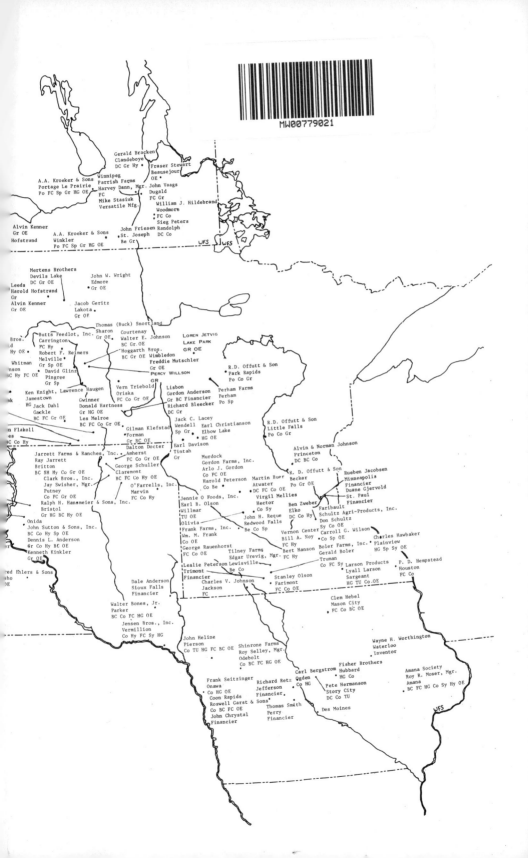

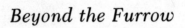

Beyond the Furrow

Beyond the Furrow

Some Keys to Successful Farming in the Twentieth Century

HIRAM M. DRACHE

Concordia College

THE INTERSTATE PRINTERS & PUBLISHERS, INC.
Danville, Illinois

Dedication

TO A FORMER STUDENT, RONALD D. OFFUTT, JR.,
INNOVATOR, MOTIVATOR, AND CHALLENGER

THE HIRAM DRACHE FAMILY on their home farm at Baker, Minnesota, pictured with some of the storage structures for their automated, confined feed lot. The feeding business is a partnership operation. Kneeling, l. to r.: David and Paul with Black Lab, Bill; standing, l. to r.: Ada, Hi, and Kay.

Foreword

No single book could be written to adequately describe the agriculture of the upper Midwest. Truly this is one of the major food- and fiber-producing areas of the entire world, and the area has been blessed with diversity from the beginning of its use by mankind.

"Diversity" is a single-word description of the soil, the climate, the environment, the production of the food and fiber, and most importantly of the people in this geographic area.

Dr. Hiram Drache has previously authored two books about agriculture and the people who were involved in the development of agriculture in this area. This previous work has obviously caused him to become more intensely interested in the several factors which have contributed to the results of this unique mix of land and people.

This book begins with a notation of some of the major inventions between 1830 and 1850 of machines and devices destined to lessen the burden of human toil on the land. In historical retrospect, one may wonder why it took so long to fully recognize the worth of these inventions, and the author helps to answer this question. Following the inventions came the railroads through this area, the adaptation of mechanical power to the farms and ranches to replace horses, and simultaneously the great increase in available and useful technology as it relates to agriculture. But inventions and railroads and machines and technology are of little consequence until put to use by people.

Author Drache dramatizes this point by assembling between these two covers a splendid array of word descriptions concerning the kinds of people who, during the period between the late 1800's and the early 1970's, manipulated resources available to them in such a way that they have made a significant and lasting

vii

impact upon the agriculture of this area. These individuals were managers and innovators, each in his own way providing early demonstration of things to come. A measure of the level of success attained by each of them is for the reader to judge.

To assist in arriving at a judgment of relative success of the individuals recognized in the book, one is pleasantly surprised to find an unusually ample supply of detail describing each of the individuals and his activities. This was made possible by the extensive research which the author accomplished not only by examining and digesting large quantities of accurate historical records and other sources of written information, but also by making countless personal contacts with knowledgeable persons still available as primary witnesses. Those who have visited with the author, or listened to one or more of his lectures, can readily visualize his aggressive, articulate, and enthusiastic manner as they read the book. It is almost as if he were present and talking.

This enjoyable history with a host of names including Kiene, Campbell, Dalrymple, Thompson, Garst, and Steiger is an excellent documentation of many of the diversities and innovations which have been so characteristic of agricultural development in the upper Midwest during the formative years of its habitation by the pioneers and their first successors. The labor of the author is to be commended.

ARLON G. HAZEN
Dean, College of Agriculture and
Director, Agricultural Experiment Station
North Dakota State University
Fargo, North Dakota

Preface

AFTER HAVING written a book on large-scale farms and another on small farms of the past, it did not take much motivation for me to attempt to find out what the future holds for farming. Farmers and agri-business friends and acquaintances from a wide area have been asking for years, what is coming next? Does an author need more prodding?

After more than 50,000 miles of driving and a total of approximately 18 months away from home on research trips, I obtained more than 3,500 pages of interview notes to support or contradict another 3,000 pages of notes gathered from census reports, agricultural statisticians, books on agriculture, and farm magazines.

Having owned and operated farms for 25 years, I found it was no problem to "get down to earth" with some of the largest and most successful or innovative farmers in the states of Iowa, Minnesota, Montana, North Dakota, and South Dakota, and in the Province of Manitoba. These men, who generally were accompanied by their wives and often other family members, revealed their deepest thoughts and convictions about why they were farmers and what they thought made them successful. Although some were inclined to hold back at first, most freely told far more than I wanted or needed to know. But that very personal information compounded to give deeper insights which helped to develop perspectives for the two books in this project.

The interviewing process offered a wide range of experiences. Spending the night in the original frame house built on the Sabe homestead in western North Dakota in 1907 and reminiscing with Oscar Sabe, an intense historian and farmer as well as a man of wisdom, helped to take me back to yesteryear. After breakfast we took a short mile drive to see a sod house still being lived in (1973), but with a twentieth-century touch—electric heat.

ix

There were nights of talking until 1 or 2 a.m., after which I was shown to a bedroom with a private bath attached. Or viewing the Big Dipper outlined in the ceiling of the Kolstad's living room, or the electrically operated drapes in the Stegner home. It was another century to be sure, and these people were thinking of the future and not of the past. Walter J. Bones, Jr., laughed as he recalled that everyone in his neighborhood had counted and stepped off the rooms of the new house they were building. That house had more bathrooms than the homesteader's outhouse had holes.

There was 19-year-old Ann Whitman, who had earned her pilot's license the previous year. She was a sophomore in college and was determined that she was going to marry a farmer when she graduated. She did. Bill and Ann have the right attitude to make a happy home and a successful farm. For many hours Ann sat on the kitchen counter and frequently interrupted her parents as they, with their typical enthusiasm, answered interview questions.

Walking on the farm yards of David Glinz, David Miller, and Ronnie Offutt or Vernon Hagen and counting up to six 4-wheel-drive tractors or forty tandem-drive and semi-trailer trucks made a real impact. What would the homesteader with two oxen have thought about this? He would not have believed that young Gregory Weiland could plow 160 acres in 12 hours with his 21 bottom moldboard plow. Nor would he have believed his eyes at the sight of the 108-foot grain drills found on some of these twentieth-century farms.

The homesteader who traveled by foot or oxcart would have been astounded, as are many twentieth-century homesteaders, to see 35 telephone numbers on a private business circuit as is the case of the Jarretts, 20 stations on a radio network as with the Offutts, or two airplanes as the Mutschlers have. Being flown to and from an interview in Charles Cannon's Lear Jet probably more than anything else impressed upon me the true technological jump in the past century.

But more important than technology is what explains the outstanding success of these people. Probably two words more than any others could best explain the farmers of tomorrow—they are *achievers*, and they have *motivation*. It was refreshing to hear positive talk about farming and its great future in contrast to the bombardment of negativism from the farm organizations, some politicans, the crossroads liquor places, or the corner store.

Gordon Anderson of the Federal Land Bank remarked, "You could expect that because you were talking to the confident people who were positive thinkers." He was right, and if there was a single purpose in my writing this story it is to show that agriculture is the growth industry of the present and the future.

When this project originated in 1968 it was intended that one book would be sufficient to cover the story. Much to my dismay the research and interviewing produced more implications than had been originally conceived, and two books became necessary. This book deals with changing technology, finance, and the farmer from the early 1900's to about 1972. A second book will deal with the attitudes and opinions of contemporary successful farmers. It will also touch on the significance of the large agri-business industry which has helped to make America the great nation it is.

The reader should keep in mind that most of the interviewing and research was completed before the great export boom that started in late 1972. Those exports and the resulting improved prices gave many farmers hopes that a new era had truly arrived. I never totally shared that optimism, probably because of my conviction that the traditional cheap-food policy of our government and the negative attitude of American consumers would not long tolerate any really great farmer prosperity.

The motivated free-enterprising farmers making constant use of innovations will continue to be the providers of plenty at a low cost to a nation of consumers who have too long taken an abundance of food for granted. However, the farmer has every right to be cautiously optimistic because, for as was true in the past, opportunities are here in greater number than ever for those who are willing to look beyond the furrow and to adapt to the times.

There is much talk today about repopulating the land, and sentimentally many readers will be opposed to some of what is found in the following pages and in the book that follows. But change, like death and taxes, is constant; and there can be no looking back if the nation is to go ahead. It is the progressive family-oriented farm, far different from its predecessor, that, as in the past, will show the way toward tomorrow's harvest.

HIRAM M. DRACHE
Professor of History
Concordia College
Moorhead, Minnesota

Acknowledgments

THIS PROJECT, which is resulting in two books, has come to a conclusion because of the efforts of many individuals. During her college career Barbara Carriere Holmquist spent three years as a devoted student research assistant and typist. Barb had the perfect sense of knowing what her taskmaster wanted next and could always track it down to the last detail. Thanks, Barb, for a super job.

Farm owner and long-time friend Howard Peet, Assistant Professor of English and Director of the Concentrated Approach Progam at North Dakota State University, served as editor and grammarian. Howie's background as an innovator in education and co-developer of a new vocabulary teaching method, as well as his knowledge of style, makes him well qualified for the task.

The administration of Concordia College and the students in my classes deserve mention because sometimes chapter deadlines had priority in my mind. For those of you who might have felt shortchanged, I apologize. On the other hand, it was students and administration who frequently were the real source of encouragement I needed to keep driving toward another goal.

Oswald Daellenbach helped in two ways. As a professional agriculturalist his concepts and perspectives were frequently important in keeping me on the right track. As a dedicated photographer it is his hours of tender loving care that have helped make the past more visible by his reproductions of faded, torn, and soiled photos. His expertise in the field of photography helps to make the future more alive to the reader. This is his third book with me, and I hope there will be many more.

Dr. Warren Kress, Associate Professor of Geography, North Dakota State University, has for the third time helped to make geography and history more alive through good maps. Dr. Kress's guidance of William Shirk, a capable graduate student, made detailed maps available. Bill Shirk spent far more hours on this project than he had anticipated, but the readers will be thankful for his efforts.

Dr. Leo Hertel, who for the third time has served me as an editor, deserves credit for being the guiding hand in reducing a massive manuscript to the size that you, the reader, can appreciate. Professor Hertel, as in the past, understood what I wanted to portray, and by standing on the edge of the forest was able to put the trees in order.

An author does not omit his family, who, probably more than any others, bear the burdens of seeing a book unfold. To Kay, thanks for getting notes typed and cataloged. David and Paul, thanks for doing the cattle chores and many other tasks while Dad was in his den. All three can attest to the fact that authors can be grouchy. To my wife, Ada—God made you a good speller, so the words were typed correctly. For you this was another book that "just had to be written."

To the nearly 200 farmers and their families who contributed directly to this project it is hard to express the correct thoughts. The emotional experience of interviewing some of the farmers of tomorrow on their farms throughout the Midwest from 1969 through 1974 is difficult to explain. The gracious hospitality, abundance of food, and unending penetrating conversations was a total experience. I finished my interviews with a feeling that the land is in the good hands of the strong farm families, and I am more convinced than ever that there is where it will remain.

Unfortunately, at least a half dozen of those interviewed are not here to see their words in print. One of those, Ferd Pazandak, had provided me with information and photos for four books. His life was motivated to search for innovations to ease the burdens for those who toiled on the land.

HIRAM M. DRACHE

Table of Contents

FOREWORD vii
    Arlon G. Hazen, Dean, College of Agriculture, and Director,
    Agricultural Experiment Station, North Dakota State University

PREFACE ix

ACKNOWLEDGMENTS xiii

  I. The Background to Change 3

  II. The Tractor 21

  III. Frank Kiene—A Loner 56

  IV. Charm, Courage, and Campbell 95

  V. The Transition of the Bonanza Farm 167

  VI. Engen, Jarrett, Schwartz, and Young—Four
     Determined Men 207

  VII. Roswell Garst—The Innovator 241

  VIII. Bert Hanson—Disciple of Alfalfa 281

  IX. Red River Potatoes 311

  X. Sugar Beets—A Bonus Crop in the Red River Valley . 358

  XI. Turkeys and Chickens—The Old Flock Is Gone 382

  XII. The Four-Wheel Drive 405

XIII. Farming Is a Business 429

XIV. The Farmer's Banker 450

ILLUSTRATIONS .... between pages 272 and 273

BIBLIOGRAPHY 499

GLOSSARY 519

INDEX 525

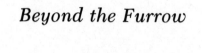

Beyond the Furrow

CHAPTER I

The Background to Change

Sᴇɴsɪɴɢ ᴛʜᴀᴛ American agriculture was on the threshold of another breakthrough in technology, W. M. Jardine, Secretary of Agriculture from 1925-1929, told a Wichita, Kansas, audience, "Could a farmer of the Pharaoh's time have been suddenly reincarnated and set down in our grandfathers' wheat fields, he could have picked up the grain cradle and could have gone to work with a familiar tool at a familiar job. And then, within the space of 20 years, the methods of crop production underwent greater changes than they had in the previous 5,000 years. At one stride, we covered ground where 50 centuries had left almost no mark."[1]

The period that Secretary Jardine was referring to was from 1830 to 1850, when many key inventions for agriculture took place. Despite Jardine's great stress on the breakthrough in technology, this era is sometimes not considered within the framework of America's agricultural revolutions and growth.

Key Inventions—1830 to 1850

The major inventions in farm mechanization that occurred from 1830 to 1850 are: the reaper, 1831; the steel plow, 1833-1837; the combine, 1836; the threshing machine, 1837; the two-row corn planter, 1839; the grain drill, 1840; the wooden hand pump using suction to lift water, 1840; the grain elevator, 1842; the first American incubator, 1844; a one-horse wheel cultivator, 1846; push-type headers, 1848; hand dump rake, 1850; tractionless but portable steam engine for farm use, 1850.[2]

Farmers frequently did not immediately adopt these machines. There were several reasons for the lag between agricultural inventions and their application. Probably the greatest of these was tradition. Many farmers believed change was contrary to the laws of nature. Others were unaware of the changes either because of

3

the lack of good farm periodicals or because of their own failure to read the ones available. Some farmers were not about to change. There were also reasons of economy as to why farmers did not adopt new equipment once it was invented. Many did not have the means to purchase labor-saving machines; others were interested chiefly in subsistence farming and saw no need to change. In many areas of the country there was also no transportation or market available to absorb any increase in production.

Nevertheless, the inventions of this 20-year period, especially the invention of the reaper and the plow, set in motion "the process of farm mechanization which ultimately was to change the economy of the entire world as well as the sociological destiny."[3]

Harold Pinches in the *Yearbook of American Agriculture, 1960*, describes the significance of this mechanical breakthrough. "Only where agricultural productivity had advanced faster than a people's needs have the economic conditions been created necessary to release larger and larger segments of the population from limited production on the land and thereby enable more and more persons to advance in intellectual, cultural, and social development above static folkways."[4] The American farmer and progressive agriculture met the task of producing ample food and fiber for the nation's constantly increasing population.[5]

The First Agricultural Revolution—1850 to 1910

Beginning with 1850 the construction of railroads gave access to growing domestic and foreign markets. In addition, the Civil War and the rise of an industrial economy were also necessary factors to create a demand and improve prices for agricultural products. These technological events caused a scarcity of labor that encouraged an unprecedented adoption of labor-saving machinery by the American farmer, who now had money to pay for his purchases. The sheer pull of economic motives overcame his inhibitions.[6]

A new system of agriculture emerged from the work of individuals who were not too different from the leaders of the growing American industry and commerce. At the same time, there were large numbers of farmers "who were unable or unwilling to contribute to, or even participate in . . ." the changing agriculture. All industries are plagued with such individuals. Unfortunately, many farmers found more reason why they should not adopt a change than why they should. But among the two million to five million American farmers of the nineteenth century some were improving their productivity through innovations. Although individual minor deviations from the established patterns were not dramatic, the cumulative effect of these trial-and-error, free-enterprising farmers was great. They changed the traditional practices and steadily increased the overall productivity of American agriculture. Basically it was the farmer on the frontier who responded most rapidly to changes.[7]

To be sure, the individual experimental and inventive farmers encountered many failures; but the energetic, the inquisitive, and the venturesome farmer usually became the most successful one in his community. In those early days of commercial agriculture (after 1850) it became obvious that there were some who faced the obstacles of change with skill, dexterity, and eagerness, while the majority retained the traditions of subsistence agriculture.[8]

The Farmer Becomes Mechanized—1850 to 1910

Several important changes in American agriculture took place after 1850, enabling it to keep up with the growing industrial society and an expanding nation. New lands opened for farmers as the Indians retreated. The Homestead Acts beginning with 1862 encouraged settlement of the then-vacant areas, increasing farm numbers from 1.4 million in 1850 to 6.4 million in 1910. New implements were available. Slavery was abolished, thereby creating a need for a new labor system in southern agriculture. The loss of a third of the labor force in the north during the Civil War and the exodus of 3.5 million farm people to industry from 1870 to 1900 was overcome by an 86 percent increase in productivity of the agricultural laborer in that period. The capacity to produce foodstuffs greatly outstripped the growing industrial and foreign demand for them. New lands opened in the West, and the application of horsepowered machinery made the big difference. Soon agriculture was straining for a new source of power. Farming in 1900 was using over 18 million draft animals and had more available horsepower than all manufacturing.[9]

As the number of farms increased, their average size declined from 1850 to 1910 because of the impact of the Homestead Act, which until 1909 restricted the granting of more than 160 acres to an individual.

If a farmer desired to increase the size of his farm he was limited by his ability to exploit his family or to hire farm workers. The relative scarcity of farm laborers caused wages to rise from an average of $120 a year in 1849 to an annual range of $400 to $600 by 1908. In addition, the work day was reduced by four to six hours.

Shorter hours and higher wages caused the farmer to invest $924 in machinery at 1908 prices in addition to the necessary horse-power required to operate a 160-acre farm. (See note 10 for prices of necessary machines.) Some writers said the farmer was no longer the man with the hoe. Fortunately for the nation, relatively few companies gained control of the farm machinery business, which tended to encourage standardization of the machinery as well as to reduce the cost and increase its efficiency. Probably the most limiting factors on further agricultural development were the lack of managerial ability of the individual farmer and the shortage of capital. By 1899 the consuming public was already saving an estimated $700 million on their food bill because of the farmers' use of machinery. At that time some agricultural writers were predicting that the day would come when farm machinery would do work "that will be worth $1,000,000,000 a year." That day was nearer than most could foresee.[10]

The Closing of the Frontier:
Hardships and Failures

The western frontier made a great impact on American society. The vast areas of the American West provided an incentive for several million immigrants to come to America to establish a new life and to gain a new opportunity. To those who wanted to farm, this provided a great speculative venture—not only to make a home, but to gain a "free farm" that they hoped would increase in value as the area was settled. To be sure, the westward movement was a high-risk venture which was often ill conceived and poorly prepared and led to many failures because of great unexpected hardships. A turnover of farm ownership of 5 to 15 percent each year during the first 5 to 10 years of occupancy illustrates the degree of risk involved. That factor, in addition to poor financial returns from farming, caused farmers to grab as much land as possible, hoping to profit from the rising value of land to offset their operating deficits. Too many farmers attempted to acquire more land than they could manage under the adversity of weather, grasshoppers, lack of markets, or ignorance of how to farm the new land. Frequently the smaller farmer was willing to bid up the price of land; and because he ignored his labor cost, he ended in failure. West of the 100th meridian farmers encountered even greater difficulties because of periodic drought.

James C. Malin, noted Great Plains historian, credits 14 individuals with basic pioneering, in determining the utility of the grasslands area. He credits Hardy W. Campbell, who settled in southern Dakota in 1879, as being the foremost of these drylands advocates. Campbell experimented widely with different machines and methods of systematic cultivation of the area. His experiments and writings covered the period from 1884 to 1916. However, it was not until after 1900 that the full impact of his efforts on how to properly work the Great Plains region were felt. The struggle over breaking the Great Plains was more than one against nature for there were many leading agricultural experts who were opposed to the idea of farming in the semi-arid region. Campbell was productive and persistent and eventually proved his point. He introduced a technique of planting grain in rows far enough apart to be cultivated, proved the value of summer fallow, and in the 1890's developed a soil packer. By 1897 he was proving the merits of his ideas by managing 43 farms for four railroads in a five-state area as part of an overall promotional effort. Once Hardy Campbell's methods were adopted, the Great Plains fell rapidly to the plowshare. Fortunately the machinery was developed by this time to make the task easier.

A great deal of credit for the available machinery has to be given to the farmers of the Red River Valley of the North, covering parts of Minnesota, North Dakota, and Manitoba. It was in the settlement of this great treeless, stoneless, fertile prairie that the farmer's vision had to be extended and newer methods of farming had to be invented and applied. The land was so flat that it was reported a man could plow out in the spring and turn around at the end of the furrow and harvest back. It is said of the Wheeler farm near Stephen, Minnesota, that "if a man started a furrow, his grandson would have to finish it." The farms were large, and commonly each section of land was treated as a single field. It was an ideal place to experiment with new larger machinery that the farmers had to use and demanded from machinery manufacturers.

A farmer's patience would not permit him to stick to the walking plow and cradle scythe when square mile after square mile of land could be operated without any natural barriers. The imagination of some of America's most professional farmers and the eagerness of the machinery companies combined to give America undisputed leadership in the development of horsepowered machinery. An economy of scale was probably reached at about 250 acres. But in the latter part of the nineteenth century the bonanza farmer, who was unique because he had extensive capital and professional management, on the western plains could operate as many multiple units of 250 acres as he desired. By specializing on one crop, real efficiency of operation was established and farms grew to many thousand acres.[11]

While the prairies of the Midwest and the grasslands of the Great Plains were being settled by the farmers, new agricultural developments took place. Lands were being transferred at a rapid rate from public domain to private ownership. This, of course, was the great attraction. Not all settlers came in order to farm. Frequently they mortgaged their lands for all they could, and if they did not succeed at farming they walked away from the land and let the loan company take it over. There is no doubt that for many of them the Homestead grant was a way of picking up some cash. North Dakota, which was settled rapidly between 1890 and 1915, reflects a pattern of farm mortgages in western agriculture. In 1900, 31.4 percent of the state's farms were mortgaged; by 1910 the rate was 50.9; and in 1920 it had peaked at 71 percent, the highest percentage of mortgage debt in the nation. At the same time the state's population growth reversed itself, for by 1915 more people were leaving the state than were coming in. Conditions were no less critical in Montana, where farm numbers increased dramatically from 13,000 in 1900 to 57,000 in 1920, only to suffer a sharp decline in the next few years. The turnabout came as a result of five successive years of drought followed by declining wheat prices. By 1923, 70 percent of the non-farmers and 48 percent of the farmers had abandoned their farms. Over 60 percent of the married farmers had quit. In one area 272 out of 550 homesteaders were non-farmers, coming from 63 diverse occupations to try their luck acquiring a homestead.

The Homestead Act on the other hand provided an ideal chance for adventure to thousands of men released at the end of the Civil War. They in turn were followed by farmers from adjacent eastern states, who desired land farther west in order to leave a farm for each of their children. Next came waves of immigrants, chiefly from northwestern Europe and Russia. To make things easier for the settler, in 1909 the government increased the size of the homestead grant to 320 acres, and then to 640 acres by 1916, and at the same time reduced the proving-up time to three years. Land was offered for sale in Montana at $10 to $20 an acre, with 15 percent down and the balance in 20 equal annual payments at 5 percent interest. With those incentives, plus the completion of railroads and the increase in markets, all the forces were in action to cause the breaking of over a million acres of virgin sod each year in Montana alone.[12]

Some Were Successful

Obviously the West would never have been settled if all who came had failed. But in all times and places, there are those who, because of stronger character, better managerial ability, more endurance, or some other good fortune, succeed. This has been the story of American agriculture from its very beginnings—failures and successes side by side at all periods of history. Let us have a glimpse at a few farmers who were successful on the Dakota and Montana frontiers in the closing days of the original settlement.

M. L. Wilson, a prominent agriculturalist of Montana in the early 1900's, surveyed the farmers of that state during the great disasters of the 1920's. He discovered that the farmers who survived the drought and the low wheat prices of the '20's had four times as many cattle as the average farm. The successful farms, he concluded, had to have 949 acres, nearly twice the state's average farm size of 484 acres. In the process of his research he took specific comments from the successful farmers about their situation. They told him: "This country is all right; you have to work hard and you can't spend money foolishly." "I don't know another place on earth where I can raise a living as easily. Everything considered, I have done well." Another stated, "Debts have hurt us worse than dry weather. Next time I'm out of debt you'll never catch me getting in again." Wilson added: "I think the biggest handicap to the 'Triangle' Country is farmers who will not farm systematically."

Charles and Jessie Grant Stick It Out

Nineteen-year-old Charles Grant came to Culbertson, Montana, in 1912 to work as a harvest hand. He liked the country so much that the next year he married a girl from back home and started farming. "We had no money when we came to Culbertson, so I guess you could say we were young, dumb, and broke." The Grants filed on 320 acres (later increased to 640 acres) which were not surveyed until the year after they homesteaded. This was near Jordan, 65 miles from the nearest railroad, "right in the middle of what has since been called 'The Big Dry.'"

In 1914, their second year of farming, the Grants sold vegetables to the cook of the government survey crews. Charles and his wife, Jessie, also subsidized their farm income by working for ranchers at $30 a month. During harvest season they were paid $10 a day and Charles had to provide a team and bundle wagon to haul grain while Jessie worked in the cook car. They needed only a little over $100 to pay for the purchase of groceries and work clothes in that year because they had cows, pigs, and chickens on the farm besides a large garden which they hand watered.

After 1916 the Grants noticed that many of their neighbors were abandoning their small homesteads. Those who did not leave by the early 1920's left in the '30's. In 1917 the Grants rented their homestead to a rancher and decided to move to the northeastern corner of the state, which was better wheat country. In 1919 and 1920 they farmed near Outlook, Montana, and received one-third share. In those two years the wheat crop on 480 acres amounted to slightly over four bushels an acre. Because of wartime demand, Charlie Grant was able to sell his wheat production for $2.20 a bushel, giving him a total gross of $2.96 an acre.

Fortunately, the Grants never relied on wheat alone to provide for them. Each time wheat failed it seemed that their sideline enterprises bailed them out. In 1919 and 1920 it was potatoes that Jessie Grant hand watered and sold for $5 a bushel and cream from the cows that survived on anything Charlie could gather for them. Charlie was allowed to keep the landlord's share of the cream check because the landlord, realizing the scarcity of renters, did everything he could to keep the Grants on his farm.

In 1921 the Grants returned to Jordan to look over their original homestead; but "The Big Dry" was true to its name, and they decided to continue renting it to a local rancher. Charlie and Jessie returned to Outlook where they rented a section from Bill Harderson, who had heard that the Grants were the kind who wouldn't "give up." He sold his machinery to the Grants, financed them at 8 percent and rented his land to them at one-third share for the landlord. Determination plus good farming paid off for Charlie and Jessie, for their wheat crop averaged 20 bushels an acre from 1921 through 1930. Wheat prices varied from a low of 71 cents a bushel to a high of $1.45, with the average about $1.00.

Charlie Grant said that with his increased income in the 1920's he purchased additional machinery. He enlarged his operation by renting more land. In 1926 he leased two sections on Montana's Fort Peck Reservation for $1 an acre. In 1926 he purchased the Harderson section for $24 an acre and another half section for $31 an acre. He could no longer get by with the 12 horses used on the first section of land, so he bought a 55-H.P. Allis Chalmers tractor for $800. This tractor could pull a four-bottom plow on breaking. This was Charlie Grant's turning point, for he expanded his farming operation and became more efficient with better machinery.

Except for a brief period during a prolonged illness in 1935 and 1936 when he was "nearly wiped out," Charlie Grant said he never once thought about quitting. He credits his wife, Jessie, his managerial ability, and his early adoption of power machinery for his success.

Gummer and Romains Hold the Line

In 1913 F. A. Gummer, a college graduate who had taught school and worked in a bank in North Dakota, decided to try his luck at homesteading at Gildford, Montana. He built a 12- by 14-foot single-wall house for $175 on his 320 acre homestead. The land was heavy, and it cost him $10 an acre to have 40 acres plowed with a 30-60 Aultman Taylor pulling a six-bottom plow. By 1917 his neighbors started leaving, enabling Gummer to buy adjoining land more suitable for wheat. Many times after acquiring that land Gummer questioned his wisdom in making the purchases, because from 1917 to 1940 there was no "good" crop in that area. Land which cost him $25 an acre in 1917 would not have brought more than a dollar an acre in the 1930's.

Gummer credits his job as a mail carrier for saving his farms, which barely carried themselves during much of the 1920's and 1930's. When he started his 32-mile mail route in 1914 there were 100 boxes on it. When he retired in 1945 there were only 30 boxes left. "People just gave up all hope, but somehow the saloons in Gildford managed to stay open. From 1928 to 1932 there just was no activity and I lost my homestead through bankruptcy, but I managed to hang on to seven quarters I had purchased. My first good crop after 1915 was 1940."

Gummer and Grant both saved their farms because they were good managers and relied on more than wheat to tide them through times of adversity. Both men had experienced setbacks, but neither ever seriously considered quitting farming. Both knew what it meant to lose land and were amazed at those around them who gave up so easily.

The story of Fred Romain is much the same. Romain homesteaded a half section near Carter, Montana, in 1913. He either worked in the Tea Pot Dome oil field, the mines, or hauled wheat for other farmers during the first 10 years that he farmed. Mrs. Romain stayed home with her large family to manage the farm chores in the off season. By 1919 neighbors in the Carter area deserted their farms in large numbers, and the Romains purchased land whenever they could. After the farm reached 1,500 acres,

Fred Romain quit working on off-the-farm jobs and purchased a tractor and a combine to do full-time farming. His operation was large enough at that point to make a living from the farm. He had very excellent crops in 1927 and 1928, just after he expanded his total operation, and this gave him a sound foundation. However, he had trouble in 1929, when the banks were short of money and put pressure on him to pay off on some of his mortgages. They knew Romain had cash saved from his two previous good years.[13]

Fingal Enger—6 feet 6 inches
and Dynamic

One of the best known rags-to-riches stories of eastern North Dakota farming centers around Fingal Enger, a Norwegian immigrant who arrived in America in 1869. He spent the next years going to school and doing odd jobs on farms and river boats and in lumber camps. This six-foot six-inch dynamic individual received double pay while chopping wood because he did twice as much work as his fellow workers. The same was true on the river boats where he was known to carry a 2-bushel sack (120 pounds) of wheat under each arm as he walked down the gangplank. Because he did not like the "pettiness and strife that applied to getting homestead lands," Enger and two friends, Iver Fecher and John Amb, decided to go beyond the limits of surveyed land. The three left Fargo in March, 1872, equipped with an axe, a gun, and some food. In early April Enger found the land he liked by a fork along the north branch of the Goose River. He marked his name on an oak tree, establishing the first homestead in present Steele County, North Dakota.

Enger made a two-story, 16- by 16-foot cabin without benefit of nails. (This building was still standing in 1975.) During construction he lived in a hillside dugout with his two pals. After the homestead cabin was finished Enger set out to find additional work because he lacked financial means to start farming. During 1872 and 1873 he worked for the Hudson's Bay Company and for government surveyors. He used his income to pay a neighbor to break his land and to purchase oxen and equipment so that he could start farming in the spring of 1874. He had little income during his first year of farming and just enough grain was flailed to provide seed for the following year and wheat for household purposes. Flour was obtained by grinding wheat in the coffee grinder. The nearest market was Fargo, which was 71 miles away. In 1876 Enger and Ellef Nyhus purchased a Buffalo-Pitts 10-H.P. threshing machine that they transported across the roadless, bridgeless prairie from Fargo. That fall Enger hauled his wheat to Fargo, making the 142-mile round trip in a week.

Fingal Enger was always in a hurry to get one job done and to start another. His marriage in 1875 to a widowed housekeeper was no exception. When B. L. Hagboe, a traveling preacher, came by, Enger immediately dropped the tugs on his plowing team and ran to the house to get his hired girl. Without either of them changing from their work clothes, they walked five miles to the nearest neighbor, where the wedding ceremony was performed. After the ceremony he returned at once to his waiting team in the field, and the new Mrs. Enger home to her household duties.

It was his aggressive spirit that pushed him into new ventures and continuous expansion. He was not afraid to plunge forward when others held back. He financially aided many of his neighbors, some of whom were in difficulty because of drinking. When these people had used up their available collateral, Enger generally received their homestead to satisfy his loan. In this manner and through excellent management his holdings grew to about 20,000 acres. However, all was not done without difficulty, for in 1900 Enger had to borrow $30,000 to pay taxes and labor because much of his crop had been damaged by drought and hail. But just four years later he was able to pay $93,000 cash for part of the Grandin bonanza. The following year he and two partners founded the Farmers and Merchants National Bank of Hatton. His family no longer lived in the 16- by 16-foot log cabin. In 1897 Enger paid $10,000 for the erection of a three-story, 60- by 48- by 24-foot, 13-room house, with hot water heat and running water. His is the story of a penniless Norwegian immigrant who amassed a farm equal to 120 homesteads, while at the same time many of his neighbors failed and left North Dakota.[14]

Tom Thompson

While Fingal Enger was settling in the Hatton, North Dakota, area, Tom Thompson left Deer Creek, Iowa, for Bachelors Grove, Dakota Territory, in 1877. He stayed in a single cabin with six other bachelors that first winter. When spring, 1878, rolled around, Tom Thompson staked out 160 acres in Fertile Township, Walsh County, west of present-day Grafton. Later he secured an additional 240 acres, which made a total of 400 acres.

The first five children in the Tom Thompson family were born in a one-room log house. After 1893 the family lived in a new frame house. Thompson's operation grew to 1,280 acres, and his family kept pace—for all five gang plows were operated by his five sons. His son Joe remembered what a treat it was when Thompson got his first Flying Dutchman gang plow with a seat. Prior to that Tom walked 20 miles each day behind the plow, between 7 a.m. and 6 p.m., with one and one-half hours out for noon.

The Thompsons were diversified farmers from the start. Besides growing wheat they always milked cows, fed hogs and chickens, and raised horses for sale. In the early 1900's Tom Thompson added potatoes to his enterprises and prospered sufficiently so that he was able to give each of his children a gold watch, a team of horses, a wagon, a milk cow, and $3,000 cash when each one started farming. After 1916 Thompson retired from farming in North Dakota and devoted full time to a 1,500 cow ranch in Cuba that he owned with two partners. Joe Thompson says he has no knowledge that his father ever really had difficulty making a go of farming on the frontier, and according to his own remembrances the most difficult years of 42 years in farming were in 1928 and 1929. "Potatoes were very low those years," Joe Thompson casually said during an interview.[15]

Garnaas and Willson Find Farming Pays

Levor B. Garnaas, who came to America in 1886 at the age of 16, spent several years working on farms and clerking in stores in Minnesota and North Dakota until he could go into business on his own. In 1895 Garnaas had saved enough to build a 20- by 25-foot one-story store building on two lots that were given to him by a lady who financed the purchase of his first merchandize. This store at Sheyenne, North Dakota, did well, and in 1899 Garnaas decided to try his luck farming. He paid Peter Gunderson, a homesteader, $1,000 for his 640 acres and assumed a mortgage for $4,500. Garnaas paid another local farmer $1,800 for seed and for planting the crop, making his total investment $7,300, or $11.40 an acre. He sold $8,500 worth of grain from the farm that year. Total return in the first year was $1,200 more than the entire cost of the land and the operation. Garnaas, in commenting on the deal 70 years later, said, "My father, Bjorn, who advised against buying the farm, replied when you have luck you don't need brains."

Where Peter Gunderson, the original homesteader, failed to make a success of the farm, Garnaas was fortunate enough to have a good crop in his first year and used that to develop a considerable estate.

Not so dramatic, but with the same end result, was the operation of George O. Willson and his son, Percy. Willson purchased his first land near Wimbledon, North Dakota, on a contract for deed, May 20, 1899. It was a quarter section of unbroken prairie that cost $1,600, with terms consisting of five equal annual payments plus interest of 8 percent. The contract specified that 50 acres had to be broken in 1899, 50 acres in 1900, and the balance of the farm in 1901. The intent of this clause was simply to improve the security of the seller. The next step Willson took was to rent a quarter section of school land for five years to guarantee himself a supply of hay. The rental contract with Barnes County specified that Willson pay $10 a year rent on each January 1 and that he could not cultivate the school land in any manner.

The Willson papers, which were available to the author, contained many of the original cost figures that present an interesting contrast to today's farming. In addition to paying $10 an acre for his first farm and $10 a year rent for a quarter section of school hay land, Willson hired J. N. Arbuckle to work for him for one year at $240 plus room and board. His taxes on a quarter section were $11.30 county and state tax and $6.85 for school purposes. He also paid $1.50 road poll and $1.00 school poll. Lumber for his first granary cost 31¢ for 2 by 4 by 12 feet and 51¢ for 2 by 10 by 16 feet at Smith and Rogers Lumber Company at Leal, North Dakota.*

On July 19, 1899, Willson entered into a contract with the McCormick Harvesting Machine Company for a binder on which he mortgaged 50 acres of crop. Three years later that note was rewritten for the unpaid balance of $144.64. Standard interest on such contracts was 12 percent. To pay for the hired labor, lumber, binder, taxes, interest, and land, Willson sold wheat to the elevator at Leal for 56¢ a bushel and barley at 35¢. His wheat averaged just over 10 bushels per acre. He also sold a cow and calf to a neighbor for $35.

The Willsons did not have an easy time struggling with the natural and economic elements, but systematically they expanded

---

*L. B. Garnaas; Percy Willson. Seventy-five years later the price of those identical pieces of lumber was $2.26 and $8.50 respectively. Wages for the year-round farm laborer would be about $500 a month plus housing and utilities. Taxes on the land would be at least 25 times more than what they were in the 1890's. It is difficult to compare the modern combine at $14,000 to $28,000 with a binder because the combine has so much more capacity and does what was formerly several individual operations. With average wheat prices from $1.25 to $2.00 and barley 75¢ to $1.25 a bushel in recent years, it becomes more obvious than ever that modern farmers are saved by volume gained through increased productivity.

their farming operation. Eventually the original quarter section became 16 quarter sections—a reasonably efficient farming operation which originally was intended to be adequate for 16 families. When their less capable or determined neighbors decided to quit, the Willsons bought them out and established a firmer basis for their farming.[16]

Three Generations on the Sabe Farm

John Sabe was a harness maker at Callendar, Iowa, until 1907, when he was advised to go to a drier climate for his health. Besides his wife, the family consisted of an 11-year-old daughter, Cornelia, and a 9-year-old son, Oscar. The family traveled by train to Dickinson, North Dakota. Mrs. Sabe and the two children rode in a coach while John stayed in the immigrant car to tend the three horses, 24 chickens, the family dog, and their machinery. They were without the family cow, which jumped the fence at the stockyards in Iowa and was then sold to a neighbor for $25.

The Sabes arrived at Dickinson in April, 1907, and immediately set out to develop their homestead.* John Sabe made 11 trips with team and wagon, hauling material the 75 miles from Dickinson to their homestead at Gascoyne. Each trip took seven days. After Mrs. Sabe and the children saw the site of their new home they returned to Iowa where they stayed until September. In the meantime, John Sabe built a frame barn for his horses, a shed for the buckboard, and started digging a 24- by 24-foot basement for the house. During this entire time he lived in a tent and cooked in the open. After 11 trips to Dickinson for material he decided that it was too much for the horses to haul all the lumber for the house and decided to build one of sod instead.

In early September John Sabe purchased a cow in preparation for the arrival of the family. A pioneer from Brainerd, Minnesota, en route to his homestead in Montana, was forced to sell a cow because she had become footsore on the trip. The family arrived at Dickinson on September 16, 1907, and after three long days of travel arrived at the farm. The 75-mile trip was very exciting for the children for they slept in tents and had to exchange help with freighting teams to cross the Heart and Cedar Rivers. John Sabe was very concerned for his horses because the grass along the way was so short from being eaten by the many horses of the wagon trains that had traveled over the same route during the summer.

---

*The Sabe farm was located-on SW 1/4-18-129-99, in Bowman County.

He became especially disturbed when on the second river crossing the children did not properly protect the food supply and it was eaten by a couple of pigs that they were hauling to their farm. On their final day of travel the family had only one coconut until late at night when they arrived at their farm and Mrs. Sabe prepared some pancakes over an open fire.

For the first six weeks on the homestead the family slept in the tent and cooked in the open. The strong fall winds blew the tent over several times and necessitated the erection of a board fence to keep the fire from being blown out. Locally mined soft coal was not the best fuel. Consequently the children gathered cow chips, hauling them to the house on a homemade sled. A stone boat was used to haul a wooden 50-gallon barrel filled with water from a neighbor's well located about one and one-half miles to the east. Water had to be hauled every day for the family, the chickens, the pigs, the cow, and the three horses. Every other day the children walked four miles to Haley to get the mail. By mid-November the family moved into their one-room sod house that had two windows facing west, one facing east, and one facing south, as well as a single door opening to the south. The walls were 3 feet thick, with the windows recessed, so that every time it rained the windows were splashed with mud. An unusual feature of their sod house was its wooden floor constructed over the basement Sabe had dug.

When the Sabes first arrived there were still a few cowboys and range cattle left in the area. Some homesteaders had to put fences around their sod houses to keep the longhorns from gouging them. Fortunately, most of the cattle had left the region by 1906, when homesteaders started to move in. Both 1908 and 1909 were good crop years, and the area filled rapidly. Then it turned dry in 1910, and in 1911 there was a total crop failure. The exodus commenced from the farms, and then from the town of Haley, that by 1911 had hit its peak of population. Those farmers who chose to stay, turned to milking, cattle feeding, and even turkey raising, besides having a few pigs or chickens. It was not until 1915 that they had the next good crop. Prices started to rise in 1916, with oats bringing a dollar and rye $2.65 a bushel. Unfortunately for those who grew nothing but wheat, rust destroyed the crop. Sabe noted, "that was the last of blue stem wheat."*

*Two miles south of the Sabe farm on the site of the ghost town of Haley is a sod house, now heated with electricity and still used, which serves to remind the Sabes of the frontier which their ancestors helped to conquer.

The decisive blow to the farm economy of southwestern North Dakota came in 1921 when there was a sharp drop in prices in addition to almost a total crop failure. Farmers moved out rapidly; but the Sabes, who had added 160 acres and were operating 320 acres, managed to stick it out. The turning point for Oscar, who had taken over the farm after his father had suffered a stroke in 1918, was in 1925, when he purchased a Twin City tractor and a three-bottom plow. Oscar, who had enlarged the farm to 480 acres, sold eight work horses and because of the feed savings was able to replace them with 17 cows to increase his income. The cream checks paid for the groceries, the coal, and the taxes. As other farmers left the area during the 1930's, Oscar was able to buy 240 acres at $2.10 an acre in 1933 and 320 acres in 1936 at $2.00 an acre. By tilling 510 acres and grazing the rest, he was able to gain more efficiency and in this manner was able to survive. In 1933 he could not borrow $480 cash from the local bank to pay for a 240-acre farm because the banker thought it was a mistake to buy. However, he did convince the banker to loan him money on 60 acres of wheat after he agreed to buy hail insurance. Oscar had to borrow additional money to pay the premium. The next time he went to the bank he took in a check for $495 collected from the insurance company for hail damage. That paid for 240 acres plus the insurance premium.

Today the third generation of Sabes farm the ever-growing homestead that has expanded from the original 160 acres to 2,000 acres.* J. Odell Sabe, third-generation farmer, says 2,000 acres is an absolute must in his area to make a living for his family. He also rents an additional 580 acres for grazing to maintain his 100 head of brood cows, which he maintains he needs to adequately supplement his income for his grain operation. The calves are carried to yearling stage before they are sold as feeders. Odell uses a chaff saver behind his combine to increase the income from his grain crop. To do all the work on this third-generation homestead, Sabe is aided by his young son and a college boy during summer vacation. The Sabe farm in 67 years time has increased fifteen-fold in land area and thirty-fold in livestock, using less labor than any time in the past. The 26 horses have been replaced by 120-H.P. and 70-H.P. field tractors and a 35-H.P. yard and loading tractor. Clearly each generation on the Sabe homestead has proven its

---

*This is in keeping with G. F. Warren's comment written in 1913: "All progress in civilization depends on having each farmer produce more than his father produced."

ability to outproduce the previous generation—typical of what many family farm operations have done through adopting the available technology.[17]

NOTES

[1]Bert S. Gittins, *Land of Plenty* (Chicago: Farm Equipment Institute, 1959), p. 5.
[2]Gittins, pp. 58-59.
[3]William J. Promersberger, *More Time to Live* (Fargo, 1972), p. 2.
[4]Harold E. Pinches, "Revolution in Agriculture," *Yearbook of Agriculture, 1960: Power to Produce*, ed. Alfred Steffrud (Washington, D.C., 1960), p. 1.
[5]J. Brownlee Davidson and Leon Wilson Chase, *Farm Machinery and Farm Motors* (N.Y.: Orange Judd Co., 1908).
[6]Ernest Ludlow Bogart, *Economic History of American Agriculture* (N.Y.: Longmans, Green, and Co., 1928), p. 111.
[7]Clarence A. Danhof, *Change in Agriculture: The Northern United States 1820-1870* (Cambridge, Mass.: Harvard University Press, 1969), p. 281.
[8]William N. Parker, "The Productivity of the American Farmer in the 19th Century" (unpublished manuscript, Yale University), Danhof, pp. 278-286.
[9]Theodore Saloutos, "The Agricultural Problem and Nineteenth Century Industrialism," *Agricultural History*, XXII, No. 3 (July, 1948), p. 162; Parker; *Statistical Abstract of the United States: 1962* (United States Department of Commerce, Washington, D.C.), p. 607; Danhof, p. 280.
[10]Davidson and Chase, Np.; Suggested machinery to operate a 160-acre farm in 1908: grain binder $125, mower $45, gang plow $65, walking plow $14, riding one-row cultivator $26, walking cultivator $16, disc harrow $30, two farm wagons $150, smoothing harrow $17, corn planter $42, seeder $28, manure spreader $130, hay loader $65, hay rake $26, light road wagon $60, and a buggy $85, for a total of $924; Parker; W. B. Thorton, "The Revolution by Farm Machinery," *World's Work*, VI, No. 3 (August, 1903) 3766.
[11]Parker; James C. Malin, *The Grassland of North America; Prolegomena to Its History with Addenda* (Privately Printed, Lawrence, Kansas, 1961), pp. 173-227; Asbjorn B. Isaacson, "Farm Mechanization in the Red River Valley, 1870-1915" (unpublished Master's thesis, University of North Dakota, 1949), p. 33; Hiram M. Drache, *The Day of the Bonanza* (Fargo, North Dakota: North Dakota Institute for Regional Studies, 1964); Mary W. M. Hargreaves, "Dry Farming Alias Scientific Farming," *Agricultural History*, XXII, No. 1 (January, 1948), 39-55.
[12]Fred A. Shannon, "The Homestead Act and the Labor Surplus," *American Historical Review*, XLI, No. 4 (July, 1936), 637-651; Louis B. Schmidt, "The Agricultural Revolution in the Prairies and the Great Plains of the United States," *Agricultural History*, VIII (October, 1934) 169-195; Carleton R. Ball, "The History of the American Wheat Movement," *Agricultural History*, IV, No. 2 (April, 1930), 48-71; Paul A. Larson, "A History of Farm Mortgage Indebtedness and Direct Farm Mortgage Relief in North Dakota from 1920 to 1950" (unpublished Master's thesis, University of North Dakota, 1963), pp. 1, 10; Elwyn B. Robinson, *History of North Dakota* (Lincoln, Nebraska: Nebraska Press, 1966), p. 372; M. L. Wilson and H. E. Murdock, *Reducing the Cost of Montana's Dry Land Wheat Harvest*, Montana Extension Service in Agriculture No. 71 (Bozeman, 1924), p. 3; M. L. Wilson, *Dry Farming in the North Central Montana "Triangle,"* Montana State College Extension Service No. 66 (Bozeman, 1923), pp. 1-132; J. D. Black, R. H. Allen, and O. H. Negaard, "The Scale of Agricultural Production in the United States," *Quarterly Journal of Economics*, LIII, No. 53 (May, 1939), 333; *The Resources and Opportunities of Montana* (Montana Department of Agriculture and Publicity, 1914).
[13]Wilson, pp. 107-122; Personal Interview with Mr. and Mrs. Charles Grant, Plentywood, Montana, May 8, 1972. Mr. Grant was born 1893, was married and started farming in Montana in 1913. They farmed until 1963; M. C. Taylor, et al.,

Prices Received By Montana Farmers and Ranchers 1910-1952, Montana State College Agricultural Experiment Station Bulletin 503 (Bozeman, 1954), p. 26; Personal Interview with John Romain, Havre, Montana, May 11, 1972; Personal Interview with Clarence Romain, Chester, Montana, May 12, 1972. John and Clarence are both sons of Fred Romain; Personal Interview with F. A. Gummer, Havre, Montana, May 11, 1972.

[14]"Fingal Enger Family History," privately printed, copy in possession of Mrs. Delmer Nystedt, (Fargo, 1961); Clarence H. Tolley, "Fingal Enger, King of the Goose River," *North Dakota History*, XXVI, No. 3 (Summer, 1959), 107-120; Records of Augsburg College, Minneapolis, Minnesota; Letter from Dr. Oscar Anderson, President of Augsburg College, Minneapolis, Minnesota, August 2, 1967; Personal Interview with Mrs. Roy Lee, Hatton, North Dakota, August 12, 1967.

[15]Personal Interview with Joe Thompson, Grafton, North Dakota, March 13, 1973.

[16]Personal Interview with Levor B. Garnaas, Sheyenne, North Dakota, December 11, 1970. Mr. Garnaas was born July 1, 1870, and at the time of the interview was very exact about details of events that took place prior to 1900; Personal Interview with Percy Willson, Wimbledon, North Dakota, June 24, 1971; George Willson papers in possession of Percy Willson.

[17]Personal Interview with Oscar N. Sabe, Gascoyne, North Dakota, April 8, 1972. Mr. Sabe was born in 1898 and vividly recalls his family's homesteading days commencing in Gascoyne in 1907. He lives in the original frame house on that homestead. Only four miles from his farm is a sod house that has been converted to electric heat and was still being used as a home in 1972.; Letter from Oscar Sabe, April 14, 1973; G. F. Warren, *Farm Management* (New York: Macmillan Co., 1913), p. 31; Gilbert C. Fite, *The Farmers' Frontier: 1865-1900* (New York: Holt, Rinehart and Winston, 1966), pp. 223-224; Parker, pp. 23-24.

CHAPTER II

The Tractor

Attempts to Develop Steam Power

ABRAHAM LINCOLN, speaking at Milwaukee, Wisconsin, on September 30, 1859, said:

The successful application of steam power to farm work is a desideratum—especially a steam plow. It is not enough that a machine operated by steam will really plow. To be successful, it must, all things considered, plow better than can be done with animal power. It must do all this work as well, and cheaper, or more rapidly so as to get through more perfectly in season; or in some way afford an advantage over plowing with animals, else it is no success.

Experiments with steam plowing had been in progress in western agriculture since 1816, but it was not until the 1850's that reasonably successful steam-powered plows in Great Britain brought many Americans to a keener awareness of this practice. Breaking the sod-bound Midwest posed a real challenge to the American farmer, and he sensed a better source of power was needed to farm. Horace Greeley, who was always seeking ways to promote the West, wrote in the 1850's that the American farmer needed a locomotive that did not weigh more than a ton, could work for a half hour steady without stopping to refuel, could travel on plowed fields, could be hooked to any machine in a minute—including carts and harrows—and could also be used as a stump puller, thresher, plow, or pump, as well as dig cellars and ditches. Greeley correctly predicted that the farmer would probably have to wait a few years for such a machine and probably realized that steam power might not be the answer to the machine he was describing. In 1855 Obed Hussey, farming on the frontier, put a steam plow into operation. This was followed in 1858 by a more successful steam-plowing outfit invented by J. W. Fawkes.

By the 1880's it became evident that steam power would not

substitute for horse power, even though big steam engines were used extensively for breaking sod on the frontier. The steam engine was too heavy and slow, besides being a fire and explosion hazard. Despite many promotional schemes and much publicity the average farmer preferred to ignore the steam tractor, and when one became stalled or broke through a bridge it was a sure cause for hilarity. Even the argument that the machine did not eat when it was not working did not help the steam engine advocates, because when it was working it was only with about 2 to 3 percent efficiency. When a better steam engine was developed, when a steam-lift plow was perfected in 1886, when longer periods between refueling stops and when definite reductions in manpower were achieved, it still did not help the steam engine to replace the horse.

Despite early failure of mechanical power, leaders in agriculture kept saying that "a new agriculture" was close at hand. Farmers would then receive their share of prosperity. Farmers realized that some kind of expandable and continuously dependable power unit had to be produced.

Faced with an ever-increasing cost of labor and land, the farmer of the early 1900's knew that he would have to find ways to reduce expenses. Farm cost studies between 1905 and 1907 reveal that labor costs varied from 12¢ to 19¢ an hour and teams from 15¢ to 30¢ per hour. A Minnesota study points out that horses averaged only three hours of work per day for 300 days a year. On one exceptionally well organized farm, horses averaged 4.9 hours of work per day for 300 days a year. The studies also point out that the average rate of speed in field work was two miles per hour. The number of acres farmed per horse varied from 9 to 28, and the cost of horse labor per acre fell to between $3 and $6. Farmers were advised to rent horses during the busy season, if possible, in order to keep their power cost down. The bonanza farmers of the Red River Valley exchanged horses with the lumber camps of northern Minnesota, which proved beneficial for both parties; and they came as close to industrialized agriculture as was possible by using horse power.

Besides the cost factor, farm horses were consuming 72 million of the 325 million acres of crop land harvested in 1910 at a time when the leaders of the nation were concerned over a food shortage. In addition, horses in the city were consuming crops from another 16 million acres. The total feed consumed by horses required 27 percent of all crop land in use.

G. F. Warren in his book, *Farm Management*, advised that the easiest way for the farmer to increase his profits was for each man to drive more horses. Ironically, he cast off all reference to mechanical power in one statement: "Engines in farming have not usually been successful . . . [and only] on large wheat farms, where the heaviest work is plowing, engines may be used in place of horses."[1]

Early Tractors and Their Opponents

While the great majority of farmers remained horse lovers and gave little thought to the need for moveable mechanical power, and steam power promoters were doing everything possible to perfect that machine, others were dreaming about the application of an internal combustion engine to the farm.

In 1889 the Charter Gas Engine Company had built six gasoline tractors by putting stationary-type engines on Rumely steam-tractor chassis, but little is said about their effectiveness. John Froelich, an Iowa farmer, mounted a gasoline engine on a Robinson running gear in 1892. "This was probably the first gasoline tractor of record that was an operating success." This tractor did a 50-day threshing run belted to a Case 40 by 58 thresher. It had to travel over difficult terrain and performed in temperatures from -3° to 100° F. In 1902 at Charles City, Iowa, C. W. Hart and C. H. Parr built a tractor designed for drawbar rather than belt work, and Massey Harris Company built the Wallis Bear, a gasoline tractor capable of pulling ten 14-inch plows. Hart-Parr became the nation's first exclusive tractor manufacturer in 1905.*

By this time many other inventors were busily working to produce an internal combustion machine to give the farmer the mechanical power he needed to catch up with the world of industry. Gas Traction Company of Minneapolis produced the Big Four (a 35-70) in 1907, with its 18-inch-wide and 8-foot-high wheels. Avery produced a 12-36 combination machine that could pull three 14-inch plows in the field or haul 3 tons on the road. Wooden plugs were inserted in the wheels for better field traction. Henry Ford, who was probably personally as concerned about getting mechanical power to the farmer as he desired a car on the road, had been working on tractors since 1883, when he produced a steam tractor for plowing. Ford produced his first

---

*W. H. Williams, sales manager for the firm, is credited as the individual who popularized the word "*tractor*" in an advertisement, instead of using the standard expression "gasoline traction engine." The word "tractor" had been patented by a Chicagoan in 1890, but Williams was completely unaware of that fact.

gas-powered tractor in 1908 and a second in 1913. By then he insisted that "tractors were superior to horse flesh in efficiency, in ease of handling and in cost per unit or work produced."

Even though production of horse-drawn wagons and business vehicles reached an all time high in 1905, it was obvious to visionaries that the day of the horse was numbered. Horse and mule numbers reached their greatest number in 1919-1920 with about 27 million, but dropped sharply to just over 16 million head by 1925, and to 3 million by 1959.* In 1910, *World's Work* had numerous articles about the farmer's future, based on the use of the gasoline engine. One such article read:

> There is good reason to hope that the gasoline engine has greater triumphs within its easy and early reach even than the automobile and the flying machine. Mechanical help is needed by the plowman more than it is needed by any other man on this planet. . . . It may free more men from the plow and the hoe than have yet been freed from muscular drudgery by steam and electricity.

The next issue of that series, in an article entitled "The Passing of the Man with the Hoe," stated:

> In the near future, to escape ruin, every farm must have its own power plant in an efficient motor. . . . As machinery multiplies rapidly for every conceivable farm need, as it cheapens in cost and grows in efficiency, we begin to understand the astonishing gist of the thing.

Other writers adding their wisdom to the above stated flatly that the tractor would lead to "a new age in agriculture." Some looking far into the future predicted that the family farms might persist but they would become "fewer, larger, and costlier." Farmers would probably be forced into some kind of effective combinations of producers among major commodities. Farmers were reminded that power shapes the world and that the tractor was replacing the farmer's greatest source of power. The tractor was touted as "the great history-making machine of the twentieth century. The reorganization of the farm which must take place will surely hinge on the solution of the power problem."

Tractor companies were quick to promote the ideas picked up from independent writers. Gas Traction Company of Minneapolis sponsored an advertising booklet in 1908 emphasizing the farmer's spending of nearly two-thirds of his time caring for and producing for his horses. Less commercially oriented publications

---

*In 1905, 644,000 wagons and business vehicles and 937,000 carriages and buggies were made.

verified that the tractor company's ads were not too far from the truth. A gasoline tractor cost about $90 for each horsepower, in contrast to $175 to $200 for each horse. The fuel cost of a tractor was about ½¢ per mechanical horsepower-hour in comparison to the more than 8¢ per animal horsepower-hour. One publisher, noting the reputed rise of 143 percent in the cost of keeping a horse from 1900 to 1910, wryly commented that the same horse "is not one pound stronger today [1910] than he was 30 years ago. . . ."

The average time spent on chores alone per horse according to government studies was 27 minutes per day—equal to 20 eight-hour days per year. The same report indicated that a 20-horsepower tractor could work around the clock and do the work of more than 25 horses at the cost of about 10. American farms were producing a fifth of the world's wheat, half of the cotton, and three-fourths of the corn, because they were mechanized, with the exception of power. "We can do neither with horses nor without them; but the steady pressure of events is forcing us away from our horsepower point of view and compelling us to take notice of trucks and tractors," concluded one of the earliest books on tractor power. The author continued with a bit of fact plus philosophy: "The horse today is an unprofitable servant . . . . [But] nothing is more difficult than to move the mind to a new habit of thought. American farmers must get out the horse habit."*

In the first decade of the twentieth century a two-way battle was shaping up in the minds of the farmers. On one side there was the resentment toward the tractor from the traditionalist and the lovers of horses. On the other side was the realization that to improve conditions in America, "agriculture had to have a reliable portable engine unit that would not only propel itself, but also furnish power to attached machines."

Many horse advocates doggedly opposed the tractor into the 1920's and early '30's. Their position was not without support, as this expensive, heavy, clumsy tractor compacted the soil and seemed to continually aggravate its owner. Even some of the most progressive thinkers in agriculture questioned whether the tractor would ever completely replace the horse, because so long as

---

*It was in the largest cities that the "horse habit" was first broken. Because of better business practices, the city businessman was much more aware of rising costs of feed, horses, and associated labor. From 1900 to 1910 there was a reduction of 25 percent in use of horses in the nation's 30 largest cities, even though overall commercial use of horses had increased 33 percent. By 1912, 50,000 motor vehicles had replaced only about 200,000 horses.

farmers needed some horses it would be more economical to have only one source of power.* Economists expressed the view that a farm had to be at least 200 acres before a tractor would be feasible. Only one-sixth of the nation's farms on that basis were large enough to justify a tractor. Not unless a farm needed more than five horses was there justification for owning a tractor.

Not all farmers who opposed tractors were small operators. Some of the most heated arguments against tractors came from large land holders of the Dakotas and Montana. Ironically, however, it was in these very areas that mechanization caught on most rapidly, resulting in the development of some extremely large farms. The Noble Foundation Farms at Lethbridge, Alberta, had developed a 16-horse hitch to pull three 12-foot drills and seed 80 acres per day. Other farmers in the prairie provinces and Montana figured out ways to hook two 2-bottom gang plows together and pull them with 8 to 12 horses. It was a particular delight of the horse farmer to acquire plows that had been abandoned by tractor farmers during the dry spell from 1919 to 1923. A bulletin authored by M. L. Wilson of Montana in 1925 took special care to point out that no argument against tractor farming was intended, but that a "majority of farmers will be interested in expansion through big teams rather than through tractors." An obvious opponent of tractors commented: "I have driven horses, bulls and even dogs, also Red River carts and Indian travois, and I know the biggest mistake the farmers have ever made was to try to farm with engines of any description."

One Miles City area farmer, who apparently considered himself quite a large operator, also supported Wilson's claim by making a typical statement:

I believe that in talking in favor of big teams you have to pick your man. Some men are rather weak at heart, or, as we say, small potatoes when it comes to trying something new and a little hard. Last year I took a young man who had been raised in town. He held a job carrying mail, using one horse. He had never driven anything larger. I stayed with him most of the first day. . . . He got along fine with my big team. In fact, he doesn't feel right . . . if he can't drive eight horses or more. . . . In my own mind I am convinced that when a man intends to farm in this country he must use big teams or quit. In fact, I get more "kick" along the big team line than I expected, because some of this country is settled up by farmers from the east, who farm "small potato style" even yet.

---

*From 1910 to 1920 there were only three and one-tenth horses per farm and little need for the average farmer to have a tractor.

Apparently not everyone agreed with the fans of the big teams. Many farmers lost interest in horse breeding because of the low horse prices. In the long run this tended to create a shortage of horses, making them even less competitive with tractors. Not only that, but as tractors continually developed into more and more efficient units the horse remained as awkward, irritating, expensive, and slow as ever. At last the farmer had a practical power alternative.[2]

The Tractor Arrives

In 1890, animals provided 15 million of the 17 million horsepower used on farms. Rapid increase in consumption of agricultural products from a growing population, expanding industry, and World War I caused total horsepower on American farms to grow to 49.2 million. Even though horse and mule numbers continued to grow until 1920, they made up only 44 percent of total available power on the farms by 1924. Truck and tractor power alone matched all animal power. In 1924 only transportation exceeded agriculture in total useable horsepower. This represented a major breakthrough because many farmers had long recognized that " . . . the producing capacity or earning ability of the farm worker is in direct proportion to the amount of power one is able to control." E. R. Eastman, writing about the mechanical power increase, stated that this was a major step for the nation because it could not rise beyond the height of its weakest industry. The tractor, which provided this power, gave the farmer the ability to substitute it for labor. This initiated a trend toward larger tractors jumping from 30 H.P. per farm laborer in 1960 to 60 H.P. per laborer by 1970, with the rate continually increasing. To the progressive farmer, mechanization saves labor.

As soon as they sensed a potential market for the tractor, manufacturers blossomed everywhere. In 1906, 11 companies produced 600 machines. The year 1910 found 15 companies with a total production of 4,000 initiating a virtual avalanche of companies and tractors, which resulted in a total production of 203,000 machines from 166 companies in 1920. Because of the increased demand for food, the advent of smaller, more versatile tractors, and the tractor demonstrations in the midwestern states, 1915 was a pivotal year. Competition among the many manufacturers forced a reduction in tractor price of about 40 percent from 1910 to 1917. Prosperous times for farmers brought about a tractor craze. Each year farmers purchased more and more tractors until 1920, when they acquired 140,000 machines, a record not surpassed until 1936. A book published in 1916 on the history of the tractor positively declared, "Whatever happens, the tractor is here with bells on and going so fast you can't see it for sparks and smoke."

With the advent of the agricultural depression in 1921, there were writers who questioned the economics of the tractor, but the progressive farmers sensed that the tractor was one way to economize. By 1926 all doubt about the future of the tractor was erased. There were 1,000 tractors recorded on farms in 1910 and no trucks, but by 1920 official census figures revealed 246,000 tractors and 139,000 trucks. The numbers increased until by 1970 there were 4,790,000 tractors and 3,185,000 trucks. Since then a noticeable decline in the rate of growth has set in chiefly because of a corresponding increase in size of tractors and trucks. Liberty Hyde Bailey, who had professionally worked in agriculture since the 1880's, noted that in the decade 1910 to 1920 horses disappeared from the highways, hitching posts disappeared from the towns, livery stables passed, and horseshoeing shops vanished. In 1927 he cautioned:

We may expect the changes from the old order to continue, and with commendable results. Farm procedure must be adjusted and readjusted. But this does not mean the elimination of the horse from farming. The tractor nor the touring car cannot take his place in any number of the common farm operations.

Others writing at the same time were more optimistic about what the tractor would eventually be able to do. Earnest Bogart, in his history on American agriculture, wrote:

The importance of this machine lies not so much in its performance up to date [1923] as in the vast possibilities which it opens up for the future. [The light, general-purpose tractor] . . . has already begun to displace horses to a noticeable extent . . . . Is there any limit to which this engine may not go in relieving the farmer, his wife, and their helpers from weary muscular effort and drudgery?

The early tractors, with few exceptions, were all very large, heavy, and cumbersome. They weighed from 20,000 to 50,000 pounds, traveled about two miles per hour, delivered only one horsepower per 600 pounds of weight, and generally were hard to start. The tractor companies promoted the bigger tractors partly because of tradition and partly because they felt the larger ones were more economical and competitive against animal power. Many of the demonstrations promoted the big engines and possibly the drama of size had some impact on the farmers. Probably the most dramatic of such exhibitions was the much-publicized Purdue demonstration where three 30-H.P. tractors were hitched to a giant 50-bottom plow. This turned a strip of sod 58 feet wide and plowed an acre in four and one-fourth minutes. Three engineers and one plowman were required to do the job that normally required 25 men with 125 horses and 25 gang plows. In 1973 a Euclid, Minnesota, farmer regularly plowed an acre in four and one-half minutes, using a four-wheel-drive tractor that pulled twenty-one 16-inch plows.

Some of the more successful big tractors which remained popular and were still in competition in the Nebraska Tests in 1920 were the Rumely Oil Pull 30-60, that weighed 26,000 pounds, produced 27.91 H.P., and averaged 1.90 mph in a ten hour test. The specifications for the Minneapolis 35-70 were 22,500 pounds, 38.67 H.P., and 2.10 mph, while the Aultman-Taylor 30-60 weighed 24,450 pounds, produced 35.09 H.P., and did 2.54 mph. The average cost of these 1910 to 1920 gigantic tractors ranged from $2,700 to $3,500; they operated at 300 to 350 rpm and had two forward speeds. A 1972 20,000-pound tractor produced 135 H.P., had 12 forward speeds, operated at 2,200 rpm, traveled up to 17 mph, and cost $13,500, a far more economical unit.

The big gas tractor boom did not last long. The big tractors had been oversold and undertested. It was common for the farmer to pay up to four times the production cost for a tractor. The tractor manufacturers prospered by taking hard cash from farmers, who for the first time had money to purchase untested machinery but who were too excited about the tractor. After a limited number of large-scale farmers had purchased tractors, the market dwindled by 1914 to 1916.

To spur tractor sales again, the manufacturers started producing smaller machines. Farmers had informed manufacturers that what they really wanted was a tractor that would do the work of four horses. Two years after the small tractor was introduced a sales rush was on. Tractor manufacturers adopting standardization techniques of the automobile industry were able to mass-produce the smaller models. Unfortunately for the farmer, many of the new model tractors were not field tested before they were sold. Because the small tractors generally were built to pull a two-bottom plow they could not be sold on the basis of being labor savers. The manufacturers stressed the value of the belt work that these tractors could do. However, most machinery requiring belt power needed more power than the small tractor provided. This led to extreme abuse and overloading of these tractors, which was not the case with the larger models. It was not long before farmers began to call for a three-plow tractor because it was obvious to most that the two-plow model was not large enough to be of any advantage.

When small tractor sales dropped sharply at the close of World War I the companies once again put on a drive to promote their product. Again they used demonstrations. One of the most notable of such tractor demonstrations was held in 1912 at the North Dakota Agricultural College in Fargo in cooperation with 26 manufacturers. To stimulate interest, 11 teams, varying in number from five to eight horses, were also on the show. " . . . Before the demonstration was over a number of horses succumbed to the extreme heat." The tractor people could not have driven their point home more emphatically.

At the demonstrations the manufacturers were also able to show the farmers the many innovations that made the new smaller tractor a much more useable farm machine. There were several innovations between 1917 and 1924 that entrenched the tractor in American agriculture. Much credit belongs to Henry Ford, who had created a tractor division within the Ford Motor Company in 1915 and produced his first Fordson in October, 1917. His tractor was called the Fordson because a group from Minneapolis, including a Mr. Ford, had pre-empted the name Ford to capitalize on it, for they knew the Ford Motor Company was working on a tractor.

The Fordson was produced on the assembly line and hit the mass market by virtue of the nationwide chain of Ford car agencies. It is the tractor that popularized tractors as far as the buying public was concerned, even though it was not a complete mechanical success and was prone to flip over backwards. Nevertheless, the Fordson did prove to the public that the day was not far off when the horse could be completely replaced on the farm. In its first full year of production Fordson captured 24 percent of the tractor market in a field of 141 companies and by the mid-1920's had as much as 75 percent of the total annual sales. By 1927 Ford officials could boast that they had produced half of all the tractors in the United States. This tractor, selling for $600 to $900 and weighing 2,500 pounds, had a 20-H.P. engine, three forward speeds, and a top speed of 6¾ mph.

The mechanical innovation for which the Fordson was credited was the cast iron unit frame that enabled it to have a rugged construction at about half the weight of comparable horsepower machines. The unit frame was rapidly adopted by most of the popular tractor makers. In 1917 Moline introduced the Universal Model D, one of the first tractors to use a storage battery for ignition, starting, and lighting. This replaced hand cranking and the magneto ignition system. This, too, became standard equipment within a few years' time on most popular tractors. In late 1918 International Harvester Company introduced a practical power takeoff mechanism. This eliminated the cumbersome pulley-and-belt system for stationary machines and enabled the tractor to transmit its power to field-operated machines.

A big year for tractor refinement was 1923. Many of the tractor manufacturers had been eliminated from competition by the farm crash of 1921, and those that remained improved their tractors considerably. Four-wheel drives appeared once again. Cletrac, Holt, and Best track-type tractors were becoming more successful; McCormick Deering introduced the 10-20 and John Deere the Model D 15-27, both of which became popular models. It was the McCormick Deering Farmall, the first genuine all purpose tractor with tricycle design, automatic braking, and power takeoff, that made big tractor news in 1924. This was the first successful row-crop tractor and was the machine that broke Ford's domination of the tractor market.

Initially there was considerable resistance to the farmer's compulsion to buy tractors. The strongest oppostion came from the financial interests who looked at the tractor as an extremely risky investment. Many bankers refused to extend credit to any farmer for the purchase of a Fordson because they felt the tractor was unreliable. Bankers and other businessmen denied credit or loans for tractors on the "grounds that tractors were a menace to farmers." They argued not only "that farmers could not operate the machine profitably, but also that if they were successful, the farmer would have too much leisure time." Studies proved that on farms of 400 acres farmers used their horses only 100 days a year and that if they used tractors for power they would be in the field only 40 to 45 days a year. Apparently many individuals felt that the farmers would not have profitable ways to use the time released from field work. Jobbers and wholesalers were instructed to give no credit to country merchants who gave credit to farmers that owned tractors.

The bankers were not entirely wrong in their opposition to financing cars, trucks, and tractors in the 1920's. For according to a survey, many farmers in North Dakota who came into financial trouble were those who mortgaged their farms for the purchase of such machines. However, some bankers had the wisdom to see the value of the tractor. The Suttons of Onida, South Dakota, were active in the horse business on a large scale, but in 1916 they bought a Case to eliminate the need of the 14 H.P. sweep in threshing. They continued to plow with 10 horses on a three-bottom plow until 1917 when they purchased an International Titan. The Suttons were financing at the Stockyards National Bank of Chicago and were advised to get rid of their horses and convert to tractor power. Edmund Sutton had been a real horse lover and probably would not have switched to tractor farming if the banker had not influenced him to do so. John Sutton, Sr., said that one switch made a big difference to the farm's ability to operate during the 1920's and '30's.

In their fight with the tractor the small farmers could not "compete with machines that represent money and lower cost of production. . . . Yet the small farmer cannot stop the coming of large machinery and mechanical power. . . ." Many wondered what would become of the small farmer who "it has long been preached" was fundamental to the prosperity and welfare of the country. Those same people had argued that the big farmer "was a detriment to good and profitable farming" because he could not cultivate his acres so efficiently as the small farmer. Now mechanical power made large-scale intensive farming possible and profitable.

One author urged small farmers to organize their farms in a cooperative so that they could successfully compete with the larger producer. It was argued that the "farmer has no more right to be independent . . . than the laborer in the city. . . ." If he could not raise the necessary capital to establish a proper unit of production he should "adopt the city's policy and employ money—borrowed money." The farmer was warned that to persist on independent small-scale farming was to "invite absolute failure. . . . For, after all, the real purpose of agriculture is not the enrichment of the man who tills the soil, but the providing of a hungry world with food."[3]

Pioneer Tractor Farmers

Time has a way of clouding memories, and because early farmers seldom kept records, it is difficult to determine who the real tractor pioneers were. Fortunately, the tractor is a new enough machine so that a few of the pioneers of the tractor age survived into the 1970's. In the Red River Valley, where the bonanzas set the tone to bigness, tractor farming had early beginnings. The bonanzas, because of the very size, were early in the adoption of steam-powered threshing, for as soon as the steam-traction engine was developed it was put to work on those farms. The broad expanse of the Red River Valley and the wide openness of the lands of the Dakotas and Montana greatly enlarged the vision of many of the settlers, and the concept of the large farm was easier to accept, thus making the region a traditional leader in mechanization.

The Reitans Break Sod

About 1903, T. S. Reitan of Comstock, Minnesota, purchased a Bull tractor that could pull a two-bottom plow. The tractor did not perform satisfactorily, but Reitan, convinced of the correctness of the tractor concept, purchased a Hart-Parr 60 from Moore Brothers of Fargo in 1905. This tractor, which cost $2,200, bore the serial number 1576 (a second Hart-Parr 60 purchased by Reitans had serial number 4400) and was an oil-cooled gasoline burner that pulled eight 14-inch plows. After the Reitans had finished their heavy work in 1906 they decided to do custom work. Oscar, age 15, and brother Walter, age 10, worked breaking sod from "daylight till dark six days a week" for the entire summer. The tractor had one forward speed that operated at 2 mph full throttle. Nearly all the plowing was done on full sections of land, and the two young boys were able to plow from 18 to 22 acres a day. The tractor had a 750-pound flywheel that cranked the starter, and by using gasoline the boys were able to get the engine going easily.

When the number of jobs increased T. S. Reitan hired two men to work the tractor nights so that it operated for several weeks continuously, with stops only for servicing. The rig required a two-man crew, one to operate the tractor and the other to lift and reset the plow, two bottoms at a time. Night plowing was made possible by attaching kerosene lanterns to both the tractor and the plow. Oscar Reitan recalls that the plowing rig traveled so slowly that not a great deal of light was needed. When the night work first started "everybody in the neighborhood thought we were going crazy and would come over and watch us, but it soon got to be old stuff for them." Soon a generator was installed along with a storage battery, and electric lights were used; but for safety's sake Reitans continued to carry the lanterns with them.

Night plowing was not without its rewards, as plowman Oscar frequently slept on the platform after resetting the plows and checking the oil drips, having the driver wake him as they approached the end of the furrow. The six oil drips on the main bearings, crankshaft, and cylinders had to be checked about every one and one-half miles, and if not dripping properly the faulty line would have to be unplugged. About every three to four miles the little cups were refilled with oil. After owning the tractor for four years the Reitans installed automatic oilers so that the job was eliminated. To ease the job of the tractor driver, a steel wheel placed on a 16-foot steel bar was attached to one of the tractor's front wheels. This wheel positioned to run in the furrow reduced the attention necessary for steering. When the end of the furrow was reached, a cable attachment enabled the operator to crank the bar off the ground, thus freeing the tractor for turning.[4]

Ferd Pazandak, an Early North Dakota Horseless Farmer

Ferd Pazandak's family moved to Fullerton, North Dakota, from Iowa in 1901. Eighteen-year-old Ferd was put right to work breaking the sod with four horses on a 16-inch walking breaking plow. Ferd soon learned how to adjust the plow so that he did not have to hold on to it while walking down the furrow. In a 10-hour day he would walk 20 miles, breaking $1^6/_{10}$ acres of sod.

Ferd and his brothers were mechanically inclined, and as soon as their father could afford to do so they urged him to buy a tractor. In 1908 the Pazandaks paid $3,000 for a Geiser Peerless Steam Tractor and a 12-bottom steam-lift plow. This rig could pull the twelve 14-inch bottoms for 20 miles a day, at 2¼ mph, to turn over 30 to 34 acres per 10-hour day. This required a three-man crew— one man to operate the engine and drive, another to fuel, and a third to haul water and coal. The engine consumed about 3,000 pounds of soft coal a day, which was fed to it from boxes on the fenders that held 400 to 500 pounds. About every four miles (or rounds) they had to take on more coal from the coal wagon. But stopping for coal was no problem for they also had to take on water every two miles. Because steam jets did the pumping, the two 200-gallon water tanks could be filled in about five minutes, but the water hauler had to work steadily to keep the engine supplied.

Even though it took three men to plow about 33 acres a day, the Pazandaks felt that steam power provided a great advantage over horse power. With an 1,120-acre farm and the 20 miles of walking per day to plow 1½ acres fresh in their minds, the Pazandaks did not find it difficult to justify the purchase of their first tractor.

By 1910 they were so convinced of the value of mechanical power that they decided to adapt it to other field work besides plowing. They purchased a Big 4 from the Gas Traction Company of Minneapolis for $2,200. The Big 4 could pull a 10-bottom Case hand-lift plow and a packer at 2 mph. The Big 4 came equipped with a furrow wheel steering device and required only one man to operate it; but because the plow was hand-lift, a two-man crew was still necessary.

The Big 4 was a gasoline-powered tractor, and the third man who hauled coal and water for the steamer was no longer needed. The Big 4 was not so powerful as the Peerless steamer and pulled two less bottoms at a slightly slower speed; but because frequent refueling stops were no longer necessary, the Pazandaks were still able to plow 24 miles a day for the same total of about 33 acres.

Getting gasoline in the early days of tractor farming was a considerable chore. The Reitans recall the Standard Oil man delivering fuel in a "brand new horse-drawn tank wagon." When he got to the Reitan farm he unloaded it into 50-gallon barrels by means of 5-gallon cans. The gas tank on Reitan's Hart-Parr held 40 gallons, but the workers always had to refill at noon to make sure they could operate all day. The Pazandaks, living in a less densely settled area, were not so fortunate as the Reitans and had to drive 15 miles to Ellendale to get gasoline. The Pazandaks owned a 500-gallon tank wagon drawn by four horses to haul their fuel. Each trip required a long day. They stored the fuel in a 500-gallon tank, which was high off the ground to enable filling the tractor by gravity hose. It was a two-man job transferring the fuel from the bank wagon to the storage tank with 5-gallon cans. Gasoline purchased in bulk in 1910 was 10¢ a gallon. After more farmers acquired tractors in the Ellendale-Fullerton area there was competition for the farmer's business. One enterprising gasoline dealer delivered to the farms in 10-gallon cans, using a flatbed wagon.

There were many shortcomings in the first tractors, and early owners with mechanical ability had a decided advantage over their fellow farmers who did not. Both the Reitans and the Pazandaks were mechanically inclined, which not only made it easier for them to keep their tractors operating and to make innovations improving performance, but also got their neighbors to call on them to help in making their machines operate. The Reitans' neighbors were frequently asking for help in timing, starting, and other technical points. Ludwig Walker had a Hart-Parr that kept getting out of time. Whenever the Reitans went over to time it and otherwise tune it up they always received $25. "That was big money for those days," commented Oscar Reitan. Because they were capable of working with big equipment, the Reitans were encouraged by their neighbors to get a feed grinder so that the horse-powered sweep grinder could be disposed of. The Reitans got a stationary double-burr grinder that could grind 124 bushels of oats or barley an hour. "The Hart-Parr always started," said Oscar, "and we would be busy many days during the winter, for the neighbors kept coming over to have their feed ground." The Reitans also operated a threshing rig and for many years spent more than 50 days each season in actual threshing. Starting about 1914 they were hired to pull road graders for building some of the first township and county roads in Clay, Wilkin, and Ottertail Counties. In spite of their faith in tractors and their ability to keep them running, the Reitans did not feel secure enough with them to get rid of their horses until 1931, when they purchased two McCormick Deering F-12's. Oscar recalls that the F-12's were able to do any job on the farm and were easy to overhaul.

The Pazandaks not only assisted their neighbors who needed mechanical help, but were always making adaptations to their tractors or tractor-powered machinery. They patented some of their ideas and sold them commercially. Their Geiser Peerless Steamer had gears that ran in the open and collected dust, causing rapid wear. The Pazandak brothers, using hand drills and chisels, made a gear box out of 16-gauge steel. After encasing the gears, the gear box was filled with heavy oil. Their Big 4, with its troublesome pencil bearings in the counter shaft, presented the Pazandaks with another mechanical challenge. The result was that they replaced the factory bearings with bearings of babbitt that they had made themselves.

The Pazandaks, after adjusting the tractors to their particular needs, started making multiple-hitch adaptations so that they could pull bigger loads. One of their first was a drill hitch that enabled them to pull four 11-foot single-disc drills. Once they got the drills properly assembled they chained 44 feet of Boss harrows behind them. By doing 20 rounds per day they could seed and harrow 160 acres in a little over 10 hours of field time. This required three men, one on the tractor, one watching the drills, and a third to treat and haul seed to the field. When the Pazandaks first started farming it would have required 14 men and 32 horses to do the same job. The hitch was later altered to pull seven 8-foot discs behind the Big 4. "We covered lots of ground, but it was clumsy. However, discing was not as fussy as seeding so that we used it several years until better equipment came along."

In 1910 Pazandaks purchased a Hansman binder hitch on which Louis Pazandak made some adaptations so that they could pull five 8-foot binders in tandem with the Big 4 tractor. A horse-drawn binder was used to open the fields and also to finish them if narrow strips were left. Five binders in tandem presented no problem on the corners, and the poles could be adjusted so that the binders all followed each other directly in line when traveling down the road. The angle of the pole was changed by a hand-cranked worm gear directly from the binder seat. A single emergency cord attaching the binders to the tractor could be pulled by any of the drivers in case of a breakdown. At the tractor end of the cord was a hammer that pounded on the fender to warn the driver of trouble. The Pazandaks kept their binders in excellent mechanical condition and except for twine problems seldom lost more than an hour each year with field breakdowns. Although this outfit required one more man than five horse-drawn binders, the Pazandaks saved the use of 20 horses. Besides, the tractor could operate all day long at 2¼ miles per hour, something that could not be done with horses unless there were 20 additional horses available for a midday change. With an 8-foot horse-drawn binder it took about one and one-half shockers to keep up, but the Pazandaks needed about two and one-half men to keep up with each of the tractor-operated machines. A wagon man hauled twine and water to the 19-man crew of shockers and binder operators.

Once the binders proved to work well in tandem, Louis Pazandak devised a hitch whereby he could pull seven wagons behind a tractor. The Pazandaks preferred to haul a large portion of their grain to the local elevator for storage at harvest. This was not only time-consuming but required more wagons than could practically be operated to keep the threshing machine going. However, this problem was solved by running a bar of iron ⅜ by 4 inches under the running gear of all the wagons, so that short poles could be used to attach all the wagons in a solid train, thus providing starting and stopping without any slack. The worst trouble came at the elevator because the Big 4 was too heavy and had lugs so that it could not pass through the driveway. Consequently, each wagon with its 125- to 150-bushel load was pulled into the elevator by the local dray team at a charge of $1 per wagon train. In the field the wagons were pulled alongside the threshing machine so that they could all be loaded without unhitching from the train. Using two wagon trains, the Pazandaks could haul grain to the Fullerton elevator four miles away fast enough to keep the machine going, except when threshing oats.

Once the wagon train was proven workable (about 1916) the Pazandaks sold their last horses. They had sold many in 1908 when they purchased their steamer, but they had to have a few for wagon work and lighter jobs. Shortly after getting rid of their last horses they acquired a lighter tractor—a Twin City 25-35, which pulled three bottoms at 3 mph. They pulled only three binders with this tractor and fewer wagons, but it was much less unwieldy than the Big 4. When rubber tires became available for tractors and wagons they adopted them at once. Ferd Pazandak said, "And because we were so close to town we never had to buy a truck."

The Pazandaks were firmly convinced that the tractor was a major reason why they were able to farm through the "desert season of the 1920's and '30's even though having three years without any crop was no joy ride." Ferd Pazandak reasoned that "even the 1920 model tractors were so much more efficient than horses." The Pazandaks' chief motive for acquiring tractors and bigger machinery was to "get out of horse chores and eliminate the need for a year round hired man. We liked to be free to travel in the winter," said the 89-year-old tractor pioneer.[5]

Max Peppel—"All I Could Do Was Manage"

Max Peppel, describing the 20-40 Case he purchased about 1910, said there were many little things which he could tell needed to be improved. As Mr. Peppel was nearly blind, he "had to explain to the blacksmith how the job should be done." Although his eyesight prohibited him from actual field work, his men would explain any tractor problems they were having and often he made adjustments by "feeling if it was right or by listening to the way the tractor ran."

Peppel, who said he always had excellent hired men—"maybe because they knew I had to rely on them to get the job done"—did considerable custom work. His 20-40 Case, with a seven 14-inch bottom plow, cost $1,800, "which was a lot of money in those days, but it paid for itself," according to Peppel, "because we got our work done earlier and then had time for custom plowing." They could plow 24 acres a day and by adding a Presto light could do nearly 50 acres by going "around the clock." Peppel charged $1.75 an acre for custom plowing.

After he determined that the tractor could do the job, Peppel built a hitch to pull two 10-foot discs side by side followed by two 10-foot drills. This enabled him to disc and drill 20 feet at one time, a job that formerly would have required 14 horses and four men. After the tractor proved itself, Peppel got rid of 12 horses. At one point he was so enthused with the potential of tractor power that he decided to quit farming and just do custom work with a full line of tractor machinery. "We were kept busy plowing, discing, and seeding but it really did not pay," said Peppel, "for sometimes you had to wait too long to get your money."

Max Peppel believes the turning point for him in farming came in 1926 when he purchased an I.H.C. Farmall with a two-bottom plow for $700. After plowing 580 acres, that one tractor was used to disc, plant, cultivate, and harvest during the rest of the season. The shift to the untiring tractor power is exemplified in the tractor-powered binder, which "was really something as it made it possible to harvest right through the hottest weather without stopping." He kept that Farmall for seven years, but also got an I.H.C. 15-30 to do the heavy work as the farm operation expanded. When rubber tires became available he traded the Farmall for two F-12's. Shortly after buying the F-12's Peppel said, "I was asked to give a speech at a farmers' machinery show and I told the crowd that one F-12 could replace eight horses and two men. Someone in the crowd called me a liar so I offered to prove it, but no one took me up! Today [1973] everybody knows it is true. Before I retired in 1940 I was operating with seven tractors."

Eighty-eight-year-old Max Peppel, commenting on his farming career, said:

Once I was asked by our minister in the depths of the 1930's how come I never complained. I told him I did not like complainers and besides my family was living good and I was getting ahead. Because I was blind I had decided to quit farming about World War I when I was able to get a job as foreman with the Great Northern. Then the local banker wanted me to take over some of the farms which others had walked away from. I agreed on condition that the bank finance some good breeding sows and milk cows. After I started farming for the second time in 1924 until I retired I never had a year that I lost money—that I swear on a stack of Bibles. And in all my years of farming I had to hire all my field help because I could never see to cultivate corn or strike out a land of plowing. I did all the yard work because I could feel my way around to fix things, feed hogs, cows, and do the milking. Of course, all I could ever really do was manage.

In the 1920's and 1930's this blind man saw clearly the secrets that were to become chief ingredients of agriculture's success—mechanical power and management.[6]

Eugene Young Breaks Sod with Tractors on Three Frontiers

Eugene Young was born at Sanborn, North Dakota, in 1893, where his father served as a telegrapher and station agent. Mr. Young, also a part-time farmer, got the urge to go farming full time in 1902. After packing their possessions and saying farewell to friends, the Youngs struck out for New Coleharbor, the frontier of North Dakota. There was no bridge across the Missouri, forcing them to make the last leg of their journey by ferry. Upon arrival, Mr. Young purchased the rights to a quarter section from an earlier homesteader for $150, and the Young family lived in the log cabin that was there until they were able to construct their own frame home. Because his sons were old enough to help him with field work, Young purchased additional land from the Northern Pacific for "about $5.00 an acre and kept doing so until 1909 when he owned 800 acres."

In 1911 the International Harvester dealer appraoched Young with the idea of selling him a repossessed Mogul Type B single cylinder 20-H.P. tractor. A homesteader had purchased it in 1910, but when his flax crop failed he left the tractor, plow, and several barrels full of gasoline sitting in the field and "skipped the country." The dealer offered this 6-ton tractor and six-bottom plow for $1,000. Young refused to buy, but the dealer persisted and offered to have him break 150 acres of prairie owned by the dealer as down payment. Young consented, but "it was so dry in the fall of 1911 that we could only pull four bottoms in the sod and that was after the rain came."

The Mogul was the first non-steam tractor Eugene Young had ever seen. That fall he and his father camped on the International dealer's land and broke his sod. They did their cooking in the open and slept in a little shack. On one damp, frosty morning the tractor would not start, and after Eugene and his father had worn themselves out pulling the flywheel, they took the ignitor off and went 12 miles to Underwood to find out what was wrong. By that time the ignitor had dried out and was "O.K." They decided to buy an extra ignitor and have it in the oven over night so that it would be dry for starting during the frosty fall mornings. Even though the Mogul had open gears and the babbit bearings needed constant attention, the Youngs soon appreciated the fact that it could do the work of their "twelve not-so-good horses," so in 1912 they decided to get a second tractor.

The 1912 slightly used Mogul was a 25-H.P., single-cylinder, 14,000-pound machine and cost $1,250. The Youngs also secured a 32 by 36 steel threshing machine. "Dad was skeptical about buying a steel machine because all the neighbors had told him there was no way to repair it. A wooden one could be repaired with boards and nails." But the Mogul and the "new steel thresher" proved to be the most efficient threshing rig in the area. "Besides, the steel machine was not bothered by rains and dampness like the wooden ones were. The owners of the wooden threshers always were fussy about covering them when the showers came." The smaller gasoline-powered tractor and steel machine made "good money" for the Youngs so that they decided to expand a bit.

In 1915 Eugene Young purchased an 11-ton 30-60 Aultman-Taylor tractor with extension rims and a steering guide, a 42-64 Aultman-Taylor thresher, and a gas wagon, for a total investment of $4,200. The rig arrived October 23 and "everybody in Coleharbor made fun of me for getting such a big outfit so late in the season. But we had fifteen days of ideal threshing weather and broke some records that had been held by the big steam rigs." Young felt good about the new records because the "steam powered people all laughed at my four cylinder gas tractor—but it had smooth power."

Confident that the Aultman-Taylor would be reliable, Eugene Young rented two sections of native prairie on the Fort Berthold Reservation near Van Hook in 1916 for 50¢ an acre. Young had to break the sod and fence the area. His equipment had to be hauled by rail because Van Hook was over 100 miles from Underwood. The Aultman-Taylor pulled an 8-14 P and O breaking plow that required a two-man crew because each bottom was on a separate lever. By working around the clock a five-man crew was able to break 40 acres a day, the work of 50 horses and 10 men. Immediately after plowing, while the sod was still fresh, an 11-foot tandem disc, an 11-foot double-disc Kentucky drill, and a 22-foot wooden Boss harrow were pulled in tandem behind the Aultman-Taylor to plant the crop. The extra-wide harrow enabled two draggings of the land with one trip over the field. The tractor and its five-man crew plowed and seeded a total of 1,280 acres into flax by mid-June. That fall the Aultman-Taylor pulled four new $125 Minnesota binders with flax attachment in tandem to harvest a crop that averaged 12 bushels an acre.

Young had to get the help of his brother and men from Underwood as well as import horses and wagons to thresh that fall because there was little available labor in the Van Hook area. Flax was $2 a bushel in the fall of 1916 when Young first sold, but before he finished hauling he was getting $5 a bushel. His entire crop averaged "about $3.50 a bushel and paid for the tractor, plow, disc, drill, harrow, four binders, and thresher besides all expenses. Only that spring I was $3,000 overdrawn at the Van Hook bank and had to mortgage the crop to pay operating expenses." Young received $5 an acre for plowing, discing, seeding, and harrowing 160 acres of flax for a neighbor who had no equipment. That fall he got 75¢ an acre for cutting it, plus 15¢ a bushel for threshing the crop.

In July, 1917, after putting in the crop at Van Hook, Young moved his Aultman-Taylor and plow to McLaughlin, South Dakota, to summer-break native prairie, which he rented for five years at 50¢ an acre per year. He purchased a 20-40 Rumely to harvest and thresh on the Fort Berthold Reservation. He was happy with both of his big tractors, but "they took tender loving care." The Aultman-Taylor with its compressed air starter "worked beautifully and it was a long time before we ever used the crank." Young had very little mechanical trouble with either tractor, and he was fortunate, for one time a crankshaft broke on the Aultman-Taylor and "it took seven days after wiring Minneapolis before the part arrived at McLaughlin."

Crops were not good on the Fort Berthold Reservation in either 1917 or 1918, but Young kept his faith and continued to expand. He purchased a Case steamer at a sheriff's auction in 1918 for $250 and used that for threshing so that his gasoline tractors could be used for fall plowing. That year he purchased two tractor adapters for Model T Ford pickups, enabling them to do field work. These adapted pickups were used to pull seeders and worked well except that the engines wore out rapidly. Hard-rubber-tired Model T trucks were used to haul grain to the elevators, with each truck hauling 50 bushels. Later Young adapted his 20-40 Rumely to pull four wagons loaded with 550 bushels at three miles per hour to McLaughlin. A long rope was used to pull each wagon into the elevator to avoid running the tractor into the driveway.

By 1919 Young's power equipment consisted of two 30-60 Aultman-Taylors, a 20-40 Rumely Oil Pull, two adapted Model T pickup tractors, and several Model T trucks. That year he was farming 2,500 acres and for the only time, ever, his neighbors using horses got into the fields before he did. Horse operators were in the field in March and got their crops in before a big snow storm in early April. When the cook car ran out of food it took six hours to drive four miles to town to get new supplies. That snow was the last moisture for the growing season. When the crop came, "the grasshoppers chewed off half of each kernel of rye that grew." Later in the season the grasshoppers returned and cleared a strip of about 20 rods around every flax field, reducing the yield to 3 bushels per acre. "When the hoppers were gone about a hundred horses belonging to the Indians broke through the fence and ate into the windrowed flax. "Somehow I did get about ten bushels of wheat per acre," Young said philosophically. He added, "Maybe that had something to do with my getting rid of the last six horses I owned in 1924."

Able to take the bad years with the good ones, Young continued to expand his operations, buying or renting land whenever it was available. He had several other ventures going, along with his farming enterprise, so that when the black rust of 1922 brought crop failure he was not too disturbed. Young's supreme test came in 1925. He had $25,000 in the bank and decided to purchase 10 12-20 Twin City tractors. These weighed 4,500 pounds each, pulled three 14-inch plows, and retailed at $1,200 each. As a dealer Young was able to buy them at $900. He also purchased a plow and other equipment for each tractor. After receiving the machinery and the invoice he mailed his check to cover the bill. While the check was en route, the McLaughlin bank went broke, leaving Young with 10 sets of equipment unpaid for and no bank account. That year Young was farming 10,000 acres with tractors, including 2,400 acres of new sod that he opened and seeded in 24 days. Besides the bank failure, 1925 also brought Young a crop failure. "My hair turned from dark to white that year, but we were able to survive and prosper," said the 79-year-old veteran of drought, blizzards, and grasshoppers.[7]

Other Early Tractor Users

Not all tractor pioneers went about acquiring mechanical power with the certainty and eagerness of the Reitans, Pazandaks, Peppels, or Youngs. There were many who had only special use for the tractor or were otherwise skeptical about buying one. Oliver Dalrymple, owner and manager of a 100,000-acre bonanza farm in eastern North Dakota, remarked when he first saw steam-powered plowing that he would leave the experimenting to others. He preferred horses because the steam tractor could do no more than thresh and plow, so that horses were needed anyway. Obviously two sets of power on a huge bonanza would have caused major financial problems. But, in 1908, Dalrymple did buy an International Harvester Company Mogul tractor. Apparently there was reluctance among his men as to who should operate it because the laborers preferred to stick with the horses rather than change their ways. After driving the Mogul for an entire day and disliking the experience, one man pointed to the "I.H.C." printed on the fender and haltingly remarked, "*In Hell Continually.*" Many hired men and even some farmers complained about the work of plowing with the tractor in contrast to horse plowing. More than one objection came from the fact that it kept them so actively occupied they were unable to "sight-see or even spit snoose."

F. A. Gummer, who farmed in northern Montana, purchased a Fordson and a two-bottom 12-inch plow in 1917 for breaking land. Even though the Fordson worked well, he remembered that on cold fall mornings he often had to use a team of horses to pull it to get it started.

A. J. Robinson, manager for the 24,000-acre Schermerhorn Farms at Mahnomen, Minnesota, in 1917, had similar problems with all of his early mechanical power units. The farm used two Caterpillar track-type tractors for breaking timber land, "and the tracks were constantly breaking on those units." At the same time, Schermerhorns had 14 Fordsons "which really gave us trouble and required two mechanics working full-time to keep them going," said Robinson. To top those problems there were three Republic trucks with hard rubber tires and "it seems like they were stuck all the time." When the Schermerhorns cleared the land and established their farm, there were no roads in the general area of the farm, forcing them to build their own to effectively use those trucks.

Oscar and Melvin Lund were not so critical of the Fordson's starting ability. "In all the years we owned our Fordson, from September, 1919, to August, 1972, it never gave us trouble starting. Of course, we were both good at mechanics and kept it in top condition," said the 75-year-old Oscar Lund. The Lunds paid $745 for their tractor which included a pulley plus $150 for a two-bottom

Oliver plow made especially for that tractor. The Lunds plowed five to seven acres in a 10-hour day, using 20 gallons of kerosene, which cost 12¢ a gallon. This was the only tractor along with four horses on a 400-acre farm that had 200 acres in crops. On one occasion when two horses were sick, the Fordson was hitched alongside the other two horses on a 22-disc drill. Even though the Lunds liked their Fordson for field work they felt that its greatest contribution was on the belt running a 20-36 thresher, a silo filler, or the saw mill. (See photo of this tractor.)

C. H. Underlee, who did not start farming until 1934, related that the biggest break he had in farming was beginning with a used Fordson tractor:

I was a greenhorn at farming and it looked to me like a tractor was a better way to farm than using horses. It would have taken six horses to replace the Fordson and I wouldn't have had enough to sell from my farm to live on. The tractor was the best quack killer because it could work in the very hottest temperature and it is easier to kill quack on such days. I never really liked horses, so the next year to operate more land I bought a John Deere Model D, which helped me replace horses for good.

From that single Fordson the Underlee farm increased its use of tractors to a track-type tractor, two 4-wheel-drive tractors and nine standard-type tractors, totaling more than 900 horsepower.

Whether or not farmers had trouble with the early tractors, the benefits of their performance and constant improvement caused most farmers to adopt them about as rapidly as they could afford to do so. William R. Page, long-time agriculturalist in the northern Red River Valley, says the first tractor around Hamilton, North Dakota, is reputed to have bankrupted its owner and "certainly kept many farmers in that area from buying one." It was not until 1919 that the Page family purchased a 22-36 I.H.C. that was excellent for fall plowing "but was constantly boiling in the hot summer." Farms in that area were small, and the urge for tractors was not so great as farther south and west in the Valley. Page said, "As a county agent in the 1920's I could see the advantage of tractor power and could sense that the farmers were waiting to try them. Farmers, except for the most sentimental horse lovers, were not against tractors but were not sure of themselves."

To Page it was apparent that the tractor would win out when he noticed that farmers were not going to the multiple-horse-hitch demonstrations put on by the American Horse Breeders Association. Instead, they were going to tractor demonstrations! "When the weather got hot and dry in the early 1930's the economy and ability of the tractor to effectively control weeds proved to be the turning point in its favor," recalled Page. "Farmers first objected to rubber tires because of cost, but when World War II came and tires were rationed it seemed that this encouraged farmers to insist on having tires."

Ole Flaat, who was to become one of America's largest potato growers, started farming at Fisher, Minnesota, in 1918. Ole and his father had 12 horses on 320 acres, but when they purchased more land in 1927 they bought an I.H.C. Farmall. Ole realized that if he wanted to become a bigger farmer he needed a better source of power. He added, "My dad, who was 58 at the time, saw the need for the tractor and encouraged me to buy it." In just a few years, the Flaats purchased three Caterpillar track-type tractors—"those cats enabled us to really expand."

By 1934 the Flaats, who by then had started using rubber tires on their tractors, "were able to get rid of all their horses except one team which we couldn't bear to part with because the whole family was so fond of them," remembered Ole. The Flaats continued to have racing and riding horses throughout their farming career.

By the early 1940's the Flaats had 20 wheel tractors and three I.H.C. TD-14 track-type tractors. The Flaats profited not only because of the efficiency of the tractor but because they were able to hire better quality men, who "liked working with good power units rather than horse drawn machines."

Clarence Romain recalls that he and his brother both left their Chester, Montana, farm home because they were made to work with horses. They both "just hated it because we yelled ourself hoarse and used wire whips on the horses, but could not get them to move in the hot weather." When their father let them buy a Model D John Deere they returned home to help farm.[8]

The Impact of the Tractor, or the Big Farm Better Tilled

Henry A. Wallace, in a radio address October 24, 1938, listed seven causes of the farm problem. Of those seven causes only two were of such a nature that farmers as individuals or a group had any control over—the opening of 40 million acres of new land during World War I and the displacement of the horse by the tractor. This eventually freed an additional 95 million acres of land for other production. When Wallace spoke there were still 16 million horses, and only about 35 to 45 million acres of land had been freed; but that amount was making a clear change in the agricultural production problem.

The surplus problem of which Wallace spoke was obvious to all Americans and appeared to be a permanent feature of the economy. Liberty Hyde Bailey, the dean of American agriculturalists, explained that in the 1920's surpluses were caused by the increase in the productive power of man. Bailey touched upon the total problem saying, "If there is consistent overproduction . . . then there are too many farmers, not too much production to the acre or to the man. . . . The situation will adjust itself in time, . . . Eventually we shall arrive at a workable program as to the number of farmers required to produce the needed supplies."

Bailey did not advocate reducing yields or output per man, nor did he recommend giving the problem to anyone running for political office, "because that would only complicate things." It was clear to him that mechanical power had caused much of the surplus problem starting with the World War I era. What was clear to Bailey in the 1920's probably was not clear to some rural Americans 50 years later, or at least many refused to acknowledge that they understood the real cause of the problem. Although mechanical power had already helped to create an overproduction problem by the 1920's, Harold Pinches, writing in the late 1930's, suggested that agriculture was the greatest industry still not sufficiently aided by industrial engineers. Pinches acknowledged that the adoption of the internal combustion engine and electricity by agriculture had caused a production revolution in the 1920's and 1930's, but at the same time he warned that future mechanical innovations would put enough farm laborers out of work so as "to constitute a major social and economic problem." Pinches' prophecy proved correct.

The burden of plowing—the greatest power consumer in agriculture—was greatly reduced with the advent of the tractor. One man with a spade can till one acre of land in 96 hours, while an ox-drawn plow can plow an acre in 24 man-hours. Using five horses and a two-bottom gang plow, it takes 1.8 man-hours to turn over an acre of ground. However, by 1938, the standard tractor needed only a half hour to do the same job, and by 1970 that task required only four minutes—an increase in productivity of over a thousand-fold.

Replacing horses with tractors produced a many-sided problem for the farmers while they were going through the transition. First of all, horses, a very valuable asset, fell drastically in price during the changeover. Horses, which in 1910 cost from $125 to $250 each, had a market value of 2¢ a pound, or roughly $20 to $25 per head by 1950. Not only was the value of horses constantly dropping, but the market for horses produced for cash sale came to an end. Prior to this, farmers had sold several hundred million dollars' worth of horses and mules to other farmers and users of horses in the city each year. Now this market was gone. It required from 3 to 10 acres of ground each year, depending on the area, to feed a horse; so when lumber camps, mines, and city users got rid of their horses and mules, the 16 million acres of land needed for feed had to be diverted to other crops. As the farmers got rid of their work animals another 70 million acres of land were also freed for other production. In all, somewhere between 75 and 95 million acres of crop used to feed work animals were added to production of commercial crops, thus creating havoc in an already depressed market. At the same time, farmers were spending up to one-sixth of their income for mechanical power in addition to the necessary cash outlay for operating expenses.

The pain of the transition lasted over a period of 40 years because during the Great Depression of the 1930's many farmers delayed purchasing tractors in an effort to avoid cash expenditures. With a horse consuming at least 40 bushels of oats and 2 tons of hay a year at a cost equal to about 400 gallons of tractor fuel, this was poor economy and forced the farmer to change or quit.

Surveys conducted by the government in the early 1930's indicated that in most areas it was not uncommon for a tractor to displace three to five families. In an extreme case nine families were displaced by one tractor. Most of the displaced farmers came from the bottom of the economical scale. This created a social problem in addition to an economic problem that continued until "the necessary adjustment," as suggested by Bailey in 1927, could be made. In the 1970's the end of the adjustment was not yet in sight. Those farmers still on the land were forced to become full-time commercial farmers, something that had not been true of American farmers, for historically most of them had been part-timers. Two kinds of farmers clearly evolved—the large-scale commercial operator who produced the bulk of the nation's food and fiber, and the industrially employed residential or hobby farmer who struggled to maintain a rural way of life. From 1910 to the 1970's American agriculture had gone through three revolutions— mechanical, technological, and managerial—which had left many farmers behind while at the same time giving a boost to the progressive farmer. Meanwhile, a nation of consumers benefited considerably from these revolutions.

A nationwide United States Department of Agriculture study in 1940 determined that horses and mules worked about 16½ acres of crop per animal. Studies had long established that larger farms were more economical in use of horse power than smaller farms, so even in the days prior to mechanical power the larger units appeared to have certain economic advantages. Individual farms studied varied from 17 acres of crop land per horse on 320-acre farms to 40 acres per horse on the Blanchard bonanza managed by George Hilstad. On bonanza farms specifically, the smallest number of acres operated per horse was 25. The largest of the bonanzas, that of Oliver Dalrymple, had 37 acres of crop per horse. Dalrymple used a system of renting horses from nearby farms and lumber camps to give him additional economy not available to the small farmers. On the average, small grain farms in the Dakotas and Montana operated about 26 acres of crop per work animal. In Iowa and Minnesota, where there were more row crops, the acreage was lower.

Some of the earliest tractor studies clearly established that tractors of 30 H.P. were more efficient than the smaller-sized machines. It was proven that a 15-H.P. tractor would be more suitable than larger models on a 160-acre farm, but on that size farm it was difficult to justify any tractor. On a 1,200-acre Dakota grain

	Plow Cost per Acre	Grain Hauling per Bushel	Initial Cost
Horse	$1.63	5.6¢	$6,000
Tractor	$1.05	2.8¢	$2,750

farm a tractor could plow at a cost of $1.05 an acre. This included a 16-percent depreciation on the tractor and a 10-percent depreciation on the plow, as compared with $1.63 an acre for plowing with horses. The tractor could haul grain to the elevator 10 miles distant for 2.8¢ per bushel, while with horses it cost 5.6¢. The initial cost of a 30-H.P. tractor to operate a 1,200-acre Dakota grain farm was $2,750, while the necessary horses and harness for the same farm would have been $6,000.

Mechanization changed agriculture from a "labor-intensive" to a "labor-extensive industry." That factor has been one of the major causes of farm enlargement as well as working in the farmer's favor in several ways. Probably one of the most significant aspects is that of timeliness of seeding and harvesting. There are many good illustrations of the value of timely work, and no good manager will dispute the point. One of the best examples of what timely work means is that of Herbert Reese of Greenbush, Minnesota, who used two Caterpillar 60's to till, plant, and harvest 500 acres of land in 1934. By working one Caterpillar 24 hours a day he was able to plow the 500 acres in 15 days, while a second outfit followed with disc, drill, and packer, putting flax seed in moist ground. Because 1934 was a dry season, getting the seed started in moist ground made a big difference. Reese's flax averaged 12 bushels an acre, while his neighbors who used horses and seeded in dried plowing got spotty stands and averaged only 3 bushels per acre. Timeliness and speed of getting planting and harvesting done in shorter periods of time have also enabled farmers to extend cropping into more marginal areas, where the days saved through the use of machinery in effect lengthen the growing season.

The actual time saved through elimination of the horse and the adoption of the tractor have resulted in some of the greatest changes in American agriculture. This, plus the fact that people involved in agriculture have been chronically underemployed, has released a steady supply of labor from the farm to other industry. A study conducted by the United States Department of Agriculture in 1938 determined that each tractor purchased up to that date had displaced three horses, and each car or truck displaced six-tenths of a horse. Because of man's increased productivity, the time needed to care for a horse for one year had dropped from 61 hours in 1909 to 50 hours in 1938. In 1938 nearly 8 million horses had been displaced on the farms, which meant a direct saving of 400 million man-hours of labor on work horses plus an additional 65 million man-hours saved by not having to take care of colts. Not having to produce horse feed saved another 530 million man-hours, and the increased speed of field work because of the tractor's use saved another 195 million man-hours. The above figures plus other time saved by not having horses to care for totaled 1,315 million man-hours in 1938 alone. Based on a 3,000-hour work year on the farm in 1938, that freed about 440,000 men from the farm.

A 1945 Department of Agriculture study established that each tractor and associated equipment saved 850 man-hours per year.

Where a tractor had displaced only three horses in 1938, by 1945 the machine had increased enough in size and convenience to displace 4.4 horses. Tractors, trucks, and associated equipment were saving over 3 billion hours a year for the farmer by 1945. Even greater savings have come since that date. Total investment in all forms of farm power (exclusive of electricity) and equipment increased from $7.3 billion in 1945 to $32 billion in 1970. Total time needed in agriculture in 1910 to supply a nation of 92 million people was 22.3 billion hours. Each farm laborer was capable of providing food for 7.1 people. In 1970 only 6.5 billion hours were needed in agriculture to provide for 200 million Americans, and each farm laborer could provide food for about 54 people. The tractor, the truck, and the automobile, as well as electrical power, had been the biggest contributors to that progress.

Examples of what the tractor did to save time are so dramatic that it is easy to overlook the importance of the truck as a timesaver to farming. Erick A. Erickson of Portland, North Dakota, using four horses to haul grain to the nearest elevator 8 miles away, could deliver two loads of 120 bushels each if four fresh horses were used on the second trip. The 16-mile road trip took six to seven hours. In 1929 Erickson purchased a new 1½-ton Model A Ford truck for $460. That truck could haul 150 bushels per load and could carry six loads in one day if there was no delay at the elevator. Hand loading of the grain was then the most retarding factor. The time needed to haul a bushel of wheat to the same elevator eight miles from the farm was reduced from three and one-fourth minutes per bushel with horses to about 50 seconds with a small-sized 1929 truck. On 1,000 bushels of grain the total saving in time was 40 hours.

The Kroeker family settled in Winkler, Manitoba, in 1876. When Abram Kroeker delivered his first grain to Emerson on the Red River in 1878 it took him six long days to make a round trip hauling 60 bushels of grain. In 1972 the Kroeker brothers, using one of their five-axle semitrailer units could deliver 800 bushels of wheat and return in five hours. The time necessary to deliver a bushel of wheat was reduced from 66 minutes per bushel to 22 seconds per bushel.

In the 1920's and 1930's farmers and other leaders in agriculture, including Liberty Hyde Bailey, were aware of the great strides made with the tractor and the motor truck, but they were also expressing the opinion that further progress along those lines was limited. U.S.D.A. studies in 1945 predicted that although each tractor displaced 4.4 horses, after 1955 a tractor would displace only about two horses. The reason was the feeling that there were some jobs the tractor would not be able to take away from the horse. An Iowa horseless farmer, speaking at the annual Farmers and Homemakers Week at the University of Minnesota in January, 1927, said of himself, "It would be difficult to operate without horses except for the fact that I was able to get the use of my neighbor's teams. . . ." Yet by 1950, 39 percent of the commercial farmers of Minnesota operated without horses, and many in the Dakotas had been doing so since the 1920's.

In addition to the adoption of tractor power, rural electrification came to the farms after World War II and enabled a time-saving similar to that of the tractor. The two factors caused total production to increase considerably faster than the cost of production. This gave the early adopters and larger units a decided advantage in net profits, for their cost per unit of production did not rise nearly so fast as on the smaller farms. In the 1950's the total cost of full-time equipment on an 80-acre farm was $75 an acre, but on a 300-acre farm that cost was about $21 an acre. There were advocates of small farmers co-operating by purchasing machinery jointly to reduce the cost of operation. Grant Mattson, one of the nation's pioneers in machinery rental, predicted that the "smaller-scale farmers must do joint purchasing or leasing of large-scale equipment if they wanted to maintain a competitive position."

The advantage in favor of the larger farmer is borne out in the fact that nationwide on all commercial farms there is a tractor for each 68 acres of crop land, on farms over $100,000 in gross sales each tractor handles 113 acres, and on farms of over $1 million in sales a tractor operates 142 acres. Robert J. Hampton, Vice President of Ford Motor Company, addressing the American Society of Agricultural Engineers in 1967, forecast that by 1980 farmers would be operating tractors in the 300- to 500-H.P. range and that possibly some would be electronically controlled. By 1973 Mr. Hampton's prediction was proving a reality.

It took the small tractor, such as the 12-H.P. Bull tractor introduced in 1913 and Henry Ford's 1917 Fordson, to make tractor farming popular with small farmers, but a 1913 publication correctly predicted what would happen:

The small tractor is coming immediately. The large tractor which is the four horse team among farm machines, looms up, threatening to displace the small farmer's power. With the large tractor making far greater profit in large scale production, there comes the question of larger farms, more capital, and possibly capitalized ownership. Effective cooperation is the only alternative. It is up to the man on the farm. . . . The larger the machine, the more economical, and the larger the farm, the greater the possible profits and the finer the opportunity for the well-trained businessman to exercise his managerial ability. The farm does not now require brute force and physical endurance so much as careful management. . . . The ideal of the "little farm well tilled" is being superseded by that of the big farm better tilled and better managed.[9]

NOTES

[1]R. B. Gray, (Compiler) *Development of the Agricultural Tractor in the United States: Up to 1919. Part I* (American Association of Agricultural Engineers, 1954), pp. 3, 4; Clark C. Spence, "Experiments in American Steam Cultivation," *Agricultural History*, XXXIII, No. 3 (July, 1959), 108; Earl D. Ross, "Retardation in Farm Technology Before the Power Age," *Agricultural History* XXX, No. 1 (January, 1956), 12-16; Warren, pp. 31-33, 344-349; Isaacson, p. 34; Edward A. Rumeley, "The Passing of the Man with the Hoe," *World's Work*, Vol. XX, No. 4 (August, 1910), 13249; W. R. Porter, *Cost of Producing Farm Crops*, North Dakota Agricultural Experiment Station Bulletin No. 104 (Fargo, 1973); William H. Cavert, "The Technological Revolution in Agriculture 1910-1955," *Agricultural History* XXX, No. 1 (January 1956), 20-24.

[2]Gray, pp. 13-18; The 1902 Hart-Parr weighed 20,000 pounds, had a two-cylinder engine that operated at 250 rpm and developed 17 H.P. on the drawbar. Top speed was about two miles per hour. Promersberger, pp. 1-8; Reynold M. Wik, "Henry Ford's Tractors and American Agriculture," *Agricultural History*, XXXVIII, No. 2 (April, 1964), 79-80; *Historical Statistics of the United States from Colonial Times to 1957*, United States Department of Commerce (Washington, D.C.), pp. 280; "New Hope for the Man with the Plow," *World's Work*, XX, No. 3 (July, 1910), 13, 116; Rumeley, pp. 13246-13248; Ross, pp. 14-16; *Modern Farming, The Passing of the Hoe*, Gas Traction Co. (Minneapolis, 1908); Herbert N. Casson, R. W. Hutchinson, Jr., L. W. Lewis, *Horse, Truck, and Tractor: The Coming of Cheaper Power for City and Farm* (F. G. Browne & Co., 1913), pp. 1-3, 10-12, 59, 121; McColly, p. 398; John A. Hopkins, *Changing Technology and Employment in Agriculture*, B.A.E., United States Department of Agriculture, p. 56; E. R. Eastman, *These Changing Times: A Story of Farm Progress During the First Quarter of the Twentieth Century* (New York, 1927), pp. 38-41; John A. Hopkins and Eldon E. Shaw, *Trends in Employment in Agriculture 1909-1936*; W.P.A. Research Project No. 8-A (1938), p. 60; M. L. Wilson, *Big Teams in Montana*, Montana State College Extension Service Bulletin No. 70 (Bozeman, 1925), pp. 29, 76-107.

[3]Eastman, pp. 42-49; Promersberger, pp. 7, 10; Yiyiro Hayami and V. W. Ruttan, "Factor Prices and Technical Change in Agricultural Development: The United States and Japan, 1880-1960," *The Journal of Political Economy*, LXXVIII, No. 5 (September-October, 1970), 1120-1128; Benjamin H. Hibbard, *Effects of the Great War upon Agriculture in the United States and Great Britain*, Preliminary Economic Studies of the War No. 11, (Oxford Press, 1919), p. 87; Gray, Part II, pp. 1-6, 10-22, 53, Part I, pp. 24-28; Arthur G. Peterson, "Governmental Policy Relating to Farm Machinery in World War I," *Agricultural History*, XVII, No. 1 (January, 1943), 37-39; By 1918 there were 200 tractor manufacturers in business, and the government took steps to limit their production. Ross, pp. 16-17; *Historical Statistics of the United States 1971*, Bureau of Commerce, p. 281, The peak year for tractor production was 1948, when 753,623 machines were made and sold. By 1950 production had eased to 693,646, and the trend has continued downward slightly each year; but tractors have become larger, so total horsepower produced each year

has become greater. At the same time, a large portion of the market has been made up of lawn and garden tractors; L. H. Bailey, *The Harvest of the Year to the Tiller of the Soil* (New York, 1927), np.; Bogart, p. 135; Barton W. Currie, *The Tractor: And Its Influence upon the Agricultural Implement Industry* (Philadelphia: Curtis Pub. Co., 1916), pp. 138, 179, 206, 223; Casson, pp. 124-129, 132-142; Wik, pp. 81-85; Larson, p. 3; Personal Interview with John Sutton, Sr., Onida, South Dakota, April 11, 1972. Mr. Sutton experienced plowing with oxen, with walking, sulky and three-bottom horse plows, and with tractors. He was born in 1898.

⁴Personal Interview with Oscar S. Reitan, Comstock, Minnesota, February 20, 1973. Mr. Reitan was born in 1891 and was actively involved in the operation of his father's farm at the time the first tractor was purchased in the early 1900's.

⁵Personal Interview with Ferd A. Pazandak, Oakes, North Dakota, December 16, 1972. Mr. Pazandak, who was born in 1883, probably has the distinction of operating the first horseless farm in North Dakota. Pazandak was very aware of the fact that he was a pioneer and recorded his inventions and adaptations to tractor farming by taking pictures of them. See the picture section for several of his photos. Prior to his death, shortly before his ninetieth birthday, Mr. Pazandak was a constant source of information about early tractor farming. It was on his farm that the jacket picture of *The Day of the Bonanza* was taken in 1910.

⁶Personal Interview with Max Peppel, Barnesville, Minnesota, February 20, 1973. Mr. Peppel, who was born in 1885, had tractors on his farm continuously from 1910. Like the Reitans, Pazandaks, Sutton, and others, he credits much of his ability to go through the 1920's and 1930's to the efficiency of the tractor over horses.

⁷Personal Interview with Eugene M. Young, McLaughlin and Rapid City, South Dakota, April 9, 1972. Mr. Young was born in Sanborn, North Dakota, in 1893 and in addition to experiencing homesteading with his parents in McClain County, North Dakota, also opened new lands in Montana and South Dakota.

⁸John Stewart Dalrymple, *No. 1 Hard: Oliver Dalrymple, The Story of a Bonanza Farmer* (Minneapolis, 1960), p. 55; Gummer Interview; Personal Interview with Oscar and Melvin Lund, Twin Valley, Minnesota, August 20, 1972; Personal Interview with C. H. Underlee and sons, Nolan and Leslie, Hendrum, Minnesota, January 19, 1972; Personal Interview with A. J. Robinson, Mahnomen, Minnesota, June 6, 1971. Mr. Robinson, who was born in 1884, served as manager of the Schermerhorn Farms from 1917 to 1940,̄ at which time he purchased part of the operation; Personal Interview with William R. Page, Grand Forks, North Dakota, February 15, 1972. Mr. Page, born at Hamilton, North Dakota, was a county agent from the 1920's through the 1960's in Grand Forks County, North Dakota; Personal Interview with Ole A. Flaat, Grand Forks, North Dakota, January 18 and February 15, 1972. Mr. Flaat, born in 1897, became one of the largest volume potato growers in the nation, planting more than 4,000 acres a year during World War II; Romain Interview. As a farm boy in southern Minnesota, the writer remembers hearing a neighbor constantly swearing at his horses to get them to move. Besides yelling, it was also possible to see that he was forever using the whip, so what the Romains said was apparently true with many others.

⁹O. E. Baker, et al., *Agriculture in Modern Life* (New York, 1936), np.; Pinches, pp. 510-513; A. N. Johnson, "The Impact of Farm Machinery on the Farm Economy," *Agricultural History*, XXIV, No. 1 (January, 1950), pp. 59-61; Martin R. Cooper, et al., *Progress of Farm Mechanization*, United States Department of Agriculture, Misc. Publication No. 630, Washington, D.C., 1947; Donald D. Durost and Warren R. Bailey, "What's Happened to Farming," *Yearbook of Agriculture, 1970: Contours of Change*, ed. Jack Hayes (Washington, D.C., 1970), p. 2; John Lier, "Farm Mechanization in Saskatchewan," *Tijdschrift Voor Economische en Sociale Geografie* (May-June, 1971), pp. 183-189. In Saskatchewan it took an estimated 10½ acres to feed a horse. Between 1926 and 1962 enough land was freed from producing horse feed to equal 54 percent of the total wheat acreage of the

province; Hopkins, p. 67; Casson, pp. 67-69, 156-159, 165-189; Drache, *The Day of the Bonanza*, see for details on horse operating expenses on bonanza farms; Herbert R. Reese, *Seventy Years Down the Road: The Life Story of Herbert R. Reese, Sr.* (privately printed, 1971), p. 47; *Statistical Abstracts, 1972*, p. 600; Erling A. Erickson, "A North Dakota Farm Auction in the Great Depression," *North Dakota Quarterly*, XXIV, No. 1 (Winter, 1971), 45; Personal Interview with Walter and Donald Kroeker of A. A. Kroeker & Sons, Ltd., Winkler, Manitoba, February 15, 1973; Cavert, p. 22; Personal Interview with Grant Mattson, Casselton, North Dakota, January 4, 1973. Mr. Mattson established one of the nation's first and largest machinery-leasing agencies; Rodoje Nikolitch, "Our 31,000 Largest Farms," Agricultural Economic Report No. 175, United States Department of Agriculture (Washington, D.C., March, 1970), p. 40; Robert J. Hampton, "What Management Expects of the Engineer," an address to the American Society of Agricultural Engineers (December, 1967); E. M. Dieffenback and R. B. Gray, "The Development of the Tractor," *Yearbook of Agriculture 1960: Power to Produce*, ed. Alfred Stefferud (Washington, D.C., 1960), p. 33.

CHAPTER III

Frank Kiene—A Loner

FROM ITS earliest days the Red River Valley of the North has been the home of large-scale farming. In the ox and horse era of farming farm size in acres made little difference, but as horse machinery developed and multiple hitches came into vogue, larger fields provided a better opportunity for greater efficiency. The vast fertile prairie of the Red River Valley, void of trees and stones, was being settled in the last quarter of the nineteenth century, when bigness was becoming part of the American industrial society. Men of vision were beginning to apply the ideas of industry to large-scale agriculture.

James B. Power, land agent for the Northern Pacific Railroad Company, devised the basic scheme paving the way for the bonanza farms of the Red River Valley. Power, whose chief concern was to promote the potential of the land along the Northern Pacific, understood that large-scale farming and its associated attention would do just that. Although promotional schemes attracted many settlers into the region, the sheer economic potential of large-scale farming was not overlooked. For example, James J. Hill, founder of the Great Northern Railroad and long-time friend of James B. Power, became intensely involved in regional agriculture because of its importance to his railroad.

Hill and Power, both innovative agriculturalists, were constantly experimenting on their own farms or working with the research stations. Power's farming operation was at Helendale, near Kindred and Leonard, North Dakota; while Hill's two locations were North Oakes, near St. Paul, and the Hill Farm near Humboldt and Northcote, Minnesota. All of these farms included large experimental programs with both animals and crops. The example set by these men and the professional managers of the various bonanza farms established a professional, businesslike attitude toward large-scale farming for some people and became a source

56

of resentment for others. Power and Hill made a good team because Power received widespread publicity for the things he was attempting and Hill used his great resources to support him. A professional attitude in agriculture demanded bigger and better equipment and frequently placed the valley ahead of the nation in ratio of farms over 1,000 acres.

Frank Kiene—The Businessman

It was in the area of these large bonanza farms near Lake Bronson, Minnesota, that Frank Kiene was born (1876) and reared.* When Frank was 15 his parents separated, and he had to support his mother. At the age of 21 he purchased the homestead from his mother, who was in ill health. Frank had worked for the Bengt Sundberg family, Swedish immigrants who farmed south of Kennedy. The family took a strong liking to young Kiene and not only accepted him as a full-fledged member but also involved him in their business activities. Kiene lived with the Sundbergs until the age of 25, when he decided to go into business for himself.

Bengt Sundberg, an extremely active man, was not only a good foster father but also a first-rate tutor for young Kiene, who early in life proved his managerial ability. Besides farming, Sundberg also bought and sold horses and was involved in politics to the extent of becoming a state senator. In 1895 Sundberg sent young Kiene to the School of Agriculture at the University of Minnesota. His practical training under Sundberg in buying and selling livestock apparently paid dividends, for Kiene scored in the high 80's or low 90's in most of his academic courses. Upon returning from the University, he was taken into partnership with Sundberg and on October 14, 1897, at the age of 21, was made a notary public to facilitate necessary business transactions.

Besides supplying the area homesteaders with horses and livestock, Sundberg and Kiene conducted an extensive business in trading horses between the bonanza farms and the lumber camps each spring and fall. Because the bonanza farms and the lumber camps used large numbers of horses, there was no problem in developing a volume business in horse trading. However, in 1900 Kiene sadly left his beloved home with the Sundbergs to take over a farm of his own, about one mile from Kennedy, Minnesota. The

---

*Lake Bronson, located in Kittson County in extreme northwestern Minnesota, was near the 33,000-acre Donaldson-Ryan bonanza farm at Kennedy, the 7,680-acre Reid farm at Northcote, and the Hill Farms at Humboldt and Northcote, which were reputed to be between 20,000 and 50,000 acres.

sadness was somewhat tempered, however, by the $25,000 bank roll frugal Frank Kiene carried with him.

Kiene quickly proved his business insight in the first transaction he made as an independent young man. Rather than placing his entire $25,000 stake in a farm and having it completely paid for, he chose to invest half of his money in a farm and the other half in a general store in Kennedy. The store became the Kennedy Trading Company, which he used to maximum advantage in securing equipment and supplies for his farming enterprise at a discount. The KTC grew into a sizeable business and gave Kiene financial strength to help him expand his farming operations. Putting money into two enterprises and only partially paying for either of them put Kiene on close terms with L. Melgaard, the local banker, who, for a time, was involved with Kiene in the Trading Company.

Frank Kiene was a "loner," but he enjoyed helping people in a quiet way and during his lifetime assisted many. At the same time it was also apparent that he liked having control over people. Although they seldom had any personal association with him, Kiene's employees felt his controlling influence in everything from family activities to political thinking. However, the one exception to Kiene's impersonal aloofness came whenever one of his employees had a problem with alcohol. Whether Kiene was opposed to alcohol for a moralistic reason or because it interfered with an employee's efficiency is difficult to determine; but, whatever the reason, alcohol was certain to bring about the wrath of Kiene.

Kiene's opposition to alcohol evidenced itself in his actions. Once while making the daily tour of his farming operations he observed a man who, under the influence of alcohol, had fallen asleep beside a barn. Kiene contacted the office to determine how much wages were due to the man, wrote out a check for that amount and a note saying he was fired, slipped it into the man's shirt pocket and drove off without waking him. At another time a salesman associated with a burlap bag company furnishing Kiene sacks for his potato business presented Mr. Kiene with a bottle of liquor during an annual holiday call. Kiene, accepting the Christmas gift, instantly smashed the bottle on the heat radiator without uttering a single comment to the astonished salesman. Ell, Kiene's teenage son, was allowed to have all the gas he wanted for the car, anything he wanted from the store, but only enough money to go to the specific entertainment event of the evening. The elder

Kiene did this out of fear that if his son had extra money he might buy liquor. Throughout his life, but particularly in his later years, Frank Kiene donated to temperance causes, and as his financial means grew so did the size of his donations in his constant crusade on behalf of temperance.

Despite Kiene's so called "loner" attitude toward life, he was involved with many individuals on public and private business ventures. In some cases these ventures lasted for several decades; and even though there was occasional personal friction in his relations with others, it is clear that in general, these associations apparently proved beneficial to all. Kiene was able to adapt himself to many ventures, and most of them were highly successful. One of his major motives for becoming involved in numerous auxiliary enterprises was to bring greater efficiency to his farming operations. Kiene never bargained when he needed supplies or equipment, for he knew the wholesale price of most items and in his own mind determined how much he was willing to let the retailer make. Likewise, when he was selling, he established a price and that was final. He was never known to cut prices once he established his position. Whatever profits his auxiliary businesses generated over and above their contribution to the farms was of minor importance as far as Kiene was concerned, for farming was his sole joy.

After he lost his second wife, Kiene devoted his life to the challenge of amassing a big farm. His social life was very limited, and most of the associations he had were with business people. After he left the Sundberg home he rarely visited with them socially even though he had continuous business relations with the entire family. He traveled considerably, but no trip was made without at least a pretense of business. Following each trip a full accounting of all funds expended was made to his long-time bookkeeper, office manager, and business associate, Ruth E. Anderson. Miss Anderson, who started with Kiene in 1917, had a great influence over all the Kiene enterprises for more than 30 years. It was she who required Mr. Kiene to itemize monies he spent on each of his business trips. She was also responsible for loaning and advancing funds to employees, for it was a standard policy of the farm to finance employees at established interest rates.

Miss Anderson's detailed records show how many small and large loans were made to employees. This was particularly true after the mid-1930's when funds were more readily available because of the financial success of the farms. Loans were made as a matter of course, but the men had a constant fight with Miss Anderson when it came to asking for advances. Generally she tried to cut in half whatever amount was asked for, but some of the more experienced hired hands learned to ask for twice what they wanted and then happily settled for half. Miss Anderson, recognized as an exceptional businesswoman, was an excellent partner for Kiene. She stayed at the Kiene home and paid a portion of the utilities and groceries for the household.

Money, if a motive at all, was secondary, for Kiene used little of it for his personal needs. In his early years he was far from generous to church, civic, and charitable enterprises, but as he became older his contributions were significant. Those who knew him best say he "liked being busy and whether or not physically working his mind was constantly engaged."

In his early years he made daily rounds of his farms with horse and buggy. Later he drove around in Model T pickups until his mid-fifties, when he retired to the back seat of his car to be chauffeured around by a local farm youth. Although always wearing a suit and tie, if an extra hand was needed Kiene knew what to do. Many times he got out of his car and donned a white protective coat to help sort cattle or hogs, or work on a broken machine, or scratch in the ground to examine seed to determine germination or disease. Regardless of the pressure of any given situation, he rarely lost his composure. Kiene had his workers convinced that he could solve any problem, and hence they had great faith in him. Many of his men stayed with him for long periods, generally leaving only to start farming for themselves. His part-time workers came primarily from small farms east of the Red River Valley.

Kiene was able to surround himself with a solid loyal core of supervisory employees. Individuals like Conrad Lysfjord, Oscar Nordling, Jonas Cederholm, Carl Hilde, his office manager Ruth E. Anderson, and a partner like M. A. Zeigler could make almost any enterprise function smoothly. These individuals were given considerable responsibility and freedom to do their work and in most cases stayed with their employer throughout their working career. Their loyalty extended beyond doing a good job, and several let Kiene treat them in an exceedingly paternalistic way, for they lived in his house, charged at his retail establishments, and borrowed from him regularly for major items such as cows, cars, and trucks. Although Kiene appeared to be lax in his financial dealings with responsible employees, they did not seem to object. Most of the years that Lysfjord, Nordling, Cederholm, and Ander-

worked for him they seldom collected their paychecks on time. Charge accounts at the store, the filling station, the garage, and the elevator, as well as outstanding bills for hogs purchased and butchered for personal use, and even the house rent were not always collected when due.*

Although he did not appear to be very religiously oriented, Kiene's conscience bothered him about Sunday work. From his first days as an employer in the early 1900's, he made it a practice to pay time and one-half to double pay for all Sunday work regardless of whether or not the laborers had worked during the other six days of the week. He paid this to all levels of employees and always paid for Sunday work in cash at the end of the day. He was quite liberal with bonuses, particularly to all supervisors and those who worked closest to him. Each Christmas members of the families of these employees were given a wholesale catalog and told to pick out a gift to their satisfaction within certain dollar limits. An employee once ordered a leather jacket that was double the established dollar limitation. When offering to pay the excess he was ignored by Kiene who paid it all.

Just the opposite happened to Winslow Larson, a boyhood pal of Ell Kiene. One day the boys were playing golf on the lawn of the Kiene home and a ball went through a bay window and shade. Winslow and Ell had to clean up, but Frank Kiene said nothing about the damage. In December when the Larson Meat Market sent its annual bill to Kiene a check was returned for the amount of the bill less the price of a bay window and shade. That was the first that Winslow's parents knew about the broken window.

The preciseness with which Kiene handled financial matters is well illustrated in an affair with Charley C. Reeves, an inmate at Minnesota State Prison, Stillwater. Mr. Reeves had operated a stage route and had become involved in a shooting incident in the area east of Kennedy. Kiene did not keep a complete file of the letters that he sent to Mr. Reeves, but the following letters relate an interesting episode in this exchange of ideas:

---

*All accounts were run until the end of the year, as were the wages, and then on December 31 there would be an annual settlement. If employees incurred outside bills, such as the Conrad Lysfjord family doctor bill in 1928 for $44.75, Kiene paid it direct and charged the employee's account. Many years the employee had more than half of his annual wages due to him on December 31, and it appears as if this was the rule rather than the exception for his permanent year-round workers. Certainly this was a boon to Kiene, for he had the use of a goodly portion of the employee's money without interest and it created a sizeable business for all of the retail firms owned by him from which the employees purchased goods. For the employee it was the supreme test of loyalty.

Stillwater, Minn.
Box 55, Reg. 1729
Feb. 20, 1911

Mr. Frank Kiene
Kennedy, Minn.

Dear Sir:

At the time I was sent to prison, if I remembered well, I owed you $23.00 which I had owed you for some time, my intention was never to avoid paying it and as you know that I sent you $25.00 on the $48.00 which I owed you just a few days before I shot Mr. Swenson you certainly do not doubt that I intended paying it in full, nor is it my intention to avoid the debt now, but the loss of the stage rout [*sic*] left me in a hard position.

Some time ago a convict earning law was passed and since then I have earned a small salery [*sic*] now you have not had the use of you [*sic*] money all those years but this coming summer I will pay the $23.00 in full and in addition to it as interest what will you take to settle the bill and consider the debt paid? In conclusing [*sic*] I will say that I am fat, hearty and feeling fine and trusting to hear from you.

Yours truly,

Charley C. Reeves

A second letter from Reeves appears to make it clear that Kiene charged one year's interest at 4 percent, probably more as a token penalty than a burden:

Warden's Office
Minnesota State Prison

April 28, 1911

Mr. Frank Kiene.
Kennedy, Minn.

Dear Sir:

Enclosed herewith we hand you check for $23.93, which we are sending you at the request of C. C. Reeves, #1729, an inmate of this Institution. Kindly acknowledge receipt of same to him and oblige.

Very truly yours,
Henry Walfer, Warden

J. E. Desantels, Clerk

Despite his inclination to be a "loner," and to operate a totally integrated farming operation, Frank Kiene was generally liked in the Kennedy community. Most people felt that he did a great deal for the community, and as time passed he became one of the individuals whom the people of Kennedy most appreciated. Because the citizens knew that they had Minnesota's largest privately owned farm and as neighboring farms became larger, Kiene farms were less of a cause for resentment. Kiene of course made an effort to be accepted, but lost little time defending himself to those who still were critical.[1]

Kiene Enters the Tractor Age

In a region of large-scale farms it was only natural that an innovative individual like Frank Kiene would observe what other progressive farmers were doing. With the model Jim Hill farm nearby using the most advanced equipment available, Kiene had the best stimulus he needed to "keep up to date." James J. Hill had progressive ideas about farming and the resources to back them, so that his farm had the best in equipment and buildings. For those who knew Kiene it was clear that he had a strong urge to have the largest farm around and a burning desire to some day own the Hill farm.

Little is known about Kiene's first 17 years as a businessman-farmer, for the records prior to the incorporation of Kiene Farms in March, 1917, are sketchy. He was an early tractor user, possessing at least two large steam tractors prior to 1910. These were for personal use on his farm and not for custom operation, as was often true of some of the early "big rigs." In 1910 he purchased a used Minneapolis tractor and a six-bottom plow in Fargo for $1,000. His next tractors were I.H.C. Titans, and by 1914-15 he is reputed to have had at least 15 Titans. In 1917 he purchased one of the first

Fordsons sold in Minnesota and an 18-36 Avery that cost $1,692.24 plus $232.63 for freight. He also purchased a threshing machine for $1,007.18 plus $105.06 for freight. The recorded purchase of a sulky plow for $35.00 plus freight when the John Deere price list for such plows in 1916 indicates retail of $52.90 is proof of his ability to buy at wholesale. The records indicate purchase of a Buick car in 1917 for $654.17, wholesale price. Eventually he acquired the Ford, the John Deere, and the Caterpillar agencies. He was not afraid to experiment with different tractors, and because of that he owned only a few of each make.

Frank Kiene capital accounts for January 1, 1917, indicate that he personally possessed gross assets of $200,000, and the Kennedy Trading Company, which he owned, had gross assets of $40,021. He had other business enterprises at that time, including a sizeable general merchandize store at Frazee, Minnesota. Very conservatively his gross assets were in excess of $325,000, with total liabilities of not more than $30,000. This meant that Frank Kiene, who had been left by his father at the age of 15 and had a net worth of $25,000 at the age of 25 in 1900, had, at the age of 41 accumulated a net fortune of about $300,000.

Kiene's rather rapid financial rise gives room for speculation that somewhere he must have had private help. His family, obviously, has to be ruled out because after his parents' separation he had no known associations with his father. Kiene married twice. The first wife died after about one year of marriage, and the second died when their only child, Ell, was still quite young. Neither of the wives came from families with any money, and they in no way contributed to any financial gains he might have had. The nearest contact Kiene had with what might be considered inside or family financing was with the Bengt Sundberg family, from whom he was able to borrow money. In May, 1920, he borrowed $3,000 on a personal note from J. E. Sundberg at 8 percent interest. At the same time he borrowed $5,000 on a personal note from Mrs. Bengt E. Sundberg at 8 percent interest, due January 20, 1922. On January 20, 1922, the note was increased to $8,000, and Kiene gave Mrs. Sundberg a mortgage on 400 acres of land.

He owned somewhere between 3,200 and 5,500 acres of land in 1917, had about 65 horses and more than 15 tractors to operate that land. Nowhere in the files is there an accounting of total number of acres of land owned at that time, but the financial statement carried land at $131,870.75. Individual farms were valued at $15 to $42 an acre, with more of them listed below the average of those two figures than above. Kiene always seemed to be in a position to pick up bargains, and because he loaned money and extended credit to customers at his retail enterprises, he knew of those who were in financial trouble. Frequently accounts receivable were used as credit to Kiene when he made purchases from debtors. During World War I he purchased liberty bonds at the rate of $2,000 to $4,000 a month. These bonds were traded to him at discount to apply on charge accounts at his retail agencies. This was particularly true of the Kennedy Trading Company, that throughout its history under Kiene maintained an extremely solvent position.

In 1917 Kiene paid himself $1,800 annual salary from Kiene Farms, Inc., in addition to funds drawn from other enterprises. For 1917, $1,800 would have been considered a good income, but it was only a part of his for he also earned from other enterprises. His chief carpenter, Oscar Nordling, received $720, and Conrad Lysfjord, a farm foreman, earned about $600. Total wages paid in the last nine months of 1917 by Kiene Farms, Inc., exclusive of Kiene's, were $10,396.78. If that amount had been paid equally to full-time men it would have meant that at least 25 were employed. Total employees were many more because most were hired on a monthly basis and many worked less than two months a year. During October, 1917, in addition to his regular workers, he paid 31 extra laborers for their participation in the grain and potato harvest. In 1918 he had 76 extra harvest workers, and in 1924 the number rose to 120. Total wages paid rose rapidly after 1917, reaching $23,481.27 in 1920 and then dropping to $13,440 in 1922, reflecting the farm depression. However, by 1928 the amount had risen to $38,336.35.

Like most farmers of the World War I era, Kiene made use of the sudden farm prosperity caused by the war. He purchased land and equipment rapidly from 1917 through 1920, which was typical of all farmers during that era. His great financial strength enabled him to buy many farms with 100-percent financing. The farm that came to serve as the main base of the Kiene operations was purchased in March 1918, for $50 an acre, with a 100-percent, 10-year, 6.5-percent interest loan from Minnesota Mutual Life Insurance Company. Kiene evidently thought that land was a good investment for by this time he owned farms at Ulen, Barrett, and Lawndale, Minnesota, as well as at Page, North Dakota.*

---

*The farms at Barrett and Lawndale were used as a source of supply for native hay, which Kiene needed for his large horse herd.

The farm prosperity of 1917, of course, was extremely profitable to anyone operating on the scale of Kiene. That year he sold hogs for as much as $14.75 a hundred pounds, Hereford cows for $100 a head, horses for $175 to $200 a head, wheat for $2.04 to $3.00, barley for $1.65, oats for 67¢ to 95¢, flax for $2.96, and potatoes for $1.20 to $3.00 a bushel. In the case of crops and animals, Kiene had high-quality products and was able to sell a considerable portion of his total output to local farmers for foundation stock or seed. This enabled him to earn a premium on those sales. He had registered livestock and generally had connections to acquire certified seeds. This was something few farmers were concerned about or were able to do. Farmers bought freely in 1917, and judging from his accounts receivable, they paid in cash. For those who did not, Kiene willingly took notes at 7- to 10-percent interest. In at least one case Kiene sold a Rumely tractor, separator, and plow for which he took a mortgage on 160 acres for $1,840.19.

Prices of things that farmers needed to purchase did not rise quite so rapidly as the products they had to sell. This is typical of such times as during a war, when food and fiber suddenly come into intense demand. Kiene was able to hire Pete Olsen and four of Olsen's horses for $9 a day to pull a potato digger in the fall of 1917, and a hired girl worked in the Kiene home for $3.50 a week. Day labor was $1 a day, and good men were available by the month for $25 plus room and board. However, during the crucial period of potato harvest these rates doubled, but the real estate tax on some of his best land was only 42¢ per acre, and a large modern 60- by 80-foot hip roof barn completely built cost him only $1,975.

Buying supplies, equipment, labor, and land at such reasonable prices and selling his farm produce at some of the highest prices in the history of American agriculture really helped Kiene to profit. Total cash flow from his farms in 1917 was $339,733.64. His large potato and livestock operations made that volume possible, increasing his total income by at least 50 percent over what he could have done through grain farming alone. His net profits for 1917 would appear unbelievable to the average farmer of that day and certainly put Kiene in the same category as many large industrialists. When Kiene Farms, Inc., was formed, Kiene made a $7,000 contribution in equipment and cash as a basis for the capital stock of the corporation. For that contribution Kiene issued $5,000 of capital shares to himself. At the end of the first year of operation, $41,000 was applied to surplus and $14,600 to undivided profits. His $7,000 investment grew to $60,600 in one year's time. He had 16 percent net profit on his gross volume after withdrawing his own salary and a considerable personal expense account. Kiene was proving to himself what many farmers have since learned, that it is easier to get ahead with volume. Part of this is because personal needs become such a small portion of the total cash flow.

Kiene had made steady progress up to 1917 and at that time felt the need for forming a farm corporation to separate the assets and business of his various enterprises. In this respect he proved his understanding of the value of the corporate tool to agriculture. As time passed he created several more corporations both for his farming and for his retail firms. His understanding of how to use financial leverage between his various businesses put him in an even stronger position to expand. Kiene thoroughly studied his personal set of law books, referring to them constantly. Locally he was regarded as one who understood legal matters well. There are few references to attorneys or attorney's fees, so it is safe to assume that he personally did much of the legal work to create Kiene Farms, Inc., early in 1917.

The 1917 records show the first references to Ruth E. Anderson, who, as mentioned, was a long-time office manager of the Kiene operation. Miss Anderson apparently made some sort of capital contribution to part of Kiene's enterprises, for frequently she collected interest on money she had loaned to her employer. Several years she did not receive her full salary, preferring to leave it in the business to draw interest. In 1926 Miss Anderson received $45,300 of capital stock in Kennedy Trading Company, which was the general department store owned by Kiene. The capital stock of the KTC was increased to $100,000 with Ruth E. Anderson's purchases. In later years several blocks of land were recorded under her name, apparently for estate and tax purposes. There was no evidence in the records indicating that she had actually made the purchase. It may have been that Kiene, during his financial decline of the 1920's, transferred some unmortgaged land to his bookkeeper as a hedge against any decline. With Kiene's financial ability and the obvious close working relationship between him and Miss Anderson, this move would have been considered the least possible risk. In any case, after the death of Kiene most of the land held in Miss Anderson's name logically reverted to the obvious owner. Such unique financial maneuvers made it possible for Frank Kiene to survive the adversity.*

For unexplained reasons 1918 was not quite so profitable to Kiene Farms, Inc., as 1917 had been. The gross cash flow of the business was down to $226,909.97, labor expenses rose to $19,290.87, and his interest bill nearly doubled, apparently reflecting the finance charge of land purchases. Kiene was, however, optimistic, and had reason to be for he increased his personal salary to $3,000 and nearly doubled the amount he allocated to what could be considered his personal business expense account, which was much in excess of his salary. Despite the sizeable amount set aside to surplus and undivided profits for the farm corporation, there was an increase in bills payable from $25,000 to $59,000. Ironically, even when he was financially distressed at a later date he did not increase the bills payable account. He properly funded his obligations to longterm real estate mortgages and financial institutions. There is little evidence that Frank Kiene ever had anyone "holding the bag" on charge accounts or personal loans, or because of adverse financial conditions. He had a reputation for paying his bills.

Kiene purchased more land each year, including some in Canada. As his land holdings increased, the need for horses and machinery became obvious. In one purchase in early 1919 he secured 21 horses for $3,075, three I.H.C. 10-20 Titan tractors for $3,315, three Van Brunt drills for $538, three I.H.C. McCormick binders and tractor hitches for $694, several plows, a harrow, a mower, potato equipment, a Model T Ford pickup for $505, and a Model T Ford car for $585. Total cost of the new equipment and horses for 1919 was $9,476.35, an amount equal to several times the power and machinery investment of the average-sized farmer in his area.

Kiene preferred tractors, but he still purchased horses to ease the transition to tractor power. An animal horsepower in 1919 cost him $146 per horse (generally a horse was rated two-thirds of a horsepower), unlike tractor horsepower that cost only $110 per

---

*Miss Anderson apparently had cash that she invested in the operation at that time. In December, 1923, Ruth E. Anderson collected $327.59 interest on a $4,096.74 note of Kiene's that she was holding. At the same time she received a note for $837.46 for uncollected labor for 1923. This appears to have been the upaid share of her $1,500 salary. The records indicate that Miss Anderson did not receive her full salary any year from 1922 through 1926. Each time she took a note for the difference from her employer. The notes apparently bore 8 percent interest, and in later years she was able to collect on all of them.

unit. He remained in the horse-trading business even though he saw the obvious advantage of the tractor for his own operations. Kiene must have been amused while pocketing the profits from selling horses to farmers who did not want to change, when he, personally, was convinced of the superiority of the tractor.

Besides Kiene Farms, Inc., the Kennedy Trading Company, and a livestock business, the 1917 to 1920 records indicate that Kiene became involved in several other ventures. In May, 1919, he purchased a hardware store at Hallock that later became Northwestern Implement Company. In Kennedy he owned an automobile agency, a machinery firm, a hardware store, and a gas and oil business (station and bulk). He also owned several houses and miscellaneous pieces of property in the area towns.

The farm profits of 1917 continued into 1918, 1919, and 1920, but at a somewhat reduced rate because of increasing costs. It is safe to speculate that the rapid expansion caused increased management problems and probably prevented the profit margins from being so wide as they were in 1917.

The good personal progress plus the very bright agricultural picture nationally encountered in the years 1917 to 1920 caused Kiene to lose his generally good perspective on financial matters. In this respect he, like most of the nation's farmers, thought the good times would never end. Each time he purchased another farm he came one step closer to realizing his burning ambition of some day owning the famous James J. Hill farm at Northcote. The larger his operations became the easier it was to comprehend the purchase of "the show farm of the Northwest." Kiene saw his opportunity and seized it, but within a matter of months realized that he had made a mistake. The purchase of that farm gave him his supreme test, and few men would have had the determination to fight it out as he did. Only by the masterful juggling of his diverse and sizeable assets plus the maintaining faith of those who were financing him was he able to survive the combined ordeal of owning the Hill farm and the farm depression of the 1920's. One of his staunchest supporters at this time was the McCabe Brothers Grain Company of Duluth and Minneapolis. Working closely with them was probably the most important external factor in his favor and eventually enabled Kiene to recoup and enlarge his fortune.[2]

Kiene Learns a Lesson in High Finance

During the period of World War I many farmers with small acreage made considerable progress and enlarged their farms.

Nationwide at this time the average farm increased from 138 to 148 acres, and in the Red River Valley the average size was well over 300 acres. At the same time the last of the bonanza farms were going out of business. The bonanza farms initially had favorable conditions, for during the late 1870's and early 1880's, weather and the price of wheat were both good. By the mid-1880's prices started to drop sharply, and a long dry spell hit western Minnesota and the Dakotas. These adverse conditions were climaxed with an extremely short growing season in 1888 when the last killing frost of the spring was in early June and the first killing frost of the fall came on August 17. There was little market for the shriveled, frost-damaged wheat, and several of the bonanza farms sold out while others changed their method of operation.

Instead of farming the land under single management, the bonanza farmers leased much of their land to share cropping tenants who furnished their own labor, power, and equipment but followed the directions of the landlord. With horse farming there was little advantage operating on a large scale because basically a man and a five-horse hitch were needed to operate at full economy of scale and most farmers in the Red River Valley were functioning at that level or better. For the bonanza farmers the only economic advantages were in purchasing and marketing of equipment and produce on a large-volume basis. But the gains from that were not enough to offset the higher cost of hired labor in contrast to the underpaid labor of the family farm. It became evident that unless management was part of the ownership there was not the devotion to the business that was found on the family farm. The advent of the tractor hastened the downfall of the bonanza farms that were not financially profitable enough to justify the conversion from horses. Besides, distant stockholders were more interested in cash dividends than in seeing long-term investments made on the farm.

With the advent of World War I, taxes of all kinds—but particularly corporation taxes—that affected most bonanza farms rose sharply, creating an even more undesirable climate. The bonanzas, which were forced to rely almost entirely on hired labor, were doubly affected by the actual shortage of manpower and its increasing cost. Land prices associated with the war prosperity gave the remaining bonanza farmers an excellent opportunity to sell. The Dalrymples of North Dakota stated that they were making about $2 an acre farming the land but that they could sell it for $100 an acre and invest that amount at 4 percent, thereby earning double the return. Kiene was well aware of what was happening. In fact, he attended the Dalrymple bonanza farm auction on March 20, 1920.

The James J. Hill demonstration farm in Minnesota was no exception to what was happening to the large farms of the area as far as the profit picture was concerned. The farm was opened by the first resident manager, Mr. Valentine, and then developed by Captain Hugh W. Donaldson, who also managed other farms. In 1896 David McCleary became the manager, and it was from then on that the buildings reputed to have cost $500,000 were erected. The farm was located in an area of Fargo Clay, a very heavy soil with poor natural drainage, and until that problem was solved there was little chance of operating economically. Because of this obstacle Hill became the father of surface drainage for the Red River Valley, a cause to which he devoted much time and money. Walter Hill, at the request of his father, lived on the farm during the early 1900's but took little interest in its operation. His reputation in the area as a rabbit and wolf chaser is exceeded only by the wild stories of his other escapades. With the death of James J. Hill in 1916 no one was left who had any desire to develop the farm. Walter Hill preferred another way of life, and since the farm was not a profitable venture he disposed of it.

Wilson H. Hubbard, president of Hubbard Grain Company of Colfax and Mason City, Iowa, was the next owner of the Hill farm. What success Hubbard had with the farm is not clear, but when Kiene made the original agreement to buy from Hubbard he was informed that there was a $125,000 mortgage against the property. That amounted to just over $64 an acre on the 1,940 acres involved in the purchase. The adverse economic outlook for the farm was obvious. For a 1,940-acre farm to support $200,000 to $500,000 worth of buildings is somewhat questionable. Jim Hill could afford a show place because of his tremendous resources. He could also justify it because originally the farm was reputed to be close to 50,000 acres, and its purpose was to show how agriculture could be improved along the route of the Great Northern Railroad. Furthermore, because of personal reasons it was probably worth it to him to keep his son Walter busy constructing buildings on the farm rather than having him in the Twin Cities. What Hubbard intended to do with the place is not clear, but whatever his interests were it is certain that he was disappointed. This gave Kiene the opportunity to realize one of his great ambitions.

The original agreement between Hubbard and Kiene was dated

December 10, 1919. Kiene agreed to assume the 6-percent first mortgage that Hubbard had with Capital Trust and Savings Bank of St. Paul. He gave a second mortgage at 7-percent interest to Hubbard in the amount of $25,000. The terms included a $12,500 payment on January 2, 1921, with the balance due one year later. The Hubbards were allowed to remove three of the workers' houses, a stud barn, and three electric lamps. The contract specified the lamps as "two in the bedroom and one in Mae's room." The farm was assigned on rental contract to Edward Florance and J. M. Lohr until November 1, 1920, at which time Kiene could take possession. Kiene's payments on the contract from 1919 were partially offset by the rent he collected from Florance and Lohr.

To finance the purchase of the $175,736.64 farm and inventory, Kiene Farms, Inc., actually did not have to extend itself greatly. Kiene traded his store at Frazee, Minnesota, and a 677-acre farm at Ulen, Minnesota, with a $12,000 mortgage on it to W. H. Hubbard as a $45,000 down payment, after which he assumed a $100,000 first mortgage and gave a $25,000 second mortgage to Hubbard. The Kennedy Trading Co. that Kiene owned purchased not only over $23,000 worth of machinery, but 51 horses and other livestock for a total cost of $33,671.50. Kiene personally purchased the furniture originally owned by Walter Hill for $4,000. The actual cost of the land was established at $35.00 an acre, which probably reflected its true worth and the buildings at $39.21 an acre. The building inventory listed 12 cottages for workers at $22,200, a water plant and powerhouse with a large Holt Caterpillar engine for $10,000, the big house, barn and silos at $65,000, a hog house at $5,000, and miscellaneous buildings at $58,000. The building site was meant to be a show place, but it certainly was not a sound investment for a practical farming operation. The machinery inventory included two Caterpillar track-type tractors, three Fordsons, five tractor plows, eight grain drills, and 10 binders.

During 1920 Kiene geared his entire operation toward raising all the money he could to meet the payment schedule on the Hill farm. He sold large quantities of wheat at harvest for $2.71 a bushel. In addition he sold grain that he had stored from the 1919 crop. His total volume of business in 1920 reached $456,042.73, and if he could have kept that up for a few years he would have been able to fulfill his commitments. But late in 1920 a sense of urgency appears in the notations and transactions of the Kiene enterprises that is in direct contrast to the well-planned smooth operations prior to this. Kiene refinanced whatever property he could and sold much land. It was not his nature to sell land that was within easy farming distance of his base operations; therefore, one must conclude that "the heat was on."

Besides refinancing and selling of land, Kiene's records, for the first time, include an entry denoting a delinquency on interest and principle payments. On November 1, 1920, he had to pay $115 penalty for delinquent payment on $20,000 due to the Capital Trust and Savings. The following year he made several trips to Fargo-Moorhead and the Twin Cities to arrange refinancing, but with no luck. In January, 1922, he paid $6,000 interest on the first mortgage, $875 on the second mortgage, plus $33.92 past due interest because payments were not made on December 20 as specified; but he was unable to make any payment on the principle. He received an extension on the second mortgage principle payment to January 2, 1923. By then he had 5,360 acres of his original land mortgaged for about 50 percent of market value, besides owing $21,000 borrowed on signature at various area banks. Kiene faced an interest bill that was double what it had been just a year earlier.

There were several factors that affected Kiene in 1921, only one of which was within his control. His personal problem was that of managing a new unit of 1,940 acres of land 16 miles from his base of operations. Machines and equipment had to be transferred by rail from Kennedy to Northcote and back when needed. There was no other practical way to move the big equipment. This added considerably to labor cost even though railroad sidings were convenient to both farms. Kiene was used to working heavy land, but the heavier Fargo Clay on the Hill farm made field work extremely difficult, causing more than the usual headaches. Then there was the problem that faces every manager in expansion—total management requirements grew more rapidly than management personnel. Because of this the best job of farming was not done, thus preventing the chance of making a good profit. In fact, labor expenses on the Hill farm exceeded the combined income from grain and hogs. Operating expenses alone were more than the farm's gross income, even without accounting for interest, taxes, and principle.

There were other factors that made the purchase of the Hill farm a bad deal for Frank Kiene. The biggest one was something he could not control—the farm price crisis of 1921. Prices of agricultural products dropped drastically between midsummer and fall.

Kiene's farm income, which was over $456,000 in 1920, dropped to $317,968.19 in 1921. Mr. Kiene tightened his personal belt and for several years did not draw any salary from his farming operations. His total income tax liability dropped from well over $2,000 in 1919 to $11.65 in 1921—not too good for one of the state's largest farmers, but typical of agriculture for that year and for some years to come.

When the national agricultural economy fell so did the fortunes of Kiene's most profitable non-farm enterprise—the Kennedy Trading Company, Inc. What happened to that store's business volume reflects what took place in the agricultural economy of Kittson County. Total sales at KTC were $54,654 in 1917, and as the war prosperity advanced, sales climbed to $74,471 in 1919 only to drop sharply to $51,904 in 1921. The decline in total business continued steadily each year, reaching a low of $26,696 in 1932 when the full impact of increasing agricultural production and declining prices was felt. One notable aspect of the KTC was that throughout the period from 1917 to 1933 only once did the charged-off accounts reach 3 percent and during most years no money was lost through charge accounts.

The message was now clear. By the fall of 1920, before he was even able to take possession of the Hill farm, Kiene realized that the challenge was too big. By 1921 it was a veritable noose around his neck, slowly tightening as the farm values dropped, cutting off all chances for refinancing. He had two choices—"dig in" or let the farm of his dreams go. Except for those closely involved in his financing, few sensed that Kiene was beginning a 10-year decline that would end in the loss of the biggest single purchase of his career. However, the experience only hardened him to face his next opportunity with more determination than ever. Kiene, "the loner," would fight his own battle, but he kept the faith of his key employees, who, except for Ruth Anderson, probably never knew his big trouble.[3]

The Long Slide Through the 1920's

The 1920's price sag forced even the most prosperous and progressive farmers into using every means at their disposal just to maintain their incomes. Tractor power and bigger equipment provided an advantage over those who did not have the financial or managerial means to make the necessary changes. The equity position of the nation's farmers decreased each year. Probably the only reason why no more than 155,000 farmers left the land during the decade of the 1920's was the fact that the alternatives were unattractive. Industrial labor was nearly as depressed as agriculture, so where could the farmer turn? The other basic economic problem of the 1920's was the difficulty in selling land during a down market. Not only were the farmers uninterested in buying land, but the speculators were turning to Wall Street.

What happened to Kiene's real estate inventory is reflective of what was happening nationwide. One basic difference was that Kiene's land was carried at a purchase price while nationwide the gross land inventory reflected the declining market value and hence dropped more drastically. Kiene's real estate inventory rose from $131,870.75 in 1917 to $176,555.02 in 1920. This figure is for land only and exclusive of the value of buildings, which were inventoried separately.*

Had adverse events beyond his control not occurred after the $183,000 purchase of the Hill farm it would not have been a strain on the Kiene enterprises. Other factors changed which made an initially sound purchase an economic disaster. First of all, some of the property Kiene sold reverted back to him because the buyers were unable to make the expected financial negotiations. Kiene held mortgages on the land, and he had to take it back or lose his remaining equity. This forced Kiene to sell more land. Each year from 1920 on he was forced to liquidate some property to hold his enterprise together. Some of the property Kiene sold during the 1920's was sold at a loss. He was forced to sell for less than he had paid just to secure cash for operating. This was particularly true of commercial property he held in the local villages. He desperately tried to sell his North Dakota holdings in Cass and Steele Counties, but buyers were not available. From a total $176,555 land inventory in 1920 his holdings steadily declined, hitting a low of $105,281.79 in 1930.

Fortunately he had enough land to sell, and in this manner he was in a position to keep himself more liquid financially than most farmers who had to maintain their basic unit or quit entirely. The

---

*This is an extremely important consideration. The purchase price of the Hill farm real estate of $143,972.75 was composed of $76,072.75 for buildings and $67,900 for land. When Kiene bought the Hill farm he traded in one $37,000 farm on the deal and sold enough other property so that his total land real estate inventory, exclusive of buildings, rose only $9,575 in 1920. When liquidating land to raise money to purchase the Hill farm, he improved his position because he realized profits on the property he sold that enabled him to pay on the cost of buildings and personal property. Net profits from liquidation sales in 1919 were $7,201.90 and $16,070 in 1920.

smaller operator had no choice but to mortgage his farm a little more each year. This automatically increased his interest load and his principle payments. Kiene was able to eliminate that pitfall. His interest bill, which was $3,656 in 1917, rose rapidly after the Hill farm purchase to $24,766 in 1922. Kiene liquidated real estate and personal property at that time to reduce the interest burden. Some of the property was shifted into possession of his retail enterprises. These businesses could stand to pay interest. Title to some land was even transferred to his office manager, Ruth E. Anderson, who from the 1920's until Kiene's death operated the land in her name.

In 1925 a large portion of Kiene Farms, Inc., property was transferred to Kennedy Trading Company, Inc. Records of the transfer from Kiene Farms, Inc., to KTC indicate that $88,869.72 of personal property consisting of $36,000 worth of machinery, $12,000 in horses, $13,000 in hogs and cattle, $28,000 in grain, plus miscellaneous equipment and supplies, was involved. Internally this was an admission that the farming corporation was bankrupt, but because KTC had sufficient assets it could assume the farm operation and tide it over until better days. Externally, except for the financial people working with Kiene, there was no evidence of the seriousness of his position. Locally it made little difference whether title to land was held by KTC, Kiene Farms, Inc., Ruth E. Anderson, or Frank Kiene personally, for everyone knew it was still a Frank Kiene enterprise. Only the financiers and Internal Revenue personnel knew what really had happened. Through such transactions the interest burden was shifted or reduced from $24,766 in 1922 to a low of $1,370 in 1932. By that time Kiene had weathered the storm and had clearly reversed his financial fortunes.

Kiene intended to amortize the Hill farm debt through the big cash crop—potatoes. The farm had a history of potato raising, and potato warehouses were among the facilities purchased. However, although Kiene won a $4.80 second prize at the State Potato Show in 1922, the heavy Fargo Clay soil would not consistently produce even second-prize potatoes.

In addition to the poor yield and quality, Kiene, like all potato growers, suffered a serious price break. When he purchased the Hill farm he was selling potatoes for up to $2.40 a bushel, but by the time he harvested his first crop from that farm he had to sell them for 80¢ a bushel. His best potato crop on that farm was in 1922, when he grossed $19,568 even though 24,000 bushels were sold at a new low of 25¢ a bushel. However, this was not the lowest recorded price for potatoes, for in March of 1923 many sales were made for 35¢ a hundredweight, or 21¢ a bushel. Total labor and fuel expenses that year were $20,948, which meant that the grain crop had to pay for all seed and all operating and financing costs. Ironically, 1922's top grain sales of $28,022.80 were not enough to cover the remainder of the expenses and the necessary principle payments.

After trying for two years to make the Hill farm pay, Kiene apparently lessened his efforts. He gradually tapered off expenditures in an attempt to reduce them to match the income potential. Obviously, if intensified potato farming was not profitable he had to try extensive grain farming. Unfortunately, grain prices fell too, but not so drastically as potatoes. Oats, barley, and wheat all took sharp drops in price from the spring to the fall of 1921 and recovered only partially in the following years. Kiene's timing for large expansion could not have been worse. During 1921 and 1922 he sold wheat for a low of 66¢ in contrast to over $2.30 in 1920. He sold barley for a low of 25¢ a bushel compared to $1.39 in 1920, oats for 19¢ in contrast to 90¢ in 1920, flax for $1.66 instead of $5.50 as in the spring of 1920. Hogs dropped $23.00 a hundred to $6.50 by late 1922.

But even with the economical use of large-scale equipment, it was nearly impossible to withstand such drastic price declines. The prices Kiene and all farmers received for their produce was beyond their efforts to control, and, unfortunately, the prices of what the farmer had to buy were also beyond his means to regulate.

The horse market affected farmers in two ways. Many of them made a business of selling horses, and, like grain and livestock, their prices dropped sharply. For those who could not afford to buy a lower-priced tractor, horses were the alternative; but the drop in the market value meant that sometimes farmers ended up owing more on mortgaged horses than they were worth. Kiene purchased horses from $185 to $265 per head during 1919 and 1920, and after mid-1921 the prices ranged from $70 to $165 per head. The drop in the value of horses produced an oddity in the price structure. Many horse breeders went out of business, so that by the mid-1930's the price of horses rose because many farmers did not have the money to buy a tractor. When horse prices dropped, so did the rental rate. Kiene paid $2.50 to $4.00 a day for the use of teams in 1919 and 1920, but in 1921 he was able to rent them from the same farmers for $1.50 a day. This was obviously a help for an operation such as Kiene Farms, Inc., which expected to hire large numbers of team and man combinations every year.

Wages, taxes, and rent all dropped sharply from 1921 through the 1930's. Labor costs, which rose rapidly during World War I, dropped almost as drastically with the end of the wartime economy after 1920. Mechanization during the wartime era worked to the disadvantage of the laborers in the postwar slump, creating a surplus of workers with a corresponding decrease in wage rates. To the small farmer this may not have meant much, but to an operation that employed at least 20 year-round laborers and over 150 in the peak season, this meant a great savings. Kiene relied on transient labor during rush periods, but whenever possible he gave preference to employing family members of the small-scale farmers living on the eastern fringes of the Red River Valley. This labor supply provided dependable and trained farm workers who needed cash income from outside jobs to maintain their family farm.

It was important to Kiene to maintain an excellent relationship with these people as they not only provided him with a labor supply but also served as regular customers for his retail enterprises. The common practice of many of these people was to work for Kiene to pay off accumulated bills at Kennedy Trading Company. This, of course, enabled Kiene to profit on the goods sold in exchange for labor. Not until the Soil Bank program of the 1950's did many of these smaller farmers, for the first time, have an alternative way of life.

While Kiene had paid from 60¢ to 90¢ an hour for labor in the years from 1918 to 1920, he was able to hire the same people after mid-1921 for 15¢ to 30¢ an hour during the peak work periods of grain and potato harvest. During the off season he had paid 50¢ an hour for wood chopping and hauling in 1920, but in 1921 men were available for a dollar a day and board for the same jobs. Instead of having a $300 a month labor cost during harvest, as in the fall of 1919, he had to pay only $45 to $70 in 1921. For spring and summer work he paid $1.25 for day laborers and $35 by the month. For potato picking alone this meant that instead of a $45,000 labor bill his costs were only $8,000 to $12,000. The wage situation failed to better itself in 1922, and for each year following there were more people looking for work, so that even during harvest season men could be hired for $30 to $40 a month. The long-time supervisory employees, such as Oscar Nordling, Carl Hilde, and Conrad Lysfjord, had their wages increased annually from 1917 through 1921 only to experience a similar drop each year after that until 1925.*

Kiene had boarding halls and bunk houses on the main farm places and a moving cook car that provided meals for a large portion of the labor force. These facilities, except for the cook car, were still in use in 1973. In the 1920's, 25¢ was deducted for each meal eaten at the boarding house. Henry Meek, who worked on the farm in 1923 said, "The men were very satisfied with the quality and price of the meals." Kiene never tried to cut corners on the food service even though in the post World War II era that enterprise resulted in a considerable annual loss.

Many of the small farmers working for Kiene part time frequently purchased their grain, potato seed, machinery, and supplies from him. Because of a lack of cash they often were forced to work off their bills. In a few cases they became excessively indebted to Kiene, and he ended up with their land. Kiene made some of his biggest profits selling hay to farmers. He had scattered hay lands in the wetter areas that produced when it was otherwise dry. He produced or purchased hay for $4.50 a ton and shipped it to Kennedy for $2.24 a ton, making a total cost of $6.74. This hay was sold from the car for $11.50 a ton in 1921, a price far in excess of what comparable feeds were selling at the time. There are several cases in the records of 1921 through 1923 where farmers worked for $1.25 to $1.75 a day to pay for $8.00 and $11.50 a ton prairie hay. Kiene's profit from his auxiliary enterprises served him more than just by reducing the cost of goods purchased by the farms.

---

*Sam Sang's experience in settling his wages is typical. At the end of April, 1923, he was credited with working 75½ days at $2.50, for a total of $188.75. However, January through April he had lived in a Kiene house at $8.00 a month rent. He had purchased 150 pounds of hog feed at $.0125 a pound, used a team to haul wood for three days at $1.25 per day, bought a 39-pound hog's head at 4¢ a pound, eaten at the farm for six days at 75¢ a day, as well as consuming 52 quarts of milk at 8¢ a quart. Total deductions were $47.84, leaving $140.91 net pay to Sam Sang, except that he had to pay a four-month grocery bill at KTC.

Oscar Nordling, Kiene's chief carpenter, drew $1,212.50 in total salary in 1920. When settlement was made on the usual December 31, Nordling had a $610.35 bill at the store and $130 house rent deducted before receiving $381.15 in cash. It must have taken good management by the employee to live for 12 months between paychecks. Fred Gulstrand, who worked on the farms in the early 1920's received a note for unpaid labor in 1922. When Kiene sold large amounts of grain from the 1921 and 1922 crop in November of 1923, Gulstrand received payment of the note for back wages and interest. At that time his accounts at the KTC were also marked paid.

Real estate taxes and land rent, major items in agriculture, also reflected the depressed profit picture. With farmers competing vigorously for land during the war boom, taxes, rent, and land prices rose sharply. Real estate taxes that averaged from 30¢ to 50¢ an acre prior to the war economy had risen to 80¢ to $1.10 per acre by 1921. Fortunately for the farmers, most county commissioners in rural areas were farmers, and they understood that when commodity prices dropped, so must real estate taxes. The counties had no choice in many cases because the farmers just did not pay their taxes and the county commissioners were forced to reduce taxes to a more realistic rate.

Kiene experienced a reduction in rate from $1.04 an acre in 1921 to 87¢ in 1923 and to 43¢ by the late 1920's. Even with the reductions in taxes, he found it difficult to make payments and for several years ended up paying a penalty on delinquent taxes. Conservatively, at no time during the 1920's and 1930's did Kiene pay less than $7,000 in real estate tax. Some years he stored his grain rather than sell it at low market prices. When he did that he was unable to pay his taxes; but that was part of his plan, for in effect he was using tax money to finance his operation when no other financing was available. It was difficult for county officials to prove otherwise, for many farmers were unable to pay their taxes during several years in the 1920's and 1930's. Kiene took advantage of this common adversity. Controlling farm sales because of low prices while taking a calculated risk on rising prices and at the same time keeping creditors happy was part of Frank Kiene's overall strategy. Such management decisions took iron nerve, but during the 1920's they were a major factor in keeping the operation afloat and in the 1930's made considerable profits possible.

Difficult as the 1920's were, it appears that Kiene never had thoughts about quitting. Even the disastrous year 1921 failed to slow him down as he simply tightened his belt and took as little salary as possible from the farm operation to reduce expenses. In 1923 he waived his salary entirely. At the same time his business expense account was reduced from $11,277.20 in 1918 to only $1,239.33 in 1923, not a large sum for one who traveled as extensively for business purposes as he did. At the same time he purchased a 30-60 Hart-Parr and a 40-65 Twin City tractor and matching plows. These were 8- and 10-bottom plow outfits, and acquiring them indicated an act of courage when the countryside was full of repossessed machines available for much less money. Ironically, during some of the years when he was hardest pressed

Kiene paid organization dues to Farm Bureau, Woodmans, and to United Commercial Travelers two years at a time instead of annually.

In 1925 after a portion of the land held by Kiene Farms, Inc., was deeded to KTC, the farm company rented the land back from KTC for $2.50 an acre. The purpose was partly to satisfy creditors and partly to put more financial strength into the farm while reducing operating cost and principal payment that the farms had incurred. The maneuver was apparently the right decision at this low point in Frank Kiene's financial fortunes. From 1925 on he attempted to get rid of the burden that the Hill farm was placing on him. He realized that the excessive overhead combined with the apparent difficulty of keeping production costs within limits made long-range-profit prospects for that farm questionable.

In 1928 his hopes were brightened when Fredric E. Whalen of Kansas City, Missouri, working through Andrew Carlson, a local realtor, made a firm offer for the farm. A contract dated September 7, 1928, indicates a sale of 2,460 acres at $74 an acre, totalling $182,000. Kiene was allowed to retain a mahogany dresser, washstand, and table and was to have free use of the elevator, potato houses, and barns until June 1, 1929. In addition he reserved the right to hold an auction on the farm prior to that date. A check for $25,000 was issued by Whalen to Kiene on September 11, 1928. On September 23, 1928, Kiene wrote to Whalen by registered letter asking why the check had not been paid. The letter was returned by postal authorities because Whalen could not be found. All hopes were smashed and an uncashed check remains in the files nearly a half century later as a memento of "Kiene's folly."[4]

A Bonanza in the Dirty '30's

After the low prices of 1921 and 1922 any improvement was welcomed. Based on actual sales Kiene made in 1925 and 1926, oats, barley, flax, and wheat prices nearly doubled, and potato prices improved even more from their lows of the earlier years. The Hill farm operation, no longer accounted for on the overall records, made it clear that Kiene was again on the move. Each year after 1926, except 1932, Kiene increased his salary and his withdrawals from the farming enterprise. The same was true for his office manager and business associate, Ruth E. Anderson, who realized sizeable salary increments. In addition, Miss Anderson had a rapid growth in income from the farming enterprises operated in her name.

By 1927 Kiene's fortunes improved enough to start new business ventures as well as increasing his farming operations. John Bogestad, who farmed land adjacent to some of the Kiene property, remembered that in the early 1930's Frank Kiene would urge him to "buy or rent all the land you can because it will never be a better bargain." Bogestad indicated that Kiene was so sure that "his theory on land was right that it was almost an obsession with him."

Kiene practiced what he preached, for at this time he rented or bought all available choice land that was near to any of his properties. If he had overlooked quality he could even have added many more thousands of acres. By late 1931 Mr. Kiene had about 5,500 acres under lease and owned 7,350 acres in seven townships in addition to 11 pieces of village real estate. The land Kiene owned was secured in 37 pieces averaging 198 acres and generally in blocks of over a section each. He had economic units of good land that were either heavy Fargo Clay or Fargo Silty Clay Loam. The land was sufficiently spread out to give him a degree of safety from adverse localized weather conditions. By 1931 the cash flow in his various enterprises exceeded the previous high of 1920 by over $80,000. People were speculating on how many acres Kiene owned. Some guesses were low and many high, but most would have been astounded if they had known that by 1931 his business volume surpassed $530,000.

Kiene's strategy to acquire the land is partly revealed in the following telegram dated December 19, 1930, to Mrs. Harold Tvede, Rio Linde, California:

Difficult to sell land at any price. Farmers terribly hard up. Grain prices very low. Conditions panicky. Taxes accumulating fast amounting to over four hundred dollars [on] this land January first. Offer . . . [240 acres of land] one thousand dollars to you. We pay all taxes. You furnish deed abstract showing good title. This for immediate wire acceptance.

Kiene's message was true and typical of many that were sent at that time to absentee landholders by people on the spot who were looking for bargains. With $400 in back taxes on a farm that was probably assessed at about $100 a year, Kiene knew that Mrs. Tvede had to either sell the farm or borrow money to pay the taxes to prevent seizure by the county. Borrowing money on land was difficult for anyone at that time, and for an absentee owner it was virtually impossible. Although Kiene was well aware of Mrs.

Tvede's plight, his $5.83 per acre offer was not out of line with typical land prices ranging from $4.00 to $7.00. Kiene recorded sales of flax at $1.19, wheat at 54¢, and potatoes at 20¢ a bushel, which obviously were no cause for rejoicing.

Frank Kiene, "the loner" who had been through the anxiety of low prices and unpaid interest and taxes, was well prepared this time and took every advantage of the general distress. In 1931 he built two large barns, two farm houses, two potato houses in Kennedy, a windmill, fences, and feed racks, for a total expenditure of $6,400, because he knew he would probably never be able to do it for less. Labor was available and cheap, and the lumber dealers were eager to slash prices to unload inventory. While he was spending $6,400 for buildings, he also invested $14,464.74 in new mechanized machinery and equipment. The amount of machinery that could be purchased for $14,000 is indicated by the following ledger entries:

John Deere sixteen foot Model A. V. combines $900 plus $165 freight each, five Massey Harris tractors, one plow, and one combine $3,270.50, three Ford Trucks and one Ford car $2,483.10, eight John Deere tractor discs $798.00.

Purchases of similar equipment and volume were made in 1932 and 1933.

Kiene was far from being solvent in 1931 to 1933, when he made these extensive acquisitions of land, buildings, and machinery, for his interest load was well in excess of $5,000 a year. However, by this time he was earning nearly $3,000 interest annually from past due accounts to his auxiliary enterprises. While others were still going broke, his farming enterprises were gathering momentum. For example, in 1931, 1932, and 1933, on a gross income of $238,344.07, after deducting $57,075.36 for new machinery purchases, $6,400 for new buildings, both of which were written off as current operating expenses, and $55,179.09 for back salary and incurred expenses by Frank Kiene, the farm showed a three-year net profit of $20,133.44.

Kiene proved that even with the low prices of those three years by operating on a large scale with good management he could make money farming. His tremendous surge of capital expenditures was an obvious indication that he sensed what was ahead and prepared for it because the machinery and building purchases were made before the profits occurred. Old timers in the area remember what almost appeared to be a mission on the part of

Frank Kiene to encourage local people to expand because of what he considered to be "the time to do it." By personal example he practiced what he preached.

Kiene, an apt man of experience, used the similarities and contrasts of the 1917 to 1920 era to his advantage during the period 1931 to 1933. His big profit year of 1917 was due in part to the fact that neither wages, taxes, nor rent had increased so rapidly as farm income. By contrast, in 1931 prices of farm goods had not fallen so rapidly as the cost of wages, taxes, or rent, so that there was chance for profit. Kiene was able to cash rent land for little more than taxes during several years of the 1930's. In 1931 he paid an average rent of about 56¢ an acre; in 1933 he paid about 85¢, and he was able to "pick his land" at that price. His taxes during those years varied from 43¢ to 69¢ an acre on his own land, much of which was the best in the area.

Experienced as Kiene was, by the 1930's he sensed a contrast that worked to his advantage. In the World War I era nearly everyone was bullish and competed vigorously to buy land, thereby forcing prices up rapidly. In the 1930's most farmers were either so broke or so extremely cautious that there was almost no competition for land. Kiene had cash reserves, was operating at a profit, and was a proven manager with loyal, experienced supervisory personnel and a reputation of being honest in his business dealings, all of which caused "land to seek him." Many landowners, including the big lending institutions, were anxious to rent or sell to any reliable person.*

Kiene's judgment of the 1932 marketing scene caused one of his master strokes of management. Because he had owned elevators since the early 1900's, his direct involvement in grain marketing had helped him establish a sound friendship with the McCabe Brothers, big grain marketers in Minnesota, the Dakotas, and Montana. In 1931 Kiene had sold them $43,013.87 of wheat, but in 1932 when wheat averaged 56¢ at Duluth, which meant about 31¢ to the farmer, Kiene's wheat sales fell to $15,115.05. He paid only $1,445.78 in real estate taxes out of a total bill on the farms of $6,000, knowing full well that a large portion of farmers were not paying their taxes and that for five years no harmful action could be taken against him for not paying his. His own feeling that the market was "too low and was bound to come up" caused him to

---

*Ole Olson of Fargo, a land agent for the National Life Insurance Company of Vermont, said at one time he personally had 165 farms in his territory that he had orders to rent to reliable farmers or otherwise get rid of at "almost any terms."

hold his grain. Besides, the McCabes had encouraged him to do exactly that. In June of 1933 he had to empty bins to make room for the coming crop and he received 71¢ a bushel; by harvest time of that year he was receiving over 80¢ for his wheat. His average sale price for wheat that year was 77¢ a bushel at Duluth, an increase of 36 to 40 percent over 1932 prices. He sold $67,586.92 worth of wheat to the McCabes in 1933, and as soon as he received the money he paid his back taxes and his delinquent first-half taxes for 1933. The calculated risk paid off. Most farmers sold their wheat in 1932 at the low price because they "had to pay their taxes to avoid the penalty." Kiene felt otherwise.

When farm profits improved after 1933, Kiene made the necessary adjustment in wages to his permanent personnel. Most of them had been with him prior to World War I. During the 1930's when Kiene was struggling financially the workers frequently took notes for unpaid wages or traded out their earned wages at the Kennedy Trading Company. Although he cut wages of permanent employees somewhat during the hard times of the 1920's, most of them fared quite well by contrast to the transient employees. Conrad Lysfjord, who started working at the farm in 1917, had an annual wage of $640 in 1920. That year his family purchased all of their food, clothing, furniture, and miscellaneous household needs from KTC for $245.88, which serves as a guide to the value of the dollar at that time. When the farm depression came in 1921, Lysfjord's wages were periodically reduced to a low of $500 in 1924, then they climbed gradually each year to $790 by 1930. In 1930 the Lysfjords had three children, and their total charges at KTC for the year amounted to $369.50. During the 1930's he received regular annual increases, so that by 1937 he was paid $1,800. A side benefit for Lysfjord was that he was permitted to raise chickens, hogs, and cows and feed them from the farm supplies. He was financed by Kiene to buy a Model B Ford truck that Lysfjord needed for his own personal ventures. He also rented it to Kiene during grain and potato harvest. This gave Lysfjord additional income. With war time economy and resulting inflation, Lysfjord's wages rose to $5,200 by 1951. Lysfjord was one of Kiene's very trusted foremen, and when he was killed in 1954 Frank Kiene was deeply grieved. Lysfjord's son, Charles, who started with Kiene Farms, Inc., in the mid-1940's, became general manager of the several enterprises in 1971.

The wage pattern for Oscar Nordling, chief carpenter, and E. F. Moberg, manager of KTC was similar. Jonas Cederholm, bulk oil station manager, was put on commission in 1933 and gained a substantial increase in earnings over 1932. Kiene was alert to the potential use of tractors and wanted Cederholm to get the new business. Ruth E. Anderson, whose recorded salary was $1,800 to $2,400 during most of the 1920's and 1930's, drew fringe benefits equal to that amount nearly every year. In addition, by 1933, farm income of over $14,000 was recorded to her account. Her participation in the income of the farm and her close association with Frank Kiene over the 40 years proved to be a cause for community talk.

Throughout 1931, 1932, and 1933, many seasonal employees, such as Ben Sorvik, were paid either 20¢ an hour or $10 a week. During those years none of the seasonal employees was paid more than $10 a week for spring planting and summer work. During grain and potato harvest, when the demand for labor was at its peak, the top pay any of the transients received was $18 for a week's work. Potato pickers were generally paid according to the number of bushels picked. This caused considerable variation in the amount received by the workers. When Kiene "loaned out" a man as a tractor driver on tractors that he leased to other farmers, he charged $1.50 a day for that man's wages.

Even though the wages paid by Kiene in the 1930's seem ridiculously low, they were equal to or better than those paid on most farms in the area. The sum total of all wages paid by Kiene was beyond the comprehension of most farmers of the nation in those years, for Kiene always had at least 20 full-time employees. The social security records for 1934 list 161 different employees. Total wages paid by KTC in 1934 exceeded $33,000; and by 1937 the total, exclusive of family salaries and Ruth E. Anderson's income, surpassed $64,000.

Innovations adopted by Kiene in the early 1930's helped him increase yields and at the same time reduce relative labor cost. His intensive use of commercial fertilizer in 1931 paid big dividends and soon he was making carload purchases. At the same time, he bought tractors and other mechanized equipment, especially combines, on a volume basis. When Roy Sundberg along with Mr. Hunt purchased what were probably the first combines in Kittson County in 1928, Kiene was on hand to watch them work. His remark to Roy Sundberg at the time was, "I'll see how yours works, but I'll thresh one more year." The burden of the Hill farm prevented Kiene from buying combines until 1931, but he admitted to Sundberg before he bought his that he could see "the labor savings."

The adoption of smaller, more flexible tractors in 1931 helped the economy of the farm operation in several ways. Diesel fuel cost Kiene, on a wholesale basis, less than 8¢ a gallon and was far cheaper than horse feed. A second benefit was the ease of hiring men to work on a farm where they would be driving tractors without having to do horse chores. The third benefit was that Kiene could rent his tractors and labor force out to local farmers in slack periods. He charged $5.00 a day for a $500 tractor plus $1.50 a day for an operator that Kiene paid only $1.00. In addition Kiene supplied the diesel fuel at a profit of over 3¢ a gallon. Roy Sundberg, Adolph Johnson, and A. A. Snare, who rented tractors from Kiene in May, 1932, were just three of many such farmers who more often than not became customers for tractors at Kiene-owned implement stores.

Through the use of fertilizer, tractors, and larger machinery, Kiene was able to do a better job on his land than was possible with horses. Total cash flow from farming operations in 1933 exceeded $576,000, and for all enterprises under his control it easily surpassed $2 million, a substantial business for a small farming community. With such a dollar volume it was not difficult to buy a dozen D-6 Caterpillar track-type tractors, enabling yet another reduction in the labor force while at the same time increasing the efficiency of the farming business. Judging from the net profits and federal income tax payments, the 1930's were a blessing to the many Kiene enterprises by contrast to the 1920's. From 1930 on everything "went right."[5]

1934 to 1953—A Joy Ride: Sixteen Caterpillars and a $100,000 Payroll

The last 20 years of Frank Kiene's life were a pure "joy ride" for a man whose ambition it was to operate the biggest farm in Minnesota. Probably there were only two privately owned and operated farms in the state that exceeded Kiene's in total acres. One was the Schermerhorn Farm at Mahnomen, that had 24,000 acres, of which only about 11,000 acres were under cultivation on a less intense basis than Kiene's. The dollar sales contrast makes Kiene Farms, Inc., the easy leader. Schermerhorn, like Kiene, wanted to be a big land owner, but, unlike Kiene, he was not so interested in doing the actual managing. The other large one was the Humboldt Farming Company, owned by the Florance family, bankers in northwestern Minnesota. The core of the Florance holdings was the Hill farm at Northcote-Humboldt, which Kiene lost in 1929. The *Kittson County Enterprise* of Hallock September 11, 1935, credits the Humboldt Farming Company with 25,000 acres, of which 15,000 were in crop. The same issue credits Frank Kiene with 17,000 acres of land.

There are many stories in the Kennedy area that credit Kiene with 30,000 acres, but because he frequently traded land it is nearly impossible to determine the exact amount he might have owned at any given time. A copy of the 1935 agricultural census in the company files together with a hail insurance company survey sheet lists Frank Kiene as owner-operator of 14,690 acres, of which 13,880 were crop land. Neither of these list land that he held elsewhere but that was not operated by his companies, nor do they list a sizeable acreage to the east of the Valley that was used for ranching purposes only. The 1935 census lists 186 beef cows, 274 beef feeding animals, 9 milk cows, 680 sheep, 58 sows for breeding, and 26 horses. The above land and livestock were listed for Frank Kiene and make no reference to Kiene Farms, Inc. or to KTC, both of which had sizeable holdings. The census sheet indicating 14,690 acres showed a listed evaluation of $161,590, or $11 an acre. That year a KTC ledger carried land real estate valued at $74,974.77. If the same evaluation is used, about 6,800 acres must have been owned by KTC. The 1945 census sheet shows this subsidiary owning 14,175 acres and lists Frank Kiene with 7,480 acres, Kiene Farms, Inc. with 5,991, Ell Kiene with 1,760, Ruth E. Anderson with 2,240, and 559 under miscellaneous. If these figures are not overlapping that would indicate a total holding of 32,205 acres, similar to the figure frequently quoted in the Kennedy area.*

Kiene was 60 years old in 1936 and was as interested as ever in the farm. His interest and participation in what took place remained intense until the day he died "on the job" at age 78 in 1954. Many larger farmers testify that "the larger they become the easier it gets," and this was obviously true in Kiene's case. The farm to him was a romance. Charles Lysfjord, who was on the payroll from age 10, first as a cow herder, then a store clerk, then a personal handyman and errand boy for Mr. Kiene, next his chauffeur, later a foreman, and eventually general manager, said, "Mr. Kiene was interested in everybody and everything on the farm.

---

*It is the opinion of this writer that there is overlap in these figures and that the total should not be much over 18,000 acres.

He was almost nervous to make his rounds each day to see that everything was under control and how the crops were doing, and he would just about jump out of the car before it stopped if he saw that the boys needed help."

Charlie added, "Mr. Kiene was interested in anything about agriculture and had a good understanding about work economy and agriculture, but he did not speak much about it preferring generally to listen and analyze what was being said."

Roy Sundberg said, "Frank was a great questioner and observer. He was always asking about things and when he found out what he wanted he was on the go."

Next to working, there was nothing that Frank Kiene enjoyed more than taking people around his farm. John Scott, Sr., recalls that Kiene would escort a person around the land for a full day, show him everything, and then take the guest out for a fine steak dinner, at which time Kiene would proceed to ask the questions. He knew the value of exchanging ideas.

Many people who knew both Frank Kiene and his son, Ell, say that the father was restless because his son Ell did not show the intense interest in the farm that his father expected of him. Although it is recognized that Ell was every bit as capable a farmer as his father, he did not have Frank's driving desire for farming. Ell had a broader range of interests. Personal friends who played golf with him testify, "the intensity was there but not as one-tracked as in the case of Frank." Frank Kiene was a "driver," of himself and those around him, but with the apparent capacity to get people to do the job without provoking them. This was one of the keys to his success, for without loyal workers he could not have weathered the storm of the 1920's nor achieved such heights in the '30's, '40's, and early '50's.

As the farm became better mechanized the Kiene operation was expanded in several directions. A beef cow herd of 300 to 400 head was maintained, and a feed lot with a capacity of 3,000 head was operated, finishing as many as 6,000 fat cattle a year. The potato operation was enlarged to 1,400 acres by the late 1940's. This required a labor force of 300 pickers and a hauling and warehouse crew of 30 to 35, using 15 trucks. To assure themselves of a more stable labor force and to make it easier for the transients, a second bunk house was erected in 1945. Over 70 beds, mostly in single rooms, were in regular use from then on until the need for big potato crews disappeared with the advent of the mechanical potato harvester. The dining hall accounts during the 1940's and early '50's varied from $12,000 to $15,000. In 1944, 11,000 pounds of pork and 4,935 pounds of beef were consumed at the dining hall. Using national average annual consumption figures, that would have represented enough meat for well over 100 persons.

The Kienes purchased four potato harvesters in 1952, one of the first years the machine was available commercially. Through the adoption of those machines, the field labor force was reduced to 20 sorters riding the machines, four tractor drivers, and 12 truck drivers. Despite the great improvement in the production of potatoes, Ell Kiene decided to discontinue growing them after 1956 because of the tremendous problem of marketing. The time consumed in that management function was more than he felt he could devote to it, particularly since he did not personally like that type of work.

Kiene first purchased Caterpillar track-type tractors in 1931. Between 1934 and 1936 he bought 16 more, which enabled him to eliminate horses and sharply reduce the labor requirements. At that time he also purchased his first combines, quickly adding more in successive years. Now the land could be tilled quicker, and in a sense the growing season could be extended by getting the crop in the ground earlier in the spring. A trucking business was established in 1935 chiefly to supply the auxiliary enterprises from the wholesalers and the factory. In slack times the trucks were used to deliver some of the produce from the farm.

A full-time labor force of 35 to 40 was maintained on the farm during the late 1930's. With each improvement in mechanization the number of full-time and part-time people needed was reduced. By 1950 there were only 13 full-time farm and 2 full-time office employees, but the payroll exceeded $106,000, exclusive of any salaries paid to members of the family. The annual payroll rose while the number of full-time employees was reduced, but the advantages of hiring better skilled workers became obvious, for in 1952 the Kienes built eight potato harvester truck boxes at a cost of $1,802.08, using their permanent labor force. Those boxes, which were nearly identical to the commercial models, would have cost $4,800, had they purchased them. There were also more than 20 full-time employees in the auxiliary enterprises by the early 1950's. Four of the non-farm subsidiary enterprises in 1954 had a payroll of nearly $60,000, and these subsidiaries were doing nearly a million dollars in business. At that time the Kienes listed eight non-farm companies and four individual farm units.

After financial success became obvious in the early 1930's and as the farm increased its cash flow and net profits, it was possible for Frank Kiene to withdraw funds for re-investment elsewhere. As the years passed this amount so diverted became substantial. The resulting investments no doubt became sizeable; but in no way were they used to enhance, nor did their withdrawal jeopardize, progress of the farm. A landmark in success to Frank Kiene personally happened in 1944 when some property of Kiene Farms, Inc., was transferred back to the farm corporation. Kiene Farms, Inc., which avoided bankruptcy only by absorption into KTC in 1925, was now fully solvent. The $7,000 stock investment by Mr. Kiene in March, 1917, was now worth $276,863.07 plus $192,724.73 in the surplus account. This represents an increase of more than sixty-seven- fold in a period of 37 years that included 18 years of the most severe and prolonged agricultural depression in western history.

In 1945 Ell Kiene Farms was created to further subdivide the operation from an ownership standpoint. In March, 1948, Prairie Farms, Inc., was organized to completely absorb the remaining farm holdings of the KTC and the personal holdings of 72-year-old Frank Kiene, who was properly preparing his estate. Even though the three operations were as one to the men working in the field, for financial ownership and income tax purposes, the office breakdown charged Prairie Farms, Inc., with 43 percent of the expenditures, Kiene Farms, Inc., with 36 percent, and Ell Kiene Farms with 21 percent. Such business practices not only represented extremely astute financial thinking, but were also indicative of the thoroughness of Frank Kiene's sense of business to the last detail.

From the early 1940's to the death of Frank Kiene in 1954, every subsidiary had interest income. There were no mortgages or long-term indebtedness. Financial liquidity was conspicuous in the form of bond accounts, actual reserve for depreciation accounts, personal loans to employees and local people, extensive funds in transit from grain-brokerage firms, idle checking accounts, and very large quantities of "non-sealed grain" on hand at all times. The dollar value of the grain alone exceeded seven digits.

Frank Kiene "loosened up" financially in his late years. Locally there are mixed feelings about how charitable he was, and the impressions about it go to the extreme either way. In his early years he was not known to take part in what might be considered civic or charitable activities. It is from this that the "loner" characteristic was applied to him. In 1923 he charged the Kennedy

Baseball Association $25 rent for land for the season. This was at a time when baseball was a very popular sport in the small towns of the Midwest and most businessmen contributed to their local teams. There is no such contribution listed on his part for any of those years—for baseball or any other charitable or civic activities. His limited donations to local churches, Red Cross, and temperance groups were precisely itemized. His membership subscriptions, outside of Farm Bureau, Woodmans, and United Commercial Travelers, were nil. In 1925 he sold potatoes to "The Pool" controlled by the Minnesota Potato Growers Exchange. It was not until the 1940's that annual donations to "just causes" exceeded a few hundred dollars. By the late 1940's that yearly figure passed the thousand dollar mark and went òver $6,000 in the 1950's. Temperance groups and his local church received the largest amounts. Since the 1930's his salary and business expense accounts surpassed the five-figure sum annually. Many years his federal income taxes of all enterprises were high in the five-figure bracket, but apparently even that was not enough incentive to cause him to "give."

The Kiene name will long be legend in northwestern Minnesota. Since the death of Frank Kiene, the founder, many of the auxiliary enterprises have been sold. Since the death of Ell Kiene, the son, in 1971, some of the land has been sold, but it is still a very large farm—large enough to gain from the economies of scale for many years to come. The fleet of 16 Caterpillars and 16 smaller two-wheel-drive tractors was being replaced by fewer but speedier and more versatile four-wheel-drive tractors of 225 H.P. or more each. Those tractors plus 16 trucks and pickups are dispatched through a complete radio communications network. Rubber-tired, radio-controlled farming no longer needs to rely, as Frank Kiene did in the 1920's, on the railroad to haul equipment 16 miles to the field where it is needed.

Frank Kiene, though not a zealous innovator, showed the people of the lower Red River Valley the advantages and pitfalls of large-scale farming. Determination to succeed, complemented by the ability to attract and hold a devoted labor force, a keen understanding of financial and legal matters, a sense of thoroughness without being bogged down by details, and a cool head under pressure were the dominant traits of the man who wanted to own the biggest farm in Minnesota. His strong temperance position, reserved mannerism, his lack of charity, his "loner" impression made this farmer who always wore a suit and tie exactly what he wanted to be in his community—a "loner." Not disliked, but not warmly liked—just respected— best describes Frank Kiene. A man who lost a mother and was deserted by his father by the age of 15, who lost two wives through death prior to his age of 48, probably had some reason to want to be alone. During his lifetime little was ever written about him or his farm because he avoided publicity, and if he had a trusted confidant that person never betrayed "Mr. Kiene."[6]

NOTES

[1]Stanley N. Murray, *The Valley Comes of Age: A History of Agriculture in the Valley of the Red River of the North, 1812-1920* (Fargo, 1967), pp. 130-164. See *The Valley Comes of Age* for a detailed account of the total growth of agriculture in the area, and Drache, *The Day of the Bonanza,* for specific information on bonanza farms; Copy of a letter written by Ell Kiene, September 12, 1969, found in the files of the Kennedy Trading Company, a subsidiary of Kiene Farms, Kennedy, Minnesota; Personal Interview with Roy Sundberg, Hallock, Minnesota, January 30, 1972. Mr. Sundberg, who was born in November, 1893, was the son of Bengt Sundberg and lived with, worked with, and had business relations with Frank Kiene throughout his lifetime; Personal Interview with Charles Lysfjord, Kennedy, Minnesota, January, 1972, and March, 1973. Mr. Lysfjord is the general manager of Kiene Farms, Inc., Prairie Farms, Inc., and P. J. & L. Farms, Inc. He has been employed by the Kienes since the early 1940's. His father, Conrad Lysfjord, worked for the Kienes for over 40 years, and in the years prior to his death in 1953, he served in a supervisory capacity. Charles Lysfjord is probably the best-versed individual on the past and current operation of the Kiene enterprises. Since the death of Frank Kiene's son Ell in 1971, there are no survivors in a managerial capacity; Records of Kennedy Trading Company, Frank Kiene, Inc., Kiene Farms, Inc., and other business accounts of Kiene enterprises. Mrs. Ell Kiene, daughter-in-law of Frank Kiene, granted full access to all journals, ledgers, and papers of Frank Kiene from 1900 until the time of his death in 1954. All available records were completely examined for documentation of data contained in this chapter or in related appendix. Mrs. Kiene has had the unique experience of being reared on a 160-acre dairy farm in Otter Tail County and then along with her children, Jennifer and Peter, becoming the sole stockholder of one of the state's largest farming operations. Her cooperation has made possible a thorough research into the company records; Personal Interview with Winslow Larson, Hallock, Minnesota, March 23, 1972.

[2]Sundberg Interview; Deere & Webber Co. price list dated February 10, 1916, in possession of Grant Mattson; *Kittson County Enterprise* (September 11, 1935), found in File 554, North Dakota for Regional Studies; Kiene Farm, Inc., records; Lysfjord Interview.

[3]Kiene Farm, Inc., records; Kennedy Trading Co., Inc., records; Murray, pp. 133-135; Drache, p. 212; *Kittson County Enterprise;* Personal Interview with Cliff Bouvette, Hallock, Minnesota, January 29, 1972. Mr. Bouvette was the editor of the *Kittson County Enterprise* for many years and a personal companion of Walter J. Hill; Contracts between Frank Kiene and Wilson H. Hubbard, dated December 30, 1920, plus other papers incidental to the transaction, all found in Kiene Farms, Inc., records. There are conflicting figures on what the buildings on the Hill farm actually cost when built in the early 1900's. Nothing was spared to make them the best. The lowest recorded figures are found in the Kiene files which indicate their cost to be $181,100. On the other extreme is the $500,000 figure recorded in the

Kittson County Enterprise, September 11, 1935, and because of the direct relationship between editor Cliff Bouvette and Walter Hill that sum must be seriously considered. It is very likely that the amount also included the many miles of fencing, road building, ditching and other developing that took place on the farm.

[4]Kiene Farms, Inc., records; Kennedy Trading Co., Inc., records. One aspect of daily life of the 1910-1920 decades that contrasts to later periods is the cost of utilities. Kiene household records show that for January, 1923, his telephone bill was $1.50 and electrical bill $3.85. The light bill for February was $3.45, for March $2.65, for April $1.45. In 1917 Kiene purchased a Delco plant for his farm at a cost of $100. In contrast to his $1.50 telephone bill, the cost of telegrams for March, 1923, was $52.36; Personal Interview with Henry Meek, Stephen, Minnesota, January, 1972.

[5]Kiene Farms, Inc., records; Kiene records; Kennedy Trading Co. records; Personal Interview with John Bogestad, Karlstad, Minnesota, February 26, 1972; Personal Interview with John W. Scott, Sr., Gilby, North Dakota, June, 1971. Mr. Scott, a large-scale farmer in the valley, was a frequent visitor with Kiene. Their mutual interest was that of a competitive nature, for they enjoyed challenging each other. Scott said in all their visits, Kiene never once told anyone how much land he operated. Scott and Ronald D. Offutt, Sr., recall the time when Kiene had shown an out-of-state visitor along with Scott and Offutt around his farm for an entire day. After dinner the visitor remarked, "You have some operation, Mr. Kiene; how many acres did you say you farmed?" Kiene's reply was "I don't believe I said," and changed the conversation; Personal Interview with Ole A. Olson, Fargo, North Dakota, March, 1970; Sundberg Interview.

[6]Robinson Interview. Mr. Robinson was the general manager of the Schermerhorn Farms; *Kittson County Enterprise,* September 11, 1925; NDIRS File 562; Mrs. Ell Kiene and Charles Lysfjord, farm manager, in an interview January 17, 1972, indicated that to the best of their knowledge the farm in the Kennedy area actually operated by the company had been 18,000 acres throughout the 1950's and 1960's. Of that amount, 15,000 was in crop and the balance in a solid block ranch unit. It is the opinion of this writer that except for brief periods when Frank Kiene was reshuffling his holdings to get rid of some poorly located, unproductive, or odd piece of land and at the same time was purchasing more desirable real estate for his operation, the base farm generally did not exceed 18,000 acres; Roy Sundberg interview; John Scott, Sr., interview; In 1972 following the death of Ell Kiene, his estate was reorganized into P. J. & L. Farms, Inc., the initials representing the first names of the children and widow of Mr. Kiene; Kiene records.

CHAPTER IV

Charm, Courage, and Campbell

IN THE NORTHERN Great Plains one of the most familiar names in agriculture is that of Thomas D. Campbell, founder of Campbell Farming Corporation. This corporation has been one of the most talked about farming operations among farmers, farm leaders, and politicians since its establishment in 1917. Tom Campbell's farm, one of the best publicized in American agricultural history, is deeply cloaked in mystery whenever the subject of profit and loss in farming is raised. The small-farm advocates have almost all been convinced that it could never have been profitable. The large-farm advocates have been dubious about its profitability because of the complexity of Campbell's economic philosophy.

For those who were close to Tom Campbell in business matters there was less of a secret, for when he lost money he told them so, and when he made money they knew it too. In that respect Campbell was like the greatest of the bonanza farm managers, Oliver Dalrymple, who had the constitution to smoke "twenty-five cent cigars" when he was broke as well as when he was solvent. Debts and losses did not bother either Dalrymple or Campbell, and, as far as the world knew, they were always "in the chips." Dalrymple and Campbell were alike in other respects—they were great promoters. In fact, it seems that it was almost an obsession with both men to let the world know about their giant farms. Few farms of the late 1800's had as many important visitors as did Oliver Dalrymple's bonanza farm at Casselton, North Dakota. Few farms of the 1920's to the 1950's have had as many national and international visitors as Tom Campbell's giant wheat farm at Hardin, Montana.

In some respects both men had a mission to accomplish, and they did the job well. Oliver Dalrymple, best known of the bonanza farm managers, had to sell the world on the value of Red River Valley soil and the advantages of large-scale mechanized horse farming. His results were mixed—he did prove the value of the fertile, treeless, stoneless prairie of the Red River Valley, but multiple-unit, large-scale horse farming with hired labor could not overcome the natural advantage of family labor of the smaller farmers. Where Dalrymple failed to prove the merits of large-scale mechanized farming, Tom Campbell succeeded. With necessary vision and proper timing, Campbell, the tall, muscular, long-striding, flashy, silver-haired Dakota farm boy, became the apostle of large-scale mechanized industrialized agriculture. He informed the world that he was a "manufacturer of wheat," a phrase calculated to convince everyone of the reliability of mechanical agriculture even in the nearly arid valley of the Little Big Horn River.

Campbell was a leader of men, and even though he made few close personal friends, he had the ability to "charm" his employees into being absolutely loyal. Like Dalrymple, he had few personal contacts at the scene of his operations, for both men spent much of their time in larger centers—Dalrymple in the Twin Cities, and Campbell in California. From the beginning, Campbell picked capable men to work with him in organizing one of the world's largest privately operated wheat farms.

Both men thrived on publicity. Tom Campbell, like Dalrymple, knew how to influence the right people at the right time. During many adversities experienced by Campbell he always managed to come up with that ever-essential financial instinct that tided him through to the next crop. He was an opportunist, leader of people, money manager, lover of farming, and eternal optimist. Campbell deserves his position as the monarch of mechanized industrialized large-scale agriculture.

Archibald MacLeish, a prominent literary figure of the 1930's, wrote in *Fortune* magazine:

> The narrative of the grass . . . is a narrative on which much history hangs. . . . There is secondly, the story of the breaking of the Great Plains sod in the years from 1900 on. . . . The actors are numerous and of all breeds and professions; one may stand for them all—Tom Campbell, the master of the wheat. . . . American agriculture has Farmer Thomas Campbell to thank for a good bit of the improvement which has taken place [from the binder to the combine]. . . . If the narrative of the grass were a simple story . . . its principal and most portentous figure would be a single man: The man who broke the sod . . . Tom Campbell. . . . The impact of Tom Campbell upon the grasslands of the Great Plains was the impact of the American passion for power, speed, and the predictable machine.[1]

Tom Campbell was never critical of the fact that the bonanza farms of the late nineteenth century had failed, nor was he overly concerned about it. He expressed his feelings about bonanzas and large-scale farming with this statement to a writer for *Country Gentleman*: "The day of the bonanza farm is gone with the buried past! It was too unwieldly, too wasteful, too costly to compete with the small independent farm unit in economical production!"

The same writer quoted an often repeated statement: "If agriculture is to endure, it must be by virtue of the small farm intensely cultivated, operated largely by the labor of the farmer and his own family, with profits dependent upon the detailed attention that only the small farm unit will permit."

The writer of the article doubted that the above statement, even though believed by most people, was true. In his opinion Campbell's operation "refutes small farming operations." Campbell commented, "Granted that most of the bonanza farms of the past have failed, was the fault with the bonanza farm or with the bonanza farmer?" He added, "Certain it is, with the exception of the invention of labor-saving farm machinery, and the newer possibilities of cooperative marketing, there have been no great fundamental changes in our farming methods as there have been in American industrial methods, since our forefathers wielded the sickle and the flail." At the time he made that statement Tom Campbell had already been responsible for some of the most decisive mechanical changes in agriculture and was on the verge of making even greater ones.[2]

Boyhood to Manhood

Tom Campbell, who in his day would visit with Wilson, Coolidge, Hoover, Roosevelt, Truman, and Eisenhower, was born on February 19, 1882, on a homestead only a few miles from Grand Forks, North Dakota. Campbell's father, Thomas Campbell, Sr., was a Scottish immigrant to Canada, where he arrived penniless to take a job in the lumber camps. After his marriage he and his newlywed wife came to settle on a farm at St. Paul, where they worked until March, 1876, after which they left by covered wagon for Grand Forks. Their cargo included five bushels of wheat, which was seeded that year and was the first of 90 crops that the father and his later famous son would plant in their lifetimes.

The land owned by Thomas D. Campbell, Sr., was so near to Grand Forks that it was not long before he became involved in real estate in that frontier village. In October, 1882, he purchased a lot there for $3,500 that he sold in June, 1888, for $5,200. Four months later that lot was sold for $11,900 to become the site for the Hotel Dacotah. Frontier farmers who dealt in village and town real estate must have had either family money or a successful farming operation to be able to risk such a venture. After farming less than six years Campbell, Sr., was able to buy a steam-powered threshing rig believed to be one of the earliest in the Red River Valley. He must have been an unusual operator and speculator, for his first farm was not big enough to generate either the borrowing power or the profits to justify such a purchase. However, he did eventually own 4,000 acres of land in the Valley.

Tom Campbell, Jr., later recalled that his father obviously was both innovative and aggressive as far as the business of farming was concerned. Young Tom Campbell, despite his being reared in a home of more than average affluence, resented the fact that members of his family had to work so hard to make things go. Besides living in a sod house for many years, Campbell remembered that his mother gave birth to her first three children "without medical assistance of any kind . . . [and] I have always resented the fact that my mother and the other women on the farms had to work so hard. It always seemed wrong to me that in this great country of ours it has been necessary for the women and children on our farms to work eight hours twice a day, all of their lives for practically nothing, in order to make the farm a success."

Tom Campbell used the word "resent" many times in his life when he referred to the hardships of his parents and his sisters. He was particularly upset over the endless chores of his mother and other pioneer women. In 1928 he commented, "My mother and father pioneered and I realized as a boy that something must be wrong in the scheme of things when it was necessary for my mother and other farmers' wives to work so hard and put in so many hours of toil." Later, at the age of 74, Tom Campbell wrote, "It was bonanza farming which was so familiar to me when a boy that caused me to have the urge to do something about the mechanization of farming. I resented the hardships and long hours of toil suffered by my mother and other pioneering women."

Tom Campbell never claimed to have been born in a sod house, for before he arrived in 1882, the family had prospered enough so that they were living in a newly constructed frame house. The endurance, courage, and fortitude that he saw in his parents as they worked to make the farm succeed was well implanted in him. At the age of 17 he enrolled at the University of North Dakota on the west edge of Grand Forks and only three miles from his father's farm. Daily he walked to and from school except in the most severe weather. Each spring and fall he took a month off from school for seeding and harvest on the family farm. He also helped with chores morning and night, but he still did not neglect his college schedule and studies. In 1903 Campbell earned a B.A. degree from the University of North Dakota. The following year he went to Cornell University in New York to take additional work in mechanical and electrical engineering, but he was forced to return home because of the illness of his father. At the same time he continued his engineering studies at the University of North Dakota and earned a graduate degree in engineering in 1904.*

The young college graduate operated his father's 4,000 acres while the latter regained his health. In 1906 Tom Campbell married a Grand Forks schoolmate, Bess Bull. Her father was one of the founders of the Cream of Wheat Company in Grand Forks. In spite of his relationship with the Bull family, Campbell had no interest in the food manufacturing business because he preferred to farm.

In 1907 young Tom Campbell was made president of the Northern Dakota Railroad Company that was organized to haul concrete 20 miles from the mine at Concrete, North Dakota, to nearby Edinburgh. The railroad was not a major success by any means, but it was part of the training that helped mold the ambitious young Campbell. In 1908 it was discovered that the young bride had contracted tuberculosis and had to seek a healthier place to live. Tom Campbell became operations engineer of all properties of J. S. Torrance, a big time West Coast financier, farmer, and contractor. The move to California turned out to be a blessing, for the job with Torrance was ideal training for the young man who someday would operate the nation's largest wheat farm. The connection proved beneficial in another respect, as J. S. Torrance was a board member of J. P. Morgan's bank. Under Torrance, Campbell did everything. His engineering of huge construction projects and his overseeing of large-scale bean production, as well as his management of massive development properties, proved him capable of any task.[3]

A Boyhood Ambition Is Fulfilled

Despite Tom Campbell's resentment about the hardships of farming, it was his boyhood ambition to have the biggest wheat

---

*It is believed that Campbell earned the first engineering degree from the University of North Dakota.

farm in the Red River Valley. Having worked with steam power on his father's 4,000-acre wheat farm, studied engineering in college, and managed the large projects of J. S. Torrance, which used big power equipment, young Campbell was now ready to let his ambition blossom. Two of the ingredients basic to the success of such an ambition—the desire and the training—were there. All that was necessary was the right timing and the capital.

Tom Campbell will long be a legend in the Great Plains, not only because he broke more sod than any other individual, but "because he broke it in the years when the breaking was most significant." The Homestead Act had been changed in 1909, 1912, and 1916 to adapt that act to the Great Plains. By 1916 it was possible to prove up as much as 640 acres in three years. At that time the work of the United States Department of Agriculture, other agencies and individuals, specifically Hardy W. Campbell, provided effective propaganda "favorable to dryland farming." Many people had the idea that with these new techniques, farming on the dry plains would be relatively safe. A wet cycle caused good crops to be produced in 1914, 1915, and 1916. That factor, compounded with the high wheat prices of World War I, encouraged people from all walks of life to pour onto the Great Plains to farm.

The stage was set for Tom Campbell to become "the most portentous plower of the plains in the history of the world . . . and in the process . . . [become] the world's greatest wheat farmer." During early 1917, Andre Tardieu, a high-ranking official of the French government, worked closely with U.S. Secretary of Interior Franklin K. Lane and food administrator Herbert Hoover in an effort to start large-scale farming enterprises in North Africa to help relieve the food shortage. Campbell, because of his connections in large-scale farming projects with J. S. Torrance, was aware of the contemplated North African project and probably would have been one of the men selected to manage such an undertaking. But he had different ideas. He wrote to the Indian Bureau early in 1917 outlining a plan for raising wheat on unused Indian lands and asked how he could lease those lands. The Bureau informed him that what he proposed could not be done.

The bureaucratic reply that "it could not be done" was all that was necessary to get the ambitious, confident young Campbell to try another approach. This time he telegraphed President Woodrow Wilson. Persistence paid off, for Wilson answered, stating that the matter had been referred to Secretary of Interior Franklin K. Lane. At once Campbell was on his way to Washington to get a commission in the Corps of Engineers to operate an African farm or to see if he would be able to "sell his idea of raising wheat on Indian lands." Lane, Tardieu, and Hoover liked his proposal of large-scale wheat production, but they preferred that the job be done in Algiers. But Campbell held firm to his idea of using Indian reservation land on the Great Plains and won his point.

There is some question over what really transpired next, for Campbell implied in a 1919 magazine story that he wanted to start with 20,000 acres but that Secretary Lane suggested that he go to 200,000 acres. Asking to operate only 20,000 acres when several million were available would not have been characteristic of Tom Campbell unless he had doubts about his ability to finance such a project.

Now was the time for the 35-year-old Campbell to turn on the charm and realize his proposal.* There were two hurdles left—get the approval of the Bureau of Indian affairs and raise the money. Because of the urgency of the food crisis caused by the war, he was allowed to select any Indian lands he wanted in either Montana or Wyoming. Campbell finally convinced some of the federal officials that there were 2.6 million of the 10 million acres of Indian lands that were capable of growing crops, so he was given freedom to pick his area.

To protect himself against a crop failure, he convinced federal officials that the Indians should be paid a percentage of the crop instead of a fixed cash rental per acre. It was agreed that for the first five years, while Campbell had to stand the added expense of breaking the sod and fencing in his grain fields, the rents should be one-tenth of the crop delivered to the railroad. After five years the rent should be one-fifth of the crop for a period of five years. The share of the rent appears nominal, but the land had to be fallowed if production was to be maintained. Consequently, the renter had two or three acres of land to work for each one he cropped. Campbell had to stand total production expense as well as pay for all the fencing. The Indians, who had been receiving from 6¢ to 20¢ an acre rent from the cattlemen and 50¢ an acre from wheat farmers, were paid 75¢ to $1.20 an acre in the first 10 years from Campbell. In later years the share rent was increased to one-

---

*"Tom Campbell was a natural born promoter, a wonderful personality, smart as hell, and not afraid of anything," according to J. R. "Punk" Taylor, who served as a foreman and then general manager for many years at Campbell Farming Corporation.

third of the crop. Eventually a cash rent basis was established for some of the land.

After arrangements were made for the Indian land, Tom Campbell had the difficult task of securing financing. What happened next was one of the best publicized phases in the entire history of Campbell's career. Secretary Lane arranged a conference with J. P. Morgan and James Stillman, president of the National City Bank, Charles H. Sabin, president of the Guaranty Trust, Charles D. Norton, president of First National of New York, Francis H. Sisson, and J. S. Torrance, who had been Campbell's employer for several years. Campbell boasted many times that he had secured $2 million in a 20-minute conference with J. P. Morgan. In later years Campbell commented that he "persuaded the greatest bankers of New York to invest heavily in their pet aversion, agriculture, and that Mr. J. P. Morgan was the banker most persuaded. . . . "

With the government urgently pressing to increase food production, the bankers had the signal that this was an important project. But at the meeting Tom Campbell had to prove its economic feasibility. He had his production costs well in hand and knew what the capital expenses would be. Some of the early news articles made it quite clear that this project was backed by the federal government. The headlines of one article read, "Uncle Sam Operating World's Biggest Farm." The article continued, stating that the United States in an effort to increase wheat production had become involved in a 200,000 acre farm in Montana. In 1922, when the wartime food crisis had passed, *Current Opinion* carried a feature story on the farm, which stated that the bankers and the government had withdrawn, letting Campbell operate on his own resources.

The Montana Farming Corporation, created to satisfy the needs of the financiers, loaned money to Campbell through that corporation. Besides Campbell there were top governmental and financial people on the board of MFC. It was governmental insistence as well as Campbell's charm and ability that convinced the bankers to make the $2 million loan. They were well aware of the great margin of risk. They resolved that in case of failure it could be written off as a "contribution to the war effort" and probably they would be able to recover some income tax. The hurdles were cleared, and the North Dakota farm boy who believed that the future belonged to the farmer with the machine was ready to prove his point.[4]

The Big Farm Gets Going

The confident, 35-year-old Campbell immediately started buying equipment and hiring the labor force necessary to operate 200,000 acres. He had convinced the financiers that machinery, labor, and seed cost would be $15 an acre and upkeep cost would be $10 an acre. If the weather held out he projected a profit at the end of the third year. His cost projections were high, but that was fortunate, for adverse weather caused the income projections to be a bit optimistic too.

Campbell's equipment order was topped by 34 30 x 60 Aultman-Taylor tractors. These were 26,000-pound machines that pulled 8 to 12 bottoms plus other equipment, depending on whether they were plowing sod or previously plowed ground.

In addition to the 34 Aultman-Taylors, Campbell purchased 40 10-foot binders, 10 threshing machines, 4 combines, 100 grain wagons, 60 grain drills, 50 discs, 50 ten-bottom plows, 10 automotive trucks, and large gasoline and kerosene storage tanks. Much later Campbell stated that the original machinery purchase totalled a million dollars. It is not possible that the above list could have amounted to more than a third of that figure, but if fencing, shop and office equipment, bunkhouses, cook houses, and grain storage were included, the original investment could have equalled that sum. Nearly 10,000 spools of barbed wire alone were used. This gives some idea of the immensity of the operation. Many magazine and newspaper articles were written about the all-tractor-powered 200,000-acre farm.

In 1918, 7,000 acres were plowed, seeded, and fenced. The following year 45,000 acres were under cultivation. By 1923, 110,000 acres were under cultivation and 150,000 acres were fenced, a feat that surpassed anything accomplished by the largest of the bonanza farms. Because of the war, both materials and skilled labor were scarce, forcing Montana Farming Corporation to rely on inexperienced labor and high school boys to plow and fence. Campbell was very much the center of the show on this giant wheat farm, but from the start he hired excellent management help to get the job done. Even though he was a "driver" he gave a surprising degree of freedom to his management force, probably because he knew that was the way to get the most out of them. He liked to supervise everywhere when he was around, but the details were left to the staff. Having the farm at Hardin, Montana, and a home in Pasadena, California, did not prevent rigorous supervision on his part. Starting with World War II Campbell was gone from his farm for long stretches and never appeared to be overly concerned.

Fred Gordon, a professionally trained agronomist, was one of the first members of his top management staff. Gordon was "a very scholarly man" and on that basis a good team member for the academically inclined Campbell. It was Gordon's responsibility to develop the system of operation and management. He devised the camp system, which made every unit a self-contained farm except for the big equipment that was moved from place to place. The camp system was a bit of a luxury because it contained an overlap in management, blacksmith, commissary, kitchen, and other facilities. Gordon was manager of the northern lands selected by Campbell on the Fort Peck Indian Reservation. Camp One was located at Poplar, and Camp Two east of Brockton, Montana, making the two camps about 30 miles apart.

Tom Hart, from Fargo, North Dakota, was selected to be "The Number One Man" on the Crow Reservation south of Hardin. Hart was a hard driver much like Campbell and had the attitude toward a man that "if you can't whip him don't hire him." Mrs. Ola B. Maddox was a very sharp-minded woman in cost accounting and financial details, which was a great help to Campbell, the "idea man," who was often impatient with details. The Campbell-Maddox association continued for over 40 years, even beyond the death of Mr. Campbell. Dan Maddox, Ola's husband, was counsel for the corporation. Tom Hanes was later made the first general manager of the entire operation, and served from 1941 to 1956. Unfortunately for the immediate success of the enterprise, Fred Gordon left to manage a large operation in California and Tom Hart took over a similar position in Wyoming. Those resignations made J. R. "Punk" Taylor the number-one man under Campbell.

Taylor's chance appointment in the early 1920's proved to be an excellent choice, for he knew farming from the ground up and had gained a few years' professional management experience before his friend, Fred Gordon, encouraged him to come to Hardin as a second-ranking manager. In his initial position Taylor was able to observe the operation from a lower level and had a different view of things than the top-level management. Taylor had another advantage; although a college graduate, he was not handicapped by being quite as academic as members of the previous management and compensated for that with a "real seat-of-the-pants judgment."

His nickname, "Punk," fit him perfectly, for in his eightieth year he was still as "sharp as a tack and hard as a nail." Taylor knew how to trim expenses and was given a free hand to cut almost whenever he saw the opportunity. Another trait possessed by Taylor, which may have been one of the greatest single factors in the long-run success of the Campbell Farming Corporation, was his mechanical ability. His boss, Tom Campbell, was a true disciple of mechanized agriculture. He was constantly telling the public that "the cheapest production is attained by the largest use of mechanical power to replace hand operation." Reducing cost of production to Campbell was the first order of business, and "Punk" Taylor's mechanical wizardry proved to be one of the best ways to reduce those costs.

Taylor, with the encouragement of Campbell, who was excited over the adoption of almost any mechanical labor-saving device, was constantly working with the blacksmiths and mechanics, such as "Curly" and "Otto," in the early days and later Pat Roach and Johnny Owens, improving, improvising, and inventing machinery or attachments. Johnny Owens was sent to tractor school but said he learned more about tractors from Campbell and Taylor than he did at school. These adoptions from the blacksmith shop proved to be a prime factor in cutting actual machine costs by 50 percent in Taylor's decade as manager.

Farm machinery manufacturers were constantly calling to see what kind of experimenting was taking place on the "big farm." Of course, the manufacturers were always more than willing to give special deals to Campbell because that was a good place to experiment and also it gave them good publicity for their machinery. Taylor's mechanical genius was not overlooked, and when he left Campbell after 11 years of employment he was hired at once by Caterpillar Tractor Company. Mr. Taylor's patented inventions made him one of the extremely successful leaders in the agricultural heavy equipment lines of Caterpillar Tractor Co.

The innovative Campbell was always searching for the latest technique, variety, machine, or idea. He had widespread contact through his constant travels, and he associated with some of the best minds in agriculture in the nation. M. L. Wilson, who later became an undersecretary of agriculture and one of the nation's outstanding professional agriculturalists, was active in Montana during Campbell's early years. Campbell and Wilson became good friends and cooperated closely on all phases of farming. Campbell's alert mind never tired of receiving new ideas even though he is remembered as one who freely and strongly expressed his own theories.[5]

Montana Farming Corporation

Montana Farming Corporation had the biggest and best equipment available and an excellent managerial staff closely backed by some of the finest agricultural minds in the region, and yet it had problems. Skeptics surmised that this was the nature of agriculture, others said that a farm of this size "just couldn't work and was bound to fail," and a few felt that the grasslands should never have been broken. Tom Campbell had an answer for every critic and after weathering some real hardships proved his point with the long-run success of the enterprise.

No one previously had farmed the northern Great Plains on such a grand scale. And with the exceptions of 1914, 1915, and 1916, there was little success in large-scale grain production. Initially, partly as a hedge against the weather and partly because of strong persuasion by a Mr. Zachary, a government official, Campbell decided to farm in two locations. His decision to do so proved wise, for even though one location soon proved to be a failure, the other saved the day. It was Zachary who convinced Campbell to break sod on the Fort Peck Reservation in northeastern Montana. However, this choice was a bad one, for there was a steady stream of disasters at Fort Peck. The area proved too cold for winter wheat, which Campbell felt had to be raised to take advantage of the moisture from the snow.

When the wheat crop failed to mature, 2,000 head of beef were purchased with the idea of feeding them wheat straw. This again was Zachary's idea, and Campbell was convinced of its merits. Too late he discovered that the water supply at Fort Peck Camp Two near Brockton was inadequate for a herd of 2,000 cattle so that they had to be driven to the Missouri River for watering. The watering place on the Missouri was 18 miles from Camp Two. While the cattle were at that location feed was hauled in. Wheat and wheat straw were transported 18 miles by wagons, with the big Aultman-Taylors providing the power. That was supplemented with sage brush hauled in from North Dakota, a very expensive process, considering the quality of feed supplied. It was not long before the feed supply ran low and the cattle had to be shipped to Texas. They were called the "tourist herd" by the employees. On the way south the train was stalled in a snow storm, and many cattle died. The following spring the herd was shipped to the Big Horn Mountains, about 40 miles south of Hardin. With such experiences it was only a matter of time before the Fort Peck operation had to be closed.

Montana Farming Corporation was not the only farm to suffer misfortune in northeastern Montana at that time. Many of the small farmers who lost out because of either frost or drought went to southern Montana to start over again. Two Legging Creek, about 8 miles south of Hardin, is an area of heavy gumbo soil that normally receives considerable snow fall, but not regularly enough to be depended upon. It was felt that with irrigation the land would be more dependable, and Campbell, not being one to overlook an opportunity, secured additional funds from J. P. Morgan for the purpose of irrigating the land. Campbell's idea was to use the displaced farmers to farm the irrigated land.

J. R. "Punk" Taylor's first job with the Company was to supervise these transplanted farmers and to oversee the plowing operations at Camp One south of Hardin. Taylor was given a Model T Ford Roadster to carry out his work. The government purchased a large number of bins for grain storage and placed them in the area of these transplanted farmers. Seed grain was supplied to them through Montana Farming Corporation; but it never got into storage, for most of the farmers took the grain directly to the elevator and sold it in their names. When seeding time came those who had not already abandoned their farms ran around the fields with empty grain drills. It was Taylor who discovered what was going on. Obviously that phase of the master plan had no future, but man and not nature was responsible for the failure this time. Art Koebbe, who has farmed in the region since the 1920's, remembered that only three farmers out of one group of 20 survived. Koebbe said, "Those who depended on horses for power just could not compete."

Montana Farming Corporation under Campbell's guidance had expanded from the original 7,000 acres of crop in 1918 to 55,000 acres by 1921. Over 110,000 acres of sod had been broken. Progress was made because of sufficient capital behind the project, but it required more than adequate capital to continue an enterprise of this scope in view of the adverse price and weather conditions that plagued farmers of that era. The years from 1918 through 1921 were not particularly encouraging to the bankers on the corporate board, who expected to see profits in the third year. The government officials involved were no more enthusiastic. Tom

Campbell, however, accustomed to the gamble of agriculture, was a bit better prepared to endure the setbacks.

The production of 1918 was not large enough to make much difference in the newly generated enterprise at Hardin, and the winter wheat at Fort Peck was a failure. In 1919 Montana suffered one of its driest years on record, and the farm produced less than 8 bushels per acre. Grasshoppers added to the problem. Weather conditions were better in 1920, but economics took its turn. The crop was planted under high wartime costs, which were partially caused by the tremendously optimistic attitude of the farmers because of the good prices they had been receiving. The crop was satisfactory, but the cost of production was out of line enough to cause a loss. In agriculture, 1921 is the "collapse year," for the bottom dropped out of the market and basically remained that way until World War II. Wheat dropped from $2.75 to $1.05 a bushel in just a few month's time, and the slide continued during the prolonged agricultural depression of the 1920's and '30's.

The MFC four-year period ended in the red, and the overall outlook for agriculture was bleak. The "citified business oriented financiers" decided it was time to quit farming. They had an obvious reason to terminate the project for it was originally established to overcome the wartime food crisis. Now the war was over, and food was in surplus. Because the project had been financed at government request and the banks were flush with wartime profits, it was decided to ease out by making "generous terms of the whole proposition" to Campbell. For Tom Campbell this was the break he needed. Campbell was psychologically better prepared to face the adversities of agriculture. He knew that farming took perserverance. Besides, a man with his pride was not about to accept defeat that easily. He was so sure that his theory of mechanized industrialized agriculture was capable of outlasting traditional farming that he never doubted the eventual outcome. Or, at least if he did, he never told anyone.

At the age of 41 Tom Campbell was about to become the owner of one of the nation's largest farms. He had pushed the bankers to the limit of their endurance. A story is told that on one occasion, a representative of J. P. Morgan was visiting with Campbell over southern fried chicken at one of the dining halls, the cook heard him say in a loud voice, "I'll give you $50,000 more, and that's all." The eternal optimist no doubt constantly pushed the bankers for a few more dollars to try another project. The representative probably felt more like a benefactor than a financier of the enterprise. He sent Joe Keeps to Hardin to inventory and oversee the finances in preparation for the anticipated withdrawal from Montana Farming Corporation.[6]

The Campbell Farming Corporation
Is Created

The Campbell Farming Corporation was the well endowed stepchild of the Montana Farming Corporation. When bankers and governmental leaders decided to abandon the project, the end was decreed to Montana Farming Corporation. The obvious thing to do was to liquidate in favor of a successor, who, in this case, was the Campbell Farming Corporation. Tom Campbell was not quietly standing in the hallways to be asked if he wanted the chance to take over. No, he was on stage campaigning at "full steam" for the opportunity. Of the original $2 million loan, over a million had been paid back with interest. Campbell was offered the $400,000 machinery inventory for $150,000 and asked to sign a note for $500,000 for other assets. At a later date the half million dollar note was cancelled. Art Koebbe, a neighboring farmer who worked for Campbell as a fence builder and in later years did custom combining on the "big farm," said that Campbell had told him that he "settled with Morgan at seventeen cents on the dollar."

Although it appears that banking interests lost about $750,000 on the Montana Farming Corporation project, Tom Campbell was far from bitter when he severed relations with J. P. Morgan. He publicly stated:

A good many people rail at Wall Street. But I would rather do business with the big bankers down there than with most people. I have never known a more reasonable set of men. I had dealings with Mr. J. P. Morgan and with men associated with him for four years and they always treated me with the greatest consideration and fairmindedness. Wall Street is just as anxious for the farmer to succeed as he is himself; for without agricultural prosperity there is no general prosperity.

What else could he say about a group who had been so generous? His day with bankers was not over yet, for Campbell Farming Corporation was destined to fall to new lows in the 1930's before it would recover with the benefit of government programs and the wartime prosperity of the 1940's. Fortunately, conditions improved somewhat, and the newly created corporation had eight years of profits from 1922 through 1929. In fact, conditions had improved enough so that the implement companies, the railroads, and the Chambers of Commerce were able to successfully promote a land boom between 1924 and 1929. Tom Campbell's corporation lost money for the first time in 1930, even though less efficient small-scale horsepowered farms had gone out of business in great numbers during the 1920's.

Tom Campbell owned about 75 percent of CFC. The remainder of the stock was distributed to managers, foremen, and permanent employees. Ola Maddox, Tom Hart, and J. R. Taylor were some of the original shareholders in the new company. Taylor, who was manager until 1930, said he had about $50,000 worth of stock, but in the 11 years that he was employed he never received any dividends. About 1940 he sold his stock. Taylor knew that the farm made money in the 1920's, the late 1930's, and through the 1940's. He lost contact with it after that. Campbell's original profit-sharing plan for the top-echelon employees was something new in agriculture, but a half century later many farms were using this technique to encourage tenure of their employees.

Once Campbell was on his own, drastic changes were made in the operation. Sifting through the information available, one is left with the impression that as long as there was money available, there was no point in attempting to reduce cost. But when Campbell became the major stockholder he recognized the need for tighter management. Precisely how Campbell did it is not known. Only he could answer that, for not even his general manager in the 1920's could say exactly what the financial results of the operation were.

Campbell broke sod at Fort Peck under persuasion, but as soon as CFC became a reality the Fort Peck operation was abandoned. It was mentioned by some of the original employees that Campbell felt he had a more suitable working arrangement with the Crow Indians than with the Indians at Fort Peck. This could have been a factor, for the relations between the CFC and the Crow tribe had been extremely satisfactory.

Big farming meant big problems. The distance from Poplar to Hardin was over 250 miles, and machinery was shuttled between farms by railroad. With two tractors to a flatcar the logistics became a considerable problem. When "Punk" Taylor was sent to Poplar to oversee the operation in January, 1923, it took him and his wife three days to make the trip by car. They lived in a "shack" at Camp One and saw to it that the grain was all delivered by a coal hauler to Poplar. Len Cothren was in charge of the Poplar operations, where the land had become so infested with mustard and tumbleweed that it could not be plowed. Consequently, the weeds were burned off by pulling burning oil-soaked sacks and blankets across the land with Model T pickups. The fires quickly got out of hand and went beyond the reservation. That year four sections of land were tilled by pulling a 30-foot single disc, a seeder, and a packer behind the Aultman-Taylors. The crop got off to a good start, but it quickly curled in face of the hot, dry winds. A potential 40-bushel crop was reduced to about 12 bushels in a few days.

Campbell was able to secure more land from the Crow Indians about that time and because he was having better luck at Hardin decided to close up the Fort Peck farm. Taylor and Cothren supervised the loading of 52 carloads of equipment to be moved to Hardin. When the Fort Peck farm was closed the buildings were dismantled and the fences taken up. Thirty miles of four-barb wire and steel posts were pulled up and reused at Camp Four, 38 miles south of Hardin.

The next change in the operation was to drop the crop-fallow system and go to a three-year rotation of cropping—two years of crop and one year of fallow. The standard practice in Montana had been to crop every other year, but Campbell believed that with mechanical power and better machinery he could preserve moisture and control weeds well enough so that he could go to a two-year crop, one-year-fallow rotation. With the improved moisture situation of the mid-1920's such a rotation was possible, but when the dry cycle returned in the 1930's there was a reduction in the crop yield under this system. But it still paid.

With two-thirds of the land in production the total potential income increased, which was an important factor to the net returns of the farm. Campbell proved to be exceptionally progressive in the area of conservation and cooperated closely with the soil conservation service, which he credited as much as any other program for being vital to the long-run profitability of the farm. He continued the two years of crop, one year of fallow until the late 1930's or early '40's, when soil conservation programs plus governmental farm program restrictions made it more attractive to crop half of the land and fallow half. Better prices and land payments via governmental programs enabled Campbell to select his best land for production and to shed some of the more marginal acres. Profits were maintained with less struggle. Having 95,000 acres under cultivation and 64,000 acres in crop in 1927 proved that Tom Campbell was not losing his courage.

In 1949, at the age of 67, he expressed the view that there

should be "legal compulsion to insure the use of conservation practices." At the age when most men quit their working career Campbell was still innovating, this time with the use of weed spraying by airplane. As confident as ever about his farming methods, he remarked that "The future depends more upon economic and political intelligence than upon the weather. . . . " He blamed his crop failure in 1919 more to his own lack of understanding of conditions than to the drought. He was aware from research that 1919 would probably be a low point in the moisture cycle for a period of seven years and gambled heavily on that fact when he took personal possession of the farm in 1922. Scientific data paid off, and Campbell's gamble with the weather worked in his favor until 1931, when he experienced a complete crop failure. Between 1919 and 1931 the poorest year for battling the weather was 1921, when grasshoppers destroyed 15,000 out of 55,000 acres. The bulk of that loss was at Fort Peck.

By 1926 the ever-confident Campbell stated, "I worry less about weather conditions than I do about the human element involved in our operations. . . . We will never have a complete drought failure again, no matter what the rainfall." Those were strong words for a Montana farmer, but they were evidence of his faith in the machine to aid man in producing a crop under the most adverse conditions. As early as 1923 he had ventured some rather daring opinions about his ability to cope with the weather, even though he had only one good crop to support his record at that time. According to Campbell:

> The old fashioned farmer took the seasons as they came, without much change in method. But we are engineers. We are constantly working out plans to make us more independent of the kind of weather nature hands us. With approved methods, and the machines with which to apply these methods, we expect to reduce the hazards of farming so that they will be no greater than in any other large industry.

Everyone at CFC knew that good soil tillage was extremely important to crop production, and accordingly from the beginning the best known methods were used. Three distinct tillage systems had been used since 1918, changing only when equipment, power, and knowledge of the soil improved enough to do a better job.

The original tillage crop rotation plan of the Montana Farming Corporation was one year of crop and one year of fallow. This meant that half of the land was idle each year. Using virgin land and assuming a continuation of favorable rainfall as experienced in 1914, 1915, and 1916, this should have worked favorably, and it was the system advocated by most of the Great Plains dry farm leaders.

In the beginning, Campbell used a combination rig on plowed ground in order to cut cost and seed the grain in moist soil. Each Aultman-Taylor pulled a 10-bottom plow, a 12-foot disc, a drill, a packer, and a harrow. This rig traveled less than 2 miles per hour and by using two men could do 30 acres each day.* As soon as Campbell took over on his own he switched to a three-year rotation, using two years of crops and one year of fallow. In this system the first-year crop land was disced and planted with 20 pounds of wheat per acre.

For the second year the land was double disced and seeded with 10 pounds of wheat per acre. Campbell seeded lightly to get the best possible wheat with a minimum of straw. After the second year's crop was removed the land was disced immediately, then left idle until May, when there was a good growth of weeds and volunteer wheat, at which time it was plowed. After that the fallowed land was harrowed lightly whenever any weeds appeared. Generally during the second week of September it was seeded to start a new three-year cycle. The three-year rotation was used by Campbell until the late 1940's, when because of government programs it became necessary as well as profitable to switch to a two-year rotation again.

There were many improvements made in tilling the soil during the period from 1922 through 1948. By the late 1940's Campbell was using nearly all track-type tractors that were capable of pulling large equipment with less compaction of the soil. Because of contractual requirements with the Bureau of Indian Affairs Campbell was forced to switch from block farming to strip farming. Campbell and most other farmers in his area preferred to block farm because it was much more economical and the danger from wind erosion was not so serious as on much of the Plains. But the BIA had "different" opinions. Both dry-basin listing to work the land and deep-furrow drill seeding were adopted to create water and snow traps to help hold moisture.

From the 1940's on, Campbell Farming Corporation used the two-year rotation. After a crop was removed their brood cow herd was allowed to graze the land for several months and sometimes all winter. In the spring a chisel plow or a one-way disc was used

---

*A half century later those 1918 methods seemed slow and cumbersome, but the Campbell rigs were capable of doing what would have required forty-three horses and nine men under horse power farming.

to break the ground, leaving a good mulch to hold moisture and prevent wind erosion. During the rest of the summer a rod weeder or a flexing harrow was used three to four times to control weed growth. In mid-September the land was worked with 60-foot field cultivators followed by 60-foot or larger drill units, all pulled by 300 H.P. four-wheel-drive tractors. In a day's work requiring two shifts of men, each unit could till or seed over 600 acres. The 21,000 acres of crop land was tilled and seeded by four cultivators and drills in about nine days. Timeliness of fall cultivation and seeding was extremely important. In the days of slower-moving and smaller equipment, seeding was sometimes not completed until January. Each third year the above system was altered slightly by the use of the moldboard plow instead of the chisel plow.

The Campbell Farming Corporation could not use the method described above on their large acreage leased from the Crow Indians, for the BIA prohibited the removal of straw and stubble from reservation lands. Because of that a fully integrated grain brood-cow operation could not be conducted on the leased lands.

In some respects Thomas D. Campbell's boast of 1923 about never having a total crop failure again can be upheld. The area he farmed covered such a great distance that part of it was bound to have a crop. His tillage methods improved and became more timely, enabling him to grow a crop with minimum moisture. Campbell was so sure of himself that he was able to instill into his staff the conviction that if winter wheat doesn't grow nothing will.[7]

A New Age for Agriculture

Tom Campbell in his confidence was as eager to prove to the world that his mechanical-powered agriculture would work as he was to make profits from it. Tom Campbell was the man agriculture needed. "He was brilliant, likeable, friendly, and the world's best promoter." He had a great innovative ability to make the new machines and methods work. He also had the ability to sell the world on his ideas. Maybe his affluent friends did not greatly profit by investing in Campbell Farming Corporation, but the point is that they were sold on the idea to invest. There is speculation that at times Mrs. Bess Campbell supplied personal funds to keep the farm going. Nevertheless, the ability to "push" the bankers just a little longer and a little harder than the average small farmer dared to do could have been a factor in his eventual success. Maybe the experimental machinery or the reduced cost of machinery because of quantity purchases was enough to give CFC an operating margin that kept it in business.

Campbell was a great promoter who thought "big" and did not let reversals stop him. Once he got those big 30-60 Aultman-Taylors with their 10-bottom plows in the fields he could not wait to break production records. With the size and the amount of equipment and manpower available to him it was easy to break the records of the horse age. An article in 1923 credits Tom Campbell with saying: "If there is any place in the world where we need engineering it is on the farm. Man-power, horse-power—they are weak and insignificant. We must have speed and greater power. I'm going to study engineering as a preparation for farming."

Since he did make that decision as a young man he was quick to see the results once he started farming. By 1923 he was heralded as "a prophet of the future," and his farm must have appeared fantastic to the five-horse gang-plow operator. CFC consisted of four units of 10,000 acres each, with 100 men on each unit at harvest. Although he employed 400 men from July 15 to October 15, he was quick to point out that this amounted to 100 acres, or five times more land per man than the nationwide average on horse-powered farms. His combines cut a 24-foot swath, and "only" seven men were required to harvest up to 60 acres a day, using the "wonderful combined harvester and thresher."

He liked to boast, and at every opportunity he stated that at CFC power machines did the work and that there were no horses or mules. The only animals were cows to supply milk at the camps and pigs to eat the waste from the kitchens. In supervising the work, his camp managers averaged 150 miles a day in their Model T pickups, checking the crews in scattered fields. Surrounded by mechanical powered machines, Campbell was called the "super-power farmer . . . and may well point the way out of the morass that threatens American agriculture. . . . "

In 1924 it was claimed that two records were set on the big Campbell farm. That year 15 Aultman-Taylor 30-60's pulling plow, disc, drill, and packer worked 640 acres in a 16-hour day without a single stop for any kind of mechanical trouble. They were even refueled on the go. It would have taken about 1,800 horses and 500 men to do the same job in 16 hours with horse-drawn machinery. That record has been broken on CFC lands many times since, always with fewer men and tractors.

There may be some reservations about the second record claimed by CFC in 1924, for every "threshing run" in the country wanted to claim that they could do the most in the least time. If they had surpassed Campbell's assertion it surely would have been published, for he claimed that one of his all-steel stationary threshing outfits turned out 4,321 bushels of wheat in a 14-hour day. Probably no other threshing unit could match that record.

Campbell was not satisfied with past performance, and he was his own best pacer to establish new records. In 1926 he harvested a half million bushels of grain from 48,000 acres. With large-scale equipment he could still make a profit, but he was convinced that more gains could be made in the reduction of costs. That year his permanent labor force was 20 men, the eight-month crew was 50 men, and the harvest force added another 250. The ratio was over 300 acres per man in harvest and over 1,500 acres per eight-month man, far above the national average in crop acres per man. It took progress like that to keep the bank ledger in the black.

In 1926 Tom Campbell became convinced of the merits of the windrow system, and the combine retired his threshing machine for good at the following harvest. Campbell was anxious to get rid of the threshing machine for two reasons: first, the excess cost of harvesting as compared to the combine, and second, his dependency on hired teams and wagons to haul bundles to the thresher.

In the spring of 1926 it appeared that the wheat at Camp Four had been winter killed. To avoid a failure it was decided to reseed the entire acreage to flax at once. Tompson Brothers, wheat brokers, Continental Commercial, and First National, all of Chicago, were financing CFC at the time and were opposed to the decision. Campbell and General Manager "Punk" Taylor, who was a fraternity brother of Warren Tompson, decided to "slip one past the bankers," and refused to stop seeding. They seeded 1,800 acres a day and got the job done before the bankers could interfere. Ironically, both the wheat and the flax grew quite well, producing a most unusual crop and causing a tough harvesting problem.

With wheat, flax, and weeds in the same field, the question arose as to what could be done to harvest it. "Punk" Taylor, working with "that German blacksmith Johnny Owens," figured out a unique harvest setup that probably was a turning point in the life of CFC. It was decided that four 10-foot binders should be pulled

behind a single tractor and that the bundle carriers and trip arm
should be removed from each binder. In this manner the grain
was free to drop directly to the ground in a windrow. But a windrow from a 10-foot swath was too small for the CFC fields. Taylor
and Owens along with Campbell built 10-, 20-, and 30-foot extension elevators and placed them on the second, third, and fourth
binders so that the grain from each binder fell on the same windrow as that made by the first binder. Then came the big
problem—how to thresh it.

Taylor, Owens, and Roach, under Campbell's prodding, concluded that it would take a large machine to handle such a volume
of straw, so a big threshing machine was mounted to be pulled
backward over the windrow.* Next a John Deere hay loader was
added to elevate the straw into the threshing machine. A grain
wagon was fastened to the side of the machine to receive the grain
as it was threshed. The blower was adjusted to minimize the dust
problem for the crew. As each grain wagon filled it was unhooked
and an empty one was attached. In this manner several thousand
acres of thick wheat, flax, and weed grain was harvested.

The harvest was characterized as "the advent of the windrow
method." It is difficult to know if others used the windrow idea
prior to Campbell; but whether or not it was a first, the drama that
surrounded CFC's windrow harvest certainly was not equalled.
Windrowing made the combine practical, and certainly Tom
Campbell's wheat-flax-weed harvest paved the way. A sample of
this unique mixture indicated that it would be discounted 9¢ a
bushel. Needless to say, after field work was over, all the grain
was run through specially rigged up combines to separate the two
grains and the weed seed.

Campbell noted an immediate reduction in harvest cost after
using the blacksmith-shop-assembled equipment and his windrow
idea. A catwalk connecting the four binders made it possible for
one man riding the rear machine to observe the working of all. A
bell cord enabled him to signal the tractor driver if necessary. The

---

*At this point there is a conflict as to whether a large Holt combine or a threshing
machine was used. One person interviewed stated that a Holt combine was modified to do the job. Two others said a threshing machine was used. An article in the
Northwestern Miller said a "large threshing machine" was used, and that information probably came directly from Campbell. It is this writer's opinion that a threshing machine was mounted and powered to be pulled backward. The fact that a
wagon had to be pulled alongside to catch the grain strengthens the idea that a
thresher was used.

other immediate major saving came from the elimination of twine and shocking.*

After its success on the wheat-flax-weed crop, CFC used the windrow idea on all of its grain. Campbell claimed that if all of the farmers would adopt the windrow technique they would save $100,000,000 a year. (Later he raised that to a billion dollars.) A windrower could pay for itself in one year on one section of wheat. Most farmers already owned binders, so that the potential savings were even greater. The windrow method reduced CFC's cost of harvest by $2 an acre. Even with the use of three to five men per rig the total labor force was only half as large as the old shocking crew had been. Grain was combined and put into a wagon for less than shocking labor and twine had cost in previous years.

CFC's machinery investment during its first full year of operation was $6 an acre. Six years later, after consolidating the machinery at a single station instead of having a full line of equipment at each camp, this cost was reduced to $3 an acre. Annual operating costs, which were $10.00 an acre under the camp system, were reduced by $2.50 an acre in the first year of management under J. R. "Punk" Taylor; and in the next six years, from 1923 to 1929, costs were further reduced to $4.80 an acre. In 1926 after the windrow method was proven successful Tom Campbell saw his production costs reduced to about half of the cost that an average wheat farmer had to pay. Annual costs ran in excess of $500,000 in 1926, and the $2-per-acre saving by windrowing on 48,000 acres of crop had an impressive impact on balancing the budget. The biggest operating cost after 1926 was for plowing. Newer, lighter, speedier, but more powerful tractors and the chisel plow later paved the way to a further reduction in costs. Taylor commented that Tom Campbell gave him and his management a "free hand to make changes" and that proved to be the best method to reduce expenses.

Fewer tractor sales received more publicity than the Montana Farming Corporation's purchase in 1917 of thirty-four 30-60 Aultman-Taylors. These 13-ton engines, pulling eight bottoms on sod breaking, covered 25 acres a day. Besides the Aultman-Taylors, there were six 20-40 Case tractors, three 120-H.P. Holt

---

*CFC had attempted to eliminate bundle-hauling labor a few years earlier by adopting shock loaders that worked all right for loading, but they tore too many bundles and caused excessive waste. They were abandoned after a few day's use.

Caterpillars, and four 75-H.P. Holts. All together they could pull 500 bottoms and plow an acre in 10 seconds. A thousand acres of plowing a day became standard. This more than any other factor familiarized the world with the "big farm" at Hardin, Montana. Campbell felt that upkeep costs of these big tractors were ruinous, "but it got things done."

Total fuel consumption was 5,000 gallons a day. The Aultman-Taylors used 80 to 100 gallons each, every 10 hours. Continental Oil Company delivered carload shipments of fuel to CFC's storage containers at Hardin. A Holt 75 pulling six tank wagons brought the fuel to the tanks at the camps. In the busy season that rig operated continuously because the longest round trip to Camp Four took three days under the best conditions. In 1924, at 3,000 gallons a trip, that Holt with six wagons made 70 trips to bring the 200,000 gallons of fuel used to produce 487,000 bushels of grain and to break 16,000 acres of sod. Fortunately, there was a storage capacity of 70,000 gallons on the farm and at Hardin. In the following years, with lighter and more efficient tractors, ground speeds increased from 1¾ to 3 miles per hour, and daily fuel consumption dropped from 5,000 to 1,400 gallons.

There are people who knew Campbell and heard him say that raising wheat was an excuse to perfect machinery needed for agriculture. What he did with tandem-hitched binders bearing extension elevators behind a single Aultman-Taylor is a good example of this statement. When pulling six binders, each Aultman-Taylor and its seven-man crew traveled two miles per hour. It was discovered by trial and error that if only four binders were hitched to one tractor it could easily travel three miles per hour. The four-binder hitch saved the use of two men and two binders and did just as much work.*

Tractors, which were the key to Tom Campbell's grand experiment, had to be constantly upgraded for ever-increasing efficiency. The Aultman-Taylors, the 20-40 Case, and the Holt caterpillars first used by Campbell were replaced with a second fleet between 1923 and 1926. At that date CFC operated with six Holt 75 Caterpillars, twenty-three 20 x 40 Case tractors, ten 22 x 36 Internationals, and ten 10 x 20 Internationals. CFC's practice of buying in large quantities without trade-ins to secure the lowest possible

---

*Binders took quite a beating behind the Aultman-Taylors; and when breakdowns occurred, the men were not careful about picking up the parts. Erwin Schnad, who farmed land that Campbell operated from 1922 to 1930 at Camp Two, said that in 1971 he was still picking binder sprockets from the fields.

cash price relegated many of its older tractors to permanent storage behind the shed.

In 1928 Campbell wrote that he was operating 56 tractors, totalling 3,000 H.P., besides 12 trucks and 10 automobiles. The inventory also listed 11 threshing machines that were no longer used, for they had been replaced by 21 combines. Still using Aultman-Taylors for show as much as for work, on May 28, 1928, he lined up 18 of them in one field and plowed 135 acres in 7.5 hours, using 400 gallons of gasoline. It would have taken 210 horses with 42 men on 42 gang plows to do the same job.*

Campbell's claim of 3,000 horsepower in 1928 was probably stretching the limit a bit, for according to records, his original purchase of tractors in 1917 amounted to 1,770 H.P. and his second purchase in the early 1920's added 1,230 H.P. It is not likely that all of the original tractors were still in use at the later date. But the ones in operation were well used, for his tractors averaged over 2,000 hours a year in contrast to only 250 hours a year for the average farm tractor in Iowa. Logging so many hours on a tractor was possible because double shifts of men were used for much of the field work.

Strange as it might appear, CFC gained in efficiency when Campbell switched from the heavier 13-ton Aultman-Taylor tractor to the lighter 22-36 International, that weighed about 4½ tons. The new tractors traveled just over 3 miles an hour in plow speed, while the Aultman-Taylor did less than two. Although the 22-36 pulled only four bottoms in contrast to 10 for the Aultman-Taylor, it cost only about one-fourth as much and consumed only a fourth as much gas. Most tractors were operated 22 hours a day during summer fallow, seeding, harvest, and plowing periods; but the heavier, slower Aultman-Taylors would mire down easily at times, and a great amount of time was lost because it usually took two tractors to pull them out.** The Internationals were less subject to miring down, were much more agile in field work, and had a

---

*In 1973 on the Gregory Weiland farm near Euclid, Minnesota, in the Red River Valley, Weiland's son, Henry, plowed 160 acres with a 3300 Steiger pulling 21 bottoms in 12 hours and using 160 gallons of fuel. One man in 1973 did what it took 45 to do in 1928 and used 40 percent less fuel.

**Owens and Wemple helped to take Aultman-Taylors to the main shop at Hardin late in the fall of 1928. The road was very muddy, and one after another the tractors became stuck. Then the weather grew colder and the ground froze, trapping the imbedded machines. A few were dug out with bars and jacks, but several were left until spring.

higher road speed, all of which tended to make them more efficient than the heavier, clumsier Aultman-Taylors.

After purchase of the second fleet of tractors, Campbell bought only what he needed to keep going until the next period of big profits in the 1940's. The profits of World War II enabled many farmers to buy tractors for the first time and provided Tom Campbell with a chance to get rid of the fleet of smaller tractors he had acquired in the 1920's and '30's and to acquire the big, powerful units he had been dreaming about since his college days.

About two dozen track-type tractors were secured from war surplus in 1946. Four of these were D-8 Caterpillars, 14 were TD-18 Internationals, 10 were TD-14's, and one a TD-20. All of these "military green" track-type power units were redecorated with proper company colors and decals. Those "caterpillars" purchased at a low cost were very economical to the CFC operation at a time when both yields and prices were good. Campbell upgraded his field equipment at once to match the new power units. The D-8's were capable of pulling 14-bottom moldboard plows or 63 feet of duckfoot weeders. The TD-20 was hitched to a unit of seven 12-foot drills, once the proper hitch was made at the blacksmith shop because nothing that large was available on the market. The TD-20 with its 84-foot drill could seed 240 acres in eight-hour shifts.

After the surplus military purchases, including trucks, were completely overhauled, repainted, and adapted to farm work, CFC sold a large volume of its older equipment. In 1949 a sale was held at which 23 trucks, 36 tractors, and a large amount of other machinery were sold. Campbell sold most of that machinery to eager buyers who had experienced wartime profits and bid feverishly for the older equipment.

Still not satisfied with these large power units, Campbell decided to try one "really big tractor" for his farm. During the war he had observed M-6 tank retrievers in action and wondered what they would do on the land. The 38-ton, 450-H.P. diesel was balanced well enough on its tracks so that it packed the ground less than the 13-ton Aultman-Taylors. A 78-foot field cultivator was made to match the power unit. With proper care from the driver, over 33 acres an hour could be tilled. By using lights, 800 acres of land could be cultivated in a day. There was only one problem. The power unit was too big for the hitch which was made to tow the machine. It was not long before the 450-H.P. tank retriever had to be retired because the "hitch came to look like an accordian."*

Campbell Farming Corporation used the war surplus fleet of track-type tractors for 20 years before again making a major change in their power units. Only a few standard rubber-tired two-wheel-drive tractors were purchased during the 1950's and 1960's. In the late 1960's CFC purchased four 4100 International four-wheel-drive rubber-tired tractors, signifying the advent of another era for "the giant wheat farm of Montana." These 125-H.P. tractors were used to pull 32-foot-deep-tillage Graham Hohme diggers. They were much faster than the track-type tractors, but were not so powerful as the management of CFC had hoped for. Because the 4100's did not "stand up" so well as expected, no further change was made in the tractor line immediately. Under the leadership of general manager Les Curnow a fleet of 300-H.P. Versatile 900's was acquired in 1972.[8]

Experimenting with Power

Although Campbell knew the merits of good machinery he was hesitant to purchase new equipment and to keep updated. The first wooden Holt combines were used machines and had come from California. It was not until 1927 that some new Case combines were procured. As general manager, "Punk" Taylor said, "We did not have a lot of new machinery, for Tom Campbell did not believe in buying new stuff." Rex Wemple and Johnny Owens jokingly remarked that a standard phrase around CFC was "the corporation seal," which meant the baling wire everyone used to fix anything that broke down.

Tom Campbell made a point to get the machinery companies to furnish him with a large amount of experimental equipment. Throughout the 1920's and 1930's and to a lesser degree even after that, there were experimental machines on the farm at all times. His workers were well aware of the fact that they were operating

---

*Tom Campbell was not the only farmer to secure a large piece of military surplus equipment. Ward Whitman, a North Dakota farmer, purchased a large four-wheel-drive 400-H.P. tractor with 46-inch tires to use as a land leveler and to pull heavy implements. Whitman's, as well as Campbell's, problem was making a hitch strong enough to stand the stress placed on it. The 27-ton power unit with a 3-ton blade handled as well as much smaller conventional four-wheel-drive tractors, and its floatation was excellent. This unit was driven by the writer while visiting the Whitman operation. Whitman, like Campbell, and many other farmers were convinced that the big but mobile power unit was the key to efficiency on their farms.

experimental equipment and accepted the more than normal amount of trouble they had in the fields because of it. The most exotic machine for its day, according to the memory of some of the early workers, was a Case tractor that had its engine hinged on one side so that it could be easily tipped to work on.

A track-type Monarch tractor was furnished to CFC for experimental work, as was a Cletrac, which was also a track type. The Cletrac apparently was quite satisfactory, and the men agreed that if it had been diesel powered it would have been very good. The model used by CFC seems to have been a 30A that was rated 30-45 and could pull four bottoms. Hart-Parr furnished another tractor. The variety of tractors and combines shown on some of the early harvest scenes of the CFC gives a good idea of the amount of different makes and models of equipment used on the farm. Emil Headrich was made experimental tractor driver and was kept busy testing new models. Representatives from machinery companies were constantly on the farm to see Campbell's experiments, and the cooperation was of mutual benefit to both.

However, it was not always beneficial to CFC to be doing experimental work with machinery, but Campbell did get some bargains because machinery companies were anxious to sell to such a well advertised farm. In 1971, CFC owned eight combines, consisting of four different makes and models. It is obvious that even with extreme discounts it is hardly economical to run a fleet of any size with such a variety, for much efficiency comes through standardization. Campbell had generally preferred to hire custom combine fleets because they offered greater flexibility at less cost. In August, 1971, there were 32 combines working on the Campbell farm. The custom fleet of 24 combines harvested over 1,200 acres a day and was more efficient than the company machines.*

Tom Campbell was never satisfied to rest on past performances. For years he was proud of the record set by his big all-steel threshing rig turning out over 4,000 bushels in 14 hours. After that

---

*The custom combines on CFC in 1971 were owned by four different operators. Some of the operators returned for 20 harvests on their run from Mexico to Canada. There were established regulations that had to be followed by the custom combine crews: these custom combines could not start until permitted by the Campbell camp boss, they had to stop when required, they had to keep their cylinders set as directed, they had to stay out of green spots, and they were assigned fields by the general manager. In 1917 the custom rate for direct cut was 13¢ per bushel for wheat and 11¢ for barley, with 8¢ for hauling to storage. Gerry Ladd, Leo Whetstone, Big Nick Trynin, Wade English, Don Snow, Darnstaedt & Reed, and Luther Lock have been the major custom crews to work the CFC harvest.

was the 33,000 bushels combined during the late 1920's when Mary Roberts Rinehart, the well known novelist, visited the farm. A new record was set in 1958 when top-ranking Soviet agricultural leaders visited CFC. At this occasion Campbell amassed 40 custom combines in addition to 11 company machines and harvested 61,340 bushels.

For the sake of watching progress Campbell continually expanded his production, and new records reflect that trend over the years. In 1935 combines averaged 18 acres and 277 bushels per day. Some of these were still the old wooden Holt models first used on the farm in 1926; others were the smaller but newer steel models manufactured in the 1930's. In 1947, working more than one shift of men because of extremely favorable combining weather, 51 combines and a crew of 165 men averaged 21 acres a day. They harvested 706 bushels per machine on the best day, sending 36,000 bushels across the scales. By 1958 the daily average production per machine rose to 1,203 bushels. Daily production in 1971 climbed to an average of 51 acres and 1,720 bushels per combine per day for the entire season. Average daily combine capacity from 1935 to 1971 increased more than six-fold, but labor efficiency improved even more because of larger trucks and better grain-moving facilities at storage locations.*

As combine capacities increased it became a constantly more difficult problem to haul grain away fast enough to keep them moving. CFC probably had as many experiments and innovations as any farm in the nation relative to improving grain handling. The efforts were dramatic and often well publicized. In the first purchase of equipment, 100 grain wagons, each capable of holding 125 to 150 bushels (later 200 bushels) were acquired. As production increased with the adoption of combines, the wagon fleet grew to 200. Wagons were attached to the combines to catch the grain, rather than rely on the combine hoppers. If all of the wagons had been filled at one time they would have held over 30,000 bushels, an amount equal to the total output of a large number of

---

*While individual daily averages do not appear to be exceedingly high, it must be considered that these figures are weighed grain yields, using total production for an entire harvest under changing daily conditions and including moves to various fields as well as using total combines in the fleet. All CFC grain is directly combined. The terrain is rolling, and the fields are scattered over a distance of 45 miles, all of which tends to reduce top performance. At one time in 1971 three of the CFC combines were idle for several days because of the inability to get parts. Obviously there are individual combine records set every year using ideal conditions that far surpass these averages, but these are realistic production figures.

quarter section homesteads, and this was only a day's production
at CFC. These wagons had to be towed several miles to storage at
camp headquarters or to Hardin, which was more than 40 miles
away from some parts of the farm. Although these grain wagon
trains reached a top speed of 3 miles an hour on the 35 miles of
company-built roads, it was considerably slower lumbering across
the field to reach the road. The trip usually included crossing several coulees. This provided a true challenge to the Holt 120
Caterpillars capable of pulling 12 wagons, but more commonly
pulling 8 wagons. After all, eight wagons full of wheat meant a
30-ton load in addition to the weight of the wagons and the tractor.
This took real power over stubble ground.*

During the 1920's and much of the 1930's, most of the grain was
hand shoveled into storage. However, there were a few John
Deere tubular paddle elevators with 1-cylinder engines at some of
the larger storage bins. It was not until 1938 that hand shoveling
from the wagons was terminated, and not until the late 1940's did
auger elevators become common enough to eliminate the scoop
shovel. This was not an act of humaneness, but one of necessity.
The machines were dumping grain on the ground at central piling
stations faster than the regular crew could hand shovel it into the
semitrailer trucks.

The concept of the famous Campbell "grain wagon train" was
apparently a team effort. "Punk" Taylor, Tom Hanes, Johnny
Owens, Pat Roach, and Tom Campbell all had a hand in developing the 12-wagon hitch in 1924. The hilly country, plus many
coulees between the storage sites and Hardin, presented a dual
obstacle. Therefore it took power and traction to pull 80-ton loads
uphill and through coulees, and it also took great strength to hold
those loads back while going downhill.

A ¾-inch-thick and 4-inch-wide strap-iron was placed on the
axle of every wagon as well as a solid strap-iron under each wagon
so that the iron took the total strain of either pulling or braking.
The tongue of each was chained to the strap-iron running from
one wagon to the next. It had to be fairly free so that the wagons
could turn. Apparently the turning factor was well considered, for
Rex Wemple remembered that on one trip he pulled 16 wagons
back to camp, and in going around a corner and through a gate the
final wagon track was just over a half width off from that of the

---

*The Holt and Best Companies, both making track-type tractors, merged to form
Caterpillar Tractor Co., which became a major producer of the popular Caterpillar
tractor.

first wagon in the train. This maneuver was important because of the narrow roads and the sharp curves.

Once the hitch was perfected, the 200-bushel wagons were used to haul loads to Hardin, a distance of 42 miles from Camp Four. By 1924, CFC had graded but not graveled most of the 35 miles of road necessary to get automotive traffic in and out of the farm under most conditions. Twelve wagons carrying 2,400 bushels of wheat weighing 72 tons, in addition to 12 tons of wagons, were pulled by the Holt 120's at a top speed of 3 miles per hour. But top speed was not average speed, for each time a coulee had to be crossed the two man crew had to set the brakes on all of the wagons and pull each wagon through the coulee separately. If no major problem was encountered at a coulee, the 12 wagons could be pulled through it in the minimum time of three hours; but sometimes it took a full day. In dry weather or on frozen ground a round trip, commencing with greasing the wagon axles at each end and including loading and unloading, took on the average five days. The quickest trip ever completed was four days.

On such a trip a two-man crew was necessary because one man could not possibly connect and disconnect all the wagons. Besides, the second man had to set the brakes and work on the rough lock chain that was used to hold the wagons back when going downhill. The men carried their lunches and traveled continuously from early morning until they were picked up by car at dark to be taken to Camp or to Hardin. Campbell encouraged the addition of a third man and made provisions for sleeping so that the train could operate continuously. The fuel tank held 120 gallons, and in good going that was enough to reach Hardin without refueling.

Because of the narrow road, schedules had to be made so that grain trains did not have to meet on the road except during periods when the land was dry or frozen. Then the empty train could pull out of the way. In the memory of long-time employees only two accidents of note happened to the wagon trains. One occurred when a rough-lock chain holding back a gasoline tank train of six wagons broke, causing the final wagon to tip and spill the gasoline. At another time a small Fordson tractor was being used to pull a train of 20 empty wagons. Some of the front wagons did not have their rough-lock chain, and when the rough-lock broke on some of the back wagons, they started to roll. The small Fordson was too light to hold 20 wagons back and was soon in a ditch with the wagons on top of it.

Mary Roberts Rinehart noted that these trains with 72 tons of wheat operating at night were the most romantic thing she had seen in the West. Campbell modestly commented that "they made the 20-mule team of Borax look like a wheelbarrow outfit." Indeed they were glamorous—but they were also economical, for two men and one Holt 120 were doing the work of 72 horses and 12 men, plus the crew that would have been necessary to supply them. The cost for hauling from Camp Four including loading, was figured at 8¢ per bushel—fuel, labor, interest, overhead, and depreciation. Actually Campbell arrived at the cost factor of ¼¢ a bushel per mile so that each camp was equally assessed its share of the hauling charge. This was a reduction from 1¢ a bushel per mile when a single small tractor was used on each wagon and 150-bushel wagons were used.

From 1924 until 1935 the Caterpillar-powered wagon train hauled most of the grain to Hardin. But as early as 1926 Campbell's crew had even tried to speed up the job by adapting a tandem-drive truck to carry a load and pull two single-axle trailers. Campbell hoped that this rig would haul 1,000 bushels at average speeds of 10 miles per hour, requiring only one man and reducing hauling costs to ⅛¢ per mile. However, he did not get his tandem-drive double-trailer rig into operation. But by late 1927 CFC was using a Mack truck with a 200-bushel box pulling a trailer with a 100-bushel load. One man with a truck pulling a 300-bushel load and traveling at average speeds of 10 miles per hour on high rubber-tired wheels could get through the coulees. Three trips could be made easily in two days, enabling one man to do the work of two, using about the same amount of fuel, and reducing the cost to less than ⅕¢ per bushel per mile. This amount appears so minimal that the average farmer would not have been interested; but if 150,000 bushels were hauled out of Camp Four, the total savings per year of over $3,000 equalled the cost of four trucks.

In 1935 CFC purchased several 1½-ton Chevrolet trucks with special transmissions enabling them to pull 300 bushels at 30 mph on the graveled road to Hardin. Hauling costs were again reduced, this time to ⅐¢ per mile. However, the greatest advantage came in the comfort for the driver and in the gaining of shorter working hours.

Alert to the opportunity presented by war surplus goods, in 1946 the "master of CFC," then an officer in the armed services, purchased 11 army halftracks and semitrailers. Some halftracks were left unchanged, while others had the halftracks replaced with tandem axles; but all were modified with a fifth wheel so that they could pull a semitrailer. Conventional truck-tractor units were also purchased from military surplus. The track-type units were used in the toughest terrain, and the conventional trucks on the road. A complete surplus semitrailer truck at this time cost CFC an average of $2,000 and was capable of hauling 700 bushels of wheat to Hardin. Because of the speed and the size of the load, the hauling cost was reduced to $1/16¢$ per bushel per mile, or, as Campbell stated, it was "a saving of $800 per day during harvest and $25,000 per year." The speed and increased size of the truck became more important than ever to combat rising labor costs, which were $5 a day in 1927 and $21 a day in 1947.*

Tom Campbell understood the importance of topography to farming and the resultant high cost of transportation both within the farm and on the road to the market. His farm, which at one time covered an area equal to 10 townships and in the 1970's was still equal to more than five townships, presented transportation problems virtually imcomprehensible to the operator of a 160-acre homestead. Large trucks and semitrailers provided means to reduce the hauling costs. Campbell expanded his motorized fleet to 40 trucks and cars in 1952, including 19 truck trailers in Montana and another 16 vehicles at the operation in New Mexico. Of all the 200 wagons used on the famous Campbell wheat train only a few relics remain.

World War I opened the possibility of the uses of the airplane in agriculture. However, a great portion of experimental aerial work done during the 1920's and 1930's was restricted to forest areas or where there were large infestations of grasshoppers on the Great Plains. After World War II many awakened to the new opportunities for the use of the airplane in agriculture.

Tom Campbell, at the age of 66, saw the airplane as a way to reduce production costs and to increase yields. Because of good moisture conditions, weeds had done better than average in the 1940's, and by 1948 CFC had 10,000 acres of crop that was badly

---

*This was an average reduction of about 5¢ per bushel over the previous cost of hauling the crop to market. On an average harvest day, 15,400 bushels could be delivered direct to storage on the railroad siding at Hardin, saving the cost of reloading from piles or from temporary storage; a large volume, but less than a third of the total combined in an average day on the Campbell ranch. In 1947 in spite of the new fleet of semitrailers, CFC had to pile 125,000 bushels on the ground between walls of wheat straw bales because yields were exceedingly good and combines harvested grain faster than it could be hauled to storage.

infested by weeds. One field of about 1,300 acres was so overrun with weeds that it was about to be plowed under. Although planes up to this time had been used to deliver parts, fly out injured men, bring in visitors, or even spray weeds, it was never done on a large scale. In 1948 a contract was made with an aerial spray company to use four planes to spray 10,000 acres of wheat at $3.25 an acre. All of the crop was saved. The 1,300-acre field that had been considered doomed to the plow later produced over $40,000 income.*

Campbell and his crew were constantly trying new methods. The general machine shop at Hardin was busy much of the time rebuilding and strengthening commercially made machinery or trying to invent something new. It was learned early that wooden parts which had stress placed on them by the power units could not stand up. Therefore, every new machine was put in the shop, and wooden parts were replaced with iron or steel before it was put into use. This was done to avoid breakdowns 40 miles from the shop. Campbell stated in 1928 that the central shop saved him $2,000 a month. A fringe benefit of the shop work in 1928 was the development of a crop dryer to dry the crop before it was either put in storage or sold. Campbell claimed a patent on that invention, which he said would give him a higher selling price on his grain and would enable him to combine under less than ideal conditions.

Many ideas worked but many failed, and a businessman like Tom Campbell understood that you must first make a few mistakes to get somewhere. He admitted that his mistakes probably cost him over a million dollars, but apparently some of his ideas must have been correct.[9]

"I Liked the Drama"

In 1928 Tom Campbell made an interesting statement:

There are but two types of farming under present industrial and economic conditions in the United States; the small farm operated by the farmer and his family without any payroll . . . and the large farm operated as a factory with high priced skilled employees, factory production methods, technical men, and industrial management. . . . The large farm will operate on less productive land with greater transportation costs and less investment per acre. Practically all the work will be done by machinery with a larger output per man and a resultant higher wage. Lands which are now marginal and unproductive will be profitably farmed through industrialization, as growth in population demands it.

---

*Bob Boles of Hardin, a professional aerial sprayer, has done most of the crop spraying for CFC. He is considered one of the pioneers in the business.

He continued with what probably was a prophetic outlook for 1928:

We have too many people on the farms now. In twenty years we will have less than 20 percent of our population and the people who leave the farm will be needed to take care of our constantly growing industrial demands. It will not be long and we will have many farming organizations, well financed, managed by businessmen with the help of high grade engineers and skilled men, who are paid as high wages as the skilled men in the cities. When this time comes there will not be any scarcity of farm labor.

Tom Campbell grew up on what was a large farm for pioneer days, but his mother, like most farm wives, still worked long hours without pay. He knew that the average farmer expected that this was the way to get ahead along with some hoped-for appreciation in land values. But Tom Campbell thought and did otherwise. He expected to make a profit from farming by hiring all of his labor and by renting all of his land. His farm practices proved his point under some of the most adverse conditions in the history of American agriculture. He went on to prove what some of the early authors on tractor power farming had stated—"the cheapest tractor is the biggest tractor." Campbell purchased the biggest power units from the beginning and basically adhered to that principle. Using the best available mechanical power, his labor force did not have to live through the traditional days of drudgery nor of low pay. Although his ideas were prophetic they were well-founded, as the history of his farming has proven. At the age of 74, he wrote, "We believe [the Hardin operations] to be the best mechanized farm in the world and have proved that mechanization and electricity have taken the drudgery and loneliness of pioneering days out of the farmer's life."

Campbell never lost his vision about the efficient use of labor on the farm. As late as 1946, when he was 64, he told Floyd Darroll Warren, a progressive young farmer at Hardin, "If I could get a few young men like you I could farm the world. We can get the financing from many sources and there is nothing stopping us." Such talk was quite contrary to the thinking of a large portion of American farmers of many different eras, who insisted that farming was not profitable if one had to rely on hired help.

Two years later Campbell acquired 415,000 acres in New Mexico, which he intended to develop. A visiting journalist described the 66-year-old farmer as "at the height of his powers, with a swimmer's physique and the temperament of a show horse.

The man is ageless. He must have been something of a patriarch at 20; he will be a youth until he dies."

Some of the long-time employees stated that Tom Campbell knew very well "the value of the human element." He worked and ate frequently with the men, and many times they did not know that he was there until he spoke to them. "A stranger could not have told him from the crew," said Wemple. "The permanent employees liked him, but most of the drifters who really did not want to work did not like him for he was a real hustler," said both Owens and Wemple, when reflecting on Campbell's work habits. Campbell seemed to have the ability of being everywhere on the farm. Owens said, "Sometimes if things went bad at a certain location, rather than get too concerned about it, he would jump in his Stutz and go to another place where things were going better."

In the memory of most former employees, CFC always had a large labor force—Mexican, Indian, local farmers' sons, drifters, or top-level, well-trained men.* Rex Wemple said he came from Illinois for a visit in 1925 and saw the big tractors breaking sod at Camp Four. He stated, "These tractors appealed to me and I decided on the spot to go to work to be with tractors." Wemple stayed with CFC until 1935 and even after his departure he maintained direct association with the farm for the rest of his life. Johnny Owens, who had been a fireman with the Burlington Railroad and later homesteaded, said he had heard about CFC, "so I came to visit and to see the operation. I liked the drama of the big farm and went to work in 1924 and stayed until I retired in 1967."

One-time general manager J. R. "Punk" Taylor liked Campbell because "he could do anything and stand up to anybody." Taylor remembered when the International Workers of the World tried to invade the CFC labor force. Campbell always seemed to be the first to "detect if there was a stool pigeon in the ranks because he spotted the unexplainable fires or breakdowns and next the person causing them." When fights broke out, "Tom would get right in there and fight anyone for he was as good with his fists as he was standing up to anyone on an argument." Taylor, who had been a star football player in college, respected the great physical strength of his boss as well as his mental ability.

---

*The permanent employees frequently did not know the part-time employees and compared CFC to an oft-repeated story of the bonanza farms with its three crews: "one coming, one working, and one going." Johnny Owens recalled one new man who the first night got into a poker game and won $300. He left at once without collecting pay for his day's work.

Tom Campbell neither drank nor smoked and disliked drinking or smoking on his farms. There was not too much of a problem relative to the vices on the farm until after 1933, when the prohibition amendment was repealed. "After that, some of the men really got loaded when they got to Hardin," according to Owens and Wemple, who agreed that the permanent staff respected Campbell's feelings about liquor and tobacco and the part-timers feared him on account of it. "He had such a powerful handshake that if he had not seen you for quite some time he would shake so firmly it would hurt your hand for an hour," said Rex Wemple. Johnny Owens remembered seeing Tom Campbell rip a "brand new deck of cards in two with his bare hands."

Henry Old Coyote, whose association with Tom Campbell dates to the very beginning of the farm, said his people looked to Campbell as one "who was colorblind and he treated the Crow as human beings. To us he was a considerate and kindhearted man. . . . I doubt that the Crows ever felt that he was forcing himself on them."

Henry Old Coyote, a staff sergeant, was very formal to Campbell, a former officer, when they met again after World War II. Coyote recalled that Campbell commented that there was no need for such formality among old friends. "After that meeting 'the general' was friendly and businesslike, but we always had time for a joke." And in continuation he said:

> When the Indians had problems big and small, he [Campbell] would always sit down with any of them. . . . He came freely to the Indian home. He never hesitated to walk into the home of a Crow even in the early days when this was far from the common practice. The Indians hoped that when they came to the corporation office they would be able to talk directly to Mr. Campbell because they felt he would be more understanding than the office employees. Whether they were poor, drunk, sober, or sick, 'the general' always had time for them. This applied to all the Crow.

"Punk" Taylor, who was with CFC in the critical period of its development, said, "Tom Campbell had the right mechanical theories to put mechanical power into agriculture. Besides being a promoter he was sold on industrialization and did a good job." Taylor, who later independently made a name for himself in American agricultural machinery and agribusiness, probably understood Tom Campbell as well as anyone and agreed with Campbell's mission of eliminating hard labor with mechanical power.

In 1928, Campbell said with good reason:

The farm is going to be a place where the skilled mechanic will find a good job, under good working conditions, and at good pay. I shall not be surprised if, ten years from now, we will have one million such men working on farms. It is bound to come. . . . On our farm the cost of operation per acre is about one-half what it is on the average small farm! This is not because we are better agriculturalists; it is because we are engineers, businessmen, and operate on the same basis as a manufacturer.

Continuing in the same line of thought:

. . . We need very few men considering the magnitude of the project. We keep them for a longer period of employment than the average farmer does. And while we pay them higher wages, our labor costs per acre are much less. . . . The men themselves are better satisfied. The work is easier and more interesting. . . . we pay a bonus in addition to fixed wages. As all of our engineers [tractor drivers] are on a mileage basis, the man who is a good operator can earn extra money.

In 1928, relative to these ideas, Campbell pointed out that wages in agriculture had not kept in line with industry because agriculture was still tied to the horse. He predicted that the tractor would soon replace the horse and farming would be reorganized through the use of new ideas. Besides improved marketing, better transportation, and lower production cost, farmers gradually would learn to make better use of capital.*

In the late 1920's, Campbell stated that CFC had increased wages every year, but at the same time the "cost of production has been lowered annually." He attributed this to the fact that higher wages speeded up their long-time employees and at the same time attracted better new men. One of the most noticeable results was the reduction of maintenance by over 50 percent in three years' time.** In that respect, "the least expensive labor is the highest paid." To Campbell, farming was like industry; it needed larger units to survive, and the use of larger units was possible with mechanization and fewer, but better paid laborers.

Tom Campbell's views were consistent in regard to wages and labor, and in his late 60's he said, "Mechanization is the secret of success in large-scale farming. With machines, output per man is

---

*In that respect, he was several decades ahead of the business of most farmers, for it was not until the 1970's that finance agriculture became an acceptable philosophy.

**When people are employed by another, there are bound to be some mishaps that cost the employer money. Even reliable long-time employees like Rex Wemple had his problems. One time the Aultman-Taylor he was driving became overheated and stalled. When Wemple could not get it started he gave the fuel injector a "good shot" of ether, which was used to start cold engines. The engine blew apart at the next try.

increased and wages can compete with industry. . . . With more and more machines, there's no limit to the chances I have to expand . . . mechanization is my answer—and every farmer's answer to the vagaries of weather." When that statement was made to a *Wall Street Journal* writer in 1949, nearly half of the 21 billion hours of farm work still involved non-machine labor. In 1959, Campbell expressed the view that farmers could no longer expect workers to labor for them for less pay "just because we happen to be farmers." One of Campbell's favorite themes was that the small farm could only compete with family help and sacrifice.

By 1920, Campbell questioned whether a farm the size of his could be economically run without tractors. Yet, the average farmer appeared unaware of the economy of the tractor, and the number of horses increased to their all-time high that year. He felt that a $6 a day tractor driver in 1928 was less expensive than a man who earned $26 a month plus board while using horses. He had evidence that showed that labor costs were 27¢ an acre for plowing, 10¢ for double discing, 7¢ for seeding, and 40¢ for harvesting. With a labor bill of about $1 per acre, his cost was considerably less than that paid by the most advanced horse operations of the bonanza farms in the 1890's. On the bonanza farms it took 10½ hours' labor for each acre of grain, or about 42 minutes per bushel of wheat—triple the time it took Campbell to do the same amount of work in 1928. His labor costs were low enough so that even with a 10-year average yield of 12½ bushels for wheat and 7 bushels for flax, he was able to make money. To exist without profits from such low yields was a miracle. No commercial operation could have done it without using the maximum available mechanical power.[10]

The Worker on the Big Wheat Farm

One of the least attractive phases of agriculture's past has been the deplorable condition of those who were hired to toil for the farmer. Room, board, and washing, plus the very minimum wage, was too often standard fare for the "hired man" and "hired girl" of the typical small farm of the past. The farmers' sons and daughters were expected to stay home and work, generally without wages, until they were married or could strike out on their own. If they "worked out" before they were married, it was commonly expected of them to "bring home" part of their earnings to support the homestead.

Low wages were such a traditional part of American agriculture that even in the twentieth century a large portion of the American farmers assumed that labor on farms had to expect substandard pay.* This was not the case with Tom Campbell, who had seen so many deprived farm workers in his youth that he was determined to treat labor well or not farm. It was part of "his mission of mechanizing agriculture" to raise agricultural labor to the dignity of any other occupation. Men wanted to work for CFC, and contrary to common conception the farm rarely lacked having a full complement of workers.

Naturally, any firm that hired hundreds of people for very short periods at a time was bound to get many transients with a questionable past. The company did not have time for nor a policy of checking the history of its new seasonal laborers. "Several times the FBI officers came to check on people and got after us because we did not know our men. We did not care as long as they did their job." On the other hand, there were many reliable farm men who came to work for CFC for many years, using the money earned to bolster their meager farm income. College students always made up a large portion of the seasonal labor force of CFC because peak labor demands coincided with their school schedule. Throughout the years the longevity of the permanent staff of the corporation is good testimony of Campbell's benevolent philosophy toward labor.

The bonanza farmers of the late 1800's had already proven that good food and good lodging paid benefits in maintaining a labor force and in keeping its morale high. Whether Tom Campbell, who grew up in the midst of the bonanza region, knew about the merit of good food and quarters for those laborers in contrast to the general treatment of "hired men" is unknown.

Campbell had a good philosophy toward his labor force. A magazine article written in 1925, although a bit flowery, seems to reflect Campbell's views in regard to the treatment of farm laborers:

> Mr. Campbell insists that all of his men shall shave every day, and encourages them to take a shower when day's work is done and change their clothes. These are not merely sanitary measures; they are also psychological devices, designed to prevent the old fashioned "farm hand" attitude

---

*This writer in his contact with farmers has frequently heard that "a farmer cannot pay industrial rates for labor." In 1970, one farmer specifically said, "If I had to pay more than a dollar an hour for labor I would quit farming." From personal observation, only the poorest type of worker ever sought employment on that farm.

from creeping into the Campbell organization. Instead every man feels himself to be what he really is, a self-respecting member of a skilled factory trade. The revolutionary effect of this attitude upon the morale of the men can readily be imagined. It helps to explain the unprecedented "per man production" of wheat on the Campbell ranches.*

Owens and Wemple remembered more than one worker who came to CFC wearing a white shirt and either dress clothes or overalls with the intentions of leaving after the first pay check. But when that time came they frequently bought new work clothes and stayed through the season. However, they recalled one man who came to the farm during harvest when a double shift was working, one plowing at night and the other harvesting during the day. The man informed the office he would work either day or night, but through a misunderstanding he was assigned to work both day and night. He thought "that was too much."

The living quarters on CFC were not the old-fashioned bunkhouses. Instead, a row of "neat modern buildings, with hot and cold shower baths, bedrooms furnished with white iron beds and honest-to-goodness sheets on them, provide living quarters undreamed of in the old West and rare enough in the New."

Unfortunately, some of the first men to work at CFC, such as Pat Roach, did not have a "neat dormitory" to sleep in. When he was 84, Pat Roach recalled one experience during his first year on the farm when the crew was sent out to open and build Camp Four. It took about a week for the big Aultman-Taylors, pulling the cook shack, lumber and supply wagons, and equipment, to make the 42-mile trip. The night they arrived, a big blizzard was in progress, and it was impossible for the men to set up their tents. The 22 men in the caravan spent the night crowded into the only structure then on the camp site, a small oil house that had been pulled in from another camp. "We were packed in like sardines but there was no other shelter because McGibinery's cook car was packed with supplies that had to be kept dry." Those 22 men also had to break sod in the Camp Four area as quickly as possible, so the buildings were not all erected immediately. A dining tent was erected first, and the men had individual or group tents for sleep-

---

*More than one farm worker of the past slept in the hayloft, cattle barn, machine shed, or granary without taking off any more than his shoes. The 25¢ bath in the small town barber shops did a land office business on Saturday nights when large numbers of farm workers were in the area. Campbell's nightly bath idea for farm workers was probably quite revolutionary in 1925.

ing. The cook drove his 1924 Star automobile to the camp because
he needed daily transportation to haul supplies.*

Both Owens and Wemple recalled their first year at the camp
before the power plant was erected and the running water was
installed. Water was heated over an open fire, and the men bathed
in a 50-gallon steel drum that had been cut in half. Wemple remarked, "There were days when the sun was so hot and the dirt
and grease from the Aultman-Taylors so bad that I swore if I ever
got to town and got steamed out and new clothes I would never
come back to camp."

Another handicap at the start of Camp Four was the lack of a
road covering the 42 miles to Hardin. And it took time to build
and gravel 35 miles of road, all on CFC land or on leased land.
Before that road was constructed, Pat Roach once had to make a
trip to Hardin for parts and supplies for the cook shack. He drove a
Dodge touring car that had been converted into a pickup. It rained
after he got started, and even though it was a high-wheeled vehicle he could not get through some of the coulees without first removing the fan belt and covering the ignition parts. It took Pat a
full week to make the 84-mile round trip. After the road was built,
the company had a small bus that operated between Camp Four
and Hardin. This was a risky business, because in the warm season fresh meat had to be brought from Hardin every day until refrigeration was acquired at camp. If the roads were impassable,
food supplies frequently got dangerously low.

Eventually the power plant was erected, and buildings were
constructed to house a maximum crew of 165 men at Camp Four.
Many visiting newspaper and farm magazine writers described
those quarters and referred to them as "neat and clean."

Henry Old Coyote, who worked during several harvests from
the late 1920's to the early 1940's, laughed as he remarked, "I
always felt that I was getting a fair break and good food. I thought
less about the pay I was getting or the working conditions than I
did about the good food that came with each meal. We preferred
to use our own bed roll and sleep in tents." Wemple and Owens
also laughed as they talked about the food at Camp Four. "There
was some griping, but generally it was very good. We really had a
couple of good cooks, but one time they forgot to put baking pow-

---

*On one occasion he was sleeping in the rear seat of his car, and the wind plus
probably a little help from some of his buddies got the car rolling downhill.
Everyone watched as the sleepy cook excitedly waved his arms from the rear seat
as the driverless Star bounced across the prairie.

der in the biscuits and they were hard. The men tossed them back into the kitchen where they banged on the bottom of the big pans." When the camp was at its full capacity of 165 men, either two hindquarters of beef or nearly three hogs were consumed daily. The main cook stove never cooled off because it was used to bake 100 pounds of flour into bread, rolls, or pies every day, which, according to Pat Roach, were all very tasty.

Men were charged various rates for room and board. In the 1930's the charge for room and board was $1.00 a day and the lowest-paid man received $2.25 a day. During the rainy spell, there was not much left of the pay check because men were paid only for the days they worked. By the 1940's the charge for room and board was increased to $1.25 a day. Some men who lived on nearby farms preferred to eat only one or two meals at the dining hall and were charged 25¢ to 35¢ a meal. Each person called off his number when he returned his empty plate to the dishwasher. In this manner, proper accounting could be made.

CFC always hired a large portion of its crew from the Crow tribe whose lands the company rented. Generally these were family men who furnished teams or teams and wagons for the bundle hauling in the pre-combine era. Their families packed tents and went from one threshing site to another. The men usually took meals with their families to save the 25¢ to 35¢ meal charge. CFC had to send a car to Hardin for supplies every day, so the Indian women provided the driver with a list of their grocery needs. Henry Old Coyote remembered living with his parents at five different sites during one threshing season. It was quite an exciting experience for the Crow children. Henry's father, Barney Old Coyote, furnished two teams and two wagons for bundle hauling, and there were between 10 and 20 other Indian families who traveled together during threshing season.*

Excitement, good food, and hot showers are not enough to attract men to work on a big farm—the pay must be good too. In this respect Tom Campbell tried to be a leader. The pay scale in terms of the inflated dollar may not appear very exciting, but by contrast to what was being paid on most farms across the country, CFC was

---

*Several of those interviewed, including Henry Old Coyote, said that the married Indian was much steadier if he could have his family with him while working. About the only problem with the married Indian was that he would quit at noon during threshing if it was too hot. E. M. Young said he made allowances for that type of turnover and hired more men than he needed because he could just about predict how long any one of them would stay on the job. The unmarried Indian males ate in the dining halls and stayed in the barracks with the regular crew.

progressive. Generally his corporation had a wage scale that was equal to, or "higher than that of any nearby industry." That was one of the reasons why so many harvest hands returned annually.

The following are a few examples of the pay scale. Rex Wemple started at CFC for $2.50 a day in 1924 and when he became an assistant foreman about two years later he was paid $65 a month for seven months a year. He furnished his own bedding so that he did not have to pay for room. He paid for meals on an individual basis. Barney Old Coyote received $2.50 per day for his labor and $1.00 for each horse he provided during harvest and threshing season. Henry Bigday, who furnished a team and wagon as well as his own labor, got $6.00 a day. Campbell frequently provided equipment for the Crow Indians so they could thresh their own crop, which was paid for by labor.

The wage schedule by 1926 was $6 a day for large-engine drivers, $5 for smaller-engine drivers, and $4 for daily labor during the standard work day of 12 hours. By the 1940's the minimum wage had increased to $7.50, and the work day was shortened to 10 hours. By the 1970's the starting minimum per day had risen to $10 plus room and board plus a standard bonus for all men who completed the season, which varied from $500 to $2,500 per employee. Both Wemple and Owens recalled that in the 1920's and early '30's their bonuses averaged about $200 per year. When working conditions became tough and more people quit, the bonus for those who stuck it out increased.

Tractor drivers were assigned to a single tractor to encourage better maintenance, and as an incentive there was a 10¢-a-mile bonus for any distance traveled over 17 miles a day with an Aultman-Taylor, and a 7¢-a-mile bonus for the smaller but faster tractors.* A counter was placed on the wheels of the tractors which recorded the distance traveled, and every man reported that figure on his daily work sheet. The bonus system reduced time in the field and at the same time lowered maintenance cost because the drivers took better care of their equipment. To stimulate additional care, a penalty system was also set up for lost tools and unnecessary breakages.

---

*The average horse-plow operation could travel 20 miles in a 10-hour day, so a 17-mile goal in a 12-hour day with the Aultman-Taylor should have been easily obtainable. "Punk" Taylor encouraged the tractor crew to work right through the noon hour to get an extra mile or two each day and discovered that most of the men liked the idea. Johnny Owens and Rex Wemple remembered nearly a half century later that they were assigned tractor numbers 1495 and 1653 respectively.

In a 1926 news article Tom Campbell is quoted, "We find it impossible to succeed without high grade, skilled men and think that the greatest field for skilled labor today is in industrial farming. Good tractor operators are very scarce." From this it is obvious that CFC's struggle to get labor occurred in the area of seasonal and not permanent employees. Throughout its history CFC has maintained a hard-core staff of 17 to 25 employees with a minimum of turnover. With seasonal employees, the problems were no different from those of any part-time employer. The seasonal employees who worked about seven months a year numbered from 50 to 65 men. During the peak season of harvest, a maximum labor force of 450 was needed in the day when the binder and thresher were still in use. After the advent of the combine, that number dropped to about 200. In recent years the Crow Indians have found alternate occupations and have become less dependent on CFC for jobs.

Like all other big employers in rural America, CFC had to rely on part-time help from among the smaller farmers in its area. These men generally were steady, reliable workers with an interest in their work. In the 1940's CFC's payroll during peak season surpassed that of the local sugar beet plant which, like CFC, depended upon nearby farmers as its major source of workers. The four-wheel-drive tractor and big combine without question have established a definite trend toward a smaller but permanent staff. During peak season of harvest the crew does not exceed a tenth of the number of the 1920's, but the total production is as large as ever.*

Tom Campbell, like most farmers, encountered difficult times in the early 1930's. He was determined to keep his regular employees with him and to make extra money for himself to get through a period of drought and low prices. To make this possible he turned to gold mining. Each fall after the crop was seeded, the stubble ground tilled, and the seasonal employees paid off, Campbell had his permanent labor force load the trucks with the necessary equipment to work a gold mine about 60 miles from Eureka, California. This was a placer mine, so much of the

---

*In the 1920's and '30's a large number of Texans (commonly referred to as "Mexicans" in rural areas) worked for CFC because they were imported by the local sugar beet company to work in the beet fields. Beet work did not conflict with the small grain harvest.

equipment needed on the farm was useable there. For eight winters from 1932 through 1939 Tom Campbell and his permanent crew worked that mine. The men were happy to have work and therefore enjoyed their activities during the winter season in a warm climate. They felt like "forty-niners" because they had to cut trees, make a road, and build their own log cabins for dining and sleeping purposes. "It was not too fancy for we used our own bed rolls," recalled Owens and Wemple. "We don't know how much Campbell was paid for his labor and equipment, but we got $60 a month and a trip to California every winter." Campbell had many business connections in California and once had actually hoped to start farming there, but instead diversified his activities by working the mine. However, there are indications that he came very close to purchasing the 280,000-acre San Margerita Ranch in southern California.

Whenever a large number of men, many of whom are single, are gathered for a length of time, there is bound to be some frolicking. Pat Roach, a bachelor, spent more than 40 years at CFC and saw about "every type of man come to work on the farm. In the first years we were too busy for much social life, besides we had no way to get to town." This meant that a great deal of jokester activity took place around the camps.

Men were not allowed to talk in the dining room during meals; however, when "Mr. Campbell came in he talked quite a bit." Both Owens and Wemple remembered that "Mr. Campbell was carrying on a steady conversation and someone passed him some real hot horseradish. Without looking at the size of helping he had taken Mr. Campbell took a mouthful of the stuff—he stopped talking for five minutes—and the men had all they could do to keep from laughing."

"Punk Taylor, the manager, and Len Cothren, a foreman at Camp Four [supervisors], were so tough that there was absolutely no noise in the dining hall if either were present. All they had to do was look at you," agreed Wemple and Owens.

But the men did not lack fun, and sometimes extreme devilment took place, such as smashing a high-priced watch on an anvil. On another occasion a foreman sent a "greenhorn" after a left-handed screwdriver, but the not-such-a-"greenhorn" went off to nap in the shade. When he was later sought by the foreman the youth replied that he was completely exhausted from his search for the left-handed screwdriver and had to rest.*

At times men were left in the field at the end of a day's work. There were several reasons for that: sometimes they were really forgotten, sometimes someone needed to be "cooled off" and was intentionally forgotten, and sometimes it was so dark that when men moved to other fields it was impossible to locate them. Rex Wemple once slept in the cab of his Aultman-Taylor because he was overlooked by the driver picking up the men. Others were stranded by heavy rains and had to stop their tractors before they became stuck and rescue cars were unable to reach them.

One hazard of dryland, small-grain farming is fire, which compounds with the increase of men and machines. CFC has had its share of fires. In the 1940's, the main shop, a blacksmith shop, and several kitchens were destroyed by fire, and in the 1930's and again in the early 1950's, the main office at Hardin burned. An elevator fire that completely destroyed the 250,000 bushels of grain in storage took place in Hardin in 1969. Those were costly experiences, but the most frightening fires were the ones that were ignited by lightning, cigarette smokers, combines, or the exhaust system of automobiles. Ironically, Tom Campbell, who during harvest was constantly on the run from one operation to the other, set as many fires as anyone. Many times, especially when the grain was cut high, the tall stubble caught fire from the exhaust of his Stutz Bearcat, which he used in his early years. Sometimes straw stuck underneath the car, and as it traveled across the field it ignited straw along its entire path. Dirt was commonly thrown on a car to smother a fire, but on at least one occasion Campbell drove his burning car to camp in order to make use of the fire hose. Generally other cars were not allowed on the fields.

Longtime employees remembered the times when they had to get out of bed to fight fires that were rushing toward camp. Owens and Wemple recall working with 35 to 40 men, using wet sacks and shovels to fight a fire until a strip could be plowed to stop it.

---

*One well remembered mishap on CFC occurred at the railroad siding at Hardin. A 20-40 Case was used to pull loaded grain cars away from the elevator so another car could be moved up and filled. One day the driver of that tractor forgot to take the chain off the last grain car he moved. When the train came to pick up the loaded cars the 20-40 Case still hooked to the train was next seen whirling backwards long the railroad tracks. The tractor was in gear and made deep marks in the ground, but fortunately the mishap was noticed before serious damage occurred.

"We've had lots of fires and some pretty bad ones," said Wemple as Owens nodded in agreement.

After World War II the company stationed surplus military spray trucks near potential fire hazards. In addition, fast-moving four-wheel-drive tractors established plow strips as quickly as possible after combining to reduce the hazard of fire. In recent years the risk of fires getting out of hand has been greatly reduced, but the danger of their being started is as great as ever.[11]

The Wheel of Fortune

In the opinion of many, Thomas D. Campbell could not have farmed on such a grand scale if he had not had strong financial support coming from his family. In addition, Campbell's self-confident air made many people, especially in Montana, critical of him. Another source of irritation was caused by the fact that his big farm was started at a period when farm failure was rampant in Montana. Because of the high rate of failure among family farms at this time, most people were convinced that large commercial mechanized farms could not succeed.

Campbell's father had owned 4,000 acres of land near Grand Forks, North Dakota, in addition to real estate in the town proper. Tom Campbell never mentioned a lack of material goods, even in his childhood during the 1880's and '90's. His wife, as previously mentioned, came from a family of considerable wealth and no doubt later contributed sizeable funds to the corporation. The exact amount is known only to his family. In addition, it cannot be overlooked that Campbell knew how to deal with bankers, and his charm and his reputation as a successful farmer probably convinced them to loan him more money than they might have done under normal circumstances. Campbell's grandiose ideas plus his ability to think in large sums would surely have overwhelmed the average country banker. Fortunately, the scope of his operation forced him from the start to bank in New York and Chicago with bankers who understood large-scale financing. For Tom Campbell that was necessary because his goal of a million-bushel crop required big financing. After succeeding in borrowing two million dollars from J. P. Morgan and Company in 1917, he was not afraid to ask for more and even larger sums.

Campbell Farming Corporation, after absorbing the Montana Farming Corporation in 1922, had several profitable years and became quite well established with some Chicago banks. When, however, the profit picture weakened after 1927, the bankers became jittery, demanded on-the-spot inventories, and had their own appraisers on the farm during harvest time.* Each succeeding year after Campbell took exclusive control of CFC he increased his crop acres. He started in 1922 with 24,000 acres in crop and expanded to 64,000 acres in crop in 1927. At that time over 95,000 acres were under plow with the two-year-fallow system. Until 1927 the new land yielded a good crop, and profits continued to accumulate. In 1925 Campbell had informed *World's Work* that he and a few of his key employees owned the million-dollar farm outright and had a "heavy cash surplus." He attributed some of his profits to the fact that he had adequate financing so that he was not forced to sell during harvest season. He cleaned and graded his wheat out of storage so that he could "guarantee a standard quality, and therefore get a better price." CFC sold direct to the millers and avoided normal marketing costs. A saving of 5¢ to 9¢ a bushel meant a sizeable amount on Campbell's earnings.

Up to 1929 accumulated profits of CFC were about $700,000. At this time a reduction in moisture occurred that persisted for nearly a decade. Campbell had always been interested in the study and interpretation of weather cycles, and when he took over personal operation of CFC in 1922 he predicted that there would be a cycle of seven years of moisture. Whether the records actually indicated such a cycle or whether Campbell was using it as a positive argument to convince investors and bankers to loan him money is hard to say, but at least this time his prediction held true. Both 1929 and 1930 were poor crop years, with a resulting operating loss; and in 1931 there was virtually a total crop failure, that caused a loss of about $300,000. A Campbell prediction in the early 1920's that he would never have a complete crop failure was shattered by the drought of 1931. However, in *Business Week* in 1949 he was quoted, "We'll never have another crop failure due to the drought." At this time he must have felt that technology he had adopted since 1940 was enough to cope with another dry spell.

The year of hardship for the ever-confident and optimistic wheat farmer was 1932. The winter wheat crop year started in the preceding fall was good, for there were heavy rains resulting in a fair stand of wheat. In addition, the tremendous volunteer growth on summer-fallow fields was good enough to harvest. The wheat crop exceeded 400,000 bushels, and it appeared temporarily that

---

*At this time general manager "Punk" Taylor parted company with CFC and joined Caterpillar Tractor Company, which knew Taylor and was anxious to get individuals with first-hand experience in mechanized farming.

Campbell might make a profit. What he did not realize at harvest time was that he would be forced to sell his crop at a time when the parity ratio was at 58, the lowest in modern economic history.* The price of wheat in Chicago fell to 60¢ and in Liverpool, according to Campbell, it fell to the lowest point in 600 years.**

The real financial crisis came because cash rent on Indian lands and a bank loan were due shortly after harvest and Campbell could not persuade either the bankers or the Bureau of Indian Affairs to wait for an increase in price. The total amount due at the time was $86,000. Much of it came from the new cash rent contract of 75¢ an acre that CFC had to pay for Indian lands, and the balance had apparently accumulated as interest on a bank loan. Given no choice, he had to sell some of his crop for as low as 14¢ a bushel. His hauling cost to market was 8¢ a bushel, so that left him with 6¢ to cover all costs of production. The irony of the forced transaction caused by the bankers and the BIA officials was that 90 days later the price of wheat at Hardin was 65¢ a bushel. Bitterly Campbell remarked that if he had been permitted to hold his crop, as he had begged to do, he could have earned $230,000 more.

After the disaster of 1932, the 50-year-old Campbell probably felt things could not be worse. Psychologically he was possibly as well equipped as any farmer on the Great Plains to withstand the drought, but he never forgot the lack of consideration by the bankers and the BIA, a feeling not uncommon among farmers. A second event of 1932 also had an impact on Campbell's future, for it was then that the BIA reduced the amount of land CFC could rent from 50,000 acres to 19,000. Other sizeable operators, such as Erwin Schnad, Floyd Warren, and Ed Kopac, secured many of the former Campbell leases at that time. In spite of this, Campbell continued to rent a large amount of land directly from individual Indians.

Drought in 1933 and 1934 kept the wheat crop at 4 bushels per acre, but in 1934 something happened that changed the wheel of fortune for Tom Campbell—he accepted a government payment of $24,000 to reduce his acres under production. Tom Campbell,

---

*For the years 1910 to 1973, the parity ratio of 58 in 1932 is the all-time low. It was the only year that the ratio fell into the 50's. In 1931 and '33 the ratio was 67 and 64 respectively, and in no other time did it fall below the 70's.

**The big spread between 60¢ wheat in Chicago and 14¢ wheat in Hardin is caused by transportation costs for the 1,300-mile haul. The continental position of Montana so far from the international and domestic market causes some of the highest freight rates of any state in the nation. Rates of 30¢ to 45¢ a bushel of wheat were common.

who favored free enterprise, was desperate enough to agree with a government program. In fact, as late as 1948 he called the "Triple-A . . . the greatest single act of legislation since the Civil War." From 1929 through 1934 CFC lost $600,000 and at the end of the 1934 harvest season was a million dollars "in the red."

Tom Campbell's dream of a million-bushel crop was now a million-dollar debt, but the wheel of fortune was still turning, and, being an optimist, he hung on. In 1935 he planted 20,000 acres and made a profit. As soon as that crop was in the bin he announced that if profits continued to return he would put rubber tires on his tractors to get higher speeds and would redesign combines and plows to do better than ever. His operating cost that year was $4.80 an acre in contrast to the average in his region of $12.00 and an average of $17.00 in the Midwest. Improved moisture conditions, better soil conservation practices, government programs, higher prices during the World War II era, and the adoption of Russian Kharkov seed wheat during the late 1930's caused Tom Campbell once more to strive for his million-bushel crop. This time he would reach his goal.

Tom Campbell was ever alert to the possibility of finding another venture that would duplicate the achievement of CFC. In 1926 he announced that he was considering similar enterprises in other states and that he had been approached to take over one of the giant Canadian farms that had gone broke after World War I. About 1935 he told a Montana neighbor that the 280,000-acre San Margerita Ranch in California was available for $3.5 million and that he nearly had the money raised. His idea of finding land in Canada, Kansas, New Mexico, or California with rainfall areas similar to that of Hardin, Montana, was not all based on drama or speculation. Campbell felt that he could farm such areas which were not being farmed and such operations could reduce the risk of farming only in one location. As early as 1935 he thought of using the same machinery for a second or third enterprise. Obviously this meant being located on farms far enough apart so that there would be little conflict during the planting and harvest seasons. Unfortunately for Tom Campbell, the U.S. Soil Conservation Service opposed his idea because they believed that grass was more profitable than grain in areas of marginal rainfall.

In the mid 1930's Campbell became acquainted with John Jacob Raskob, a General Motors official and national financial figure. In 1936 they became involved in the purchase of the La Jolla Grant near Socorro, New Mexico. There was considerable opposition to the purchase of the lands in New Mexico, and the news media carried many articles about it. Senator Dennis Chavez of New Mexico even indicated that the sale involved subterfuge. An Albuquerque paper stated, "Chavez alleged they bought land worth $2.00 an acre for 35¢ per acre, and they haven't grown any wheat on it." This was a reference to Campbell's statements that the land would be used for dryland wheat farming. Campbell broke his silence on the controversial purchase of 126,000 acres. He announced that after a conference with New Mexico's Governor Miles, he would not purchase the 10,000 to 12,000 acres owned by settlers even though there were "$22,000 in back taxes on their land." The United States Department of Agriculture and the Federal Security Administration had publicly opposed the Campbell-Raskob purchase on the grounds that it would dispossess people who had lived on that land for generations. Campbell was undaunted and proceeded to call for bids for building a fence around the land.

After the initial controversy quieted down, *Business Week* carried an extensive article on Campbell's efforts to raise wheat on his New Mexico lands. The article implied that although Campbell had liquidated his cattle business in Montana, he felt it was necessary to combine sheep and cattle with crops in New Mexico to get a profitable operation. At that time irrigation of the newly acquired land was being discussed, and Campbell's answer to his critics was, "Anytime you get enough water on land to grow wheat you can grow other crops to far better advantage. Irrigated land is too expensive for wheat, generally speaking." It was Campbell's idea that if he struck water in New Mexico he would raise alfalfa, for one acre of irrigated alfalfa could equal 60 acres of dry grazing land. However, this was one of the few times of his life he publicly expressed a pessimistic view, as he warned that water might not be found. His pessimism must have been justified, for not much more occurred in New Mexico, and the land was leased out.*

The publicity Raskob and Campbell got on their large purchase in New Mexico was probably no more than any other well-known individuals would have received. Any purchase that disrupts families of impoverished, tax-delinquent settlers would get considerable negative publicity. The news media and some political

---

*Campbell was a real fan of dryland grain farming. After his death some irrigation was used on the cattle operation near Fort Smith at the south end of CFC where 800 acres of alfalfa ground were irrigated.

figures fanned the fires whenever they could. The transaction may not have been completely free of irregularities, as Senator Chavez implied, but eventually as a result of a tax sale clear title to the property was given and Tom Campbell became the sole owner.

For many years CFC did not own most of the land it operated. Working on the assumption that if a farmer cannot afford to rent land that he operates he cannot afford to own it, Tom Campbell did not rush to buy real estate. In later years, as the CFC profit picture brightened, especially during the World War II era, land was acquired. Eventually CFC came to own about 15,000 of the 41,500 acres of crop land it operated. After Campbell's death a ranch containing 68,000 acres was purchased under the management of Floyd Slattery, and the company brood-cow herd of 2,000 head was established. The ranch, designated as Grapevine within the corporation, served to offset the risk of marginal dryland grain farming. Its acquisition was well timed in the livestock cycle, and Tom Campbell himself could not have done better.

Once the challenge of drought and low prices was overcome and the farm started to progress again, Tom Campbell turned to other endeavors, leaving the Hardin operations in the hands of Floyd Slattery. The really profitable years were still ahead. By that time Campbell "had proven his point—agriculture could be industrialized."[12]

The Philosophy of a Wheat King

The goals and ambitions of Tom Campbell do not always coincide with his professed philosophy. Basically he was a free enterpriser. His first conflict with his general philosophy came into the open in the late 1920's and early 1930's when he advocated protective tariffs for the farmers and at the same time encouraged more trade with Russia. Later in the New Deal era of the 1930's, his free enterprise philosophy had to compromise with government subsidies from the farm program. In neither case could he hide his actions from the public for he was too well known nationally. In the case of his advocacy of protectionism for the farmer while at the same time asking for more world trade including trade with Russia, the conflict appears to have arisen from oversimplification and confusion in his own thinking. In the case of the farm subsidies, Campbell, an ardent free enterpriser, readily accepted government funds because he was financially insecure and he understood that somehow production had to be curtailed. One article which appeared at this time stated that his economic and social principles were those of the New Deal; but if that were true then this would have been a great subterfuge, for it was inconsistent with most of what he said before and after that era. Nor was it consistent with what he did.*

Tom Campbell, the man who succeeded in producing a bushel of wheat in 14 minutes, above all else wanted to be thought of as a real down-to-earth farmer. It is said that one of the few people who aroused his anger was Senator Burton K. Wheeler. Campbell once stated that the farm bloc in Congress did not truly recognize the "interests of the Western grain-grower." Wheeler's retort was that Campbell was "a promoter, not a farmer." The "wheat king's" reply was that his family had just produced its 66th consecutive wheat crop, "in the West, land of boom-bust-blow, that makes . . . a farmer."

There was little of Campbell's life that was hidden from the public, as the following article illustrates:

As the acknowledged "wheat king of the world" and one of the most theatrical figures in public life, he has been a favorite of the writers for thirty years. . . . Praise and castigation have been equally intemperate (a woman correspondent once addressed him as "Dear 95,000-acre Hog") and few have been able to make sense out of his complex character.

At a time when many farmers were being extremely critical of transportation (because one-third of the proceeds of a bushel of wheat went to freight), Campbell had words of a more objective view. "Railway companies have always been friendly to agriculture. It is to their best interest to be so." It was his impression that perhaps rates should be revamped, there should be more regulation in the grain trade, and farmers should have greater representation on all national boards; but if the farmer really wanted to find help he had to do it by adopting better methods. With bigger equipment and larger crops the farmer needed more storage and a better form of transportation to avoid selling his grain on a depressed market during harvest time. Campbell understood the problem of transportation, especially from the isolated position of his farm in south central Montana, and therefore he did not complain much about freight rates or boxcar shortages.

Because of the great size of CFC, its owner often was subject to criticism and was questioned about the future of large-scale farming in contrast to the position of the family farm. Using good

---

*A *Denver Post* article in 1948 noted that he was "formerly a loyal New Dealer," which implies that he may have been a supporter of some of the ideas of Franklin Roosevelt, whom he knew and whose guest he was at the White House.

common sense, Tom Campbell tried to avoid antagonizing the defender of the cherished family farm, but at the same time let it be known in what direction agriculture in his opinion was headed. He voiced his opinions on this matter throughout the period from the 1920's to the 1950's.

In 1928 Campbell remarked, "It is positively absurd to think that . . . the old homestead entry of 160 acres of land is applicable to modern times. Practically all corn farms in Iowa could be successfully operated with a unit of 640 acres." This was at a time when both the national and Iowa average farm was less than 160 acres. But he also made it clear that his own position was an exceptional one when he added, "I have always believed in industrialized farming, based on economical units, but this does not mean that all farms must be large farms." He recognized the social indispensability of the small family farm and felt that "we will always have the small farm with us for truck gardens and intense units near industrial centers and they will serve as a home as well as a business." In the 1920's he clearly foresaw that a large portion of American farms would someday become rural residences rather than the headquarters of a commercial farm venture. He was certainly looking 50 years into the future when he predicted:

> The average tenant or farmer has little hope or enthusiasm. His children will leave the farm as soon as they are educated or feel their obligations to their parents have been fulfilled. No woman in any other industry puts in as many hours of toil as the farmer's wife. . . . We have too many people on the land now. Less than 20 percent of our population will be on the farms in another twenty years. This farm population will drift to the city to meet our ever growing industrial demands.*

Even though many people had already left agriculture, he predicted that many more would do so because there were still too many people engaged in farming. He cautioned that his prophecy did not mean doom for agriculture but rather a progress. In his opinion:

> Agriculture is stepping into what you might call its machine age; mechanical appliances are taking the place of hand labor; scientific knowledge is being substituted for unscientific habit, and, in short, INDUSTRIALIZED AGRICULTURE, offering greater opportunity than any business I know of, is just around the corner. . . . To begin with, farms will be bigger . . . the investment in machinery will become greater. . . . Thus more people . . . are bound to leave the rural districts, but those who

---

*Campbell's prophecy made in 1928 was five years off because in 1943 the farm population declined to about 20 percent of our total population.

remain will be rated as skilled laborers whose wages will be as high, and whose working hours as . . . [short], as in the manufacturing industries in the cities.

The writer of an article in *Colliers* magazine said he was frightened by Campbell's predictions and wondered what would become of the "traditional American farmer." To this Campbell replied:

The traditional farmer disappeared from the American rural scene a long time ago . . . and if I may say it, I'd like to add that the pretty phrase about farming being "a way of life" is not nearly as appealing to the farmers as to the politicians who coined it. . . . [Farmers] would vastly prefer to have their business considered a money making one, rather than have it sniffed at as a mere living.

In the 1920's, Campbell, a prophet of industrial agriculture, expounded that complete and revolutionary changes would reduce the cost of production beyond the belief of the average person. He foresaw diesel engines for tractors and electricity on every farm to operate the equipment, and at that point "farming will . . . be recognized as a dignified business . . . and it will attract the smart, ambitious young man." When in 1959 a reporter for *U.S. News & World Report* questioned Campbell, "Is the answer in the end going to be fewer farmers and bigger farms?" Campbell replied, "It's a transition that's coming. We can't make it too abruptly, or a lot of people will get hurt."

To the average farmer of the 1920's and 1930's, Campbell must have appeared as wealthy as the mythical Wall Street banker so often blamed for the farmers' problems. In relative terms, Campbell did appear better off than the average farmer; at least he could live on a higher level. But Campbell, in spite of his higher status, had as many money problems as the 160-acre homesteader who, according to Campbell, had little chance for survival. Campbell probably had more problems than the small-scale farmer because most bankers were more accustomed to the small family farm and its demands than to a large commercial farm that used expensive tractors and more hired labor.

If Campbell ever had a complaint it was the lack of understanding the financial men exhibited about agriculture. Tom Campbell once remarked: "I learned as I grew older that farming was the oldest business in the world, that it had survived and continued to grow under the most unfavorable conditions, and I felt that if it were possible to apply the ideas which were used in other industries, farming too could be made successful."

Among ideas useful in industry that he felt should be applied to agriculture were better financial policies: ". . . grasshoppers, the hot winds, cold snows, cyclones, and market jigglers that are postponing the farmer's millennium . . . so much as it is the general feeling that farming 'jist natchelly' can't succeed with a big S. It is impossible to interest capital in agriculture. [I] . . . doubt if there are a dozen big bankers in the whole country . . . who believe that farming can be successfully done."

And in continuation of this, Campbell once told a writer for *Colliers* in 1930 what many rural Americans knew too well: "On all sides one hears it said that the farmer just cannot make money, and with that kind of talk filling the air he can't. The bankers hear it, the manufacturers hear it, the implement and seed men hear it, the senators hear it—and make the capital dome ring with it—the farmer hears it himself, and the first thing you know they all believe it."

Unfortunately, according to Campbell, the farmer was afraid to ask for money, and if he did the banker would not give him what he needed anyway. "He struggles along as best he can, fights his dirt with antiquated, horse drawn machinery. . . . After a few years in spite of his love for this country life he gives up and decides to make a living some other way."

Financing, in Campbell's mind, was the biggest obstacle to the development of industrial agriculture. It was his belief that when the general public, bankers, and manufacturers would change their attitude toward agriculture, conditions would improve. Should that day arrive he envisioned a United States Farming Corporation that would be bigger than General Motors or the United States Steel Corporation, "for food is the most necessary commodity of all."

Tom Campbell, an ardent student of agriculture and agricultural economy, knew that bankers had greater difficulties financing farmers than any other group because farmers did not keep good business records. This was not true of CFC, which not only had records, but also made annual projections of its operations. Looking back on his first decade of farming Campbell said, "We [my associates and I] are satisfied that over a period of 10 years, capital invested in farming enterprises, properly chosen, will earn a higher rate of interest than capital invested in any other conservative business." He lived to prove his point and also to see great progress in agricultural financing.[13]

Bold and optimistic as Campbell was, he wanted to leave the

impression that he could solve all of his own problems, but during much of the 1920's and 1930's the national agricultural situation was so serious that not even he could escape being pessimistic. From a purely economic view Campbell's start in farming in 1917 could not have been more ill-timed. He entered farming near the top of a boom, and when his farm was finally operating at full scale the agricultural depression set in. However, from a technological standpoint Campbell's timing was good because he was in on the initial stages of farming with tractor power, and on this basis he pulled through.

In the early 1920's, Campbell suggested that if agriculture would adopt industrial engineering principles and scientific practices the cost of production could be lowered and the natural hazards of farming could be overcome. He was so convincing in his arguments that when a writer for *Country Gentleman* questioned him, "Does the real remedy for agriculture . . . lie just around the corner, on our blind side, where the industrial secrets of America are openly revealed?" his answer and argument seemed quite logical. Yet the mechanical efficiency and science of which he spoke were some of the very factors creating an agricultural surplus. It was not long before he realized that the surplus problem could not be solved by mechanization.

An agricultural expert, Liberty Hyde Bailey, had warned in the 1920's that the agricultural problem should not be turned over to the politician, or the solution would never be forthcoming. A reason for this statement appears so evident from an observation by Owen P. White after he had spent several weeks attending Senate debates on the farm problem in preparation for an article he wanted to do on agriculture: "After a long, long period of this martyrdom I was no better informed than I had been prior to the beginning of the oratorical deluge. In other words, I didn't know anymore about the farm problem than I had known before all the talk started. Hence I came to the conclusion that I am either naturally stupid or the United States Senate knows not (and knows not that it knows not) what to do for the aid of American agriculture."

In this article White added that Tom Campbell had given him more information in one hour than he had gathered in several weeks of Senate debate. Campbell commented on the farm bill of 1930 that Congress had just passed:

No, as I see it, the bill for farm relief hasn't solved the farm problem and isn't going to solve it completely. In fact . . . special legislation at its best is only a makeshift device to carry any industry over a period of stress. If general conditions are not right they cannot as a rule be made right by any kind of legislation. Therefore, it would be a mistake . . . to assume that because a farm relief bill has been passed it has actually relieved the farmer. To some extent it has relieved the politicians of their much discussed obligations to the farmer, but it hasn't so far relieved the farmer of his obligations to his bank, his family or his implement dealer.

By 1930 Campbell had altered his original thinking that mechanization and science alone could save the farmer. Now he suggested first that the Federal Farm Board could back the creation of farm cooperatives, which would be a favorable factor for agriculture. Secondly, production should be limited to domestic needs only, and Congress should establish a tariff to stop importation of grains, because the farmer deserved as much protection as the industrialist was receiving. In the third place, the biggest improvement would come if the defeatist attitude about agriculture held by farmers and other groups could be eliminated. Tom Campbell's constant optimism toward farming, even in one of its most depressed periods in history, kept him from defeat.

In spite of his recommendation that production be curtailed to domestic needs, Campbell felt that if a farmer was really willing to mechanize and operate on a large scale he had little to fear. Citing his production figures he pointed out that the total operating cost including depreciation and replacement was $8 an acre. He prophesied to be proven right only 40 years later:

The apparent competitive disadvantage, due to cheap labor elsewhere, which American agriculture has to combat in its search for a world market can be overcome without the aid of debentures and subsidies, and without lowering the standard of our American farms. In fact, in my opinion, based on a large practical experience, it is possible to raise the standard of living and still win the world markets if we go at it the right way . . . is there anything to prevent an industrialized agriculture in this country from putting its products on the world market at as low a production cost as any country on earth.

At the same time, Campbell presented a powerful eight-point program for better farming. He advised, "Let farmers think more about economics, less about politics." The two must be drawn closer together, for the business world had more to offer to the farmer than the politician. He expressed it in this way:

The farmer feels that business is unfriendly to him. The businessman is not directly interested in farming and has given little attention to the agricultural problem. Meanwhile unfair politicians, prejudiced radical leaders, have told the farmer that business, especially big business, is trying to destroy him. . . .

On the other hand, the average businessman feels that the farmer is lazy, has a tendency to paternalism, and has not ordinary business integrity. Both are wrong, and a closer relationship will teach both that their interests are mutual. . . .

As depressed conditions in agriculture continued in the 1930's the combination of drought and low prices forced Tom Campbell, then a million dollars in the red, to accept the government subsidy that he had firmly rejected in his previous philosophy. Because of the size of his farm, he immediately became a target of publicity and criticism. In an article in *Time*: "Farmers: Something for Nothing," Tom Campbell was second on the list of seven wheat farmers to receive over $10,000 from the government in 1936. His payment was over $50,000. The article called attention to the fact that CFC rented 16,500 acres of government land in 1933 and 1934 for 50¢ to $1.50 an acre and received an average of about $7.00 an acre from the government program for the acres diverted from production.*

From that day on in the annual report of farm payments, listed for the largest in each county or state, CFC was always on top or near the top of the list. Only the large cotton and tobacco operations could consistently top the payments to CFC. From 1933 to 1959 Campbell had taken out loans annually and sealed his crop and in all but 5 of those 26 years had redeemed his grain to sell on the free market. His open market sales yielded him 17¢ to 70¢ more per bushel than support price.

There was resentment in the office of CFC when the newspapers and magazines published articles about the big government payments to the company. Headlines in the *Billings Gazette* in early 1956 announced the $312,998 payment to CFC well ahead of Bill McCarter's $124,146 and E. G. Onstad's $102,740, other leading farmers in Montana. Even among the biggest, CFC was well ahead.** The sharpest critique came when a *Life* photographer spent about a week on the CFC lands in 1957 under the pretense of doing a story. He took many pictures of the entire operations, but when the "story" appeared in *Life* it was a two-page picture showing manager Floyd Slattery on top of a bin full of grain with the words "This farmer was paid $330,268 by the government."

---

*Secretary of Agriculture Henry A. Wallace made a defense of the large payments that obviously and fairly had been made on the basis of the amount of land owned or operated by a farmer.

**The CFC loan was made on 172,159 bushels of wheat, while McCarter had 69,631 bushels, and E. G. Onstad 55,050 bushels. In 1958 CFC placed over 320,000 bushels under loan at $1.59 for a total payment in excess of $510,000.

Again and again Campbell warned that the farm program must be better managed "before the taxpayers revolt and deprive us of any farm support law." Later, when attacking the idea of $3.1 billion surplus wheat stored in government bins, he added, "I'm not trying to build up any sympathy for the farmers, but I am trying to acquaint the public with the true story of this extravagant program that only a country like the U. S. could afford."

To eliminate that huge stockpile of wheat that served as a constant threat to the farmer, he advocated giving the wheat to the world's needy to save $385,000 a day in storage cost. At the same time he suggested a 50 percent reduction in acreage for two years and a reduction in price supports. After that he supported a two-price plan.* The large-scale, mechanized American farmer would thereby be left free to compete on the world market. Campbell knew that he would have a favorable advantage under such a program.

Tom Campbell certainly enjoyed the publicity given to him, but he resented the emphasis on the farm subsidy aspects. He pointed out that of the $15.725 billion paid out by the USDA from 1933 through 1959, the farmers received only $3.6 billion—an amount that was only 1/39 of what the government had paid to other industry during the same period. "A drop in the bucket compared to the billions of war surplus goods that were given away or dropped in the ocean after the war," he remarked.

After having the benefit of the government program for so many years, even Tom Campbell was skeptical about returning to a completely free economy. Nevertheless he criticized the support programs because they encouraged surplus production.** He could produce wheat at $1.10 a bushel, and the support price at Hardin was $1.59 (national price $1.80).

Campbell also noted that even though he liked the theory of the soil bank it was not satisfactory because farmers increased their fertilizer use on the acres still in production. With an annual pay-

---

*The farmer would receive 100 percent of parity on the domestic allotment and would have to provide one year's free storage for that amount of grain. The balance of his crop would then be completely free for movement into the world market.

**Campbell, like other progressive farmers, was causing the government endless trouble with his tremendous rate of mechanization. In a *Wall Street Journal* article entitled "Super Production from Mechanical Farms Gives Price Proppers Trouble" he explained his theory as follows: "Labor saving devices are sowing a pack of problems for the government's program of supporting agricultural prices. . . . As one legislator put it, 'with mass production methods, federal support prices are still tied to the old mule and plow day.' "

ment of about $450,000, Campbell should have been attracted to the soil bank, but there was no challenge to him in idle acres.

Tom Campbell was very concerned about how much of a farm program the nation could afford. In a letter to Senator Milton Young of North Dakota in 1956, he wrote: "I believe that most farmers will support the flexible loan program as they realize the disposition of surplus wheat has practically gotten beyond the government control and the cost of buying and storing these surpluses will ultimately destroy the entire program of price support based on parity."

Campbell pointed out to Senator Young that the consumers would eventually vote an end to the program. He added, "Most farmers think we have been well cared for by our government and that we must look at the flexible program as suggested by President Eisenhower from the viewpoint of our national economy." Without some kind of program Campbell felt wheat might fall to 50¢ a bushel and "we would all be bankrupt again."

Shortly after World War II, Campbell stressed his interest in the human aspect of the farm economy, realizing that America suffered from a surplus of farmers as much as from a surplus of crops. "The real problems now," in Campbell's view, "are problems of social engineering. The preservation of free enterprise is in the hands of the capitalists themselves, who must define profit in terms of human welfare." Without such programs the transition would be too rapid.

The most criticism Tom Campbell received relative to farm programs and the price of wheat came immediately after W. W. II when the government attempted to move mercy grain to alleviate the starvation in other nations. The greatest obstacle to the government's mercy mission was that farmers were holding title to most of the wheat and because they were aware of the worldwide shortage were not anxious to sell at existing prices. In an attempt to get farmers to release wheat abroad general appeal was made and Tom Campbell contributed to it.

According to his appeal the farmer was permitted to release his wheat anytime between May, 1946 and April 30, 1947, but he could select the date he desired as the actual selling date, enabling him to choose a high spot in the market. In that manner he could deliver his wheat in 1946 to the government, but he could select either 1946 or 1947 as sale date for income tax purposes; plus, he was given a 30¢ bonus over market price, and the government assumed full responsibility for storage. In addition, the farmers were asked to sell for humanitarian reasons. The big appeal was made because U.S. relief wheat was 313,000 tons short of its goal to reduce famine in other parts of the world. He offered to sell 425,000 bushels to get the mercy grain moving, and he offered his fleet of trucks to any community that needed help to transport grain to the elevators. The community of Climax, Minnesota, delivered 8,000 bushels in one day "when all the big shots were around." Hardin, Montana, pledged that they would triple the volume of the Climax rally for one day, and Campbell's 14 trucks would deliver enough to fill 10 rail cars. A ¾-inch rain on the ungraveled roads prevented CFC from delivering its full pledge, so Hardin fell short of its 24,000-bushel goal.

The farmers held their grain in spite of an appeal by the government for the sale of 570 million bushels. Campbell, the world's largest private wheat grower, then visited President Harry Truman to explain why. The market was $2.94, and the farmers wanted $3.50. Letters to the editors came from everywhere after *Time* reported the Campbell visit and showed him leaving the nation's capital. Those letters condemned the farmers for holding the wheat and the government for subsidizing them. One headline read, "Time for Tom to See Harry Again???" and continued, "Do you think Harry [Truman] will arrange to give you another BONUS or $7.00 for your wheat??"

A letter to Campbell in CFC file reads:

> Oh, no, our Montana wheat king never gives anything away,,,? He will be making another trip to Harry and try to get another BONUS for his wheat. The past and the present ARMY BRASS has put our country in the SHAPE that it is today. For a good ROTTEN DEAL lets nominate for President and Vice President Bennet Meyers and Tom Campbell..?::,,?? An Ex Service Man from Montana.

Early in 1948 the market broke, and wheat fell to $2.47 a bushel, but Campbell had sold 60 percent of his crop just before the drop. When asked why he had sold he retorted, "Wheat was higher than it had been in 70 years and—what the heck, it looked like a break was coming. I wish I had sold it all." The *Cincinnati Post* was not so kind as the *Omaha World Herald* had been, for the editor of the *Post* wrote the following: "Tom Campbell never got his $3.50 a bushel. It looks as though he never will. And we cannot say that we are sorry for him and other farmers who wouldn't sell wheat last year because they wanted to hold it over to this year to save on taxes. . . . As it ends up, the greedy may lose but the hungry eat."

President Truman told reporters that he did not blame Campbell for wanting to get as much as he could for his wheat. Campbell defended himself in a letter to the editor in *Time* saying he could not understand why a farmer should be criticized for holding his commodities. Having sold wheat for less than 50¢ a bushel in the 1930's, Tom Campbell was somewhat irked at the criticisms of the fickle consumer.*

Initially Tom Campbell did not start farming with the intention of being able to benefit from farm subsidies, but when the opportunity presented itself, he, like a majority of the farmers, took what he could. But he received more criticism than most farmers on account of the size of his subsidies. No doubt he, like other large farmers, benefited to a greater extent from the programs than did the small-scale farmer. That might have been some of the intent of the programs, for America had had a cheap food policy dating back to 1910.[14]

Ah-Wa-Go-Da-Ay-Goosh

Many have been critical of Thomas D. Campbell, and as can be expected Hardin's citizens lead the list of critics. The Campbells did not spend a great amount of time at Hardin, Montana, and this caused more than normal resentment against a successful local figure. But one honor and recognition he rated very high was his adoption into the Crow tribe.

Campbell's Crow name, Ah-wa-go-da-ay-goosh, means "known all over the world." The Indians were proud of the fact that they had good relations with a well known leader in agriculture, who also had brought international attention to their reservation. They knew that he, an "associate of presidents and princes," had never been too busy to visit with any of their tribes who wanted to see him. Campbell had endeared himself to the Crow. He was escorted through the full adoption ceremony in August, 1946, by William Wall, Ben and Calvin Jefferson, Joe Mountain Pocket, Bob Wolf, Joe Knows Ground, John Old Coyote, Freddie Bird Hat, Ernest Holds, Adam Bird in Ground, and George Old Elk, all veterans of World War II. After the ceremony there was a feast of

---

*Because of his fame and his connections in high circles many people wondered if Campbell might have had pre-knowledge about the wheat market, especially when he managed to sell a large block of grain just before the price break. When asked about price prospects just prior to the drop in prices he said, "If I were speculating, I'd buy wheat. But I'm not buying. I don't believe in paper wheat." Maybe just at that time he was not active in the commodities market, but Campbell was a regular speculator. He used the futures heavily in the 1920's and early '30's.

barbecued buffalo on the tribal dance ground near St. Xavier, south of Hardin.

The purpose of the adoption was to make sure that Thomas D. Campbell would be known to "the great mysteries" in the hereafter. For, to the Crow, Campbell was more than a farmer; he was a recognized worldwide servant of mankind. He and CFC had hundreds of leases with the Crow, and at times it took supreme patience in dealing with them; but as Campbell wrote to General of the Army George C. Marshall, "We are happy to say that we have always received fair and cooperative treatment from the Indians and the Bureau of Indian Affairs."

The Indians had not misnamed Campbell, for he was known all over the world. Joseph Kinsey Howard's article, "Tom Campbell, Farmer of Two Continents," in *Harper's Magazine* of March, 1949, was translated and reprinted in six foreign editions of *Reader's Digest*. Twenty years before that article was written Campbell was asked to perform one of the most important missions of his life—a trip to Russia and consultation with the Russians on how to establish and operate large-scale mechanized farms. Campbell's work served as a basis for collective farming, believed by the Russians to be the most direct route to modern farming.

When Campbell was first asked to come to Russia in July, 1928, he declined because he was busy with harvest and "I was then very much prejudiced against anything which the Soviet Government was trying to do." *World's Work* of January, 1929, reported about the Russian invitation that the "Russian government has offered him all the land he wants on his own terms." Campbell said in another article that he was offered one million acres of the Red paradise for his own use and profit if he would consent to stay there and personally operate a farm, just to show the aspiring agriculturalists of Russia how he does it. . . . I could have cleaned up two million dollars a year if I had accepted that offer."

But the Russians were persistent, and the man who had pioneered mechanized mass-production techniques in agriculture under the free enterprise system now was called upon to pioneer mechanized agriculture for Soviet Russia. As Campbell wrote at this time, ". . . the leaders of Russia have realized that agriculture must be made prosperous, and have built all their plans of progress from that base line. . . . Because I have been identified almost all my life . . . with efforts to grow wheat more cheaply and with less waste of human energy, I . . . was brought into the confidence of men and women in all social departments. . . ."

A poor crop caused Campbell to reconsider the Russian offer, and on January 12, 1929, Mr. and Mrs. Campbell and Miss Jean Henry, a nurse, left New York on the first of Campbell's three missions to Russia. On the second day of sessions with the Russians, Campbell gave a technical talk to over 800 specialists from all over Russia. That session lasted until 2 A.M. On a later date, to the surprise of everyone, he was asked to meet Joseph Stalin and the President of the University of Moscow who acted as interpreter. At the onset of the audience, Campbell said to Mr. Stalin that he was not interested in Communism and added: "Nevertheless, I am much interested in your agricultural development, as I am an agricultural mechanical engineer, and have spent most of my life trying to develop mechanized agriculture in the United States. . . the work your Government has offered me is interesting. I will not, however, make any kind of working agreement with your Government if it cannot be done absolutely independent of my political beliefs and strictly on a business basis."

After Campbell had finished, Mr. Stalin rose from his chair, approached Campbell, took his hand and said, "Thank you for that, Mr. Campbell. Now I know I can believe you. Now I know we can respect each other and perhaps we can be friends."

It was Campbell's feeling that the Russians were completely flexible and willing to adopt mechanized agriculture of varying degrees into every local situation. According to his estimate, 14 million farming units were organized into 200,000 collective and state farms. Their problem was simple; all they had to do was buy the equipment from America and hire technicians to show them how to operate the machines. To save Russia from hunger, food had to be mass-produced.

Their original intention was to work with 75,000-acre farms, but by 1932 they were nearer to a 150,000-acre average, according to Campbell, who did his planning on a 500,000-acre model farm, or the "Giant Farm," as the Russians called it. That farm was to have 400 tractors, 278 combines, 314 drills, and 2,000 plow bottoms, which made even the machinery inventory of CFC seem small by comparison. The Russians, who had cut 80 percent of their grain with sickle or cradle and threshed with flail or corrugated log chain in 1918, were influenced by Campbell to bypass the binder and threshing machine and go directly to the combine.

On June 17, 1930, the Campbells and two of their daughters left for a tour of Russia. During this trip Campbell saw the fruits of his earlier mission. On the "Giant Farm" he observed 235 combines harvesting nearly 200,000 bushels of wheat a day. He remarked, "It's feasible, it's practical. . . . This tract had been thoroughly organized and industrialized. . . ." When he turned down a chance to farm in Russia the Soviets sent 200 farmers to his farm for a month to study thoroughly his methods.

In August, 1958, Russia's Vice Minister of Agriculture, Demetri Omelyanenko, and Mikhail Krylor, an agricultural engineer and nine other top Russians spent a day on CFC.* "They forgot their stiff protocol. . . . They marveled at the weight of a 50 plus bushel per acre yield from only 20 to 30 pounds of seed to the acre. . . ." Campbell assembled 51 combines in an attempt to harvest over 60,000 bushels in one day. At 76 he still had not lost his touch for a big show.**

Tom Campbell, all his life a Republican, who headed the national campaign of farmers for Hoover in 1932, said his work with the Russians in 1929 had cost him the nomination for Secretary of Agriculture in the Hoover cabinet.

Although he had social acquaintances with Presidents Coolidge, Hoover, and Franklin D. Roosevelt, Tom Campbell's most active period of governmental service came during the Eisenhower administration. Much of that stemmed from his acquaintance with General Eisenhower as a liaison officer between the War Department, the U.S.D.A., and the United Nation's Relief and Rehabilitation Administration.

Later recognized by the American government and invited as an advisor by the Russian government as an authority on agriculture, Tom Campbell found himself in demand by several other nations. Envoys from those nations visited his farm, and he was called

---

*Wheat, long a magic word in Russia, has not lost any importance today. Senator Hubert H. Humphrey learned from his visit to Russia in 1972 that the big wheat purchase of that year was carried out under the specific direction of the Council of Ministers. Senator Humphrey feels that the wheat purchase had to be made to avert a possible political crisis in Russia similar to the revolt of the Polish factory workers in 1970.

**Tom Campbell profited in several ways from the Russian missions. He adopted Russian Kharkov wheat, which enabled him to sell seed to Europe at a premium for several years. He also influenced Russians to buy American equipment, approving as many as 1,000 tractors in a single order. However, many modifications were made on American equipment, such as tapered bearings, dust filters, covered carburetors, and pressured lubricating systems, at Campbell's suggestions.

often to foreign lands to consult with their agriculturalists. To Mr. Campbell, this must have provided great satisfaction, for it was evident that his mission to mechanize and industrialize agriculture had proven its validity.

Thomas D. Campbell may not go down in history as Montana's most admired farmer, but his adopted state will recognize him as one of its best known citizens and one of the greatest innovators in American agriculture. Even today the farmers of America and the world have still not fully comprehended what Tom Campbell envisioned could be done in farming. Not until they do will the consumers of the world reap the full benefits of his dreams.[15]

NOTES

[1]Archibald MacLeish, "Grassland," *Fortune*, XII, No. 5 (November, 1935), 60, 65, 186; Personal Interview with Elizabeth Ann Campbell Knapp, Hardin, Montana, August 8-12, 1971. Mrs. Knapp, one of the three daughters of Tom Campbell, has been the most active in the farming operation since his death in 1966. It is through her that the contacts with some of the original employees were made for information on the early-day operation of the Campbell Farming Corporation.

[2]Malcolm C. Cutting, "A Manufacturer of Wheat," *Country Gentleman*, XCI, No. 8 (August, 1926), 18-19.

[3]Knapp Interview. Mrs. Knapp says even though her father was a registered Republican, he admired Franklin Roosevelt. He made many visits to the White House during several presidential administrations; *The Grand Forks Herald*, December 29, 1943. This article reported the burning of the second Hotel Dacotah and also gave a history of the real estate dating from June 3, 1875. In a period from 1875 to 1892 the price of that land rose from $1 to $13,385.45; Edward Angly, "Thomas Campbell: Master Farmer," *The Forum*, LXXXVI (July, 1931), 19; "Wheat Harvest Ends at Campbell's Farms," *The (Billings) Herald* (August 26, 1948); Thomas D. Campbell, *Russia: Market or Menace?* (New York: Longmans, Green & Co., 1932), p. 2; Thomas D. Campbell, "What the Farmer Really Needs," *Magazine of Business*, LIII (June and December, 1928), 724; Thomas D. Campbell letter to Leonard Sackett, File 876, NDIRS, April 11, 1956; "Montana Claims the Greatest Wheat Farmer in the World," *Current Opinion*, LXXV (November, 1923), 543; Joseph Kinsey Howard, "Tom Campbell: Farmer of Two Continents," *Harper's Magazine*, CXCVIII (March, 1949), 62; *The (Fargo) Forum* (September 29, 1957). A clipping found in NDIRS 876; *Fortune*, XII, 186.

[4]Stuart Mackenzie, "The Greatest Wheat Farmer in the World," *American Magazine*, XCVI, 37; *Fortune*, XII, 185-187; *The (Billings) Herald*, August 26, 1948; Robert H. Moulton, "200,000 Acres and Not a Single Horse," *Everybody's Magazine*, XLI (July, 1919), 47; *The Forum*, LXXXVI, 19; Personal Interview with J. R. Taylor, Memphis, Tennessee, September 11, 1971. Mr. Taylor, who was known as "Punk" to everyone on the Campbell Farms, was born in 1893 and had college training in agriculture as well as professional farm management experience before he was employed by Campbell in 1919; Wilfred L. Lingren, "Superfarmer," *The Northwestern Miller*, CCXXVI, No. 13 (June 25, 1946), 24; *Current Opinion*, LXXV, 542; Robert H. Moulton, "Is This the Biggest Farm in the World?," *The Scientific American*, CXXI (August 23, 1919), 183; *The (Fargo) Forum* (April 12, 1920); NDIRS 876.

[5]*Fortune*, XII, 187; Personal Interview of Johnny Owens and Rex Wemple, Hardin, Montana, August 10, 11, 1971. Both men started working at the farm in the

early 1920's and spent most of their working life with Campbell; Taylor Interview; *Everybody's Magazine*, XLI, 47; *Current Opinion*, LXXV, 542, 543; *Harper's*, CXCVIII, 60-62; Northwestern Miller, CCXXVI, 24; "Thomas D. Campbell, Business-Farmer," *World's Work*, XLIX, (January, 1925), 256.

⁶Taylor Interview; *Northwestern Miller*, CCXXVI, 25; *The Forum*, LXXXVI, 19; *Fortune*, XII, 187; *The (Billings) Herald*, August 26, 1948; *American Magazine*, XCVI, 39; Personal Interview with Erwin Schnad, Hardin, Montana, August 11, 1971. Mr. Schnad is a long-time resident of the Hardin area and has farmed neighbor to the Campbells since the 1930's; Personal Interview with Pat Roach, Hardin, Montana, August 12, 1971. Mr. Roach was a mechanic at CFC from April 25, 1922, until his retirement.

⁷*Fortune*, XII, 187; "The World's Greatest Wheat Farmer," *The Buffalo (N.Y.) Courier Express Pictorial* (October 12, 1947), p. 29; Personal Interview with Art R. Koebbe, Hardin, Montana, November 7, 1973; *American Magazine*, XCVI, 37, 38, 166; *The Forum*, LXXXVI, 19; Taylor Interview; Owens and Wemple Interview; Personal Interview with Henry Old Coyote, Hardin, Montana, August 11, 1971. Mr. Old Coyote is an interpreter and liason with the National Park Service in understanding Crow culture and tongue. He has a lifetime of first-hand knowledge of the CFC and Crow relationship; *Northwestern Miller*, CCXXVI, 25; "Soviet Agriculturalist Studies Montana Farming," *Great Falls Tribune Montana Pride Magazine*, August 17, 1958; *Harpers*, CXCVIII, 61; *Country Gentleman*, XCI, 18-19, 44, 61; *The (Billings) Herald*, August 26, 1948; Personal Interview with Les Curnow, Hardin, Montana, August 8-11, 1971; and several letters on file since that date. Mr. Curnow, a professional farm manager, was general manager of the Campbell Farming Corporation at the time of interview.

⁸Taylor Interview; *American Magazine*, XCVI, 37, 38, 39, 166; *Current Opinion*, LXXV, 543; 545; *Country Gentleman*, XCI, 18, 19, 44; *The Butte Miner*, July 5, 1926; *World's Work*, XLIX, 256-258; Roach Interview; *The (Billings) Herald*, August 26, 1948; *Fortune*, XII, 64, 66, 187, 188; *The World Herald Magazine*, May 9, 1948; Owens and Wemple Interview; *Harper's*, CXCVIII, 59, 60; Personal Interview with Henry Weiland, Euclid, Minnesota, March 17, 1973, son of Gregory Weiland; Thomas D. Campbell, "The American Farm Problem," *Mechanical Engineering*, L, No. 10 (October, 1928), 751-752; Thomas D. Campbell, "What the Farmer Really Needs," *Magazine of Business*, LIII, (June, 1928), 726; *The Forum*, LXXXVI, 20; Alfred D. Stedman, "Montana 'Know How' Sets Wheat Record," *The St. Paul Pioneer Press* (August 25, 1947), 1, 8; *The Chicago Sunday Tribune* (October 9, 1949); Personal Interview with Ward Whitman, Robinson, North Dakota, March 7, 1972.

⁹Taylor Interview; Owens and Wemple Interview; Curnow Interview; Roach Interview; *Fortune*, XII, 64; *The Buffalo Courier Express Pictorial* (October 12, 1947); *The Great Falls Tribune* (August 17, 1958); *Harper's*, CXCVIII, 56, 58, 59, 61; Tom Campbell letter to Leonard Sackett, April 20, 1956; in NDIRS 876; *Mechanical Engineering*, L, 752; *St. Paul Pioneer Press* (August 25, 1947); *Magazine of Business*, LIII, 725; *World's Work*, XLIX, 258.

¹⁰*Magazine of Business*, LIII (June, 1928), 725, 730 (December, 1928), 656, 657; Personal Interview with Floyd Darroll Warren, Hardin, Montana, February 21, 1972. Mr. Warren in 1972 was a very large-scale farmer on land adjacent to CFC and as a young man had worked on Campbell's farm. He knew Tom Campbell very well; *The (Billings) Herald* (August 26, 1948); Campbell letter to Leonard Sackett, April 11, 1956; NDIRS 876; Owens and Wemple Interview; Taylor Interview; Henry Old Coyote Interview; *The American Magazine*, XCIV, 166, 167; *Country Gentleman*, XCI, 18; *The Forum*, LXXXVI, 20; *World's Work*, XLIX, 257; Edward Hughes, "Rural Robots," *The Wall Street Journal* (January 19, 1949), 6; Article by Campbell (April 12, 1920), NDIRS 876; Hiram M. Drache, *The Challenge of the Prairie* (Fargo, 1971), p. 72; *Mechanical Engineering*, L, 752; "Crisis in Wheat— Can It Be Ended?, An Interview with the Nation's Largest Grower—Thomas D.

Campbell," *U. S. News & World Report*, XLVI, No. 22 (June 1, 1959), 67.
[11]*World's Work*, XLIX, 258; Owens and Wemple Interview; Roach Interview; *St. Paul Pioneer Press* (August 25, 1947); *Denver Post* (October, 1948); Henry Old Coyote Interview; *Harper's*, CXCVIII, 60, 61; Warren Interview; Young Interview; *Country Gentleman*, XCI, 44; *Mechanical Engineering*, L, 751; Curnow Interview; *The Butte Miner* (July 5, 1926).

[12]*The Northwestern Miller*, CXXVI, 25; Taylor Interview; *Country Gentleman*, XCI, 19; *World's Work*, XLIX, 258; *Fortune*, XII, 188, 189; *Handbook of Agricultural Charts 1969*, United States Department of Agriculture, Agricultural Handbook No. 373, p. 7; *U. S. News and World Report*, XLVI, 67; *Harper's*, CXCVIII, 62; *The (Billings) Herald* (August 26, 1948); *The Great Falls Tribune (August 17, 1958)*; Warren Interview; *Magazine of Business*, LIII, 727; "Farm Project in New Mexico," *Business Week*, No. 1033, (June 18, 1949), 52, 56, 57; *Billings Gazette* (July 19, 1949); *The American Magazine* XCVI, 167; *Albuquerque Journal* (July 19, 1941; July 15, 1946; March 22, 1949); *The Albuquerque Tribune* (October 12, 1949); *Lockhaven (Pennsylvania) Express* (October 6, 1941); Curnow Interview; Clippings and land maps in the Campbell Company files.

[13]*Harper's*, CXCVIII, 56; Glenn R. Winters, "Wheat and Justice," *Journal of the American Judicature Society*, XXXII, No. 2 (August, 1948); "World's Greatest Wheat Farm," *Minneapolis Sunday Tribune* (August 15, 1958); *Mechanical Engineering*, L, 745, 747, 750; *Magazine of Business*, LIII, 656, 724, 725, 728, 752; *The Denver Post* (October, 1948); Campbell, *Russia*, p. 2; Owen P. White, "Such a Relief: An Interview with Thomas D. Campbell," *Collier's*, LXXXV (March 22, 1930), 10, 11; *U. S. News and World Report*, XLVI, No. 22, 71; *The (Fargo) Forum*, (July, 1931).

[14]*Country Gentleman*, XCI, 18, 19; *Collier's* LXXXV, 10, 11, 72; *The (Fargo) Forum*, LXXXVI, (July 21, 22, 1931); Thomas D. Campbell, "The Industrial Opportunity in Agriculture," *Magazine of Business*, LIV (December, 1928), 656; *Magazine of Business*, LIII, 752; "Farmers—Relief Rebus," *Time*, XI (January 9, 1928), "Farmers: Something for Nothing," *Time*, XXVII (April 20, 1936), 18, 19; *U.S. News and World Report*, XLVI, No. 22, pp. 66-71; "Program on Crop Support Blasted by Wheat King," *Albuquerque Tribune* (1960); *Billings Gazette* (April 30, 1946; February 12, December 18, 1948; April 20, 1949; April 20, 1956); Copy of Campbell letter to Senator Milton Young, *The (Fargo) Forum* (February 28, 1954), NDIRS 876; *The (Billings) Herald* (August 26, 1948); *Wall Street Journal* (January 19, 1949), pp. 1, 5; Address to English Agricultural Leaders, *Survey Mirror* (Red Hill, England: May 9, 1941); "Bonus to Spur Wheat Sales," *Minneapolis Star Journal* (April 30, 1946); "Montana Urges Farmers to Sell Wheat and Save World," *Ipswich (Massachusettes) Chronicle* (May 9, 1946); *Washington Post* (April 6, 1946); *Hardin Herald Tribune* (May 1, 1946) "Commodities—Freedom at Work," *Time*, L, No. 20 (November 17, 1947, 91; *(Lodi, New Jersey) Messenger* (January 15, 1948); Clippings and notes in CFC files; *Hugoton (Kansas) Herald* (November 14, 1947); *Cincinnati Post* (February 7, 1948); "Taming of a Bull," *Omaha World Herald* (February 14, 1948); Letters to the Editor, *Time*, L, No. 23 (December 8, 1947), 11, 12; Taylor Interview; Personal Interview with Amelia Harris, Hardin, Montana, August 8-11, 1971. Mrs. Harris has been employed by the CFC since 1956.

[15]Henry Old Coyote Interview; *Hardin Herald Tribune* (August 15, 1946), *Harper's*, CXCVIII, 56, 58; "Really Big Harvest Show Ready for Visiting Soviets," *Great Falls Tribune* (August 9, 17, 1958), p. 4; Owens and Wemple Interview; "Apostle of Bread," *Rocky Mountain Empire Magazine Denver Post* (October 10, 1948), p. 5; Thomas D. Campbell Letter to General of the Army, George C. Marshall, May 22, 1951; Curnow Interview; *Great Falls Tribune* (December 26, 1949); Campbell, *Russia*, pp. VI-VII, 1, 15, 16, 44, 53, 74, 77, 93, 94; *World's Work*, XLIX, 259; *The (Fargo) Forum*, LXXXVI (July, 1931), 20, 21; Jonathan Mitchell, "Trade with Russia Becomes Respectable," *Outlook and Independent*, CLII, No. 11 (July

10, 1929), 407; *Colliers*, LXXXV, 63; Owen P. White, "Wheat on a Grand Scale: An Interview with Thomas D. Campbell," *Colliers*, LXXXVI (December 20, 1930), 63; "Like Schoolboys on Picnic," *Salt Lake Tribune* (August 10, 1958); *Fortune*, XII, 188; Letter to the author from Hubert H. Humphrey, Senator from Minnesota, August 4, 1973; *New York Journal American* (March 22, 1949); Knapp Interview; *Time*, X (December 5, 1927); *Time*, XI (January 9, 1928); "My Day, by Eleanor Roosevelt," *Lexington (Kentucky) Herald* (May 14, 1944); *Northwestern Miller*, CCXXVI, 29; *(Bradford, England) Yorkshire Observer* (April 30, 1941).

CHAPTER V

The Transition of the
Bonanza Farm

THE TREELESS, fertile, stoneless prairie of the Red River Valley of the North, overlapping the Province of Manitoba and the states of Minnesota and North Dakota, is one of North America's richest agricultural regions. The valley provided a natural setting for about 100 large-scale farms which existed in the last decades of the nineteenth century and early years of the twentieth.

Easterners with a variety of backgrounds, who had invested in stocks of the Northern Pacific Railroad, became owners of large blocks of Red River Valley land. Horse-drawn machinery, such as the twine binder, seeder, harrow, and riding gang plow, were used on a large scale on these huge bonanza farms. The bonanzas, some of which were as large as 100,000 acres, practiced wheat monoculture and in their day were looked upon as "factory farms."

Professional management, availability of large amounts of capital, cheap land, technological changes, an adequate supply of labor, and a good demand for grain were the underlying causes of the development of these huge farms. Agricultural technology had not advanced far enough, and labor was still too expensive to enable farming to be competitive with more industrialized businesses, but great changes were made.

The bonanza farms were never intended to be permanent, for the major objective in creating them was promotional—to farm on a large scale, to draw settlers, and to improve the price of land. Strict accounting procedures made the owners very aware of the profit-and-loss potential of large-scale farming.

However, the bonanza farms of North Dakota and Minnesota in the late nineteenth century presented one of the really dramatic elements in American agriculture prior to the age of mechanically powered farming. They were created to promote an area that had previously been considered marginal for farming, and by design they were some of the best-advertised farms in the period from 1870 through 1910. In their heyday on the northern plains only the giant wheat farms of California and the Pacific Northwest attained similar publicity.

The bonanzas brought about a definite change in agriculture, for on them the use of large-scale animal power combined with newly invented farm machinery was developed to its maximum. The presence of 1,000 men, 800 horses, and 150 binders on a single farm had its impact. But the bonanza managers soon learned that outside of volume buying and marketing they had little advantages over the family-operated farms that were large enough to be economical.*

The bonanzas, which in their day were considered to have obtained the maximum use of animal power, were nothing more than multiples of 250-acre units. For every 250 acres, five horses, a gang plow, a harrow, a seeder, a binder, and a full-season man plus additional help were needed. Bonanza management, therefore, presented little more than the willingness to accumulate acres, men, horses, and machine units to the limit of one's capital, patience, or ability. When the bonanzas were established, land was a relatively small cost factor, for most of it was purchased at prices from 16¢ to $1.50 per acre. Fortunately, all odd-numbered sections were available, making it possible to amass as much as 11,560 acres within a single township.

With their efficiency, the bonanzas were able to reduce the production time for growing a bushel of wheat from 255 minutes down to 42 minutes. But once the average farmer saw the advantages of mechanization in crop production, he was quick to copy the methods of the bonanzas and was able to gain almost as much efficiency. By the adoption of the bonanza techniques the family-operated farm once again became a serious competitor in an economic sense. To meet this competition the bonanza owner had two choices—sell land on a contract for deed to the small farmer, or subdivide his big farm and lease it to many tenants.

The bonanza operations which benefited by this transition were the Amenia and Sharon Land Company, the Dalrymples, and the Larimores, all of which leased a major portion of their land to ten-

---

*This economy of scale in the Red River Valley seemed to lie somewhere between 240 and 320 acres for a completely diversified and successful farming operation.

ants on a crop-share basis. Those bonanzas, which combined farm management with their elevators and machinery agencies, brought benefits to themselves and their tenants. This permitted them to retain the advantage of volume buying and selling. Tenant leasing also permitted the bonanza owners to overcome the great disadvantage of employing farm labor.

It was not uncommon for labor cost for one year to exceed the original cost of the land. In addition to the actual wages, there was also the cost of room and board for the men, which made the total labor bill even higher. Bonanzas needed cooks and attendants to maintain the rooms for its labor force, and this had to be added to labor costs. The third-highest operating cost on the bonanzas besides wages and board was that of animal power. For the bonanza farmer as well as for the homesteader the cost of animal power frequently exceeded the original cost of the land.*

In an attempt to minimize cost and maintain a steady labor and horse supply, the bonanzas exchanged horses and men in the winter season with the lumber camps of Minnesota and Wisconsin. The iron mines also proved to be a steady source of labor for some of the bonanzas, even though the season of the open pit mines and farming presented conflicts for men. Horses were also shipped back and forth between the forests and the prairies. Sometimes they were owned by the bonanzas and rented to the lumber interests, and sometimes the bonanzas rented them from the lumber firms.**

When lumbering declined and many of the lumberjacks became farmers, it resulted in a shortage of labor and at the same time tended to increase land values. These two factors, plus strong demands for labor in industry, finally caused the bonanzas to lose out. They could not pay industrial wages using animal-powered equipment.***

Information about labor costs for small farmers in the days before records were required for income tax purposes is virtually

---

*For a more detailed study of costs of labor and power for both the bonanzas and homesteaders in Minnesota and North Dakota, see *The Day of the Bonanza* and *The Challenge of the Prairie*.

**In the case of at least one bonanza, the Askews of Mapleton, the men and horses were maintained as a unit and worked in the woods during the winter under direct supervision of the Askews.

***Art Askegaard, whose father, David Askegaard, had a bonanza at Comstock, Minnesota, said, "Those big Minneapolis tractors even though they were clumsy made farming a lot easier and more profitable."

nonexistent. Most small farmers were not too concerned about labor cost because generally it was provided by the family.[1]*

The Dalrymple Bonanza

The largest of the bonanza farms created in the Red River Valley in the nineteenth century was that of Oliver Dalrymple. Originally formed as the Cass-Cheney-Dalrymple Bonanza in 1874, it eventually became the sole possession of Oliver Dalrymple, a St. Paul lawyer. Dalrymple had enlarged a 50-acre farm purchased in 1862 to 2,600 acres by 1873 and had become Minnesota's largest wheat producer. Low prices, grasshoppers, some inferior Odessa wheat seed, and a $100,000 loss in commodities speculation had bankrupted Dalrymple.

To recoup his fortune Dalrymple accepted the challenge to manage a bonanza farm for George Cass and Benjamin Cheney, officers of the Northern Pacific. He carefully wrote a contract that enabled him to share in the profits of the farm, and eventually he became a full-fledged partner. This was above a fixed salary paid to him by Cass and Cheney. Dalrymple insisted on adequate bookkeeping, so he knew what the farm operations were capable of, and, as he said, "Less thought and doubt, pitch in and act."

As Oliver Dalrymple became more aware of the profit potential of farming in the Red River Valley, he immersed himself in acquiring additional land "to the point that he sometimes lost all social and business contact." His efforts in this line were so determined that his brother William was constantly surprised how Oliver managed to avert economic disaster.

Dalrymple, like Tom Campbell of a later era, was totally convinced that "labor was the key in the world wheat market. . . . America's technological genius would have to overcome the vast labor resources of Asia and South America in order to survive rapidly lowering wheat prices." Even though he could produce a 16-bushel-per-acre crop of wheat for 45¢ to 50¢ a bushel, there were 5 out of 20 years between 1880 and 1899 when prices were lower. The drought of the 1880's kept yields below his needs and expectations. But Dalrymple was a genius as a manager in agriculture besides being skilled in what he called the "lax title busi-

---

*The 1,400-acre Kingman farm of Fargo had non-family labor costs amounting to 22 percent of its total cost. In the 1890's its hired-labor figure was 17 percent of the gross income, which would be prohibitive for the tractor-powered similar small-grain operations of the 1970's.

ness." Drought and low prices did not stop him from accumulating land.

Dalrymple was continuously increasing his acreage, but he knew that it could be done only as long as his farms were profiting. His management to the smallest detail had to be "systematic and orderly. . . . Watch the small economies, little leaks sink the ship," was his management program written in 1880. Using such a program he was able to crop about 40 acres per man in the 1890's, double what the average farmer in America was doing at that time. Dalrymple also knew that only by using the latest and best machinery could he successfully compete with the farmer who relied on his family for labor.

Oliver Dalrymple's sons, John and William, in a very businesslike manner improved the financial basis of the farm after Oliver's death in 1908.* In 1916 they decided to sell some of the land to capitalize on the inflated land prices caused by wartime prosperity. They sold 60 farms during the boom period at about $100 an acre average. The original farm of 2,000 acres established by Oliver Dalrymple in 1876 was sold for $151.50 an acre to a German farmer who kept pigs on the tennis court. He held the farm for about four years. When it became obvious that most of the farms that had been sold would revert back to the Dalrymple family because of nonpayment, John S. Dalrymple, Sr., decided to personally secure them because no one else seemed interested in the land. By the early 1930's land values had fallen to $17 to $25 an acre in the Cass and Traill County area, and between 1925 and 1942 John Dalrymple had regained 23,680 acres of the family's former farm.

Charles Schutt is a good example of what happened to the buyers of the Dalrymple lands during World War I. Schutt, who farmed near Manila, Iowa, until 1916, sold his 311-acre home farm for $230 an acre. He came to Cass County, North Dakota, and purchased 1,120 acres from Dalrymple for $75 an acre and 160 acres from John B. Sinner for $65 an acre. In 1918 Schutt sold a second farm of 120 acres in Iowa for $250 an acre and used the money to buy 560 acres in Berlin Township, Cass County, at $53.57 an acre. Shortly after that he purchased 320 acres in Rush River Township at $100 an acre.

In all, Schutt sold 431 acres in Iowa for $101,530 and purchased

---

*In 1906, after farming in Dakota for 30 years, he was awakened at 2 A.M. one morning to see one of his elevators burn. He remarked to his wife, "Too bad, Mary; we won't pay your bills this fall," and went back to bed to sleep.

2,160 acres in Cass County, North Dakota, for $195,200. Assuming the Iowa farms were paid for, the Schutts had 50 percent equity in their North Dakota farm, in some of the finest parts of the Red River Valley, and now had a much better base for farming. But after farming in 1919 and 1920, with the farm crash of 1921 the "roof fell in." The Schutts were in trouble at once and to that came more problems when Murray Brothers and Ward Land Company, who did some financing for Schutts, went broke.

In 1923 Schutts asked the Dalrymples to take back some of the land. The Dalrymples refused initially, but after more pleading they agreed to repossess. By that time the price of land had fallen so far that the combined down payment, principal, and interest payments for five years, plus the land itself, did not equal the initial contract price. Charles Schutt's son George remarked: "Fortunately the Dalrymples were good to us and did not hold us to the face of the contract."

Next they lost the 560 acres in Berlin Township which they had paid for in full and which they had used as security on other land. They were unable to liquidate any further because no one wanted to buy the land. The Dalrymples carried the contracts on the remaining 1,240 acres at 6 percent, and whenever Schutt was unable to make a principal or interest payment Dalrymples carried the delinquent amount at 10 percent. In 1937 Schutt was able to refinance his farm, but he still felt the strain of having too large a mortgage. In 1943 he sold 160 acres of the John B. Sinner land, which had cost him $65 in 1916 for $50 an acre. He also sold 160 acres of Dalrymple land which had cost him $75 for $50 an acre. Even though this represented a loss of $6,400 on 320 acres, it was the first time in 22 years that he had found buyers.

The Schutts retained 1,280 acres of the total purchase of 2,160 acres until 1943 when the sale of 320 acres enabled them to clear off the debts on the remaining 960 acres. George Schutt farmed that land until 1967. At that time those 960 acres were worth $100,000 more than the total 2,160 acres had cost in 1918.

Of the many who bought land from the Dalrymples, only three—the Askews, Scherweits, and Sinners—had managed to hold on to all the land they had purchased. Another three—Bill Austin, Bill Meyer, and the Schutts—held onto at least part of their land. The remaining 54 out of the 60 buyers let all of their land revert back to the Dalrymples. The grandest of all bonanza farms, which in the first decade of the twentieth century contained over 100,000 acres under Oliver Dalrymple's management, had decreased in size to about 30,000 acres in 1916. At that point the Dalrymple family, even though their operating cost was only $9 an acre, decided that the profits from the farm did not equal interest on their investment, so they decided to sell.*

By 1919 the best known of all bonanza farms had been almost completely dispersed. But before the Dalrymples could leave North Dakota, the land started to return to them. When the worst was over the family once again had 30,000 acres on its hands. To reobtain clear title to the land, John S. Dalrymple frequently had to pay several years' back taxes. In addition, he had to spend money on buildings which had fallen into a state of disrepair and also had to finance any renters who were willing to operate the farms.

John S. Dalrymple, II, who followed his father's and grandfather's footsteps upon graduation from college in 1936, said:

Father seemed like a pessimist but he had so much faith in the soil. We pulled all the brakes on and just didn't spend. In 1932 and 1933 even with interest and mortgage payments from those "lucky six" who were still hanging on, we ran behind a few hundred dollars. Those six men all knew what they were doing especially Al Sinner and Joe Askew who bought feed for cattle in the midst of the drought. In 1932 we got 27¢ for some wheat. In 1935 we had bad rust and lost about $1,000 on the total operation. In 1936 we had a profit of $9,000 from over 30,000 acres. The land was never mortgaged but the family had to liquidate some of its securities to pay for sending me to Yale 1932-1936. Dad [John S. Dalrymple, Sr.] would not have had to do so except that he forgave so many on their payments. William Dalrymple had a grain brokerage firm which helped on grain prices for the farm. But that went broke and took some land with it.

The liquidation and then repossession of the Dalrymple bonanza and the personal case of Charles Schutt, one of the 60 buyers of land, clearly explains why that large farm is still in existence. John S. Dalrymple, Sr., unlike his father, was a very cautious businessman. He took over the three sections in the early 1900's, including the original Dalrymple farm east of Casselton, North Dakota, and leased the rest of their lands to 35 renters.

The Dalrymples maintained close control of their renters. The major purpose of the three-section base farm was to do experimentation and seed raising for the benefit of the tenants who were

---

*The $9-per-acre operating cost that the Dalrymples had on their Red River Valley farms using horses in 1915-16 contrasts well with Tom Campbell's $8-per-acre operating cost in the 1930's. Campbell included the cost of tilling summer-fallow land against his cropped acres. This meant he was working land for less than one half the cost of Dalrymple.

operating the land. This policy was established by John S. Dalrymple, Sr., and was continued by his son and grandson. Eight different crops were raised on the home unit in 1973. In 1959 elevators were rebuilt on the farm to provide direct marketing of the crops raised on the operated and leased farms.

Initially there were 48 renters, but by the 1960's the number was reduced to 35, and in 1973 there were 32. Each time a renter left his unit the remaining renters were permitted to increase the size of their lease. Of the 32 renters, only one did not own land of his own. The Dalrymple renters on the average have about 1,000 acres of their own land. The largest of their renters has about 5,000 acres of his own land.

John S. Dalrymple III, the fourth generation of the family to operate the farm since 1876, assumed management of the farm after the death of his father in November, 1971. When 25-year-old Dalrymple in his second year of managing the farm was asked how it felt to be the manager of the largest bonanza still in existence, he replied:

My whole life I lived with the knowledge that I could be happy at farming. On the other hand I sort of resented the burden of knowing that I could some day be responsible for this big operation. . . . The feeling I had when I graduated from college was that I did not want the farm to cut off any other opportunities that might arise in my life. . . . [After trying other jobs that had appealed to me I decided] that they did not have the depth I was looking for. . . .

My father had always been a shining example of the fact that he never doubted that he wanted to farm even though he had many attractive business opportunities. I tried working with Dad but he was very strong so as long as he farmed I could not be there. [He knew I wanted to farm and] I came to the farm to take over four days after Dad died.

John Dalrymple III said with all emphasis:

. . . [I want] above all else to be independent. There are so few people who I think are really independent. I think every farmer is president of his own business. Next I want to be in a career where things are happening . . . in farming things are happening. In this business we are really making changes—from the seed to the consumer. Everything is involved in farming—biology, chemistry, management, finance, physical layout, people—they are all important. . . .

I find it intriguing to walk into a going operation. It is easy to get emotionally attached to the farm's history. My dad was a little this way and could not destroy a set of vacated farm buildings. . . . Dad talked to me about this. . . . I have a lot to live up to and I am anticipating a fabulous challenge. . . . Our records indicate that each new year [production-wise] is probably going to be another record year, the trend is so firm that we can almost project this.

Operations on the Dalrymple farm today are intensified to meet the rising value of land and to provide a sounder base for full-time employment of a quality labor force.

The oldest and the largest of the Red River bonanzas has functioned as a single bonanza longer than any of the other major holdings. The Dalrymple farm represents one of the most solidly financed and soundly operated farms in American agriculture.[2]

The Amenia and Sharon Land Company

The Amenia and Sharon Land Company was incorporated by 40 individuals from eastern United States in 1875. Under the leadership of E. W. Chaffee it grew to become probably the most thoroughly integrated of all of the bonanza farms. At one time or other it had 33 subsidiaries which ranged from a bee-keeping and honey-producing company, to a ranch, to machinery firms, and real estate in two communities. The company started with a $92,000 original investment in 1876 and grew to gross assets of $3.4 million in 1922 with a surplus account of $2.4 million, but the combined net assets of the parent company and its associates easily exceeded $7.1 million.

In its 42 years of operation it showed profits in 30 years and losses in 12 years. Four of those years with losses were the final years, when there was a deliberate attempt to "milk" the company as part of its liquidation policy. Another three loss years were the initial three years of the company and are accountable to the normal starting expenses of any business. The three years combined had total losses of only $24,040, of which $18,708 came in the first year. The greatest loss year in the company's history was $74,471 in 1888, when drought and early frost prevented the grain crop from maturing. In three of the other loss years sizeable land purchases and other capital investments were made that were charged against the current year's operations, resulting in a loss to appear. For the rest, only three years, from 1877 to 1923, apparently had unpreventable losses.*

At all periods of its history sizeable dividends and substantial salaries were paid to family members. But it is clear that the good margin of profit of the Amenia and Sharon Land Company was caused by its being a well integrated and diversified agricultural

---

*The largest net profit shown by the Amenia and Sharon Land Company was $236,008.68 in 1917. The profits from that year nearly equalled all of the combined losses in the 42 years of operation, including the deliberate losses of the last four years.

conglomerate. It benefited from the many savings that normally accrue to an integrated operation. The stability was there and the stockholders came to expect a steady high standard of living. As H. L. Chaffee, grandson of the first manager, said, "It was our property so we could use it the way we wanted."

For reasons of efficiency in 1893 the Amenia and Sharon Land Company went into a tenant lease operation. It retained very tight control of its renters. In a sense they were almost like a company-run operation except the tenant furnished all the labor. In addition, the Amenia and Sharon Land Company funneled its buying and selling for the farms through company-owned elevators, stores, implement shops, service stations, etc. The 42,000 acres of highly productive land operated by tenants provided a substantial business base for these subsidiaries. The members of the stockholding families, chiefly Chaffees and Reeds, provided managerial leadership for the farms and operated all of the subsidiaries.*

Like the other bonanza families, the Chaffees and Reeds did extensive traveling and lived at a high level in contrast to the average farmer's family. Being served by Twin City dressmakers as well as professional clothes buyers was an accepted practice with the ladies of Amenia and Sharon Land Company. Honeymoons in Europe, traveling on the maiden voyage of the *Titanic*, and a week at the grand opening of the Blackstone Hotel in Chicago, together with winter residence in the Twin Cities or the South were all part of the life of the Chaffees. Living this high in the social parade obviously took considerable capital.

The family of Walter R. Reed, last president of Amenia and Sharon Land Company, drew $22,370 in salary from the farm in 1919 and spent $20,244.66 on personal family living. Such sums, even for the prosperous year of 1919, indicate the grand style to which the second- and third-generation bonanza families had become accustomed. Walter Reed, whose net worth was $105,945 in 1902, had a conservatively listed net worth of $474,443.24 in 1919. His family, therefore, was living at the rate of about 4.5 percent of its net worth, obviously far less than the percentage needed to support the family of a 160-acre homestead.

But the spending habits of the families would eventually be

---

*In 1900 the Amenia and Sharon Land Company purchased an adding machine from the Burroughs Company, style No. 3, machine No. 33299. When the machine was delivered, a full set of tools and instructions accompanied it because Burroughs felt that 50,000 machines in the West would fill the needs and the demand and they would never have to provide a repair service in the region.

harmful to the Amenia and Sharon Land Company. Mrs. Carrie T. Chaffee, whose husband went down in the sinking of the *Titanic*, did not seem to have sound financial judgment and was overly charitable to worthy causes as well as to her children. Walter R. Reed as president was not in a position to counteract Mrs. Chaffee's wishes in order to maintain the Amenia and Sharon on sound fiscal policies. There were too many diverse family interests which, combined with discriminatory corporation taxes, caused the company to be dissolved into individual farms in 1922 and 1923.*

During the 1920's and 1930's the Chaffees continued their spending habits acquired during the profitable years prior to 1921. Tutors, stablemen, governesses, lawn keepers, housekeepers, and maids were part of their way of life in some of the most elaborate homes in the Red River Valley.** After H. F. Chaffee's death in 1912 Mrs. Carrie Chaffee tore down their nearly new and luxurious Manor House at Amenia and built a new house in its place.***

Mrs. H. F. (Carrie T.) Chaffee was extremely active in W.C.T.U., mission work to the Chinese, and her special project, aid to unwed mothers. Apparently she was well enough known so that she was able to make regular calls at the office of John D. Rockefeller in New York. Being involved in such work naturally took a great amount of time and money.

Such spending by different members of the family led to financial difficulties for the Amenia and Sharon Land Company. This was quite in contrast to the financial policies of E. W. Chaffee, who was known for the thoroughness of his financial records and his economy in money matters. Through the dissolution of the Amenia and Sharon Land Company in 1912 Mrs. Carrie T. Chaffee became uncontested controller of the lands that accrued to

---

*When the Amenia and Sharon Land Company was dissolved the mortgage on the original land was $19 an acre, which represented about 50 percent of market value but was 6 to 19 times the original cost of the land in 1876. It is apparent from that how rising land values enabled the Amenia and Sharon to finance more land and its string of subsidiary organizations, besides high living.

**One of the few residences to be more elaborate than those of the Chaffees or Reeds is that of Wallace Grosvenor of Casselton, North Dakota. This 53 x 58 foot, three story house was built in 1909 at a cost of $54,000 and had nearly every convenience and luxury available in the area at that time. In just over a decade it was lost by the family because it had placed too much of a drain on their finances at a time when economic conditions of Dakota agriculture became depressed.

***Prior to his death H. F. Chaffee was swindled out of $25,000 in what locally became known as the "gold brick" fraud. The man who did the swindling apparently was located in California but was never prosecuted.

herself and her children. No longer was there an H. F. Chaffee or Walter Reed to manage the farms in order to achieve the best income and at the same time hold down needless expenses.

When Carrie T. Chaffee died in 1932 holders of large mortgages on the Chaffee farm agreed not to foreclose on any additional lands if the Chaffee children consented to give William Guy, who had been C. T. Chaffee's manager since 1926, full management control of the land. Under Guy's management the Manor Farms, which was the Chaffee estate, was the only farm corporation in North Dakota to show a consistent profit during the 1930's. The mortgage companies apparently realized that if anyone could make the farms pay, Mr. Guy could, and they certainly had no desire to repossess any land. Under his management the Chaffee farms fed 6,000 or more sheep each year besides a limited number of cattle and hogs. It was those enterprises which enabled Guy for years to show a profit on the estate of Carrie T. Chaffee.*

Guy eventually assumed management of most of the land of the Carrie T. Chaffee estate and the Chaffee children, but could still not restrict the unnecessary spending by them. Of the younger generation of the Chaffees, only two sons, H. L. and Lester, retained their land until late in their life, but most of the rest of the land of that once-famous bonanza farm had been sold.

Some of the largest blocks of the Amenia and Sharon Land Company are still in the hands of members of the Reed family. One fourth-generation member of that family, Jack Dahl, along with his brother-in-law, William Carlisle, has revitalized family interest in active farming by combining a sizeable ranching operation with a feedlot and grain farm. The significance of dual management and an integrated unit is readily apparent in their operation.**

The Amenia and Sharon Land Company was in some respects the most profitable of the large bonanza farms because it was highly integrated. Its dissolution in the 1920's was caused as much by family lack of interest in farming and family conflict as by any

---

*William Guy, one of the outstanding farm managers of the 1920's and 1930's, was one of the organizers of the North Dakota Wool Pool and along with President Shephard of NDAC organized the first intercollegiate Livestock Judging Contest at the International Livestock Show in Chicago.

**Dahl, who has computerized performance testing records on his beef cows dating back to 1958, is one of the most progressive farmer-ranchers in North Dakota. But because there was little left of the Amenia and Sharon's bonanza, Dahl's operation must be considered basically a new venture.

other factor. Once the land was turned over to individual heirs, the famous farm became so fractionalized that by the 1950's there were only a few significant blocks still recognizable as part of that large bonanza. The sociological significance is important.[3]

The Grandin Farms

The Grandin Brothers of Tidoute, Pennsylvania, who acquired land in the Red River Valley by means of an $88,600 note drawn on Jay Cooke, had one of the largest bonanza farms. The Grandins, who were also involved in banking, oil, and lumber, eventually converted that nearly worthless note into an initial purchase of 23,040 acres. Through an additional purchase by the use of Indian scrip and other railroad securities, the Grandin bonanza was enlarged to between 75,000 and 86,000 acres.*

The farm was worked under close and progressive management at all times. Except for the Amenia and Sharon Land Company, the Grandins probably had the best record system of the bonanza farms. Diversification was important to the Grandin operation, and probably that factor helped maintain the profitability of the farm throughout the years in contrast to those bonanzas which practiced wheat monoculture. However, after the decade of the 1880's had passed, with its drought and low prices, the Grandins started to sell some of the more remote parts of their holdings. As a reflection of their faith in the diversified livestock company, the Grandins, in selling parts of their farms, gave much better terms to farmers with livestock than to cash grain farmers.

Fingal Enger, who started farming in 1874 in Steele County, North Dakota, threshed his first grain crop in 1875 with the use of a flail and used a coffee grinder to grind wheat to flour. Later he purchased a large portion of the Grandin bonanza. He wrote a check for $93,000 to pay for several sections of land which he secured about 1900.**

In 1883, by virtue of a management contract, Oliver Dalrymple had also become an equal partner of part of the Grandin lands. In

---

*It is not possible to determine if the Grandins possessed all of their land at one given time, but at various times they owned a total of 85,924 acres. Probably the most land held and farmed by them at one time was about 75,000 acres.
**There is dispute as to when Enger made this large purchase from the Grandins. In November, 1894, the Grandins advertised 54¾ sections of land. After 1901 they again sold a fairly large acreage. Clarence Tolley in a published article indicates the purchase was made between 1901 and 1904. The family history mentions the check of $93,000 but does not give the date.

1901 a division of the Grandin farms took place, in which J. L. Grandin and Oliver Dalrymple participated. The land was to be divided equally in acres by the two men; but in order to get the privilege of the first choice on the land Charles Grandin acted as auctioneer, and the right to select the first parcel of land was won by Dalrymple on a high bid of $12,500.

By the 1890's, labor costs, increasing machinery costs, rising taxes, and a high freight rate in relation to grain prices were the reasons given by the Grandins for selling their land. By then they were handicapped by the lack of ability to get men who would do a good job of managing when members of the family were not around. After a period of depressed years the Grandins selected times of rising prices to dispose of some of their land. A final effort was made to sell the remaining parcels of the once-great bonanza in late 1919, when William Leazanby of Mt. Moriah, Missouri, purchased all but two sections. Again timing was right, for the Grandins were able to capitalize on the high land prices of the World War I era. But by 1923 the Grandins had to repossess the land purchased by Leazanby, who was financially overextended when the depressed prices of 1921 occurred.

The Grandins, like the Dalrymples, Chaffees, and Reeds, enjoyed the best of two lives. During the growing season they spent their time on the farms and throughout much of the rest of the year attended to their other social and financial interests in the East. The first owners of the Grandin bonanza in the Red River Valley liked farming and intended to make it a way of life. Because of that they developed large building sites and conducted livestock-oriented farming in contrast to the wheat monoculture of most of the other bonanzas. The attractive farmsteads with very extensive single- and multi-row tree plantings were a distinctive feature of the Grandin bonanza. Their Mayville Stock Farm, managed by R. S. Wilson, was one of the show places of diversified farming in the Red River Valley.

Mrs. Orlando Serum worked for the Grandins from 1912 through 1920. It was her job to cook and to do housework for the Holstroms, who managed Grandin Number One, and that also meant taking care of the members of the John, Charles, and Bert Grandin families when they were in residence. Generally the Grandin families came at different times, and seldom were they all there at the same time. Mrs. Serum's pay ranged from $20 to $36 a month for the years she worked there, in addition to room, board, and uniform. During regular work days she was instructed to wear a black dress with white collar and cuffs. When there was company she had to wear a white uniform dress.

The house on Number One had 42 rooms and contained seven chimneys. There were at least seven space heaters besides the cook stoves in operation at the height of activity. A 6-H.P. steam engine pumped water from the Red River to use in the house and to water the extensive hedges, the golf course, a lily pond, and a very large garden. Two men were employed just to keep the garden and lawn on Grandin Number One. On farms Number Two and Three, pipe lines were extended nearly three miles to get water to the building sites. Meat was kept in a 20- by 60-foot cooler, with walls 2 feet thick and filled with sawdust. The ice stored in the ice house next to the cooler provided Grandins with every convenience while they lived at the farm. But that was not enough to maintain the interest of the third generation, who preferred the greater opportunities in life, comfort, and wealth in the East.

When the land purchased by Leazanby was repossessed in 1923, the Grandins did not return to farming directly. Instead, they hired managers and returned to share renting. During the years from 1927 through 1938, under the management of Allen C. Sulerud, the 3,520-acre remnant of Grandin Number One never had an operating loss. The poorest year was 1932, when the area had less than 8 inches of total precipitation and prices were at their lowest. However, the Grandins saw an opportunity to capitalize on adversity. Lumber, shingles, paint, and other supplies, as well as labor, were available at very low prices, and they did extensive repairing and painting on their buildings. There was no cash return to the owners that year even though the farm did have an operating profit. The best year under Sulerud's management was 1934, when "there was a sizeable net and an excellent return on investment."

In the management rental contracts under Sulerud the renter furnished labor and all the operating equipment. The Grandins, in addition to land, provided the seed and determined the crops and the tillage methods, as well as the timing of the operations. This gave them complete control of the crops at all times. This was their protection against share renters who were completely free to farm as they wished and sometimes disposed of the crop and left the farm without reimbursing the landlord.

D. E. Viker, who started farming in 1924 by renting some of the Leazanby land, was able to succeed where Leazanby had failed.

Viker, who farmed from 1924 until 1967, "made some money every year except 1933. That year I actually lost money—but I had several outside activities like stallions, cattle buying, and custom work. I have always fed cattle and they have been profitable. . . ." The results of both Viker and Sulerud, farming through two of the most severe decades in American agriculture, give some indication that management had to be a factor too often overlooked in the past.

Viker attributed three factors to his success. Timing was one. He said, "You bet 1924 was a good year to start farming. I made good. Equipment was cheap and I think I got $1.67 for my wheat." A second factor was the willingness to do just "a little extra something that worked in with farming to help cut cost and bring in money—like custom work."* His third reason was that "J. L. Grandin must have known how to pick good land because much of the farm had up to five feet of black top soil."**

Viker overlooked a fourth factor in his success—that of being an innovator. There were many innovations. In the early 1960's a homemade four-wheel-drive tractor converted from surplus military equipment provided the largest power unit on the Viker farm. The tractor was a creation of Bruce Viker, who was farming with his father. Through the Vikers a sizeable remnant of one of the great farms of the past is being brought back into the new era of agriculture.

The history of these large bonanza farms of the past and their potential in almost every case was directly dependent upon the desires and interests of one or two individuals. The interests of each generation are different, and until there is a basic change in concept of land holding, the resale and fragmentation of economic units of production will be an ongoing process. Sociologically this may have desirable effects, but the economic costs may some day force a revaluation of the process of farming. The farm which was once viewed primarily as a private home has now become a significant productive commercial unit in the total of American agriculture.[4]

---

*Viker liked horses and had well bred animals on his farm. He worked them hard, for on that heavy land he expected each work animal to provide sufficient power for 30 acres of crop.

**Viker made reference to J. L. Grandin's scouting trip along the Elm and Goose Rivers in 1875, when he, with the help of Colonel H. S. Back, selected the original 40 sections of the Grandin Bonanza.

The Larimore's Elk Valley Farm

The Elk Valley Farm near Larimore, North Dakota, was probably the only bonanza farm to thresh any of its grain by horsepower sweep. Its first crop, harvested in 1881, was ready to be threshed before any steam engines arrived at Larimore, North Dakota. Even though the owners claimed that they averaged 15 percent return on their investment, based on the high appraised value of $25 during the first two decades, the Larimores felt differently by the early 1900's. By 1902 the Larimores complained:

> The greatest difficulty the management finds is in securing competent men in the number required on a farm of this size. Comparatively few are needed during the winter months, and thus in order to secure them when required large wages must be paid, for men who are at best transient farm hands. In this particular case the owner of the small farm has a very decided advantage, and it is this fact which had led to the breaking up of many of the [large] farms in this vicinity.

"In 1914 the pressures of manpower, taxes, and management caught up with the Larimores," according to Jameson Larimore II, "and they reorganized the land letting it out to tenants." Management costs alone were running about $1 an acre on the 6,000 acres that were farmed directly. Charles R. Wright, who worked with Larimore, said farm managers had a difficult time to compete with farm owners who work "a third as hard again as the farm manager can expect his hired men to work."

Some land was sold, and the remaining 10,000 to 15,000 acres was leased to 23 tenants. The Larimores kept tight control on what was produced by furnishing all of the seed. By operating their own elevators and handling the tenant's grain, the Elk Valley Farm was able to pull through the 1920's and 1930's without ever showing an operating loss. Their lowest net income year after depreciation was $8,000 on 8,000 acres in 1937. Before harvest it appeared that the 1935 crop would be a complete failure because of rust. But a load of grain from each farm threshed on one central location showed better results than expected, so the entire crop was threshed. In 1936 the company averaged 7.48 bushels of wheat on its 23 farms, the lowest in its history dating back to the 1880's.

The Larimore management avoided borrowing money during those two decades. Their worst years were 1930, 1934, and 1936, but by careful operations they were able to develop a full year's crop carryover as a reserve. This practice was maintained by the company even during the wartime prosperity of the 1940's. Jameson Larimore II expressed the belief that after working closely with two dozen or more tenants over a period of four decades the success of any farm is more dependent on the man than on the times. He also recalled that one of the most difficult tasks in his career was to convince his father, uncle, and renters to give up threshing and convert to the combine. The farm bought some Model P Case combines with 16-foot platforms, and before the first year was over even the staunchest supporter of shock threshing had stopped his protest against the combine. He added:

When World War II came there was no labor available and combines took over completely. The next big break came with the self propelled swather which was an instant hit with everyone. Motivating renters to adopt fertilizer was both difficult and slow. By 1954 only seven out of twenty agreed to use it. From 1954 through 1956 average wheat yields increased eight bushels per acre so we just forced all tenants to fertilize according to our recommendations in 1957. Our average wheat yield increased another four bushels that year. Except for some real weather reversals our average yield has increased about one bushel per acre per year since 1957.

Jameson Larimore III assumed management control in 1967 and used records intensively to determine the course of the farm. Mr. Larimore stated, "Our records indicate that each year because of adding more fertilizer, using a new chemical, or by better tillage we have increased our overall production. This process has been so predictable that it has almost taken the gamble out of farming. I'm not sure when those increases will stop."

In 1944 the Larimores added potatoes to their cropping program in order to diversify and stabilize their income. They credit a great deal of their success to Walter Reitan, who was the superintendent of schools in Larimore at the time. "Adding potatoes and irrigation to a dryland grain and cattle operation in one year required some real challenge to management and Walt knew how to do it," said Jameson Larimore II.

When the Larimores first started tenant operations with 23 renters, they set a minimum of 800 acres per renter. Many of their renters owned other land, but the minimum acreage rented by them from the Larimores has increased to 1,120 acres. As renters retired the Larimores have parceled out the land to remaining renters. By 1970 they were down to twelve. It has always been company policy to use pool buying for all the operating needs of the farms except for the tenants' personal machinery. The Larimores agreed that the fun in farming came from trying to keep up with new crops and technology that appeared each year.*

The Larimores of the 1970's are consolidating their units and helping their tenants to streamline the total farm operation in every way possible. The chief concern of the farm management over the years has been to overcome the marketing problem. The Larimores established their own elevator in the 1920's, which serves as both a marketing and a buying agency for the farm. In recent years they adopted irrigation and added the growing of potatoes, which led to the development of storage and marketing facilities for that crop.

By switching to the 50-50 tenant system prior to the "boom and bust" of the World War I era, the Larimores kept themselves in sound financial condition and avoided the risks of direct operating losses during the 1920's and 1930's. This gave them the distinction of being one of the few bonanza farms still in operation after one century of ownership.[5]

The Baldwin Farms and Ranches

The Baldwin family, which owned large paper mills in Wisconsin, purchased most of their 65,000 to 75,000 acres of land near Ellendale, North Dakota, during the agricultural depression of the 1880's and 1890's. In the period of the Russian thistle infestation in south central North Dakota, this land was purchased for $200 to $500 a quarter section ($1.25 to $3.25 per acre). Originally the Baldwin farms relied on cash grain farming. In 1919 they engaged J. W. McNary as manager to improve their profit picture. McNary, from Rice County, Minnesota, was one of the highest paid county agents in the country with a salary of $3,500. The Baldwins offered him $4,500 and gave him a free hand to operate the farms. He remained with them until 1933, when he resigned to become a farm manager for the Metropolitan Life Insurance Company.

---

*Elk Valley bonanza of the Larimores received considerable criticism in its early years because of the large number of Blacks who were brought up from St. Louis to farm and homestead near Larimore. The criticism came when the Blacks gave up their homestead rights and sold out to the Larimores when they decided North Dakota was too cold for them. In contrast, fame came to the farm when 300 dignitaries from about 40 foreign countries, who were attending the Chicago World's Fair in 1893, came to Larimore to observe 43 binders cut 640 acres of wheat per day. Officials from International Harvester claim it was one of the greatest promotion events they have ever held.

By thoroughly diversifying the farm with milk, meat, and seed grain production, McNary improved the profit picture even though his labor costs averaged about 45 percent of all expenses on all four ranches.* Return on investment for the Baldwins after all expenses, based on appreciated land values, was 3 percent. Because of a good rotation program made possible by diversification, McNary was able to maintain a crop index of 125 in contrast to 100 for the area. Obviously the average small-grain farm typical of that area would have netted a considerable loss at that period.

During the prosperous years of 1908 to 1910 the Baldwins had advertised 46,840 acres for sale, but were able to sell only 5,760 acres at the asked-for price of $25 an acre. During the prosperous years of World War I they sold about 35,000 more acres to over 40 individuals at prices ranging from $45 to $62 an acre. (Most of it went for the $62 figure.) Later, when these parcels were repossessed by the Baldwins, they encountered real losses. By 1935 they had most of it back from teachers, doctors, lawyers, and other absentee speculators who had purchased the land during the years of the agricultural boom. By contrast the hard-working family farmers held on a little longer, but eventually many of them lost their land too.

A letter dated April, 1926, from George Baldwin to his nephew, George Baldwin, a college student, explains the predicament of the large-scale farm caused by the cost of labor:

I fear you have the wrong idea when you say "all other farmers seem to live and prosper." They may live, but many farmers are only eking out a bare existence. The farmer, since the end of the war [World War I] has suffered more than any other industry because of the falling prices of commodities they raise, while wages have dropped very little; ... The farmer in a dairy country ... may make a pretty good living ... but farmers in the northwest who depend upon raising small grains and corn have

---

*Baldwin real estate taxes averaged 12 percent of all expenses for the several ranch units in 1923. McNary had divided the holdings into 6,000-acre farms to get the most efficiency. Taxes on the Adams farm in Iowa in 1899 were 2 percent of gross income. The Amenia and Sharon Land Company experienced their top labor bill in 1919, which amounted to 25 percent of all expenses and was one of the reasons why the Chaffees and the Reeds decided to subdivide. In 1899, well before the days of World War I labor increases, the Adams Ranch at Odebolt, Iowa, produced 215,000 bushels of corn and 20,000 bushels of wheat for a combined value of $74,500. The labor bill on that crop was $14,921.98 or 20 percent of the gross income, far in excess of what labor costs are to the industrialized corn producer. The Adams crew averaged about 75 men during the crop season. By contrast the labor bill for all American farmers in 1971 averaged 8 percent of all farm expenses.

not been doing so well, they may have been going backward and getting deeper into the mire. . . .

Practically the only farmers who make money are those who live on their farms and do their work because they have enough help in their own families, and who are dairying and raising hogs. The farmer who has been depending upon grain has been losing even though he did his own work.

One of the chief reasons that we are losing money on our farms is we must depend on hired help to do our work, and you know people are not as interested in making money for others as for themselves. . . . Farming can never be put on a manufacturing basis. . . .

When one hires labor on a farm the owner has no way of checking up to see that they do a full days work. It is very easy for them to shirk their work, lay down on a hay or straw stack and go to sleep. A man working for Henry Ford . . . has hardly time to turn around. . . .

Now, if it is difficult for a farmer who lives on his farm and works it himself to make money, you can see how much harder it is where it is necessary to hire help. These are some of the reasons why our farms have lost money.

We are endeavoring to switch over our farming operations to a rental basis on a 50-50 scale, sharing one half of the profits with the renter. We hope that this will work out better and that we will lose less money. . . . A farmer is never sure of the amount of the crop, he is never sure of the price it will bring, and he is not sure his help will be honest and give him a good days work.

George F. Baldwin was a good businessman and understood the limitations of animal-powered pre-industrialized farming. His letter emphasizing the labor problem explains why bonanza farms could not compete with family-operated farms. But probably the Baldwins were no worse off than the farmers who bought from them. In December, 1921, they sent letters to 49 individuals who had not paid on their principal, interest, or taxes and had their farms for sale. Many replies from the purchasers asked for extension on these past-due obligations.

The 1920's and '30's was a period of struggle for all farmers, and even with the large acreages at their disposal the remnants of the bonanza farms had a difficult time showing a profit. If the large farmer could not make a profit, the chance of survival was even less for the small family farmer.[6]

The Dwight Farm

The Dwight Farm and Land Company was started in the 1870's by Jeremiah W. Dwight, a New York Congressman.* It was man-

---

*Jeremiah W. Dwight, New York Congressman from 1877 to 1883, apparently started the farm, but his son, John W. Dwight, New York Congressman from 1903-1913, was also involved.

aged by John Miller, who later became the first governor of North Dakota. The farm grew to 27,000 acres in Richland County and 32,000 acres in Steele County, North Dakota. After 1900 the land was gradually sold by the stockholders, who preferred to use money for other purposes.

John Miller, who was made president of the corporation in 1896, retained his stock in the land and by 1909 was owner of the remaining 11,520 acres, of which 6,720 were in Richland County. Those 6,720 acres held by Mrs. Anne Dwight Tyler and Mrs. John Miller were managed by Reuel Wije. Wije served as the last manager of the Dwight Farm from 1931 to 1935. At that time half of the farm was operated as a unit, and the other 3,500 acres were leased to nine tenants on a share basis. Reuel Wije said the Dwight Farms, although not exceedingly profitable, during his management never had a loss from 1931 through 1935, five of the most difficult years in American agriculture. Monthly operating statements based on complete cost-accounting procedures were submitted to the owners. Wije attributes the fact that he was able to maintain a fully hired labor force and still show a profit to five factors: (1) There was no indebtedness; (2) income from sheep, hogs, and milk cows improved the value of farm-grown crops*; (3) purchasing new labor-saving equipment as soon as it came on the market helped continually to cut cost; (4) the production of certified seeds, especially sweet clover and alfalfa, gave a higher cash return than grain; and (5) the farm had the ability to hold cash grain up to the annual traditional market highs.

Holding grain, which was probably the most important factor in avoiding operating losses, was possible because the owners were not dependent upon the farm for immediate income. The elevator, which had been a major factor in the Dwight Farm's success, could hold most of one year's production. Wije recalled that in 1932 wheat was 20¢ a bushel in the fall and the following spring it was "about a dollar." Good seed oats which brought 10¢ to 12¢ a bushel at the same time rose to 55¢ in the spring; barley did likewise.

Admittedly taxes went unpaid in the fall because there was no money, but in the spring taxes plus penalty were paid with ease after a five-fold increase in prices. Wije acknowledged that the owners did not like seeing their names on the tax default list for the first time, but when they later saw profits occur because of it

---

*Whole milk was sold to consumers at Wahpeton.

they no longer disputed the wisdom of that practice. To make withholding of the grain less of a strain on the budget, Wije avoided drawing salary until the crop was sold. He felt impelled to do so to convince the owners that he had faith in his own judgment.

When Anne Dwight Tyler died no stockholders were interested in acquiring the property. Wije purchased much of the land, and the rest was sold at auction in the spring of 1936. With that sale the once-respected Dwight Farm and Land Company went out of business.[7]

The Schermerhorn Farms

The success of a farm relying on the interest of a single person is well portrayed in the history of Schermerhorn Farms of Mahnomen, Minnesota. The father of James B. Schermerhorn, founder of the farms, had been a missionary among the Indians in Oklahoma. Because of his interest and loyalty to them, the Indians had given him a gift of 160 acres of land. Years later oil was discovered on that land.

James B. Schermerhorn, the missionary's son, was a sales manager in the agricultural department of Swift and Company. He was also very active in the oil business after oil was found on family properties. After James B. Schermerhorn had amassed a sizeable fortune he became interested in agriculture and acquired a large farm because of his strong attachment to land. Schermerhorn was also motivated by the government's plea for greater food production during World War I.* Schermerhorn started buying land at Mahnomen in 1917 and was intensely involved with its development until his death in 1929, just "when he was really beginning to enjoy the farm."

Schermerhorn looked for a manager as soon as he had decided to establish his big farm. In 1917 he hired Arthur J. Robinson, who in his youth had broken native sod at Morris, Minnesota. Robinson was 33 years old at this time and besides managing a sizeable farm had also worked as a building contractor. Schermerhorn had learned about Robinson through a banker at Chokio and was very interested in him because of his combined background in farming and in building construction, for it was Schermerhorn's wish to erect very elaborate buildings on his farms.

---

*Arthur J. Robinson, his manager, mentioned that James B. Schermerhorn was not nearly so interested in making money as he was in having rural property so his children could enjoy country life.

A. J. Robinson was hired as manager at an annual salary of $5,000 plus living facilities. His immediate task was to buy land on the White Earth Indian Reservation and to plat it into five separate units. Land east of Mahnomen, where the farms were located, was a combination of rolling hills, prairies, and swamp crisscrossed by Indian trails; it was also heavily covered with brush and trees. Of the land which eventually was secured 6,700 acres was all prairie and 17,000 acres alternate brush, swamp, and prairie.

Robinson's job was that of a later-day pioneer, for he had to clear brush from the fields, build fences and roads, and erect buildings. But he went a step further than most pioneers by erecting a private telephone line, which was connected to the Mahnomen Telephone Company in 1918. As soon as each farm was established a Delco light plant was installed in each one. The biggest plant was a 110-watt unit at the Ranch at a cost of $2,000. Each night 5 gallons of gasoline were allocated to operate the lights. When the gasoline was used up it was time for the men to retire. These Delco plants remained in use until R.E.A. was installed in the early 1940's. Schermerhorn merely gave general directions about what he wanted done and let Robinson have a free hand to implement the action.

By experience he learned that the best way to clear the woods was to hire entire families of Indians. The men cut down the trees, while the women and children cleared the brush. At times over 100 Indians were used in a single location and apparently were very efficient. It took an Indian family about 10 days to clear 5 acres of land. The pay varied from $15 to $25 an acre. Pay was not given until a designated plot of land was cleared. This proved to be the greatest incentive to get it cleared in the least possible time.* After the fields were cleared of brush, trees, and roots, two Caterpillar track-type tractors were used to pull a 24-inch breaking plow 8 inches deep in the new sod.

At the same time that the land was being cleared, 40 miles of telephone line and roads were constructed. Over 100 miles of fence were built and elaborate farmsteads constructed. The first buildings were started on the St. Pierre farm in August, 1917. From then until 1922 a steady construction program was in progress. There was an office in Mahnomen which during the height of

---

*Robinson doubted that the cost of clearing land in the early 1970's, even with mechanical power, could have been any cheaper.

the farm's activity had rooms for three bookkeepers and one room for the manager. Later it became the R.E.A. office building. Stockyards, a bulk oil plant, and a potato warehouse with 100-carload capacity were also erected in Mahnomen, as was a home for the manager. The climax of the building program came with the completion of a spacious private house for the Schermerhorn family. This building, complete with a three-car garage and living quarters for servants, was never used by the family more than two months a year. Some idea of the size of this home can be gathered from the fact that in 1971 it was used as a rest home for 28 men.*

At the height of the construction program as many as 250 carpenters were employed on buildings, fencing, and tiling. While this development was still in progress, J. B. Schermerhorn, motivated by an intense pride to build a showplace in northwestern Minnesota, hired Harry Franklin Baker, well known Minneapolis nurseryman, to landscape the five farm sites. Baker was given complete freedom to design and landscape the farmsteads. Besides his having five carloads of shrubbery shipped in, large quantities of paint were used. The farm sites were exceedingly attractive and practical, and in spite of the cost involved Schermerhorn was happy because he loved the farms and he wanted them to be appealing to his sons.

Schermerhorn's efforts were not entirely appreciated by the people of the Mahnomen area, and the Schermerhorns themselves remained quite aloof from the local people. But the Robinsons not only lived in Mahnomen but made a determined effort to become associated in every respect with the community. Robinson, who was active in the Northwest Farm Managers Association, once said that two group visitations by members of that organization plus personal visits by some of those farmers were most appreciated because "the people of Mahnomen did not take to us too well."

Eventually the local sentiments changed when it became obvious that Schermerhorn Farms was the community's largest employer. During the winters at least 40 people were kept busy in the livestock enterprises, and in the peak farming season as many as 100 were hired. The high rate of winter employment resulted from Schermerhorn's great interest in breeding and showing of registered livestock. Two herds of purebred Herefords and two

---

*Included in the construction program was a very elaborate family summer cottage on White Earth Lake, which was located only a few miles from the farms. This was the major attraction for the Schermerhorn family after the death of James B. Schermerhorn in 1929.

herds of purbred Angus cattle, besides registered sheep and hogs, were maintained. Livestock grooming for major shows, plus sales to 4-H club buyers, required a great deal of labor.

In the peak labor seasons Minneapolis employment agencies could be relied on to send as many as a dozen men up on a night train in response to a phone call for help. Many of the lumber workers in Mahnomen and Clearwater Counties worked for the Schermerhorns. These men were under the supervision of the managers of each of the individual farm units. Employees were paid from the central office every two weeks. A. J. Robinson said that it was his impression that over the years about one half of the floating labor force of Mahnomen County worked at the farm.

The economic impact of Schermerhorn Farms on the region was significant. About $2.5 million was made in capital developments on Mahnomen properties and the 24,000 acres of land. Much of the land was purchased from the Indians in 80-acre plots at $10 per acre. Some developed land near Mahnomen cost as much as $75 an acre. Overall, the 24,000 acres averaged about $30 an acre. The other $1.8 million was used to clear the land, build roads, telephone lines, and farmsteads.

In addition to the capital investment figure, Schermerhorn Farms spent sizeable sums locally in its annual operations. Figures were unavailable for the period from 1917 through 1930, but from 1931 to the selling date of the farms, operating expenses exceeded $1.3 million. Of that figure more than 40 percent was labor. In the 30-year history of the farm over $1 million was paid to local labor.

J. B. Schermerhorn did not understand farming but was greatly influenced by his uncle, A. C. Baker of Thief River Falls. Baker was a land speculator and probably caused Schermerhorn to be more optimistic than he should have been about farming. "But when the land was being purchased in 1917 and 1918 it looked like a good investment for wheat was $2.00, corn $1.00, oats $.75, flax $4.50 a bushel and hogs were $20 a hundred." Many other people were making the same decision. Schermerhorn's purchase of "idle practically worthless land . . . [and transforming it] into one of the most productive tracts of land to be found anywhere in Central North America" was a definite contribution to that rather marginal agricultural area. Newspaper accounts occasionally may have been a little too glowing, but the farm under A. J. Robinson's management had 500 acres of potatoes, 4,000 acres in small-grain crops, 2,600 acres of hay (mostly alfalfa), and several hundred acres of corn each year, besides 14,000 acres of developed and fenced grazing land. Alfalfa and potatoes proved to be two of the most consistently reliable and profitable crops.

The farm had as many as 3,000 hogs and 1,500 cattle at one time during the 1920's. In 1924, 28 carloads of cattle, 24 carloads of hogs, and 47 carloads of potatoes were shipped from the farm. In 1929, after the death of J. B. Schermerhorn, less effort was placed on purebred livestock. Robinson reduced cattle and hog numbers and expanded the sheep operation. In cooperation with the Agricultural Credit Corporation of Minneapolis, 4,000 ewes were brought to the farm from Montana in either 1930 or 1931. Besides those ewes about 8,000 lambs were fattened in what proved to be one of the most profitable years for the Schermerhorns. From that date until liquidation, sheep continued to be a major enterprise. Because there was a lack of experienced sheep men in the Mahnomen area sheepherders with their cook cars and sheep dogs were imported from Montana. These men knew the sheep and how to manage them, and this proved to be a decisive factor from the profit standpoint. When asked in his eighty-sixth year if Schermerhorn Farms was economically feasible A. J. Robinson replied that it had survived during the 1920's and 1930's with little contributed capital, once they were established.

The family lost interest in the farms and attempted to sell them. The asking price to a firm bidder on April 13, 1931, was $400,252 cash. This represented a $2 million loss on initial investment. Unable to get a cash offer at that time, the family retained the farms until the 1940's. In 1940 A. J. Robinson purchased what was called the Terminal (sometimes Brown) Farm and farmed for himself but continued to manage for the Schermerhorns. Robinson felt it was most unfortunate that the family decided to sell when they did, for the farms had made money during the 1930's and into the 1940's. For the 20 years from 1931 to 1950 covering 76 farm unit years, there were 27 unit years of loss and 49 unit years of profit. It is possible that Schermerhorn Farms was never really profitable, but records show a net profit of about one half million dollars. This was after over $565,000 was paid out in labor and $55,000 in depreciation on equipment was applied. Considering that the average small farm used unpaid family labor and did not figure depreciation in its profit-and-loss tabulation, Schermerhorn Farms had an astonishing record.

Ole A. Olson, an agricultural agent for a large insurance company, supervised 160 farms during the 1930's and 1940's. In 1934 not a single binder was operated on any of those 160 farms.* In 1936 Olson's company paid out $30,000 more in real estate taxes than it took in from those same farms. The area of Schermerhorn Farms was part of the district covered by Mr. Olson. From such operating results it becomes clear that had J. B. Schermerhorn not erected such lavish buildings on his farms and had he avoided becoming so heavily involved in the purebred business, his dream farm could have been economically sound. Whether or not greater profits would have made the family more interested in retaining ownership in the showplace farm of northwestern Minnesota will never be known. Obviously, none of them loved the land as did J. B. Schermerhorn, and that proved to be the decisive factor.[8]

The Adams Farm of North
Dakota and Iowa

Two of the largest farms of the past in midwest agriculture belonged to the Adams family. The first of these farms was the 9,600-acre Adams or Fairview Farm near Mooreton, North Dakota, which, except for a brief period, belonged to the Adams family from July 1, 1884 to 1929. The second farm owned by them was the Fairview Farm or the Adams Ranch at Odebolt, Iowa. This 6,500-acre farm became Adams property in 1896 and was retained by the family until 1962.

John Quincy Adams, founder of the Dakota farm and a distant relative of the former president, was a man of considerable wealth and was one of the first members of the Chicago Board of Trade. Members of the family also owned the Atlantic Hotel in Chicago, had property in Florida and California, and had enough stock in International Harvester Company to merit a position for W. P. Adams on the board of directors.** He also was a director of the Illinois Trust Company of Chicago.

---

*Ole A. Olson recalled that in 1934 one farmer near Landa, North Dakota, not associated with his company, hauled in a filled 125-bushel wagon box that weighed out 37 bushels of wheat. Test weight was 20 pounds. Olson bet Andrew Sharp, the manager of the Peavy Elevator at Glenfield, North Dakota, that there would not be a single load of wheat delivered to Sharp's elevator from the 1934 harvest. Olson won his bet.

**William Sebens reports W. P. Adams purchased International Harvester stock at $30 a share in the mid-1920's and sold it in 1929 before the crash at $189 a share.

On July 1, 1884, John Q. Adams purchased section 23-132-49 in Richland County, North Dakota, for $6,500. In a short period after that he added more land until the farm included 9,600 acres. Most of this land had been part of the Northern Pacific land grant. On December 17, 1884, John Q. Adams' son, William P. Adams, married Nettie Moore of Hopkinton, Massachusetts, and the young couple moved to North Dakota to develop the 9,600-acre farm. They remained in Dakota until 1897. On February 4, 1888, William P. Adams acquired title to 4,649 acres of that land from his father for $1,000.

The Dakota Fairview Farm became quite a dramatic operation under W. P. Adams. It had as many as 35 binders in a single field under one binder boss. Sheep were a major part of the farm's business. To house them three sheds 76 feet by 240 feet were built to hold up to 10,000 head at one time. The Northern Pacific built a 3-mile spur track to the 60,000-bushel elevator which was called Fairview Junction. From July 1 to November 30, 1885, 61,220 bushels of grain were shipped from this elevator, indicating a rapid start for the farm.

In 1889 Thomas D. Parsons, a traveling salesman for the Walter A. Woods Harvester Company, was employed as a foreman by W. P. Adams. By 1891 Parsons was given management responsibilities in preparation for the time when the Adams family would leave the farm. Apparently they did not like to live in North Dakota. Under Parsons' management the farm employed about 50 men in the spring and fall and 125 during harvest and threshing. The fields were fenced with woven wire so sheep could graze without being herded. Sheep were Adams' method of controlling weeds which became serious with continuous grain farming.

At one point during the 1890's, three years' crops were accumulated because Adams refused to sell at the current market prices. Over 150,000 bushels of wheat were held in their elevator on the Northern Pacific siding, in sheds and bins and in piles covered with hay and straw. When the market rose W. P. Adams put 100 men to shovel and haul wheat in order to cash in on the higher price.

The sheep barns were not the only large buildings on the Fairview Farm; it also had large horse barns. The largest of its horse barns was 68 feet by 220 feet and held 250 mules or horses. This barn had four tracks in the haymow for loading in the hay and had two driveways running through the barn for cleaning behind the

four rows of horses and mules.* On August 9, 1900, a $20,000 fire
started by lightning consumed the building, harnesses, hay, and
122 head of horses and mules. The building and contents were
insured by eight companies. Three men sleeping in the barn
"barely escaped as . . . the hay burst into flames in a dozen
places." To continue the harvest after the fire several carloads of
horses were shipped from the Adams Farm at Odebolt, Iowa.

After William P. Adams left North Dakota in 1897, Parsons managed the farm completely until John Adams, a son of W. P. Adams,
came to Mooreton in about 1908. John Adams remained in Dakota
until about World War I, when the farm was sold to John L.
Mathews and A. L. Bonjer.** The undeveloped land, which had
cost less than $10 an acre in the 1880's, sold for $80 an acre during
the inflationary period of World War I. Bonjer and Mathews paid
$20 an acre down, subdivided the farms into 160 and 320 acres,
sank wells, erected buildings, and sold on credit at very inflated
prices. Because the Adamses held first mortgage on this land by
virtue of their contract for deed sale, Bonjer and Mathews had to
be satisfied with second mortgages. In 1925 Mathews reported
that he had between "$250,000 and $300,000 in second mortgages
that were not worth the price of a breakfast." Shortly after, the
Adamses had the deeds in their hands on 6,900 of the 9,600 acres
they had sold a decade earlier. In January, 1926, Edward Sebens
was hired to manage the repossessed land at a salary of $6,000 a
year. Sebens employed his brother, William P. Sebens, who later
became a prominent agriculturalist in North Dakota, to help with
the management. When the Adamses started farming again in
North Dakota in 1925, they used tractor power exclusively. The
once-famous strings of 100 to 200 mules in a single field had become obsolete. Tractors, plows, drills, harrows, and binders were
shipped in by the carload from the International Harvester plants
at wholesale prices.

Because the land had not been well farmed by the former owners and had lost money, the Sebens brothers decided to seed 3,000

---

*There are conflicting reports on the size of the barn. One source gave 84 feet by
176 feet but stated the same number of animals. In either case the building was in
excess of 14,500 square feet. Quite likely the one set of figures applies to the original barn and the other set to the replacement barn constructed after the 1900 fire.
There were several fires on the Adams farms. In 1905 a cook house and dining
house were lost at Odebolt. Again in 1919 a $100,000 fire consumed a mule barn,
cattle barn, elevator, water tank, and blacksmith shop.

**There are conflicting reports about the time of the sale, but it occurred be-
tween 1913 and 1916.

acres to sweet clover to put it back into better condition. The sweet clover seed grossed over $33 an acre, which was considerably more than any grain crop would have yielded. Now they had hay for animals and cleaner land for the following year's crop. The farm yielded a profit in its first years of the new operation, even in the depressed 1920's. The Sebens brothers managed the farm for four years. Before the stock market crash of 1929, the Adamses sold the land again. William Kube, one-time manager of the Union Stockyards at West Fargo, purchased the largest single block of 2,240 acres.* After the land was repossessed in the 1920's the Adamses never completely re-established themselves. It appears that none of the family came here to live after the land was first sold during World War I because they apparently preferred to be around their property in Chicago and Iowa. Other than Edward and William Sebens, who managed the land locally, all other administrative personnel were stationed at Odebolt, Iowa. All records and financing was handled by commuters from the Odebolt farm office.

In the 1860's western Iowa had been opened to settlers, and late in the decade the railroads pushed hard to sell their land-grant properties. Two of the earliest land purchasers in Sac County were H. C. Wheeler and Charles Willard Cook. Each of the two transactions amounted to over 7,000 acres, and the *Sac Sun* commented, "This makes two 7,000 acre farms in Sac County. Better this size than none at all!"

H. C. Wheeler, a San Francisco realtor and a Chicago grain elevator owner, paid less than $18,000 in 1871 for his 7,000 acres (about $2.50 per acre), for which title was posted October 2, 1875. In 1887 the Wheeler farm at Odebolt produced 60,000 bushels of corn, 12,500 bushels of timothy seed, and large quantities of oats, flax, and millet, besides having 700 head of stock on feed. Wheeler was acknowledged as "the greatest farmer of all Northwest Iowa." He was also an innovator. In 1882 he experimented with steam-tractor plowing and with stationary engines using cables to pull the plow. He was an early user of milking machines, which were less than a decade old; and he pioneered in growing popcorn, which was important in making Odebolt the popcorn capital of the world.

---

*The Kubes came to own the land which had the farm home of the Adams family on it. Just across the road was the old Adams main farmstead owned by Lloyd Fleischauer.

In the 1890's H. C. Wheeler's good fortunes faded, and between 1894 and 1896 he sold his famous Odebolt farm. On September 3, 1898, W. P. Adams paid a reported $185,000 for about 6,500 acres of land.*

William Phipps Adams was a great promoter for the village of Odebolt and apparently much happier with the Odebolt farm than he was with his North Dakota property. Young Adams was "a real business man who loved the land and hated publicity."**

When he first moved onto the Wheeler Ranch it was so weed-infested, particularly with wild sunflowers, that he did not even try to farm it the first year. Learning from his experience with sheep and their ability to "clean up the land," he bought several thousand head and grazed them on the weed crop. His idea worked, and in the second year the Adamses were back at producing popcorn, a crop which had production costs similar to regular field corn up to harvest time, when cost tripled because of the added care. From the start the Adamses sold their corn through their elevator in Chicago.

The Adams Ranch at Odebolt resembled the bonanza farms of Minnesota and the Dakotas. The Iowa farm had well built and beautiful buildings. Visitors often commented that the white corn cribs must have been a half mile long, which was not an exaggeration, for if the cribs to hold all the corn produced on the farm were placed end to end they would have stretched out nearly one mile. Just as in North Dakota, the land was fenced in with woven wire; and within a few years concrete posts became a trademark of the farm, as did many rows of trees which bordered every road and most section lines.

The ranch had 240 mules to power the 18 gang- and 17 single-bottom plows, 18 manure spreaders, and 80 wagons. The 140 four-foot drag sections requiring 140 mules and 35 men covered 65 acres in one sweep across a section. The sheep barns in Dakota were dismantled and moved to Odebolt where, according to re-

---

*Wheeler donated part of his original 7,000 acres for the townsite of Odebolt. The first post office was called Wheeler Ranch, but when the town was platted in 1877 the name became Odebolt. During Wheeler's time in the 1880's and 1890's it is reputed that more popcorn was grown within a 15-mile radius of Odebolt than in all the rest of the world combined.

**A familiar site in Dakota or Iowa was W. P. Adams being driven around his land by chauffeur, Walt Masters. Masters also accompanied Adams to Chicago and Florida and is reported to have drawn his pay only once a year because he was so good at poker that he could make his expenses that way. Sebens reported that Adams once remarked, "It was a good idea to have one or two excellent poker players around—to keep the others so broke they wouldn't quit."

ports, from 10,000 to 100,000 sheep were used to graze the corn fields and clean up the fence and tree lines. Employment varied from 45 men in the winter season to 150 in the busy harvest period. Every implement and work animal was numbered and inventoried. Men were checked frequently to avoid loss of tools or materials. They could draw their pay on Saturday night if they wished.

Most of the labor force lived on the farm. The bunkhouses and other buildings were heated by corn cobs. Each week two beef animals were slaughtered for meat, and one dozen cows were maintained to supply butter for the workers. There was not a chicken or hog on the place. Each winter ice was put up in an ice house on the farm to help preserve food in the warm season. The farm had its own water tower to supply water to the network of buildings.

During the slack periods men were put to work cleaning and repainting equipment and machinery. In this manner every item was painted each year. The buildings were kept in equally good repair. Machinery was inventoried at $17,773.98 in 1899, and the building figure was $43,021.64—obvious proof of both quality and quantity. Blacksmith, harness, and repair shops were busy year around keeping everything in top condition.

A financial report indicated a profit of $50,855.22 in 1899. This represents a very respectable net on an apparent total investment of $260,496.83 (the total of land, buildings, livestock, and machinery). The labor bill was $14,921.98 for producing 215,000 bushels of corn and 20,000 bushels of wheat. The corn was raised on 3,600 acres for an average of about 60 bushels an acre. It took a crew of about 60 to 75 men using 76 wagons and 27 four-mule teams 60 days to harvest that crop, for an average of 60 bushels of corn per man per day. Once the corn was dry enough for safekeeping it was shelled at the Adamses' elevator at the rate of 1,000 bushels per hour.*

Corn on 3,600 out of a total 6,500 acres was clearly the dominant crop for the Adamses. However, as time passed the rotation changed somewhat. Reputedly Henry C. Wallace, Secretary of Agriculture in the 1920's and father of Henry A. Wallace, advised his good friend, W. P. Adams, to plant one-sixth of his farm to clover each year. The eventual rotation was two years of clover, two

---

*Because large cylinder-operated corn shellers were not produced until about 1902 the reference to shelling corn at the rate of 1,000 bushels an hour very possibly meant by using several shellers at the farm's elevator at the same time.

years of corn, and two years of grain. No doubt this enabled the Adamses to maintain a high fertility level and at the same time control weeds as effectively as they did in the days before chemical agriculture.

The efficiency of the 6,500-acre Adams farm was clearly demonstrated by the fact that, up to the death of W. P. Adams in 1937, the farm never had a mortgage on it and never had a year in which it operated at a loss. This was a singular accomplishment for having operated through the disaster decades of the 1920's and early 1930's.

Even though the Adamses consented to change over to tractor power on their North Dakota Fairview Farm in 1925, they were reluctant to convert to tractors in Iowa. W. P. Adams had a very strong attachment for animal power, but did buy 15 to 20 International tractors during the 1920's. As late as 1929 horses and mules were still a very important part of the power supply, and in 1933 it was reported that 15 tractors stood idle while 191 mules did the work.

Robert Brakke Adams took charge of the farm after his father's death. He was "extremely active in breeding, selling, and showing saddle horses and became a director of the American Horse Breeders Association. His love of horses caused the farm to retain horse and mule power past 1945." Within 30 days after Robert B. Adams' death in 1956 the new owner, W. P. Adams II, sold the remaining work animals and tore down the home that had served his namesake, W. P. Adams, from 1896 to 1937. At once W. P. Adams II turned to building a herd of registered Hereford cattle. He later became president of the American Hereford Association.* His interest in livestock breeding eventually led to the purchase of a large ranch at Valentine, Nebraska.

The Adams family avoided publicity, but it is difficult for an operation the size of the Adams Farm to go unnoticed. In 1937 it made the headlines when W. P. Adams sealed 300,000 bushels of corn stored in 20 cribs for $135,594. It was believed that this was the largest corn sealing loan made up to that time.

W. P. Adams II, apparently concerned about a potential trust entanglement with a Chicago bank, sold the family's famous farm on

---

*The Adamses came to develop an excellent herd. In 1954 Margot Adams received first prize on summer yearlings at the International Livestock Exposition. This was after the Adamses had already taken 13 first places at the Iowa State Fair. In 1955 14-year-old Margot Adams had the reserve champion steer at the American Royal Livestock Show.

October 7, 1962. The buyer was Charles E. Lakin, who operated implement and grain companies in Emerson, Iowa. Lakin also became interested in land development and acquired a large acreage during the 1960's. The new owner received over $100,000 in government payment since 1964 for diverting approximately 2,500 acres of land from production. In 1966 the now-famous Lakin Ranch, of which the former Adams Farm was a part, received $241,670 in government payments, placing it well ahead of the next largest recipient in Iowa, the Amana Society with $155,006.

The 6,510-acre farm, which first sold for $2.50 an acre in 1871 and was purchased by the Adams family in 1896 for $30 an acre, was sold to Lakin for $384 an acre. The total price was $2,500,000. This was the largest land transaction in Sac County in 70 years. Lakin sold off small parcels of land not suitable to large-scale machinery and under the management of Roy Selley put the remaining 6,200 acres into field corn, popcorn, and sorghum. The *Odebolt Chronicle* of October 5, 1967, reported that Odebolt's most famous farm had been sold by "developer" Lakin to Frances G. Bridge. In January, 1968, the "biggest farm auction in Northwestern Iowa" was staged by Lakin. The machinery list included four track-type tractors, 30 standard-wheel tractors, 20 trucks, 12 cultivators, 22 plows, seven combines, 18 harrows, and 12 corn pickers. Only the bonanza farms of the past could have matched such an inventory.

Frances G. Bridge and her husband, William O. Bridge, a Detroit trucking executive, formed the Shinrone Corporation for the purposes of owning and operating the farm. Shinrone paid over $4,000,000 for the 6,200-acre Iowa showplace. Roy Selley, manager for Lakin, continued to operate the farm under the new ownership.

The Shinrone Corporation and its new owners were the target of considerable publicity in early 1968 when it was reported that William O. Bridge could not pay his $594,398 tax bill to the Internal Revenue because he had only $10,768 in assets. Eventually the IRS settled for about $250,000. At the same time that Mr. Bridge was in conflict with IRS, his Shinrone Corporation was receiving a quarter-million-dollar shipment of machinery from five Massey-Ferguson factories on a single train in Odebolt. This purchase consisted of 14 1130 MF tractors, four combines, and other machinery.

Shinrone Farms continued to make news because of its size and the business-philosophy of its owner. In mid-1968 it launched a program of building a feedlot that had a capacity for 5,000 animals in 16 pens. There were to be two and one half turns of cattle annually, so about 12,000 to 13,000 head would be fattened each year. Three large trench silos 75 feet by 250 feet were constructed to hold 4,000 to 5,000 tons of silage. It was part of the overall goal to feed all of the crops raised on the farm. It was suggested at the time of the first expansion that other enterprises would be added as soon as feasible. Beef cattle started to flow in and out of the Shinrone lots. In February, 1970, the farm shipped 5,000 cattle to the Sioux City stockyards, which reported it as the largest consignment ever received from a single owner. Some of those cattle had been purchased as feeders from former President Johnson's L.B.J. Ranch in Texas.

By 1972 Shinrone Farms had achieved its goal of 12,000 cattle per year through the feed lot and had met other expansion goals as well. By putting up 3,000 acres of silage each year it was also able to stock nearly 3,000 beef brood cows. The cows utilized corn stalks left in the field after combining and were fed a limited ration of silage. A swine-raising and -feeding program was added and expanded to farrowing 300 litters per year. Shinrone under the Bridge family, as under the Adams family, continued to be a place for raising fancy horses for personal enjoyment and commercial purposes.

Rich Iowa soil, a very excellent facility, and a full-time crew of 26 men under topnotch management gave Shinrone the potential of being an extremely profitable farm, if a policy prevails to give a free hand to the on-the-farm manager. In the more than six decades under the Adams family this was practiced and the results were profitable. Whether the same results can be achieved under absentee ownership relying completely on hired management has still not been settled. Personality, delegation of power, and the assumption of leadership between owners and management more than any other factors will determine the ultimate answer.[9]

The Tilney Farms

Lewisville, Minnesota, in the midst of some of America's best corn and soybean land, is the home of Tilney Farms, which dates back to 1885. The farms were started by John S. Tilney, a New York grocer who became a stockbroker about the time of the Civil War. Through the purchase of scrip Mr. Tilney was able to amass several thousand acres within six miles of Lewisville. Tilney

Farms is still known in the Midwest for its tight management and steady profit records.

For much of its existence Tilney Farms has been operated on a tenant basis because the Tilney family apparently never desired to become too directly involved with farming. In 1944 Edgar M. Urevig was hired as farm manager. Observing the trends in agriculture, Urevig and the owners concluded that a better return could be secured through personal operation of the land. "Large-scale equipment used on a large area of land under direct central management were more efficient than many smaller machines and it seemed each year the equipment was becoming larger," thus providing the basis for the decision to consolidate in 1962.

The Tilney family change from closely managed tenant operations represented a reversal of the historic pattern in large-scale Midwest agriculture that had become established during the difficult decades of the 1880's and '90's. According to manager Urevig:

> It is more profitable to employ capital than people. New technology. . . . [has] made it wise to use machines rather than men wherever possible. Machines can be depreciated through a fast write-off, which is a gain. Men have to be pensioned which is a cost. Machines keep improving; men haven't changed much."

Through the more intensive use of capital and machines the management could more accurately predict the outcome and anticipated cost of farming. In 1962 the Tilneys started to operate 320 acres under central management, and they increased the personally farmed acres annually to 1,500 acres by 1972. They intended to continue this policy until a reversal of the profit trend appeared. Based on the experience of larger operators with similar crops, Tilneys probably did not have enough land to exceed the optimum limits. The expansion was done entirely on self-generated capital. The Tilney Farm Company bylaws prohibit the borrowing of capital, but that did not appear to be a great handicap, for, besides expanding, the normal dividend was paid and a large holdover crop inventory was accumulated.

For many years Tilneys have used experimental plots on their farms to keep up with the ever-changing use of fertilizers, chemicals, and tillage systems. In the early 1970's they had 48 1-acre plots on various soils, with 15 levels of fertilizer. Through experimentation average corn yields increased from 29 bushels per acre in 1945 to 130 bushels in 1969. The most rapid increase in yield came in the first five years after adopting the experimental plots. The Tilney Farms also cooperates closely with many university research programs and the National Field Research.

An unexpected benefit of the experimental plots and the shift to direct operation of the land has been the stimulation of the tenants towards doing a better job. Records proved that the personally operated land using the latest methods, the largest equipment, and only hired labor was not only competitive but generally more profitable than the smaller individual units operated by tenants. The exceptions were encountered in livestock units where the tenant did not figure his cost so precisely as the Tilneys did. Tilneys had top-notch tenants who in their opinion were doing better than average farmers, but apparently needed the competition of the company-operated farm to motivate them. After discovering this fact a system of awards was established among the tenants to encourage self-improvement.

Very tight cost accounting enabled the Tilneys to know where they started and where they were going. To accomplish this an accountant and two secretaries were needed to maintain the records, and obviously the procedure paid dividends. The management personnel were convinced that Tilney Farms operates as efficiently as the best of the typical father-son operations, which are in general portrayed as the most effective economical unit. Tilney records prove this point. A good job of labor relations has given the Tilney Farms a labor force of a quality far superior to the rank and file of agricultural laborers. This has been a positive factor in their operations. It is the opinion of Tilney management that the greatest handicap to large-scale agriculture is the lack of trained management, a commentary frequently voiced by financial supporters and large-scale farmers.[10]

NOTES

[1]Drache, *The Day of the Bonanza*, pp. 109, 110, 111, 166, 214; Drache, *The Challenge of the Prairie*, pp. 186, 321; Interview of Clarence Askew, Fargo, North Dakota, June 20, 1972; Interview of Arthur D. Askegaard, Comstock, Minnesota, May 15, 1967; Dalrymple, *No. 1 Hard* . . . pp. 43, 52, 53, 55; Agricultural Situation: *The Crop Reporters Magazine*, LVII, No. 7 (August, 1973), p. 5.

[2]Interview of John S. Dalrymple II, Casselton, North Dakota, March 27, 1967. This interview with Mr. and Mrs. Dalrymple, the third generation of this famous Dakota farm family, took place in the house that Oliver Dalrymple built in the 1880's; Interview with George Schutt, Harwood, North Dakota, February 10, 1967; Interview of John S. Dalrymple with Leonard Sackett, August 28, 1953, August 9-18, 1955, found in NDIRS File 549; John S. Dalrymple III, "Oliver Dalrymple and His Bonanza: An Essay on a Western Enterpreneur and the Operation of a Bonanza Farm," Yale University, 1970, pp. 30-32, 33-36, 37, 44, 71, 86, 101-105, 112; Interview of Homer E. Dixon, Detroit Lakes, Minnesota, February 2, 1967;

Interview of John S. Dalrymple III, Casselton, North Dakota, March 7, 1973.
³Drache, *The Day of the Bonanza*, chapters on the Amenia and Sharon Land Company; Interview of Mrs. Max Dahl, Chaffee, North Dakota, February 24, 1967. Mrs. Dahl was born Eleanor Reed, daughter of Walter R. Reed, the last president of the Amenia and Sharon Land Company; Financial statements of Walter R. Reed from 1900 to 1925 in the possession of Max Dahl; Interview of Earl T. Carley, Casselton, North Dakota, January 6, 1967. Both Mr. and Mrs. Carley were well acquainted with the Chaffees and saw first hand the manner of their lifestyle; Interview of Mrs. H. L. Chaffee, Bismarck, North Dakota, July 27, 1966. Mrs. Chaffee was the daughter of a music professor in California. She met Chaffee at Oberlin College in Ohio. After receiving her B.A. she earned an M.A. degree. She also spent one year working on a Ph.D degree. Gertrude Auld Bacon Chaffee came to North Dakota in 1915; Letter from Governor William L. Guy, August 14, 1967, son of William Guy, manager of the Chaffee farm; Interview of Jack Dahl, Gackle, North Dakota, February 10, 1973. Dahl is the grandson of Walter Reed.

⁴Drache, *The Day of the Bonanza*; "Fingal Enger Family History," published 1961, copy in possession of Mrs. Delmer Nystedt; Clarence H. Tolley, "Fingal Enger, King of the Goose River," *North Dakota History*, XXVI, No. 3, Summer, 1959, p. 119; Interview of Allen C. Sulerud, St. Paul, Minnesota, March 20, 1970. Mr. Sulerud, who was born at Halstad, Minnesota, was general manager of Grandin Number One farm from 1927 through 1938; Interview of D. E. Viker, Halstad, Minnesota, June 12, 1967. Viker started farming Grandin land in 1924, in 1946 purchased 3,300 acres from the Grandin Trust, and eventually secured over 5,000 acres of former Grandin land; Interview of Mrs. Orlando Serum with Leonard Sackett, August 25, 1955, NDIRS File 645.

⁵Interview of Jameson Larimore II, Larimore, North Dakota, August 20, 1971. Mr. Larimore has been actively involved with the Elk Valley Farms since his return from college in 1930; Interview of Jameson Larimore III, Larimore, North Dakota, December 1, 1970. Mr. Larimore has excellent farm records, which date back several decades. He uses them to help make his projections for each new crop year and thereby greatly reduces management error; *Larimore, North Dakota 1881 Diamond Jubilee 1956*, published by the Larimore Diamond Jubilee Booklet Committee, p. 17; *Grand Forks Herald*, Green Section, June 8, 1967; Louis M. Wangberg, "The Historical Geography of Selected Farms in the Larimore, North Dakota Area," Unpublished Master's Thesis, Department of History, University of North Dakota, 1964, pp. 46, 63, 65-66; Interview of Charles R. Wright, Larimore, North Dakota, with Leonard Sackett, March 22, 1954, NDIRS File 242.

⁶Orville M. Fuller, "Farm Organization as Applied to the Baldwin Farms, Dickey County, North Dakota," Unpublished Master's Thesis, North Dakota Agricultural College, 1924, p. 44; William G. Hunter, "The Baldwin Farms," *North Dakota History*, XXXIII, No. 4 (Fall, 1966), pp. 401-405; 410-414; 418; Letters and other documents in NDIRS File 1116C entitled Baldwin Farms—North Dakota Lands 1918-1930; Albert Coffman, "The Story of the Famous Adams Ranch," Unpublished Master's Thesis, Michigan State University, February 1968, p. 23.

⁷Drache, *The Day of the Bonanza*; Donald Berg interview with Reuel Wije, Moorhead, Minnesota, July 6, 1966. Mr. Wije was the last manager of the Dwight Farms covering the years 1931 to 1935.

⁸Interview of A. J. Robinson, June 22, 1971. Mr. Robinson was the only manager of the Schermerhorn Farms from 1917 until dissolution; *The Fargo Forum*, June 7, 1925, September 8, 1963; *The Minneapolis Journal*, Editorial Section, November 22, 1931; Photo book and record book of the Schermerhorn Farms; Interview of Ole A. Olson, January 12, 1972. At that time Mr. Olson presented a prepared statement entitled, "Some Hazards of Farming During the Dirty Thirties." This paper was based on his experience as an agricultural representative responsible for the management of 160 farms for his company during those years.

⁹Letter from Frank Vyzralek, October 25, 1967. Mr. Vyzralek is archivist for the

State Historical Society of North Dakota. His letter is based on legal records of the state; Albert Coffman. "The Story of the Famous Adams Ranch," dated February, 1968, Detroit, Michigan, from the files of Michigan State University; Interview of William P. Sebens with Leonard Sackett, May 11, 1955, NDIRS File 520; *Odebolt (Iowa) Chronicle*, March 8, 1951, September 9, 1954, December 2, 1954, October 27, 1955, March 5, 1964, October 5, 1967, March 21, 1968, February 5, 1970; Drache, *The Day of the Bonanza*, pp. 75, 115, 123, 138; *Farmer-Globe*, Wahpeton, North Dakota, January 25, 1965; *Compendium of History and Biography of* North Dakota, Geo. A. Ogle and Co., Chicago, 1900, p. 220; Interview of Mrs. Robert Parsons, July, 1970. Mrs. Parsons was the daughter-in-law of Thomas D. (Dan) Parsons, long-time Adams manager; *The (Wahpeton) Globe*, August 8, 1900; Interview of Mrs. C. D. (Florence) Ballard, Odebolt, Iowa, June 15, 1972. Mrs. Ballard started as a bookkeeper on the Adams Farm in 1905 and later became superintendent of the Odebolt Farms; *Des Moines Register*, January 28, 1963, April 12, 1964, August 27, 1967, January 18, 1968, August 11, 1968; Interview of Roy Selley, Odebolt, Iowa, February 23, 1972. Mr. Selley was the manager of the Adams Farm after it became the Lakin Ranch in 1964 and continued in that position for the Shinrone owners. He had a background of a practical farmer, 10 years in the United States Department of Agriculture prior to assuming his management position; *The Minneapolis Star*, March 18, 1968.

[10]Interview of Edgar M. Urevig, General Manager, Tilney Farms, Lewisville, Minnesota, March 29, 1972; Interview of Joe Metz, Operations Manager, Tilney Farms, March 30, 1972; Edgar M. Urevig, a prepared manuscript entitled *The Decade Ahead* used for a farm management talk; Dick Hagen, "The Setup That Works Year-Round . . . and the Price Is Right," *Farm Journal*, April 1964, p. 55.

CHAPTER VI

Engen, Jarrett, Schwartz, and Young
—Four Determined Men

In the development of industrial America probably no segment has gained so much from the motivation of individual owner-operators than has agriculture. Yet in no industry has there been so much skeptical thinking as in agriculture. The opportunistic and skeptical trends are deeply ingrained in agriculture, and frequently both moods are apparent at the same time. It is no secret that some farm factions have deliberately used a relatively negative approach to develop their farm program. The frontier to a degree was responsible for both attitudes. Vast free land encouraged recklessness in farm management and promoted the attitude that if a man could do nothing else he could always farm. Once the frontier came to an end this attitude was destined for doom, but it took farmers a long time to change their thinking.

Today successful agriculture more than ever before depends on first-rate thinking. Many success stories in agriculture are based on the motivation and the ability of first-rate individuals. The rise and decline of the bonanza farms as described in the previous chapter illustrate clearly that the aims, the drives, and the desires of one individual were more responsible than any other factor for the success or the decline of those farms. Agriculture, an industry of many independent and fragmented units, has conspicuously provided many failures, but also many success stories.

Eugene M. Young, Otto Engen, Ray S. Jarrett, and Earl Schwartz were four of a kind who became successful farmers. All were born about 1900 and started farming during the decades of despair—the 1920's and 1930's—but they survived to experience the successful decades of World War II and after. In spite of obstacles they never quit, and as rugged individualists refused to admit defeat.

These men became conspicuous in their home communities because of the size of their farms. In spite of some of the misgivings of their neighbors, they dared to be different and eventually succeeded.

Eugene M. Young

Eugene Young's father, Edmond Young, moved to Sanborn, North Dakota, in 1884 to work as a station agent for the Northern Pacific Railroad. While working as station agent he also started farming in Barnes and Benson counties. Edmond Young decided to leave the railroad and become a full-time farmer. In March 1903 the family moved to Coleharbor, McLean County, where they lived in a log house until "something better" could be built. In McLean County, Young purchased the rights to a homesteader's 160 acres for $150 and purchased another 160 acres from the Northern Pacific for about $800. By adding more land to those original purchases he had 800 acres by 1909 and was renting another 480 acres.

Edmond Young was a progressive farmer and instilled that attitude in his children. He purchased his first tractor in 1911 and kept buying new ones as major improvements were made. His son Eugene started driving a 20-H.P. Mogul tractor at the age of 18 in 1911. This was a difficult job because he was so small that he had trouble turning the flywheel to start the engine.

In 1915 Eugene Young, now 22 years old, started farming on his own. He seeded 100 acres of flax and did fairly well. However, his biggest profit came from a threshing run that he organized among the farmers in his area, using a newly purchased single-cylinder Mogul tractor and a steel 32 x 36 Case thresher. He had so much business threshing that he was able to buy a new 30-60 Aultman-Taylor tractor and a 42 x 64 thresher. This rig proved to be superior to the old steam-powered outfits. Initially, Young never owned horses and relied on tractors and trucks to do all the farm work.

In 1916 Eugene Young and his brother Guilford leased land near the town of Van Hook on the Fort Berthold Indian Reservation. Cash rent was 50¢ an acre, and the renter had to break the sod and fence in the crop. In the next three years the Young brothers broke over 3,000 acres of sod in Mountrail County. The first year yielded the best and only good crop at Van Hook. Their flax did well and sold for an average of $3.50 a bushel, enabling the brothers to pay all expenses and for all of their machinery.

By 1917 Dwight and George Young joined company with Eugene and Guilford. In July 1917 they decided to enlarge their farm and acquired a lease on the Standing Rock Indian Reservation near McLaughlin in Carson County, South Dakota. The rental contract was for five years at 50¢ a year per acre. In addition, the brothers also rented land near Bison, Lemmon, Timber Lake, and Firesteel, South Dakota.

The tillable acres of the rented farms in South Dakota increased from 450 in 1917, to 960 in 1918, to 2,500 in 1919. Eventually the Young brothers broke a total of 20,000 acres of virgin sod in Carson and Perkins Counties. Their experience there was one of trial and error and was not significantly profitable.

While farming in North Dakota the four Young brothers relied chiefly on the sons of neighboring homesteaders for help. But after they moved to South Dakota they actively sought to employ Indian laborers. From 1919 through 1931 the Youngs used 18- to 22-year-old Indians almost exclusively.* With the exception of the harvest season the Youngs needed only 20 men. In addition they had a waiting list of about 50 available workers and "seemed to work a sort of rotation among themselves." During the farming season most of the men worked about the same number of days without any particular scheduling program on the part of the Youngs.

The Indians were paid at the same rate as any other available farm labor. They were very good with the tractors, but "whenever they could get their hands on any tools they always wanted to tamper with the governors." After the 1930's it became more difficult to employ Indians, and by World War II the Youngs once again switched to using the neighboring farmers' sons or purchased labor-saving machinery. During the 1920's they had 10,000 acres in crops, requiring 40 employees.

Horseless Farming

The Young brothers were very mechanically minded and were determined to farm without horses. To do this they had to improvise equipment formerly used with horses to make it applicable to tractors. They made a hitch to pull three 11-foot tandem discs behind a single tractor followed by three 11-foot grain drills. Drills with unique 3-inch spacing were used to distribute the flax seed

---

*The Youngs had a cook car, with two women cooks, and a bunk car, on wheels, which were easily moved from one farm to another. Each car was built to accommodate 24 men.

more uniformly. When harvest was ready in the fall of 1917 a four-cylinder Model T Ford motor was mounted on a 12-foot McCormick header to give the Youngs their first self-propelled header. Grain was hauled to the elevator by a big tractor pulling four wagons. Trouble was encountered at the elevator because the tractor could not go into the driveway. This obstacle was overcome by a long rope that was used to pull each wagon through the unloading driveway from the opposite side of the elevator.

Because of the labor shortage during World War I, Model T Fords were converted to pull grain drills while the bigger tractors opened up more land. In 1918 a special hitch was made so that pusher-type headers which required four to six horses could be pulled by the converted Model T's. In 1920 the Young brothers built an evener for pulling 60 feet of harrow with a 20-40 Rumely. Traveling 3½ miles per hour one man could harrow over 200 acres a day in contrast to about 35 to 40 acres for a four-horse outfit.*

Despite their innovations and ability to gain real efficiency with the use of the tractor in extensive grain farming, the Youngs once were told by their banker at McLaughlin that they could not succeed without using horses to do the work and having to raise livestock for diversified income. Following the advice of the banker they purchased six excellent horses and also started raising corn. Horses were used to plant and cultivate the corn and to pull the corn picker in 1920. That fall after shelling their corn with an Avery threshing machine the Youngs shipped the first two carloads of corn ever sold out of the elevator at McLaughlin, South Dakota.

To be able to thresh continuously while the tractor was pulling loaded grain wagons to the elevator, Young built a 500-bushel portable grain bin on an old tractor chassis in 1921. The Youngs' grain-hauling rig consisted of four wagons, each with a 125-bushel capacity. When the wagons returned from the elevator they were pulled alongside the portable bin, which was higher than the wagons, and the grain was loaded. To reduce harvest cost even more, three grain headers and three header wagons were attached to a single hitch and pulled by a 30-60 Aultman-Taylor tractor. In this manner four men, one for each header plus the tractor driver, could cut and load grain faster than six men and 18 to 24 horses had done the year before. The tractor worked faster and more

---

*In 1973 commercial 80-foot harrows were available which could do up to 720 acres in nine hours.

hours per day. In 1923 because they were dissatisfied with the use of horses they hitched two double-row corn planters together and pulled them with their new International Farmall tractor. The Farmall, believed to be the first one used in South Dakota, was just what the Young brothers had been waiting for. They had wanted a small tractor but did not like some characteristics of the Fordson. So they purchased the Farmall as soon as they learned about its adaptability for row-crop work. On January 1, 1922, they started an implement firm at McLaughlin which handled several lines, including International.*

The Farmall worked so well pulling two double-row corn planters making rows "as straight as any ever planted with horses" that the Youngs decided they could cultivate with it too. So one year, before mounted tractor cultivators became commercially available, the Youngs hitched two double-row horse cultivators together behind the Farmall and cultivated their corn. That fall the brothers rigged two horse-drawn corn pickers together so that they could both be pulled by one Farmall. The elevators from these pickers were modified so that both were used to fill one wagon. This was done by pulling the wagon behind the lead picker and beside the second one.

The 1923 adaptations of tractors to row crops were not the only innovations on the New Era Tractor Farms as they now called their operation. That year International Harvester put two 12-foot combines with mounted engines on the Young farm for experimental work. They proved to be an almost instant success because of the greatly reduced labor cost. After completion of the wheat harvest the flax crop was swathed into windrows, and a pickup attachment was made for the new combines. With these modifications the flax crop on the Young farm was probably the first in South Dakota to be swathed and combined.**

Innovations to reduce cost could hardly come fast enough to keep the New Era Tractor Farm profitable. Real estate purchases

---

*Besides the International-McCormick line, the Young Brothers Implement Company also had John Deere, Rumely, Oliver, Huber, Holt, Nichols-Shepard, and many short lines. This gave them access at wholesale prices to almost any innovation taking place in agriculture.

**Grasshoppers were a frequent problem to the Youngs, and to get ahead of hopper damage they purchased the New Way Harvester made in Fargo, North Dakota, in 1927. This harvester enabled the farmer to cut grain before it was completely ripe. A stack about 7 feet in diameter and 6 to 7 feet high was made. The grain cured in these stacks and was then hauled by a buck loader mounted on the rear of a Model T truck to the threshing machine. This was a much cheaper way of harvesting than the binder-thresher method.

at the wrong time hurt the Youngs just as it did many others who became too optimistic during the price boom of World War I. After renting Indian land for many years from the Bureau of Indian Affairs, the Youngs started buying land in 1919. The first purchase was 160 acres at $31 an acre, quite a contrast to renting for 50¢ an acre. In 1922, because of encouragement from their banker, the Youngs purchased a well developed livestock facility for $125 an acre.

Before 1922 was over the four brothers had taken a couple of severe jolts. That year black rust caused an almost complete crop failure. Land at any price was too high under those circumstances. The livestock enterprise, although giving them some diversification, did not pay satisfactory dividends. It conflicted with their basic desire for extensive grain farming with maximum mechanization to get the lowest possible cost of production. Within a few years both the cattle and hog operations were discontinued.

The Young's implement business was not profitable from a cash flow standpoint because South Dakota farmers were not doing well during the 1920's and the accounts receivable caused headaches. The chief benefit of having their own implement firm came from having a good shop in which to make improvements on their farm machinery plus having access to the newest types of equipment. One of the greatest benefits of the repair shop came with the chance to build a power plant at McLaughlin, which provided enough income so that they could remain solvent. The Youngs were primarily farmers, but they were not hesitant to get involved with other enterprises to keep their farm going during the difficult periods.

Poor timing on land purchases and crop failures were not the only setbacks suffered by the New Era Tractor Farm. In 1925 it was decided to put 10 Twin City 12-20 tractors along with plows, discs, and drills into operation. These units cost about $1,200 each at wholesale prices. When the check to pay for them was on its way to Minneapolis the McLaughlin bank went broke, taking $25,000 from the sale of the light plant with it. In one stroke their cash was lost and they had a bill for $12,000 of new machinery. In addition, the summer of 1925 in the McLaughlin area was dry and windy, causing another total crop failure. Eugene Young, 33 years of age that summer, experienced "a hair change from black to white, it didn't stop at being gray."

As conditions worsened there was little choice but to reduce operations. By reducing expenses and keeping only the best land it was hoped that some profit could be realized. From farming 10,000 acres and employing 40 men in the early 1920's the operation was gradually reduced to only 750 acres just prior to World War II.

In 1923 the Youngs made an effort to raise money by borrowing $10 an acre against a 560-acre unmortgaged farm. In the following years neither taxes nor interest were paid, but the Youngs "were unable to dump it in the laps of the Federal Land Bank." When the Land Bank proposed to refinance the total original loan, including back taxes and interest at the rate of 3.5 percent and 20 years, Eugene Young commented, "I almost refused to take the offer." It took a few good years in the 1940's before "people had faith in land again," said Young, "for in 1945 I was still able to buy good land on a section basis for $23 an acre." This was a turning point in the fortunes of E. M. Young.[1]

Otto Engen

Arne and Betsy, the parents of Otto Engen, settled on a 160-acre homestead near Tagus, North Dakota, in 1904. The young homesteaders started to break 320 acres of sod on their farm with four wild three-year-old steers that had cost Engen $40 each. Once these animals were broken to work they could plow 1½ acres a day. While the oxen rested, Engen's father piled rocks to be hauled away, picked up buffalo bones, collected arrowheads, and gathered cow chips.* This farm was never larger than 320 acres, and one-half of that was always in pasture.

In 1914 Otto Engen, who was then 19, purchased a 160-acre farm for $20 an acre. Living with his folks, he used their machinery and horses to operate his farm while also helping at home. In 1917 Otto enlisted in the navy and served until April of 1919. While in the service he missed two good crop and price years and consequently he was unable to reduce his mortgage. He left the navy with $150 in cash that he used for buying machinery. Later, with his father's help, Otto was able to buy more machinery and 26 cows. Borrowing four horses from his father and a grain drill from a neighbor, he was able to produce a crop in 1919.

---

*Both Otto Engen and Mrs. Otto Engen, who were young children at the time the sod was being broken, remember having to pick up cow chips for the cook stove. Whenever it looked like rain the children all had to rush to stockpile cow chips in the shanty where they would stay dry. Using cow chips enabled both the Harris and Engen families to avoid spending money for fuel during the summer months. The Engens lived in an 8- by 10-foot house the first year and in a side hill dugout the second year.

At this time Otto Engen's total machinery investment was less than $400. Besides his own 160 acres, he rented 160 acres from the Tagus Bank for one-fourth the crop. His rye crop averaged 30 bushels and sold for 80¢ a bushel, while wheat made 12 bushels and sold for $1. Beside farming, he worked on a threshing run, furnishing a team and a rack along with his labor to make an extra $2.50 a day. In the winter Otto worked in his father's blacksmith shop for room and board. But even with all of his extra efforts Otto Engen was unable to meet the mortgage payments on his purchased 160 acres, and it was foreclosed by the Federal Land Bank. However, he was allowed to continue renting that land until 1940 when he repurchased it for $400, in quite a contrast to the $3,200 he had paid for it in 1914.

Like the Young brothers, who were expanding rapidly despite poor crops and other setbacks, Otto Engen ignored his failures and went ahead. In 1920, by renting 80 acres from his father-in-law, Otto was farming 400 acres with eight horses and a hired man who received $45 a month for five months. Engen saved all of his straw to feed his expanding cow herd. In winter Mrs. Engen often fed the cows while Otto hauled grain to Tagus. Using four horses he could haul 85 bushels of grain in a day to the town nine miles away, leaving at 5 a.m. and returning at 7 p.m. In the intense cold of the winter Engen walked beside the team for the entire trip to keep warm. Sometimes he would haul coal for the local school or for his own house on the return trip.

In spite of all their efforts, the Engens, like most of their neighbors, would have failed if it had not been for their milk cows and hogs. The livestock enterprise kept them busy in the wintertime, but it also kept them solvent. During most of the 1920's and early 1930's husband and wife had to hustle to milk 30 cows, raise 80 hogs for market, and care for a herd of beef cows. One section of land was in crop and one section in pasture, but even with all their efforts they could not afford to hire a man except during the five summer months.*[2]

Ray S. Jarrett

When the Engens were settling near Tagus, North Dakota, Ray S. Jarrett's parents moved to Britton, South Dakota. The senior Jarrett had been a machinery dealer and livestock buyer in Iowa; but in 1904 after a severe fire he moved to Britton to become a

---

*Engen recalled that during most of the 1920's and early 1930's it took seven to eight months to raise a hog to 225 pounds.

"sod buster." There Jarrett bought land, which he broke, and sold some of it to raise enough money to get his own farm going. In 1905 the Jarretts had 2,000 acres of rented land, of which 300 acres were in potatoes and 40 acres in cabbage. Ray Jarrett, then seven years old, started sleeping in the bunkhouse, a practice which he continued until he started farming his own farm in 1924. In the bunkhouse Ray learned how to get along with other men who were working on his father's farm and learned to understand their way of life.

In 1910, the Jarretts diversified their operation further by adding a 50-cow dairy herd. To secure additional income they bottled milk for the people of Britton. The Jarrett farm grew rapidly in size, and 40 full-time men and 60 horses were required to operate it prior to World War I. In 1921, Jarrett had 300 acres of potatoes and decided to speculate on the sale of potatoes by paying 50¢ a bushel down on contract to local farmers. Then unexpectedly the price broke, and when the time came to sell the potatoes the price had fallen to 25¢ a bushel.

"That setback took the heart out of him," said Ray Jarrett, "and when Mother died in 1922, Dad decided to sell out. In spite of Dad's personal setbacks, the night of his auction he put his arm on my shoulder and encouraged me to farm and to think big about it. I bought $600 worth of machinery at his auction in 1924. The same year I was married, and started farming on my own."

As Ray Jarrett later related:

Dad was an alert farmer who worked closely with the county agent. An excellent innovator in crops and livestock and was good with men. I learned to work with men and I wouldn't trade the best ten sections of land for the experience I gained from Dad. I had promised my mother that I would take care of my younger brothers and sisters and I made a vow to myself that I would farm big. The first year I farmed I don't think I saw over $15 cash.

In his first year Ray Jarrett rented 480 acres of land and ran 20 sheep. But he kept dreaming about the day when he would become "the state's biggest farmer." To reach his goal he had to overcome many obstacles, and nearly 50 years later he remarked:

. . . any man who has broken the prairie in South Dakota knows that Mother Nature . . . can be as mean and nasty as she can be good. If your crops can survive the spring frosts . . . if you can get enough rain . . . if you aren't hailed out . . . if the grasshoppers and bugs don't eat up your crops . . . if the tornadoes miss your house . . . if the sun doesn't parch your crops and the hot winds dry up every last bit of moisture . . . you've had a good year.

Even in his seventies and after many successes the weather-beaten Jarrett still adheres to the same thoughts about new ways to innovate and expand his farm.[3]

Earl Schwartz

By 1904 Minnesota was getting "too crowded" for Jacob Schwartz, the father of Earl Schwartz, so he decided to move to Kenmare, North Dakota. Earl's father was accompanied by his father, Mathias Schwartz, who came to Kenmare just to file a claim to 160 acres so that a larger farm could be developed. In this manner 320 acres were claimed under the homestead law, with a single house being built over the boundary line of the two quarters to satisfy the letter of the law. Earl's grandfather had no intentions of staying in North Dakota, but Earl's father wanted to go into the cattle business and needed every acre he could buy or rent.

Earl Schwartz's father kept expanding, and in 1919 he purchased his first tractor, an International Titan. The only other tractors around Kenmare at this time were steam-powered rigs used to open the sod, "but the farmers who stuck with them went broke." In 1927, he added a John Deere Model D tractor and in 1928 purchased a 20-foot Avery combine, another first in the Kenmare area. Schwartz made money every year during the 1920's because he had a diversified farm. In 1929, when his son, Earl, wanted to start farming for himself, his father encouraged him. Earl rented 560 acres of land on a 50-50 crop share basis and had a "fairly good first year." But he would have many obstacles to overcome before he would achieve success.[4]

The 1930's: A Test of Endurance

Young, Jarrett, Engen, and Schwartz were all in their thirties when the low prices and drought conditions of the 1930's encroached on the nation and the Dakotas. Probably Schwartz, who had just started farming in 1929, was in the best financial position because he was the least encumbered of the four and had suffered the fewest setbacks up to that date. When the stock market crash of 1929 hit the nation, these Dakota farmers were little affected by it at first because in general they had nothing left to lose. The early 1920's had taken care of that, at least for Young, Jarrett, and Engen.

Engen lost a total of $200, an amount equal to one year's living expenses, in 1921 and 1922 when the banks at Tagus and Berthold were closed. E. M. Young's big loss came with the bank closing of 1925, while Jarrett had no money in the banks to lose. Apparently he did not even have a checking account, because he either by choice or by necessity was financed by the various firms with which he did business. As he once said, "I learned to do business with postal money orders because I had more faith in them than I did in the banks. I was so broke so many times that people will never know. Not even the bankers knew my condition because none of them had any money to loan me. But somehow I always managed to keep my credit in order."

In retrospect, each of these men could look to one factor that they thought had pulled them through, but there was a greater attribute which helped them to survive—their own attitude. Each was willing to try something different when all else failed and others were quitting. Jarrett expressed the thought that all four must have had at various times: "Many mornings when I went out I felt like quitting, but I knew this could not last forever, so I hung on."

Experimenting with cost-cutting equipment was a key factor in the eventual success of all four of these men. The survival of the Young brothers testifies that mechanization was the solution for overcoming difficult times. Engen credits the purchase of a 15-30 International tractor and a three-bottom plow for $1,300 in 1928 as his lucky break. As he once commented:

It was the first really modern tractor in our area. With that I could get up at 4:30 a.m. and break nine acres of sod a day—quite a contrast to one and one-half acres with four oxen. On old plowing I could do eighteen to twenty acres a day. Eight horses could only do ten or twelve acres. By putting lights on the tractor I could do as much with two men as three men using twenty-four horses and three plows could do. From 1928 on I always used two men on that tractor and worked it around the clock until I traded for two Hart-Parrs in 1935. We broke some new sod every year and I continued to do that right down to 1972.

Jarrett, who started farming with six horses in 1924 and rented land throughout the 1920's, felt that his sheep, milk cow, and chicken enterprises on a small scale "kept food on the table and the family living." Buying a threshing rig at a foreclosure auction enabled him to make "a little extra money without a big investment," because some local people helped the financing by having their threshing done.

Jarrett had a total crop failure in 1926 because of bad wind erosion, but that only seemed to make him more determined. In 1929 he had the foresight to trade some of his horses for an F-20 International tractor and equipment. That year he had a good crop in prospect when a hail storm completely wiped it out. Fortunately, Jarrett had taken precautions to avoid a complete disaster. "I don't know what I would have done without the milk cows and the chickens because the big deal was to keep food on the table, you didn't worry about anything else because it was so far gone."

Before the hail storm in 1929 he had bought 200 head of feeder cattle for 9¢ a pound and got started in the beef business, which proved to be very practical for the type of land he was farming. Beef eventually became a major enterprise for him. Psychologically the beef business in 1929 helped him to survive, in contrast to many of his neighbors, who quit farming in those years.*

When the loan companies found the land abandoned and returned to them they had no choice but to seek other farmers willing to operate it. They approached Jarrett, who was milking 25 cows and "peddling crops to make a few extra bucks." And Jarrett, who had "all kinds of chances to rent or preferably buy land from that date on," made good use of this opportunity. In 1930 Ray Jarrett purchased his first 160 acres for $500 after he had rented it for six years:

I was able to buy a great deal of land for $3.00 to $4.00 an acre and I often secured a quarter section for $50 down, 4 percent interest and almost any terms I could name. I bought a few quarters at court house auctions. At one time I could have bought 60,000 acres at a very low price per acre. I did buy one block of 10,000 acres from Sargent County (N.D.). At that time I reasoned if I had credit to buy land I could also get money to buy cattle and that's how I really got going in cattle.

By 1928, the Engens had enlarged their farm to 900 acres of crop besides a large acreage of pasture, most of which was rented. They had 30 milk cows, 70 brood sows, feeder pigs, and 125 horses that Engen raised for the market. There were eight hired men and one or two hired girls. Because of the inability to get water on their place, all wash water had to be hauled in by 50-gallon barrels and the drinking water had to be carried in 3-gallon

---

*Some pulled out at night and left everything to the loan companies. This was an experience similar to that of Mr. Van Nice, the banker at McLaughlin, who many mornings in the 1920's and '30's found notes under his door from various debtors. The notes usually stated that although they had abandoned their farms, the chickens, pigs, cows, and horses were still alive and the banker could pick them up. Van Nice told Gene Young that this happened quite a bit to the little farmer "who stayed until everything from the garden was gone and the last cow had dried up."

Red Wing stoneware crocks from Bill Bauer's farm 1½ miles away.*

The purchase of a 1928 15-30 International tractor at this time proved to be a great asset to the Engens. The year of 1928 was wet, and the crop was good. Ordinary ground-driven binders would never have cut the heavy crop, but the power takeoff on the new tractor enabled the Engens to get their crop harvested. Two carloads of cattle were finished on that grain, and the proceeds enabled Otto Engen to purchase a Model A Ford and a gasoline-powered washing machine for Mrs. Engen.

Poor crops were encountered in 1929 and 1930, so that at the end of 1930, after 16 years of farming, Engen recalled: "I had lost all the land that I ever owned and the cattle and machinery were mortgaged for more than they were worth." He experienced another crop failure in 1931, "not even prairie hay to cut." The army worms ate the green weeds and "traveled over the house instead of around it." That fall the Minot Credit Company offered to pay the freight for moving a carload of household goods, two carloads of horses, and three carloads of cows to a farm near Davenport in eastern North Dakota if he was willing to take over that farm. The credit company saw that Engen was not a quitter, and it wanted to help him. A farmer at Davenport had abandoned his farm, leaving a silo full of silage and a loft full of hay. Engen accepted the offer. He sold horses to finance more cows to step up the milk income. In the spring of 1932, Otto Engen returned to Tagus and put in a crop, while Mrs. Engen, the children, and the hired man stayed at Davenport.

After seeding his crop at Tagus, Engen returned to Davenport and took a contract to plow 1,600 acres of land near West Fargo for the Union Central Life Insurance Company at $2 an acre. It took a 22-36 International tractor to pull three bottoms in the heavy Fargo clay soil. At that time he was told that the next lowest bid was for $3.75 an acre. But he fulfilled his contract by working with two tractors and four men. They got the job done and there were "quite a few dollars left after expenses." In the fall of 1932, he was able to sell 60 steers to a butcher in Minot for $22 each, which was $2 more than the government was paying at the time.

Crops were no better in 1933 and 1934. It was so dry in 1934 that after planting, the hired men were released because it became obvious that there would be no crop. A string of 134 horses

---

*Using a brass scrub board, Mrs. Engen or one of the girls washed clothes for the four Engens, the two hired girls, and eight hired men.

were sold gate run for $11 each. Engen retained 12 head to perform needed farm work. In 1935 the 15-30 International tractor and the remaining 12 horses were traded in on two Hart-Parr tractors. The dealer agreed to take a $1,660 note for the balance of the payment. Rust ruined the wheat crop, but some late flax sold for $2.50 a bushel, with all proceeds going to the Minot Credit Company. A total crop failure in 1936 and 1937 forced Engen to herd cattle in order to make a living. In the fall of 1937 he was asked to pay the two-year-old $1,600 tractor note, on which he had not paid any of the principle. The machinery dealer had gone broke, and the Oliver Company settled for $400 (of the $1,600) because "they already had more repossessed tractors than they could sell at any price."

Earl Schwartz's father had purchased a combine in 1928, and except for "threshing enough to make a strawpile," they used only the combine, and left the threshing machine standing idle. With fairly good returns in 1929, Earl Schwartz was optimistic enough in the spring of 1930 to buy a 20-35 Allis Chalmers tractor with a four-bottom plow. In the fall of 1930, he purchased 320 acres of land for $11,000. He paid $2,000 down and made two annual payments of $1,000 each "before I let it go because I could buy other land cheaper by then."

Schwartz was very depressed in the fall of 1932 when he was forced to sell wheat at 28¢ a bushel in order to pay off a $165 seed loan. According to Schwartz: "I was so mad I wanted to quit when the government supervisor said because I was a single man I could not be allowed to extend my loan. My dad, because he was a family man, did not have to sell his wheat to pay his loan." In the winter of 1932 and 1933, Schwartz got a job working at the mill in Columbus, North Dakota, and debated all winter about quitting the farm. He commented, "I remember Dad sold a carload of pigs and the proceeds were not enough to pay the freight. He just killed the rest of them rather than waste the feed." At this early date Schwartz could not foresee the good fortune that was ahead for him.

With continued depressed farming conditions throughout the 1920's, the Young brothers' New Era Tractor Farm continued to reduce in size. Their machinery business was no better, for Young Brothers' Implement Company had purchased 15 Holt combines in 1927 and had ordered 70 more for the 1928 crop year. Since sales were slow, George Young decided to go to Stockton, California, where the Holt combine was made, to be trained as a field service man. Crop failure in South Dakota that year enabled the Youngs to sell only 25 out of the 85 combines ordered, so the implement business came to a near standstill at this time. George Young learned of an opportunity to go to Russia as a mechanic instructor through the Caterpillar Tractor Company, which had sold many tractors in that country. George had an excellent background to help the Russians get started in mechanized farming because he had both in-the-field and in-the-shop experience.

After training with Caterpillar Tractor at Peoria, Illinois, he went to Russia at $300 a month plus all expenses paid. He was the only American on a 350,000-acre farm that had well over 100 tractors in operation. There were 55 Case four-plow tractors, 70 Olivers, and 30 Caterpillar 60's, and no trained tractor drivers. Young's chief job was to show the drivers how to operate the tractors. This gave him little time to repair those that broke down. The farm was 250 miles southeast of Novosibirsk, 72 hours by train from Moscow.

After 15 months of his 24-month contract, Young quit because he did not get his American food allotments. In retrospect he said, "Had I known that the depression was on in America, I would have gladly fulfilled my two-year contract."*

While George Young was in Russia trying to earn extra money, Gene Young and his brothers Guilford and Dwight were attempting to hold things together in South Dakota. Adverse as the 1920's were, Gene Young thinks that in some respects the weather was worse from 1928 through 1941:

The only thing that kept me going was that we won the contract to build and operate the power plant at Fort Yates, North Dakota, and even then I had to beg Van Nice, the banker, for $50 to pay for a carload of fuel. I had $700 in the bank and the bill was $750 and I could not touch it until the bill was paid in full. George Metzer, a farmer, helped me at that time by making some money available at the bank.

The Young Brothers' Implement Company was closed down in 1930 because the customers "just disappeared leaving us holding their notes and worn out machines. . . . We as dealers had guaranteed the farmers' notes to the manufacturers, which was quite a burden to overcome."

While building the power plant in 1932, using his mechanics and farm workers, Gene Young discovered that he not only "had

---

*Young commented that he was in Moscow at the same time as Tom Campbell and was aware that the Russians had great respect for Campbell's reputation.

no cash but also no credit." He was turned down by both bankers and lumber merchants until finally the lumber yard manager at McIntosh, South Dakota, agreed to let him charge $800 worth of materials to put a roof on the power plant.

In 1932, Gene Young received his first government seed loan. He borrowed the maximum of $400 and purchased 400 bushels of seed wheat with it. That fall he returned 400 bushels of wheat to the elevator and received $100, which he applied to his government loan. Young felt he had returned pound for pound, and with the price at 25¢ a bushel he was bitter enough not to give in too easily.* His banker also insisted on collecting interest and some principal payment on a loan in the fall of 1932. Although Young felt that the market at 25¢ a bushel was too low and was sure it would be higher within a year, he could not get the banker to wait. He was correct, for by May, 1933, wheat was selling for $1.07 a bushel.

Out of the Depths

Between 1933 and 1940 Engen, Jarrett, Schwartz, and Young were all able to rise out of the depths of the depression and to start on a new road toward eventual success. Of the four, it was Earl Schwartz who had thought most seriously about quitting farming throughout the fall and winter of 1932 and 1933. However, he was the first to pull himself out of the financial depths. He was the youngest of the four and the only one unmarried, so he was probably the freest to venture.

In the spring of 1933, Earl Schwartz started a flour mill on his father's farm, that was capable of turning out four sacks of flour an hour. This mill was operated during the summer of 1933 whenever Earl or his help were free from farm work. After field work in the fall, Earl and one man operated that mill 24 hours a day for six weeks. Schwartz remembered, "We ground wheat for about every farmer in the radius of sixty-five to seventy miles. And we ground some relief flour for the government too."

It took three bushels of wheat to make 100 pounds of flour, and the milling fee was 50¢ a hundredweight of flour. Besides grossing $2 for each hour for the grinding, there were also 320 pounds of by-products from the wheat, such as bran, midds, and shorts, for each hour of milling. These wheat by-products became available

---

*The next two years were total crop failures, but he was able to talk the government officials into a compromise on payment of the balance of his seed loan. Many farmers never repaid theirs.

for livestock feed on the Schwartz farm. From that point on until the mill operation was discontinued in the fall of 1937, the Schwartz farm always had ample feed to maintain a sizeable number of cattle, sheep, or hogs. In retrospect of this operation Schwartz later said:

> The milling business was a real break—I pulled through the 1930's without a single loss year. It gave me the *cash, confidence, and enthusiasm* to try other things. The by-products of the mill enabled me to have sheep, beef, hogs, and dairy which gave me the boost I needed to buy land and get an efficient unit. When the break came in weather, prices and available land, I was ready to take advantage of it. Because I was an innovator, always having the latest possible equipment right down to the four-wheel-drive tractors with cabs and air conditioners caused people to laugh at me exactly when I was not only having fun, but getting ahead of everyone else by being first.

With the additional cash from the milling income and extra money from the livestock enterprises, Schwartz was able to generate a sufficient cash flow to purchase land. In 1935 he purchased 480 acres, a "brand new 1929 combine that had never been used," a used combine (both for $350), a repossessed 22-36 International tractor for $175, and two new press drills, "the first in our area." The purchase of the additional 480 acres increased the Schwartz farm to 2,080 acres, large enough to be efficient, support a family with the technology and living standards of the 1930's, and "still pay the taxes."

The crop of 1935 was light because of rust, but Schwartz had two good combines and "did a lot of custom combining on 50-50 shares. The wheat averaged eight bushels an acre, and there was enough of a crop left, so lots of cheap screenings were available which he got on trade for his combining or for little cash. In 1936 another crop failure occurred, but one could find "lots of weeds in the country," so Schwartz decided to purchase five carloads of sheep from Montana.

He was refused money by the local banks and at Minot, but he was able to borrow some through the Central Livestock Association at South St. Paul. A total of 1,780 sheep were procured and about one-half were grazed on the weeds while one-half were fed on the screenings purchased in 1935. "Even though the winter was against me," Schwartz said, "I made about $2,700 on those sheep and that money got me started in the dairy business in 1937."

"The C.C.C. came into the area and created a demand for milk," recalled the ever-alert Schwartz, "and I found out that if I was willing to set up a Grade A dairy and bid for the business I could sell milk at $.13½ a quart bottled in half-pint bottles and $.08½ in quart bottles delivered to the towns. We were in business at once with ninety-two milk cows."

From 1937 to 1943 seven men tended the dairy business, including three delivery routes, and eight other men worked in the fields and cared for 200 hogs and 200 beef cattle. Most of the 15 men had families and were employed for $75 a month plus housing, utilities, milk, meat, and potatoes. Schwartz had no difficulty keeping a labor force because there were "always people waiting for a job." With the advent of World War II, the labor supply became less plentiful, and Schwartz, anticipating a call to serve in the armed forces, decided to sell the lucrative dairy business, which, he said, "really got me into the business of buying land and farming on a bigger scale."

A Power Plant, a Seed Farm, and a Government Loan

Eugene Young attributes his long-term success to the ability "to work with men to get the job done." But he overlooked the fact that it was costly to keep those men employed and paid at a time when many farmers were leaving the land. He, like Schwartz, was alert to new opportunities in the depths of the depression that gave him the ability to sustain himself in business until the agricultural recovery of World War II. His power plant at Fort Yates, built entirely with borrowed capital, plus charge accounts, gave Young the necessary cash flow. Young was emphatic that the steady trickle of income from the plant "kept things together at the very depths and when recovery came on the land that income financed land purchases."

Even though 1932 was by far the worst year for Young's New Era Tractor Farm, other years left even less for living. After a complete crop failure in 1934, not even seed for the following year was available, but Young dared to go ahead. He purchased a 320-acre farm at Russell, Minnesota, on a contract for deed in an effort to guarantee himself a seed supply. He operated this farm completely with hired labor for five years. At the same time he rented a 640-acre farm on a crop-share contract at Buxton, North Dakota. This farm was also operated completely with hired help. Both of these additional units were significant to Young's recovery. He got seed from the Russell farm. Grasshoppers destroyed his South Dakota crop, but he harvested a crop at Buxton and secured his first government loan on that crop of wheat. He experienced rains in September in South Dakota that enabled the Russian thistle to grow. When it was about 8 inches high he cut it for cattle feed. That was his total South Dakota crop in 1934.

With the government loan money he was able to start buying land. His first purchase was in 1937 at the courthouse in McIntosh, Carson County, South Dakota. Young bid $1 to $5 an acre and over a period of three years purchased several thousand acres of abandoned lands at an "average price of about $3.00 per acre." He added, "It seems like I paid about $.75 an acre down and had the balance at 4 percent for five years." About that time his brothers went off to other ventures, and Gene Young was left alone to manage a farm that was just starting to grow.

Young, who had long employed Indians and had great faith in them, "hired prime Indian men to operate for me. We had a lot of tractors like skeleton wheel Olivers and even Caterpillar 60's which were old but ready to go. These men stayed with me until the military took them. Then I had to use older Indian men and part-time help from neighboring farmers." In 1941 Young raised some of the first winter wheat in Carson County with considerable success.

That was the year I started to roll, and in 1942 the weather, crop, and prices were good, our costs were way down, and I knew I had it made. In 1945 wheat was only $1.30, but we had big acreages. The machinery was old, but large, so we could really turn out the grain and we made it big. I wanted to put money on C.D.'s that year, but by then banker Van Nice had so much on deposit that he would only pay 1.5 percent interest.*

In 1949, at the age of 56, Gene Young and his brother George, who had rejoined him in the farming operation, could no longer buy abandoned farms, but they were able to purchase land in Sully and Potter Counties that was still in native sod. At this time new farming techniques had made some of his marginal land profitable to be farmed. This newly broken land, added to what they had turned in the past, amounted to a total of 33,000 acres, all of it plowed in a period of 40 years.**

---

*To help handle the first big-volume crop since the early 1920's Gene Young made a portable grain auger to move the grain from the trucks into his storage bins. He said, "When the neighbors saw it they all wanted to use it or have me make one like it."

**The Young brothers' father had broken land in three different areas in North Dakota in the 30 years prior to 1910.

Always Something Extra to Do

Ray Jarrett sold wool for 4½¢ a pound in 1932 and thought "that had to be the bottom." Wool at that price hardly paid, but he had faith in sheep and stuck with them. Eventually he had one of the largest sheep flocks on a crop farm in South Dakota. But more important than keeping faith in any single enterprise at that time, Jarrett maintains, was keeping faith in himself. He had this to say about the depression era:

I never felt I had any break; if there was a break it was not quitting and each time I saw someone else quit I became a little more determined. It seems that when things were toughest I always looked for something extra to do—hay buying, cattle buying, farming, threshing, trucking, or peddling. I had learned to sell on the road during World War I and the early 1920's. I knew every little town in Iowa, Minnesota, and the Dakotas. I was exposed to all types of people and I knew that farmers were leaving the land in every state. While traveling I had lots of time to think about that and that impression stuck with me in the 1930's.

After buying his first land in 1930, at the beginning of the depression era, Ray Jarrett had courage enough to enlarge his farm when he noticed that people with land for sale came to him. Every time the wind stirred the topsoil, neighbors got discouraged and quit. In his own words Jarrett described what took place:

Neighbors who were quitting just seemed to come around to tell me because they knew I'd be interested. I often thought it was just another break for me. I kept taking on more land. Many times I think I was too cautious. . . . I heard people talking so much about Jarrett going broke someday that I almost came to believe it myself. I had no one to fall back on and many times I passed up deals. . . . But man's biggest problem is to solve himself—then you go on and maybe you make lots of mistakes, but you have to be more right than wrong. . . .

W. M. Scott, who was up and down in the cattle business, once told me, "It's no sin to go broke, but it is a poor stick who stays broke." This always stuck in my mind. Walton Thorpe gave me lots of good advice too. . . .

Now that I have lived nearly seventy-four years I believe that failure and success go together. If you don't fail sometimes you never will really succeed and I have had plenty of failures.

In 1934 when cattle were selling at the established government price of $4 for calves, $12 for yearlings, and $18 for mature cattle, Jarrett bought hay to avoid selling his livestock. This indirectly encouraged him to go into the hay buying and trucking business, and both helped him develop the cash flow he so desperately needed if he wanted to continue expanding his farm. In 1935 Jarrett started buying more tractors while looking for hay because he discovered that there were "lots of good buys in both new and used tractors." He bought two more tractors in 1935, and from that time on he added two tractors annually; and each year he noted, "They seemed to get a little bigger, so the natural thing to do was to just keep growing." This he did in spite of the fact that, except for hay, 1935 was a year of total crop failure.

In 1936 he increased his trucking business again because he had a complete crop failure, including hay. That year he used a Model T chassis to make a trailer for his 1931 Chevrolet truck. It gave him a large-capacity outfit for hauling loose hay. The trucking business seemed to him to be compatible with his farming operation, that from the start had evolved around the "jockeying" of both livestock and hay.

The following year Jarrett proudly displayed a completely motorized threshing rig. He had a John Deere "A" tractor, a 28-inch thresher, three trucks to haul bundles, four bundle pitchers in the field, and two at the machine. Each farmer who hired Jarrett's rig only had to haul the grain away from the machine and have a man in the straw pile. Although this motorized rig required less than half as many laborers as the old threshing rigs using steam and animal power, it was already obsolete, for the new combines had already proven their merit. This made little difference to Jarrett, for after paying all expenses plus paying for the equipment, he pocketed $1,000 that year from custom threshing. He grinned as he remarked, "And that went a long way toward buying land in the thirties." The work on the farm was also done while Jarrett was threshing and hauling because "By then," he recalled, "the kids were a big help. My wife and children always have been real workers."

Each year more land and tractors were purchased. Millet was the only good crop in 1937. The next year was just a bit better, but 1939 proved to be the turning point, and the Jarrett farms "really started to roll."

Two significant events on the farm that year repaid Ray and Lena Jarrett for all their hard work and sufferings of the previous 15 years. Using four Model A John Deere tractors to plant and cultivate, plus some bigger old tractors to till the land, Ray Jarrett raised 3,000 acres of corn in 1939, and, he remembered, "That was the year it started to rain and we were in business because we had lots of land and a good labor supply." Once corn harvest started, Jarrett looked for a market and discovered that while whole ear corn was 25¢ a bushel at Britton, it was 55¢ west of the Missouri, just 100 miles from his farm. "That got me thinking," he smilingly said, "and I bought two new International trucks and started hauling corn."* With operating costs of less than $20 a load, Jarrett was able to gross $60 more for each load of corn that he hauled to West River Country. The net profit on just that corn hauling was enough to pay for a farm.

Later, the 74-year-old Jarrett reminisced, "I bought lots of used tractors at that time and each year got better. When you could pay for another farm in one year it made things easy. Others were still giving up so it was no trouble adding land." This was the same man who had slept in a bunkhouse until he was married, who had started with machinery purchased at his father's auction, and who had operated on the rented sandy lands north of Britton that nobody else wanted. Now he was over the hump. Then he noticed that people were becoming critical of his success. He began to understand what his dad had told him, " 'There is something wrong with a man who does not have enemies.' I have come to learn this and it hurts, but I have come to live with it. I find that I am recognized by people in other areas who have made a success of themselves. Our family likes to work hard, but we are great party people and we like to play hard too. My greatest joy in life is looking into the future for all the challenges that are still around."

These were the impressions of a man who had stuck it out when things were toughest and, once he had made a success of himself, discovered that many in his community resented him for it. Jarrett was not yet through with growth, for he was only in his early forties, but true to his philosophy he had "solved himself." Once he learned how to use the light sandy soils, Ray Jarrett knew he had found his new challenge.[5]

"The Anticorporation Law of 1932
Was My Big Break"

When Otto Engen was asked what he thought was chiefly responsible for his success in pulling through the disastrous decades of the 1920's and 1930's, he was unable to answer. Mrs. Engen, who had gone through those years with him and remembered every day's struggle, quickly responded: "Otto never gave up and

---

*To better understand the prices of the 1930's, a listing of card prices sent to the Kindred Farmers Elevator at Kindred, North Dakota, by the Minneapolis Grain Exchange October 29, 1932, will suffice: No. 1 Hard Dark Northern Wheat 60# 30¢, Amber Durum 60# 23¢, Flax No. 1 86¢, Oats No. 3 26# 5¢, Corn No. 2 7¢, and Barley No. 2 46# 11¢.

once the ball started there was no stopping." In Otto Engen's words, "I managed to keep myself busy breaking new land and keeping the plowed stuff from blowing." Of course, good moisture conditions in later years and wartime prices were also factors in the success of Engen.

Even though he had poor crops from the early 1920's to 1940, except for "a spurt of a flax crop," which he sold for $2.50 a bushel in 1935, he remained optimistic.

The Engens recalled the late 1930's:

There was a total crop failure in 1936, and 1937 was just about as bad and in 1938 we summer fallowed because there was no hope for a crop. In 1939 I had prospects for a big crop and then hail came so thick that it laid in the ditch for a week and where it rolled off the barn there were big piles. But that was sort of a turning point because we had moisture then.

In 1940 Engen sensed that moisture conditions were getting better than in any of the previous years, so he was prepared to put in a big crop. In addition, the Anticorporation Farming Law of 1932 proved to be a big break for him. When the Federal Land Bank, the Bank of North Dakota, county governments, and other finance agencies were forced to dispose of land by 1942, they looked to men like Engen as prospective buyers. In Engen's own words:

It seems like there were many who still wanted to quit farming and then these finance corporations were being forced to sell so it was easy to pick up land. The land I had bought in 1914 for $20 an acre and lost to the Federal Land Bank was sold for $2.50 an acre in 1940 when the bank knew they were going to have to unload their land.

Analyzing the Anticorporation Law, Otto Engen felt that here was the opportunity of a lifetime. He was beseeched by finance companies in 1940 and "picked up land nearly every chance I could." As he described this chance:

The Bank of North Dakota offered to rent 1,440 acres to me with the option to buy. I could rent for one-fourth crop, but if I could see that if I could pay the full $6,000 [$1.16 an acre] in the fall they would give me their fourth of the crop. When the crop was getting close I got the elevator man out to look at the crop and he agreed to advance me the money. I paid the Bank of North Dakota for the entire 1,440 acres with their one-fourth of the crop. That was not all for the Federal Land Bank practically forced a section on me for $1,600, [$2.81 an acre] and I paid for that farm with less than one-fourth of the crop.

In the fall of 1940 Engen had acres and acres to harvest, so he purchased three new combines and hired several custom machines. At the same time, North Dakota farmers were leaving their farms at even a faster pace than during the 1920's or 1930's because World War II gave them opportunities to work elsewhere. Otto Engen and his brother decided to expand as fast as they could. He explained what took place next:

> Everybody thought I was crazy because I was buying land as fast as I could, but I realized that the counties were more concerned about getting the land on the tax rolls than they were about the price they got for it. I didn't care if they [the neighbors] thought we were crazy because many who quit during the past fifteen years had told us we were crazy just to want to keep on farming. Many told us we didn't know what we were doing. Early in 1941 there were two court house auctions on the same day so my brother went to Minot, the Ward County seat, and I went to Stanley, the Mountrail County seat. That day at those two auctions we bought 3,200 acres for an average of $1.50 an acre. We paid 10 percent down and had terms of five years. If you paid in advance they discounted the purchase price. At one time I got quiet title on sixty-two quarters (12,800 acres) and only lost two by contest.

When he was able to purchase so much land in the spring of 1941, Engen had no other choice but to expand his tractor power. Machinery was not quite so easy to buy as land, but the cash flow from the crops was so much greater than he had ever experienced that it gave him courage to expand even more. A few days after a courthouse purchase of 3,200 acres, he bought four LA Case tractors for a price of less than 2,000 bushels of wheat. Fortunately, his cost of operating had not gone up much, so he was experiencing a wide operating margin. The 1941 crop was also good, and he just kept growing. For the next 32 years the Otto Engens had an easier time with farming, and they bought more land and faced every new challenge with greater confidence than during the first 20 years. They did not know how to quit and saw no reason to do so.

The Challenge of Bigness

Engen, Jarrett, Schwartz, and Young had major traits in common: they knew what it meant to put in long hard days of work, they were willing to risk making mistakes by trying new things, they thought big, and they had fun facing challenges. The tougher the obstacles were, the harder they tried. None of them was afraid to do something extra to keep the farm going. Once they started to grow they all were quick to grasp new opportunities and to use the combination of big equipment plus large acreage to surpass those around them. But there were also other characteristics of these men that helped them in their success.

Robert Peters, who farmed near Jamestown, North Dakota, during the same period, remembered that after the economic recovery many farmers were tired of battling debts and interest payments and had no desire to expand. Mr. Peters said, "I missed my chance in the mid 1930's when I bought my first tractor. My 480 acres was too small all at once and that is when I should have expanded."

Percy Willson, who started farming in 1920, recalled that during the next two decades when people left the area "they laughed at those of us who stayed." Then Mr. Willson added, "And when they came back after World War II and saw what those of us who stayed had done they criticized us for having taken an unfair advantage."

Walter E. Johnson grew up on a farm, but left to improve his education. At the conclusion of his education he had to make a choice—whether to return to the farm or remain in the city:

Farming seemed to have no future to it as far as I could see. I decided to take a trip and bummed rides to Chicago, the Twin Cities, and the West Coast. I looked at the cities and country both and when I got back home to North Dakota I decided my opportunities there were better than any place I had seen. That was the right decision, but for several years I wondered about the merits of my choice.

In general, this was the thinking of many Dakota farmers who lived and farmed through the same period as Engen, Jarrett, Schwartz, and Young. Peters, Willson, and Johnson, all capable men, had different desires and motives, but in the end they overcame the social and economic pressures and succeeded in their own way.

Direct Selling

One of the weakest aspects of most farming operations is that primary efforts are placed on production and little concern is given to marketing. Historically, many of the more successful farmers are those who have stressed the value of market management. In this respect Engen, Jarrett, Schwartz, and Young were among the pioneers in their respective areas. Three factors that helped them to become successful were that they liked to bargain with people, they knew how to sell, and they had large volumes of commodities, so that buyers were willing to work with them.

Eugene Young was the first to become involved in direct marketing. His father had done so prior to World War I, and when Young started farming on his own, first at Van Hook, North Dakota, and then during World War I at McLaughlin, South Dakota, he went into direct marketing with the terminals at Duluth and Minneapolis. In his first years he received 5¢ to 25¢ a bushel of grain more than he would have obtained from the local elevator.

I never got less than a 1 percent dockage at the local elevator but seldom ever had dockage on sales to the terminal. This irked me because I was a real crank on having clean grain. . . .

Father loaded 400 bushel cars by the hand shovel method, and us boys did our share of the shoveling up at Coleharbor. He knew about margin, dockage, and the inability of the local elevator to take grain from us fast enough. When I started farming I got the elevators to bid for my grain and if they didn't come high enough then I bargained with them on handling fees for direct marketing. Benson & Quinn or Cargill were our major buyers. In 1938 I started with GTA—I continue to sell direct even though I am a member of the local farmers' Co-op which is not a member of GTA. Too much of the local price depends on the ability of the local manager.

During the lifetime of an average farmer it might happen that he will produce a bumper crop which can be sold at good prices. In 1917 to 1919 the nation's farmers experienced this pleasant occurrence. It happened again in the late 1940's, especially in 1947-1948, and then in 1972 and 1973. In those years the crops were good, and there existed a large domestic and export demand. Eugene Young was farming on a large scale by 1946 and had to store his flax crop. In April 1946 he had 26,500 bushels of seed flax for sale at a market price of $10 a bushel. But because he had already sold so much other grain that year, the additional income tax on his sale of the flax would have amounted to 84 percent and would obviously have worked to his disadvantage. So Young went to the Internal Revenue Service to find out whether there existed a possibility for an arrangement for deferred payment. His contention was that the farmer was drastically penalized by a tax system which did not make exceptions for the "once in a lifetime bonanza year." He persisted and overcame early obstacles; he was able to get a favorable opinion from the IRS and obtained one of the first such marketing arrangements in South Dakota. He also received permission to sell on a deferred payment plan with an interest bearing non-negotiable note.

Negotiations with the IRS proceeded slowly, and Young, anticipating this delay, made other arrangements to dispose of his $10-a-bushel flax without being penalized by an 84 percent tax. He made arrangements with other farmers through seed houses to provide seed flax on a one-fourth crop basis. By distributing the seed throughout five states he greatly reduced the risk of total loss because of weather. When the 1947 crop year was over, Young had realized up to $16 a bushel for his seed even though the market price for flax was down to $5.50. But more significant than the improved price of his flax was that he discovered that some of the best returns per acre in the five-state area came from around Onida, Sully County, South Dakota, an area very near to where he had farmed since 1917. At once he sought and secured a large acreage of relatively virgin land in that area, which has since become the major part of his farming operation.

Interest in direct marketing has continued on the Young farm under the management of Dennis L. Anderson, who sells a large volume of feed grains to feedlots in Texas and Oklahoma. The truckers for the feedlots load the grain at the farm, thus eliminating the usual cost of hauling the crop to market. Further efforts to improve their marketing took place in the 1960's when Anderson, with 32 other men, formed the Owahe Grain Company at Onida. This highly successful company, with more than two million bushels of capacity, has served as a model in efficient grain marketing. The objective of the stockholders is to have the grain company pay its owners standard market prices; and if the firm makes money, the stockholders profit from their investment.

Earl Schwartz and Otto Engen followed the traditional pattern of farmers who became large enough to do their own marketing. They purchased or built their own elevators in order to sell not only their own grain, but also the grain from their combined 59 renters. The annual volume potential from the farms of these two men easily surpasses a million bushels. By having their own elevators they are able to sell their own grain as fast as their combines can harvest it, even when they have as many as 20 machines operating at one time. They are also able to haul their crops from the field directly to their own storage-on-rail siding. In this way they eliminate the expense of having to move the grain again, once they decide to market it. Both men capitalized from the government storage programs that made it very profitable to build large warehouse facilities. This also gave them the advantage of holding their grain for the traditional market highs. A further advantage in having their own elevators came from the installation of a certified seed-cleaning plant used to clean their seed grains. This also provided them with a screening supply used to good advantage in their livestock enterprises.

Marketing was the first order of business for Ray Jarrett. His father had taught him to sell vegetables from the farm as soon as he was capable. Jarrett stressed that it was his exposure to many people on those selling trips that enhanced his ability to deal with people. There is little doubt that much of the reason for the tremendous success of his farm can be traced to Jarrett's ability to sell. From selling vegetables, retailing milk, and peddling hay and corn to large-scale direct selling was a natural step.

Jarrett often mentioned that he was probably one of the largest hay producers in the United States. In 1972 he stacked over 3,000 five-ton stacks of hay, plus a large acreage of straw. In his own words: "That is 15,000 tons of hay alone, who can top that? We have sold as much as forty carloads of hay to Texas in a single year and for many years I furnished all the hay for the Milwaukee Road from central Minnesota to Billings, Montana." Hay remained a profitable crop for the Jarretts because they adopted the latest in agricultural technology as soon as it became available. "I think buying equipment has been the biggest labor-saving device on our farm," said Jarrett, "It seems each year we put up more hay but have bigger equipment and need fewer men." The Jarretts use large-scale mowers and dump rakes on some of their hay land, and self-propelled swathers on other. All hay is "put up" with three totally automated Haybuster stackers that pick up the hay from the windrow and form it into a stack. Two stack movers and two tub grinders have to work every day of the winter season to keep up with the hay consumption on the Jarrett farm.*

Around Britton, South Dakota, the rumor is that Ray Jarrett got into the elevator business because he bought the town of Newark, which bordered Jarrett's land. It had a tavern which was frequented by many of his employees and caused him so many problems that he bought out the town and closed the tavern. Jarrett smiles at the story but leaves the impression that there may have been some truth to it. He admitted that he became involved in the direct marketing of grain when he purchased the elevators at Newark (now a ghost town) and nearby Kidder. These elevators were used chiefly for his own grain, but he also bought some grain from area farmers. The railroad branch line running north from Britton through Jarrett's land accommodated about 75 carloads of freight annually for the Jarrett farms. Jarrett grinned:

---

*Ward Whitman, who uses very simple but large-scale haying equipment, probably has established some kind of production record for hay harvest on the drylands of central North Dakota. Using two 9-foot mowers on a tractor, then using a 42-foot hydraulic dump rake and tractor-mounted hay stackers, he is able to "put up" a ton of hay with 14 minutes of labor. This compares to 35 hours per ton with the hand methods of the 1870's.

Grain marketing direct has been a boon to me. It has netted us five cents a bushel over the local market. I know someone has to pay for the elevator operation, but we need the crew anyway so it is just another way of supporting year-round employment. This way we can clean all of our own grain and save the shipping cost on screenings; besides, we need them for our sheep. It helps, of course, that we were able to buy the elevators at the right price; but they weren't paying off very well for the previous owners.

When Ray Jarrett's livestock volume became very large, a natural flare for showmanship appeared in him, for he decided to make an exhibition of his cattle shipments. The headlines read, "25-Car Cattle Shipment Makes Marketing History." This represented the first special livestock train ever made up by the Milwaukee Road for a single shipper east of Aberdeen, South Dakota. The details of Jarrett's first big cattle shipment of January 8, 1948, indicate that the 25 cars were loaded at the rate of one every five minutes. There were over 600 cattle, which averaged 855 pounds, and sold for from $25 to $27 a hundred. The cattle train was hooked to a passenger train after it reached the main line to be speeded to the stockyards at South St. Paul.

The *Minneapolis Tribune* and the *New York World Telegram* saw fit to run a United Press wire story of the 1949 shipment of 22 cars with 612 head of cattle and 28 cars containing 2,719 sheep. Two locomotives rushed this special 50-car train with over $170,000 worth of livestock to the market in South St. Paul.

The January 20, 1950, shipment had "only" 23 carloads of cattle and hogs and 22 carloads of sheep, creating a freight bill of over $3,500. Six carloads had to be sold a week earlier because of adverse weather. The cattle for the shipment spread out over three miles of road as they were driven to the Jarrett stockyards at Newark. The division superintendent of the Milwaukee Road sent Jarrett a letter thanking him for the excellent publicity their company had received because of his large shipment of cattle.

On February 23, 1951, Jarrett's annual shipment consisted of 29 cars of cattle, 3 of hogs, plus 7 of sheep in the previous week. Although it was the smallest of his annual shipments, it ranked first in gross receipts—$250,000. His final trainload exhibition on February 1, 1952, consisted of 30 cars of cattle and 5 of hogs. The news articles, which were less dramatic than in previous years, commented simply that Jarrett was the "biggest single farmer shipper year after year on the Milwaukee Road." That year, as in previous years, the consignment was made to five commission firms at the South St. Paul market.

Not content to hold single-shipment records at South St. Paul, Jarrett, following the suggestion of his longtime friend and fellow sheepman, Art Moyer, decided to consign a shipment of livestock to West Fargo. In January, 1955, the *Western Livestock Reporter* noted that Ray Jarrett had made one of the largest consignments in the history of the West Fargo Union Stockyards. His shipment of 3,400 sheep was delivered by 9 semitrailers on Monday and 10 semitrailers on Wednesday of the same week. In retrospect, the 74-year-old Jarrett said:

I no longer ship train loads of cattle to market. We are big enough and have bred quality animals so there is a steady demand from a large area for our bred heifers and yearling steers. Our lamb crop is sold on the place to packers—as many as 6,000 head in one crack. Direct selling like this has eliminated our trucking expense and reduced sales cost and has given us a very reliable limited risk program. We live in a feed deficit area and this year we were able to sell our corn production from 2,700 acres. We just combined it, ran it over our scales, and got paid by the feeders who bought it.

But after the glamour of the shipping show disappeared, Jarrett realized that there were few economic gains to be made from spectacles and the risk of loss was too high.[6]

Social Resentment

In most communities social resentment seems to build up against those who are successful. In small rural communities this feeling probably is stronger than in the larger, more metropolitan areas because of the closer relationship among individuals. Engen, Jarrett, Schwartz, and Young were all highly motivated men, and antagonism arose against them quite naturally, even though they sought to avoid it. Earl Schwartz describes this reaction:

When I bought four quarters of land in Burke County for $4.00 an acre at a $1.00 down and four years on the balance, there was little competition for the land. That year the crop from one quarter paid for all four quarters. Those four quarters went on to generate enough income to pay for fifty-five quarters in Burke County. That year I also purchased a big Caterpillar which pulled six bottoms on sod-breaking, besides a disc and a press drill. I used it every year on new sod land that I bought until I quit buying land in 1969. On all the land that we acquired we only inherited two sets of buildings so apparently most of it had never been settled and if it had ever been broken it was left to go back to sod.

Everything was fine until the late 1950's when we noticed that people were getting downright angry against us. They started to refer to us as the land hog, so with that kind of social pressure, even though I liked the challenge of continued growth, I decided to stop buying land and looked for other investments. I know this has not reduced the community resentment even though we have tried to lessen it by being boosters of the community.

Mr. and Mrs. Otto Engen did not change their ways of frugal living even after their farming became a great success. They both liked to work but also enjoyed the fruits of their labor:

> After we got to a certain size it seemed that the social pressure hit us from neighbors who had not expanded either by buying or renting when we did. We feel that pressure now from those who told us in the 1930's and early 1940's that they thought we were crazy for farming and that we didn't know what we were doing. Now that we are successful they think we should let them rent our land. Maybe we were lucky because we were never really turned down by a bank and the Minot Credit Company always worked with us. We made good deals with the Federal Land Bank and Bank of North Dakota once they found out we were not afraid to stick our necks out.

At the age of 77 Otto Engen had purchased land and broken new sod every year for 32 years and saw no reason why he should curtail his activity. He learned to ignore the social resentment of those who once thought he was "crazy for farming."

Eugene Young had always been an active civic leader in his community, and the few who were left when good fortune returned to agriculture remembered the struggle he had endured. One advantage for Young in overcoming resentment was the fact that he had his farming operation in more than one location. Another was that he had farmed with his brothers, and this lessened the impact of size somewhat. A third factor was that the largest holding of land was acquired in a community where many outsiders were entering and building large units at the same time. These newcomers were less tradition-bound and simply took advantage of an opportunity available to them from which the natives had walked away. The Youngs encountered less envy because they did not live in the community and hence were less exposed to local social pressures.

Eugene Young, when interviewed at the age of 79, was still looking for new ventures. Besides buying land he wanted to build a museum to house antique agricultural machinery at Rapid City. "And then," he said, "when I get old I would like to build a home for the aged on this hill where we are sitting." At heart he was still young and not ready to quit.

Ray Jarrett and his family would provide for any sociologist an ideal topic in a study on community attitudes. Jarrett frequently commented that since the age of 10 he dreamed that some day he would be the state's biggest farmer. That ambition alone must have sustained him during the trying years in the 1920's and early 1930's. As the 74-year-old Jarrett reminisced:

Dad farmed with sixty horses before the twenties so I was used to thinking big. I had over a hundred horses myself in the thirties even though I had a Farmall by 1929. Over 10,000 acres of crop land were operated on this farm with Model "A" John Deeres. By the 1950's we had as many as seventy men working for us. The boys really pushed me to get bigger equipment about 1955 to be more efficient. Now with more crops and more livestock than ever we get by with thirty-five men. Of course my wife and children have been real assets, the boys are both good managers and my daughter, Fern, could really handle men and knew every detail about the farm.

One of the best-informed businessmen in Britton said about him:

There is jealousy in the community, but it is mostly petty stuff. To the best of my knowledge Ray has never taken advantage of anyone in the area who is in trouble, but he has been a tough competitor for the prosperous farmers who want to buy land. His operation may be a mixed blessing to our community. He trades with as many merchants as possible. It is difficult to determine whether or not the buying power of thirty-five separate farms would be greater than it is at present with thirty-five families employed by the Jarretts on one farm.*

It seems that those who are most critical of him are those who understand him least. Jarrett makes his feelings known, but he is a real community booster and wants to be the friend of everyone in town. Generally it seems that people like working for the Jarretts. One of the reasons that some don't like him is that he makes a point that except for the Wool Growers Association he has little time for farm organizations. He is a loner, but his sons and their families are active in the community and are respected as good managers and widely recognized for their significant civic contributions state-wide even more than locally. There is no community concern about corporate farm inroads from people such as them.

When asked about community criticism of either himself or his large farm, Jarrett thoughtfully remarked:

Are we too large? I don't know how large a farm should be. This depends upon the individual. If I had twenty more years to live I would probably double what we are today. I am sure I could do it. Labor and real estate and income taxes are my personal concerns. There are too

---

*It is the opinion of this writer that the standard of living of the thirty-five families working for the Jarretts is higher than it would be if the land were subdivided into thirty-five farms. The economy of scale, the excellent management of this farm, and the knowledge of employee income in contrast to what they could make on thirty-five individual farms is the basis of that conclusion.

many programs to help those who are really only half way interested in making a go of it. I believe there is nothing like the challenge I had to get started, but too many of those who criticize me don't know that story.

In a *Minneapolis Tribune* article of the early 1950's is found the following statement:

Jarrett's farm simply does not fit the conventional argument over rights and wrongs of large landholdings. It is the biggest in the area, but it is no corporation farm, with an absentee landlord who has no particular concern for the land and the community. Jarrett cares a great deal. . . .

Jarrett says he has learned to take criticism because of his size. "We've never frozen anybody out . . . and if we have grown it's because we never spent our rainy days in town. We saw the possibilities in this sandy land when nobody else did, and we've done a lot of good for this country." . . .

Farmers nearby resent Jarrett's size, but the townspeople of Britton show no antagonism. They joke about the mistakes he has made but commend him for being a good community man. . . . He has a good reputation of being a good fund raiser for community and charity projects.

A. E. Stoa, Assistant Vice President of Northwest Bancorporation, wrote to Jarrett in regard to the above article as follows: "I would certainly take exception to one statement made in the article concerning the resentment of the people in your community because of the size of your farm. From my experience in visiting with people in your area, I have heard nothing other than good concerning Ray Jarrett as a neighbor."

Ray Jarrett, a born entrepreneur, had lost little of his spirit up to his eighth decade of life. But he also was a man who intensely wanted to be liked by people in spite of his natural aggressiveness. At 74 Jarrett was still saying, "There is an opportunity in anything if it is handled right—farming is no exception. I would like to start all over again."[7]

NOTES

[1]Interview of Eugene M. Young, Rapid City, South Dakota, April 9-10, 1972, August 26 and 29, 1973, at Fargo, North Dakota; Typed manuscripts of the high points of the "Life of Eugene M. Young" in possession of author; Interview of Dennis L. Anderson, Rapid City, South Dakota, April 9, 1972. Mr. Anderson is the manager of the Young Farm properties and the owner of D. L. Anderson, Inc., a farm management firm in South Dakota, the largest stockholder in the 33-member Owahe Grain Company, and a founder of a bank specifically oriented toward agriculture; Interview of George Young, Fargo, North Dakota, August 29, 1973.

[2]Interview of Mr. and Mrs. Otto Engen, Minot, North Dakota, April 3, 1972.

[3]Interview of Ray S. Jarrett, Britton, South Dakota, March 2, 1972, December 14-16, 1972; Interview of Donald Jarrett, Britton, South Dakota, March 2, 1972; *The Wool Sack*, August, 1952, found in the clippings of Jarrett Farms and Ranches office files.

[4]Interview of Earl Schwartz, Kenmare, North Dakota, April 4, 1972; Letters from Earl Schwartz dated March 30, 1973, and January 26, 1974; *Land of Plenty*, p. 58;

Post card prices Minneapolis Grain Exchange, October 29, 1932, in possession of Mrs. Ralph Scott, Jamestown, North Dakota.

[5]For a detailed exploration of the Anticorporation Law of 1932 and the exodus of the farmers from North Dakota since 1920 see Elwyn B. Robinson, *History of North Dakota*, Nebraska Press, Lincoln, 1966; Interview of Robert Peters, Jamestown, North Dakota, July 11, 1973; Interview of Percy Willson, Wimbledon, North Dakota, March 2, 1972; Interview of Walter E. Johnson, Courtenay, North Dakota, March 8, 1972; Interview of Charles Card, editor *Britton Journal*, Britton, South Dakota, December 16, 1972; Interview of Larry Gustafson, attorney-at-law, Britton, South Dakota, March 2, 1972.

[6]A six-page hand-written article by Eugene Young on his working with the Internal Revenue Service to obtain the deferred payment arrangement, in possession of the author; Interview of Ward Whitman, Robinson, North Dakota, February 2, 1974.

[7]*Agricultural Markets*, January 14, 1948; *Minneapolis Tribune*, January 14, 1949, March 4, 1951, August 18, 1957; *New York World Telegram*, January 18, 1949; Letter from Division Superintendent, Milwaukee Road at Aberdeen to Jarrett in the files of Jarrett Farms and Ranches; *Milwaukee Road Magazine*, April, 1951; Interview of R. D. Olson, Fargo, North Dakota, April 28, 1972; *St. Paul Pioneer Press*, February 25, 1951; *Western Livestock Reporter*, January 17, 1955; Letter in Jarrett files from A. E. Stoa, Assistant Vice President, Northwest Bancorporation, August 20, 1957; Letter in Jarrett files from A. C. Miller, Kennebec, South Dakota, March 7, 1951; Considerable additional information about the attitudes of the people of Onida and Britton, South Dakota, and Minot and Kenmare, North Dakota, was gathered by conversation with citizens of those communities who preferred not to be identified but in every case were well acquainted with Otto Engen, Ray Jarrett, Earl Schwartz, or Eugene Young. The insights of these people were a valuable contribution to this chapter.

CHAPTER VII

Roswell Garst—The Innovator

Roswell GARST, interviewed during his seventy-third year, was asked who in his opinion was the most influential man in American agriculture. He replied, "I have probably influenced agricultural people as much as anybody, but I am really not trying to lead someone." For those who knew Bob* Garst the answer was not surprising, for he will be remembered as an innovator, supersalesman, and motivated teacher of farming. Trained in the college of experience, the outspoken advocate of corn, cows, and beef seldom took a back seat to anyone.

The offices of the Garst & Thomas Company and the Garst Farms at Coon Rapids, Iowa, are so much a center of agricultural activity that a visitor gets the feeling that he is near the nerve center of American agriculture. Undaunted by a throat mike, Garst's voice dominates every conversation. His great achievements, bold innovations, and even bolder language have made him a powerful spokesman in the nation's agriculture. He has earned his position by being a founder and partner in a firm that possesses several seed corn processing plants, including the largest one in the world at Coon Rapids, Iowa. In 1971 their corn production amounted to 2.5 million bushels from 17,800 acres of crop land. Of this amount enough seed was selected and sold to plant 7 percent of the nation's corn-growing acres. In addition, the company processed enough sorghum seed from its Garden City, Kansas, plant to supply 15 percent of the national requirements for that crop. At the same time Garst's farming enterprise had sprawled over a half township, which contained a home-grown cow herd of about 4,000 head and a feedlot for the calf crop. The integrated farming enterprise employed 35 men the year around.

Life had never really been difficult for Roswell and Elizabeth

---

*Roswell Garst is commonly known by his nickname, Bob.

Garst. Viv Bell, who farmed for Garst for 20 years, said that when he first knew Garst, "he really had to scratch." What Mr. Bell didn't know was that Roswell Garst freely admitted that even though he occasionally complained about hard times he never experienced them. (He paid inheritance taxes on "a substantial estate" in the 1920's.) Garst was never near insolvency; and it is to his credit that even though he was financially secure, he lost none of his enthusiasm to build a better agricultural business. This makes him an exception to the common "rags to riches" myth.

Early Years

Roswell Garst's grandfather, Dr. Michael Garst, had started a store in Boone, Iowa, in 1866 and failed. Michael Garst's son, Edward, was graduated from the University of Michigan in 1869, and after securing the unsold inventory of the defunct store of his father, he started business in Coon Rapids, Iowa. Edward Garst complemented his store with his income as the postmaster in Coon Rapids from 1871 to 1885. In 1878 Edward Garst married Bertha Goodwin, a graduate of Northwestern University in 1878, and started to buy land.

Immigrants from Illinois, Indiana, and Ohio moved into the Coon Rapids area in great numbers after the completion of the Milwaukee Road in 1881. The new settlers purchased land for $5 an acre by paying $1 an acre down and the balance on contract. When some of them encountered difficulties Edward Garst would give them clothes or other store supplies for their equity in their homesteads and assumed their land contracts. In this manner Edward Garst amassed "many thousands of acres."

Son Roswell, who was born in 1898, was "raised the son of a prosperous merchant who owned much farm land." Roswell commented, "I did not know farming, but somehow was encouraged to become a farmer."* His father's store continued to prosper, and the land he had accumulated increased in value.

Roswell and his brother Jonathan gained their first farming experience on the Apple Valley Farm starting about 1915. This 200-acre farm that Edward Garst had owned for 30 years had never netted him over $500 a year. When Roswell and Jonathan started farming there they worked with eight horses and had "twenty and thirty head of cows, fed out a couple of carloads . . . of steers and kept maybe ten sows and fattened their pigs." Jonathan wrote, "In

---

*Roswell Garst's uncle, Warren Garst, was elected Governor of Iowa in 1906 and later held the governorship from November, 1908, to January, 1909.

1915 Thomas Jefferson would have felt very much at home on the farm at Coon Rapids. . . ." The local corn-picking champion at Coon Rapids could pick 100 bushels a day. World War I came, horses were needed for the war, and farmers started buying tractors to replace the horses.*

Jonathan and Roswell both liked farming, but both also wanted an education and an exposure to other ways of life. Jonathan entered the service in World War I and then went to Canada to farm. Roswell stayed on the Apple Valley Farm until the fall of 1919, when he had an auction and entered Northwestern University. Later, Roswell Garst commented, "By not farming during 1920 and 1921 I missed two very bad years by luck."

During his nearly three college years Bob Garst said he "took everything from blacksmithing at Ames to philosophy at Northwestern, so I am not highly educated but broadly educated." When he decided to start farming again in 1922, he returned to Coon Rapids and married Elizabeth Henak, to whom he credits much of his success. Bob Garst said, "Nothing exciting happened—I milked cows and sold whole milk and I learned that agriculture was a very tradition-bound industry."

After four years of farming, restless Roswell and Elizabeth Garst decided to try their luck at selling real estate in Des Moines. That proved to be a turning point in their lives, for quite by accident they became friends of Henry A. Wallace, who became Secretary of Agriculture under Franklin D. Roosevelt. Roswell grinned, "We became very good friends because we both had such an intense interest in farming." Wallace had been working on the development of hybrid corn since about 1915, and Garst was a receptive listener to Wallace's ideas. When Garst farmed in the early 1920's, he had heard much about the corn borer invasion and its damage to the corn plant. As a precaution he planted some sunflowers. They were high in protein and oil and made good chicken feed, but they were no substitute for corn silage. It was this experience that made Garst so interested in Wallace's ideas on hybrid corn. As an afterthought Garst remarked, "We were scared of the corn borers, but they did not get here until 1949; my worries of the 1920's were in vain."

---

*Jonathan Garst once asked a neighbor how he liked his new tractor. The farmer replied, "I used to be able to turn the horses into the furrow and sit on the plow and watch the hawks and chew tobacco. Then at the end of the field I'd turn the horses around and that was it. Now I have to drive this damned tractor every foot. I don't get time to spit."

In 1930 Elizabeth and Roswell Garst decided that farming in Coon Rapids was more challenging than the real estate business in Des Moines. With three young children, they returned to the Apple Valley Farm to start farming once again. Elizabeth and Roswell preferred to farm; but they also had the feeling that the years ahead might be lean, and they "desired to sit it out on the farm as a peasant rather than as a pauper in town."* Things were gloomy indeed in the Coon Rapids area in 1930. Armour and Company made the announcement that it was closing down 13 cream, egg, and poultry stations in West Central Iowa. An article published at this time by the State History Department, entitled "Abandoned Towns, Villages, and Post Offices of Iowa," indicated that nine towns had already disappeared in the Coon Rapids area. Of the 1,046 farms in Carroll County only 413 were mortgage-free; the 634 that had loans were mortgaged at 59 percent of their value. It was a very serious situation, considering that this was one of the best farming regions in the nation. Number 2 Yellow Corn was down to 31¢ a bushel, oats 20¢, and top hogs were $4.60 a hundred.

Even the Chicago & Northwestern Railway was selling much of its remaining land as fast as it could because it was no longer able to realize a fair return from it. By 1930 the price of land was down to less than half of what it had been worth prior to 1921. At farm auctions the farmers were paying cash because they had little faith in the banking structure. One large farm sale, which had 237 head of livestock, grossed $12,500 and received the total payment in cash except for $200 by check. At that sale "a Des Moines politician was actually applauded when he shouted to a group of farmers that they are already reduced to peasantry."

Elizabeth and Roswell Garst were in the middle of all this uncertainty, but kept moving ahead as if the opportunity would never be better. Viv Bell, a tenant for many years, said, "Garst never gave up and really knew how to get the job done." Garst, always alert, had a half dozen experiments under way most of the time. In the early 1930's he received a shipment of four dozen

---

*Roswell Garst had a favorite story about his father, which he often told. Edward Garst had purchased considerable land in the late 1800's, and in the 1890's he tried to increase the loan on some of it. When the representative of the loan company stopped at the farm, Jerry Croak, the tenant, degraded the farm so seriously that the representative considered asking for the immediate payment of the existing loan rather than granting an increase. When Croak later was asked why he had run down the farm he answered that he thought the man was the assessor. But even after an explanation of the tenant's false statements was made Garst was unable to get a loan increase.

Kiki Camel duck eggs from England. These eggs were in incubation, and Garst wanted to raise them because they were known for their high laying ability.

Garst started headlong into tractor farming because, as he figured it, "horses cost five times more per unit of horsepower than a tractor did." Besides, for each colt he raised to working age on his farm he lost the chance to feed four steers for two years. He could not afford this because he needed to sell steers to raise his income. By spending the same amount of money for tractor horsepower as he did for animal horsepower, he could do more than five times as much work. Roswell also needed to be freed of some of the sheer drudgery of working the soil because he became too much involved in other activities. Cooperating with Farm Bureau and Mr. Vial of Iowa State College, he helped to promote the feeding of minerals to livestock to improve feeding efficiency. At the same time he was also being promoted as a Republican candidate for the governorship of Iowa. The newspapers of many parts of the state "felt he was a winner"; but Garst did not accept this challenge, saying, "I am definitely not a candidate." Besides, he was already quite involved in the development of hybrid seed corn, for which he saw a greater future than in politics.[1]

Garst—Sells Hybrid Corn

Mr. Henry A. Wallace was pioneering in the breeding of hybrid corn; and since the Garsts knew the Wallaces, it was only natural that Garst should buy some of Wallace's hybrid seed as soon as it was available. Garst invested in his first bushel of hybrid corn for personal use in 1927, increasing his purchases to two bushels in 1928 and four bushels in 1929. When the Garsts returned to full-time farming in 1930 they had the good fortune of acquiring enough foundation corn seed from Henry A. Wallace to plant 10 acres. That foundation seed was the beginning of a success story for Roswell Garst and the dawn of a new era for agriculture.*

In 1930 Garst became allied with Henry A. Wallace to produce and distribute hybrid seed corn. That was a hot, dry year, but he produced 300 bushels of seed, which he stored on the front porch

---

*Garst noted that the hybrid was a very high eared corn and yielded about 10 bushels per acre more. In 1930 Garst had 50 bushels of corn allocated as experimental programs with cooperating farmers. The corn was planted three kernels per hill 42 inches apart in rows that were 42 inches wide. A bushel of seed planted 7 acres, and even though no fertilizer was used the average yield increase for the 50 bushels seeded was 11 bushels per acre.

of his home. The total national output of hybrid seed at this time was not enough to plant one township. But through promotion, hybrid corn sales increased and the national average yields increased correspondingly from 26 bushels in 1930 to about 40 bushels by World War II. Nearly all of the increase was because of the use of hybrid seed.

Roswell Garst knew how to farm and to promote farming, but he lacked two basic premises to make his ideas pay off: (1) the detailed mind of a production manager and (2) enough money for such a sizeable enterprise. This led to another major event—the beginning of a 42-year partnership with Charles and Bertha Thomas. These two men, although never close social or personal friends, proved to be an ideal business combination. They understood each other's strengths and in partnership created one of the world's largest seed corn farms as well as extensive personal farming operations.*

John Strohm, a well-known agricultural journalist, wrote "Bob Garst . . . farms big, thinks big and acts big . . . [while fellow farmer] Charles Thomas was extraordinarily insistent about perfection in production," which made him a natural partner for promotion-minded Garst. The partners were both innovators and were interested in improving their product. The best way they could improve their seed and convince the public of its value was through widely scattered test plots. In 1931 the newly formed Garst & Thomas Company had hybrid corn plots with cooperating farmers in every township in Humboldt County and several in Webster County. Those two counties were reasonably close and were better corn-producing areas than the region around Coon Rapids. The results were gratifying.

But those early years of farming and producing hybrid seed corn were frustrating for Bob Garst and Charlie Thomas. The west-central district of Iowa, which normally has a precipitation of 29 inches, had three years from 1933 through 1939 in which there was less than 21 inches of precipitation and another with just over 23 inches.

The drought-depressed yields of those years combined with low prices forced Garst into extensive participation in government programs. In spite of being an agressive, free enterpriser he became deeply involved in the government programs, which were contrary to his basic philosophy. Garst, a staunch Republican like

---

*The Thomas Charolois herd was reputedly the largest in the United States by 1959.

his friend Henry A. Wallace, took an active role in formulating the farm program of the first administration of Franklin D. Roosevelt. Roswell Garst was one of 25 farm leaders from Iowa to sit on the original committee to formulate the Federal Farm Program in March, 1933. Ray Anderson, prominent middle-western agricultural journalist, first came to know Garst as a member of the "original corn-hog committee." Garst "agitated for a processor's tax to finance the farm programs and was one of those who encouraged Wallace to destroy little pigs" to keep them off the market.

Because of his alertness and ability, Garst continued to remain in the forefront of government agricultural programs. As his business grew and his circle of political acquaintances enlarged, his conservative Republicanism became modified. By the 1940's, Garst had a desire to avoid political controversy and preferred to be called a political independent, even though his deep-seated defense of free enterprise and rugged individualism remained with him. In early 1940 he was quoted, "There isn't a chance the farm program will be done away with. . . . We should hold on to what we had and improve it as we go along." Like many farmers who might have preferred a different solution, the guarantees of the government payments were too tempting for Garst to resist.* Besides, the programs were causing a virtual technological revolution in agriculture that increased hybrid seed corn sales enormously.

Drought and low prices of the 1930's were not the only cause of frustrations for Garst and Thomas. Even more significant was the farmers' refusal to buy hybrid seed. In one of Dave Garst's speeches he repeated an oft-quoted statement of his father that Roswell credits to John Stuart Mill, "The despotism of custom is everywhere a standing hindrance to human advancement." Probably no other technological advancement so significant to mankind has better shown the resistance of man to change than did the farmers' refusal to buy hybrid seed corn in the 1930's. Interestingly, Iowa State College faculty members were shocked by the farmers' refusal to accept a new product that was almost certain to increase profits. Some significant sociological studies on

*The editor of the *Coon Rapids Enterprise* understood the political significance of the government payments on the farmers. On October 12, 1934, he wrote "Carroll's [County] Bonus Checks—We don't bet on politics but we are willing to hazard one little guess: we predict the corn-hog bonus checks will arrive before November 6 [election day]." On October 19, 1934, the headlines read, "$384,310 Corn Hog Checks Arrive in the County." It continued and said that farmers were paid 15¢ for each bushel of corn not raised and $2 for each hog not raised.

traditionalism have evolved from that research about the farmers' hesitancy to adopt hybrid corn.

Scientists had worked on the development of hybrid seed corn since 1913. It was not until 1926, however, that foundation seed was sold, and it was not until 1930 that hybrid seed became commercially available. That year Garst and Thomas produced 300 bushels. A newspaper account stated that "Inbred Corn Promises to Give Real Farm Relief." The story reported that Garst and Charles Rippey had visited the Henry A. Wallace farm at Grimes, Iowa, and that they were "enthusiastic about the possibilities of breeding drought resisting corn. . . . Garst has thirty acres of inbred corn which he has carefully detasseled."

Wallace, Garst, and Thomas thoroughly understood the value of hybrid corn to the farmer, but they were also aware of the farmers' resistance to change. Even Apple Valley Farm, one of the first farms in the world to produce hybrid seed corn, did not greatly impress the neighbors. The key to success would be the salesman, and because of the personalities of Wallace and Thomas in the total venture, Bob Garst had to be that salesman. He was a natural for that difficult job. With determination Garst gradually found people receptive to his product; and their example awakened others, until a new era eventually came to agriculture. That promoting job alone would indelibly stamp the name of Roswell Garst in the pages of agricultural history. The job was slow, and by 1936 only 10 percent of the corn acres nationally were planted to hybrid seed. In the next decade, however, figures grew to 80 percent of all acres planted to corn. It took nearly 15 years in Iowa alone to convince most of the farmers of the value of hybrid seed.*

The time lag of acceptance of hybrid seed corn encouraged Neal Gross and Bryce Ryan of Iowa State College to do intensive research about the causes for that lag. This original study has been followed by many others on the same topic. Through the research it was discovered that the innovative farmers on the average tried hybrid seed about 19 months after they first became aware of its existence, while the laggards required about 9½ years from awareness to trial. As late as 1943, after entire communities had proven the worth of hybrid corn, many of the last to try it would do so only on an experimental basis to convince themselves. This

---

*Total hybrid seed corn sales in 1930 were less than 4,000 bushels out of the 20 million bushels of seed necessary for the 101 million acres of corn planted. That was 2/100 of 1 percent.

was true even though they were surrounded by neighbors who had successfully used hybrids for years.

Garst knew that hybrid seed could produce at least 10 bushels more corn per acre, using the standard cultural methods of the 1930's. Because of that he could not understand the reluctance of the great majority of the farmers to buy the hybrid seed.* He tried every sales technique possible to get farmers to buy his seed. The most common device used was to get farmers interested in accepting a bushel of seed and paying for it in the fall with the increase in yield from one acre. He knew that a bushel of seed would be enough for planting from 4 to 8 acres. Garst seldom took corn in trade because by fall the farmers could visibly see the increased yield and were willing to pay in cash.**

From 1930 to 1938 Roswell Garst spent almost all of his time traveling the dirt and gravel roads of Iowa trying to deliver the message of hybrid corn to the farmers:

I did not know anything about the genetics of corn and neither did any of the farmers, but I did know how to sell corn hybrids by demonstration. When I had 90 percent of my seed sold I broke up the remaining 10 percent of my bushels into seven bags of eight pounds each. Then I gave each farmer enough to plant about one acre. Then I followed through with a yield test on those acres and the next year I knew each eight-pound gift bag would produce another customer. Of course, we had to work on the best farmers first. . . . The rewards go to those who look ahead. Historically, it is the first farmers who adopt a new practice who reap the greatest rewards. This was particularly true with hybrid corn because the price of corn was based on the lower yield of open pollinated corn.

Garst was not betting his entire future on hybrid corn alone; there were other improvements to be made to increase the farmers' income. In 1930, he and Jack Chrystal experimented with flax and had satisfactory results.

---

*The average national yield increased from 20.5 bushels per acre in 1930 to 28.4 bushels in 1940, while the Iowa average jumped from 34 bushels to 52.5 bushels for the same year. Almost all of the increase was because of the adoption of hybrid seed. Even more significant than the mere increase in the decade of the 1930's was the fact that the national average corn yield per acre had been relatively static for about 40 years prior to that date. This made the hybrid breakthrough even more conspicuous to those observing national production trends.

**Nearly one-half of the farmers indicated that seed corn salesmen were their first source of knowledge of hybrids, and it is apparent that expensive sales campaigns were also conducted by the corn companies to promote their product. Ironically there was no organized resistance to hybrid seed corn because there was no one who had personally lost by exchanging hybrids for homegrown seeds.

Henry Brown, one of the first farmers around Coon Rapids to have a mechanical corn picker, reported that in 1929 his two-row tractor-powered picker had harvested 22,000 bushels of 100-bushel corn in 21 days. Brown must have been an innovator like Garst and had high corn yields at that date; he also owned a two-row P.T.O. corn picker the year after they were first manufactured. His example was well received, however, for in 1930 Roswell Garst and Charlie Thomas were among 23 farmers in the Coon Rapids area to harvest with mechanical corn pickers. The labor cost of hand picking had forced them to change to the cheaper mechanical methods.

The rapid adoption of the mechanical corn picker also helped Garst and Thomas sell hybrid seed, for the picker required a corn variety with a stalk that would stand erect for the mechanical gathering chains. At the same time, it made the farmer more aware of his potential for growing corn, for he was no longer limited to 100 bushels a day per man picking capacity.* Where the Indian had required about 400 minutes of labor to produce a bushel of corn, the average horse-powered Iowa farmer needed only 70 minutes, and the best could do the job in 30 minutes. By adopting tractor-powered equipment the time requirement reduced this even more. All this helped to increase seed corn sales, which grew to 135,000 bushels for Garst and Thomas by 1940. After overcoming the initial obstacles even greater growth was just ahead.**

In 1932 corn prices fell to a low of 12½¢ a bushel, and everything seemed at a standstill. But in 1933 prices started to rise again, and Garst and Thomas became optimistic because they had a feeling that the worst of the depression was over. Pushing the sales, they were able to sell out the total production of what was then called the Hi Bred Corn Company. By February, 1934, the entire 10,000 bushels of seed produced was gone and late orders were being returned. A little note was included stating that the company hoped to have more seed available for 1935.

---

*The corn husking champion in Carroll County in 1938 was able to husk for 80 minutes in the contest at a rate equal to 168 bushels in a 10-hour day. Bert Hanson, who was Minnesota State Champion twice in the 1930's, was able to husk 3,655 bushels of corn in 33 days in 1933. His best single day was 159 bushels in 10 hours and 20 minutes.

**Bob Garst, Herman Riis, and John Walsh all had an unexpected profit from their 1930 corn crop. A Dubuque, Iowa, firm was in the market for buying 5,000 tons of stalks annually to make wall board. After the cattle had grazed the leaves from the stalks the stalks were harvested and baled. They averaged about 1 ton per acre, which sold for $7.20 a ton. Garst was still a believer of humus in the soil, but to earn the extra income he was willing to sell the stalks.

During the summer of 1934 a new drying and grading plant was erected in Coon Rapids. The driers had a capacity of 100 bushels an hour. Garst and Thomas had 70 men detasseling 500 acres around Coon Rapids and another 110 men on 750 acres in scattered areas of Iowa. By August it became obvious that the new plant would not be needed because of drought conditions. Both 1933 and 1934 were dry years, and the yield was seriously reduced. Open-pollinated corn in the Coon Rapids area was expected to average less than 15 bushels per acre.* The disappointed partners needed only 35 men each on two shifts for 10 days to process their 10,000 bushels of seed. They had hoped to produce at least 30,000 bushels of seed.

The 1934 crop was less than one-half of the 1933 crop, and the price of corn rose from 35¢ a bushel in December, 1933, to 90¢ a bushel in December, 1934. After that, prices steadily dropped to 46¢ in December, 1935. The drought of 1936 sent prices "soaring" again to $1.24 in April of 1937, but good moisture conditions in 1937 brought the second largest crop of corn in Iowa's history (498,690,000 bushels). The result was a low corn price of 33¢ a bushel by harvest time of 1938, not much better than the price of 1932 and 1933. It was not until December, 1941, that the price of corn rose above 60¢ a bushel and remained there. The impact of hybrid corn was obvious by 1937. Since then with few exceptions the Iowa average yield has climbed steadily, reaching 100 bushels per acre in 1971.

In 1937 Garst and Thomas carried the sale of hybrid seed corn outside of Iowa into five states to the south and west, an area that was "nearly all vigin territory for hybrid seed corn." In 1937 the company's field plots took first place in the Iowa yield contest. This achievement gave them good publicity, and seed corn sales skyrocketed. The new low prices for corn had forced farmers to seek ways to reduce the cost of production, and hybrid seed was the best available alternative. By early 1938 seed sales surpassed 60,000 bushels in 13 states, and orders were still coming in strong.

Plans were made to increase foundation seed planting by 1,000 acres and to build a new plant that could process 4,000 bushels a day, four times the volume of the plant built in 1934. The new plant had a capacity of three times that of any other plant in the nation and was the largest corn plant in the world. In September 1938 it had 500 men on the payroll, who handled 8,000 bushels of

---

*Iowa average corn yield in 1934 was 23 bushels per acre. This compares with 40 bushels in 1933 and 43 bushels in 1932.

corn a day from the 3,700-acre crop. Eight furnaces consuming 200 gallons of fuel an hour were needed to dry the corn to 12-percent moisture. In 1938 Roswell Garst was caught short, for production was up again. The state average yield reached 46 bushels, which surpassed all previous yields except in 1912 and 1920. But more important to Garst was the average yields realized by the leading corn farmers in the Coon Rapids area. Wallace Knapp had 226 acres that averaged over 100 bushels per acre. Hosea Heath, from Carroll, had 25 acres of hybrid corn, from which his county agent measured 120 bushels per acre.

Fantastic yields like that proved what Garst had been expounding for a decade. It took only two extra bushels of corn per acre to pay for the seed, but hybrid growers were averaging 10 to 20 more bushels per acre. There was another factor to consider—the hybrid strains stood so much better because of strong stalks. This became increasingly important each year as more mechanical pickers were introduced. In 1938 over 60 percent of the corn in the Coon Rapids area was mechanically picked. The International Harvester dealer that year sold 19 new McCormick-Deering pickers and eight used machines besides 18 new Farmall tractors. Garst and Thomas processed about 160,000 bushels of corn to acquire enough seed to double the amount sold for the 1938 crop.

After 1938, because of more restrictive government programs, corn acres in Iowa dropped from 10.4 to 9.4 million acres. The total impact on corn production was the opposite of what the program had been designed to do which was to reduce the total crop. The farmers cleverly used their best land and purchased more hybrid seed, probably using money from the program to purchase the seed. Average yield for the state in 1939 broke all previous records by over six bushels per acre and total production for all previous years but 1932 and 1937. The farmers had found a way to satisfy their government restriction on acres as well as their own desire to increase production. Hybrid seed sold better than ever.[2]

Commercial Fertilizer—A New Challenge

By 1938 the challenge of selling the hybrid seed corn had lost its appeal to Bob Garst. He needed something new to motivate him because after eight years of aggressive promotion most farmers were using hybrids. The rest would follow or perish; Garst knew the results were inevitable.

When Garst started farming for the second time in 1930, as mentioned previously, it was just another drab year for Iowa farmers.

But Garst and Thomas witnessed another exciting happening in the development of corn growing. Henry Lubkeman and Son, of Coulter, Iowa, and DeWitt Mallony of Hampton experimented with commercial fertilizer. Mallony had applied 125 pounds of 20-percent phosphate to his corn ground and produced a measured yield of 95 bushels and 13 pounds of 15-percent moisture corn. Just a few miles to the west the Lubkemans did what was then considered a very scientific job of corn growing. They applied 200 pounds of 20-percent phosphate before fall plowing. The next spring they disced the land three times, harrowed once, planted, and harrowed twice after planting. They check planted four kernels every 36 inches. Lubkeman's yield was 96 bushels and 46 pounds of 15-percent moisture corn, an unbelievable yield to the average farmers of that day.*

Besides watching those Iowa farmers and doing fertilizer experimentation on the Garst and Thomas lands, Roswell Garst was keeping close contact with his elder brother, Jonathan. Jonathan Garst was running an experimental farm on the Hebrides Islands of Scotland and was working with fertilizers. Roswell did not want to be outdone by his brother, whose ideas he valued highly. Garst and Thomas also experimented with the use of fertilizer on their corn acreages, and at one time in the 1930's one-tenth of all the fertilizer used in Iowa was shipped to Coon Rapids, where it was used on one-tenth of one percent of the state's cropland.** By 1940 Garst and Thomas were using balanced applications of nitrogen, phosphorús, and potash at a time when the great majority of the nation's farmers had not considered using any of these basic elements in fertilization. That year Charles Thomas had a 180-acre field that produced over 150 bushels per acre by wagon-box measure and was expected to weigh in excess of 125 bushels. Two things were exciting about that yield. One was the result of the total application of fertilizer; the other was the fact that it had been harvested by rubber corn roll designed by Ralph Wheeler. This contrasted to the average yield for Iowa in 1940 of 52 bushels.

By 1938 Garst and Thomas had discovered the value of potash on high lime soils. It was Jonathan Garst who, Roswell says, was

---

*This corn was weighed on the basis of 70 pounds per bushel for ear corn. Corn pickers were still the vogue, for the corn combine that harvested only the shell corn did not come into existence until the 1950's.

**The Garst Company, owned by Stephen and David Garst, a business firm apart from Garst and Thomas, has probably been the largest fertilizer dealership in the state of Iowa.

"a real soils expert" and noticed that the corn was both taller and later in the low spots which had once been potholes and ponds. Jonathan reasoned that this was because of the high lime content from the snails and other life in the old water bodies. He suggested fertilizer to counteract the high lime content. Roswell Garst credits John Coverdale, an Iowa Farm Bureau leader in the use of fertilizer, for getting him to use potash. Through such experiments it was also learned that proper fertilization caused the corn to develop more evenly. This meant a great savings, for the detasseling trips were reduced from 16 to 5. In addition there was a more uniform maturity, which enabled the saving of a higher percentage of the ears for seed.

By 1940 Garst and Thomas knew the value of the balanced application of commercial fertilizer and were using large quantities of it. They purchased their nitrogen and phosphate from the Consolidated Mining and Smelting Company in British Columbia. Using nitrogen prior to the outbreak of World War II proved to have tremendous dollar value because nitrogen was restricted during wartime to essential industry and to previous users of it in agriculture. The annual carload each of nitrate and phosphate enabled the partners to gain valuable experience in corn production.

World War II prices encouraged farmers to do a better job in production, and seed corn sales grew rapidly. Garst and Thomas sold 150,000 bushels of seed in 1942 and announced that they would increase their seed corn production to 5,000 acres to meet the even greater demands anticipated in 1943. Corn Belt hog feeders were happy in 1942 when hogs hit $14 a hundred, the highest since 1926. Bob Garst sold some 1,300-pound steers for $12 a hundred, the best price for cattle since September of 1928. As Garst said: "Those of us who were in fertilizer early, like those who were first with hybrids, had an advantage for the price of the grain was based on the lower yields of unfertilized corn; the market price had not yet found the lower cost of production per bushel associated with fertilizer."

In their experiments with fertilizer Garst and Thomas discovered that with the proper fertilizer one could raise a crop in potholes if there was proper surface drainage. The land around Coon Rapids is very rolling; in fact a good percentage is steep enough so that it probably is most profitable as permanent sod and should not be tilled. Getting rid of water in the low spots with tile is a costly process and needs continuous attention. In 1947 the spring season was very dry, and Garst saw an opportunity to do some work on his low spots. A large dirt scraper was employed to move land that was more than 4 feet above the tile lines. The excess dirt was used to fill the low spots at a cost much below that of deep tiling. From that time on the former potholes have produced a crop every year. The partners also learned that by using proper fertilizer applications no reductions in yields occurred where the topsoil had been removed. Garst emphasized:

For years the agricultural schools and farm papers have taught the virtue of humus. But I have discovered since that there are places where as much as 300 tons of manure per acre have been applied and it was not noticeable the second year. Commercial fertilizer could do the job just as well—this taught me that humus was over-rated. I know I can remove my top soil and with proper farming can avoid a reduction in crop. That surface drainage was the single most profitable expenditure that I have ever made in farming.

The 1947 crop was short, averaging only 30 bushels for Iowa and resulting in $2.60 corn by January, 1948, a price not equalled until February, 1974. It proved to be a virtual bonanza to corn growers, especially those with big acreages.

After World War II Garst felt sure enough about the possibilities of fertilizer that his "goal was to teach the other farmers how to use it." Once he was assured of a steady supply of nitrogen, Garst decided that it was up to his salesmen to be teachers. Sales meetings were used to explain the value of commercial fertilizer to the salesmen, and each of the 600 men was given 1 ton of fertilizer for every 100 bushels of corn sold. As Garst recalled:

All at once I had 600 fertilizer demonstration plots in Iowa, Kansas, Missouri, Nebraska, Colorado, and Oklahoma. All of the plots had nitrogen and phosphate and in some places potash too. This gave proof of the value of fertilizer to lots of farmers in six states. In the next ten years the use of commercial fertilizer in Iowa alone increased over 1,000 percent.

The grand sales pitch of 1948 was the climax of Garst's crusade to "teach" the value of fertilizer, for immediately the farmers grasped the meaning of his message. By 1950 average national corn yields had climbed to over 71 bushels an acre because of the use of nitrogen, phosphate, and potash. In 1950 hybrid seed corn sales at Garst and Thomas surpassed 350,000 bushels, and projections were still pointing to greater demand.

Because of the wide variety in quality of land in the Coon Rapids area, it became advantageous to grow continuous corn on the best quality land. Garst, a leader in corn monoculture, soon learned that continuous cropping did not have an adverse effect on production. By 1959 he was able to prove that fact because some of his land had been in continuous corn for 15 seasons. A five-year average from 1954 to 1958 on over 1,000 acres of continuous corn was just over 100 bushels an acre. The Iowa average yield for that same period was just over 56 bushels. His 1958 average on 3,600 acres of corn was 108 bushels per acre.*

After going to continuous corn, about 40 percent of his land was seeded to permanent pasture which was fertilized as heavily as any crop on his farm. In 1962 Garst stated that he was using a balanced fertilizer application of 200-54-46** pounds of actual ingredients on corn and over 150 pounds of actual nitrogen plus whatever phosphate or potash was necessary on pasture. He boldly asserted, "Fertilization of pasture brings as high a return per dollar invested as fertilization of grain crops." Experience has proven him to be correct, and by the 1970's he had increased his fertilization of pastures to even higher levels.

To Bob Garst the use of nitrogen was one of the greatest developments in modern agriculture. His brother Jonathan, who often influenced him, wrote in 1963:

> The whole field of agricultural production has exploded. It cannot be much of an exaggeration to say that in the last ten years the potentials for food production have increased as much as in the whole ten thousand previous years of farming in the western world. . . . And the end is not in sight—how far can it go? Production potential has not been reached,—one ton of nitrogen in fertilizer is equal to fourteen acres of good cropland. To keep up with our population growth we need only to increase our nitrogen output by 200,000 tons a year.

Jonathan Garst understood precisely the value of nitrogen in agriculture, for it was under his guidance as a Special Assistant to the U.S. Secretary of Agriculture in 1951 that production of nitrogen was greatly increased. Production of about 1 million tons annually in the early 1950's increased to 10 million tons in the next decade. This happened while the capital requirements for a ton of annual production of nitrogen were reduced from $300 to $60 a ton. Industry was helping to increase the volume of agricultural commodities by mass producing a basic ingredient and at the same time keeping the costs down. Fertilizer was supplementing the land. From 1955 to 1967 national average corn yields increased at the rate of about 3 bushels per acre per year. From then

---

*Garst also grew continuous hybrid grain sorghum on some of his land.

**Fertilizer rates are generally designated in order as nitrogen, phosphate, and potash and in the trade are symbolized as N-P-K.

to 1972 the rate of increase was about 5 bushels per acre per year. Roswell Garst for once did not seem wildly optimistic about a continued rate increase because he thought that moisture might become a limiting factor. But that did not slow down his efforts, for he and Thomas in theory and practice were recommending "maximum use of fertilizer as the best way to cut production cost and getting increased yields of better quality feed."

The American writer John Dos Passos interviewed Garst in 1948 and indicated to him that he was fully aware of leadership in many aspects of American agriculture:

Even here in the heart of one of the richest farming sections in the world . . . you can see on every hand the results of poor farming practices. . . . Farming has become the most extinct occupation in the world. . . . Something new all the time. What we are seeing . . . is a revolution in agriculture. . . . Improvement in corn has started off improvements all down the line. It is my impression that the success of hybrid corn had opened people's minds to the possibility of all kinds of new advances in farming. . . . In my opinion . . . the main thing about all these new developments is they stimulate people's minds, make them more alert. They are more careful with their crops, they take more pains . . . [to do a good job].

In 1972 Bob Garst, still confident, recalled:

By 1955 fertilizer had become common knowledge and like peddling hybrid seed corn ideas after 1938 I lost interest in promoting fertilizer. I have always had one project started before I finished my last one to keep ahead of the crowd. I guess that's the fun of it. In 1946 quite by accident I got all excited about feeding by-products of feed grain and seed corn production to cows. And that brought about my next challenge to stay ahead of the crowd.

At the same time Bob Garst was working on another innovation, which in thought had disturbed him for many years. "I don't claim to be very smart when I must admit that it took me eighteen years to discover that something I was working with for each of those years would have great commercial potential to corn farmers."

The new innovation was the adaptation of the corn drier to commercial corn production. Garst and Thomas had installed a corn drier to process their seed corn as early as 1930.* The female seed corn must not be frozen in the field, so it has to be picked early and artifically dried. The rest of the corn could be left to be harvested later in the traditional manner. In 1948 Garst, realizing

---

*It is believed that only Henry Wallace and Funk Brothers Company had driers prior to that time.

the potential of the commercial corn drier, installed one for his crops. Recalling that event, Garst asked himself, "Why didn't I start using the drier on all of our corn years before? We could have saved lots of corn." And he added, "I harvested some commercial corn early and dried it to sell on the local market for $1.85 and within thirty days the market had dropped to $1.25 after the others started to harvest.

In 1949 a corn borer infestation hit the Coon Rapids area. Bob Garst had worried about borers in the 1920's and finally they had arrived. But because he had corn driers he was able to get most of his corn harvested prior to the biggest stalk breakage. Garst grinned: "The one field we harvested late had a 40 percent loss. With my experience of 1948 and 1949 I concluded that every sizeable corn producer would have to have a drier. I started advocating the commercial drying of corn. Obviously this had an impact on the next step in mass corn production—the adoption of the field picker-sheller."[3]

Corn Cobs and Cows

The sum of Garst's expectation for American agriculture is well expressed in a statement he made shortly after World War II:

> In some ways the advance of agriculture has just begun. Farming . . . [is] a well-heeled fast moving industry ready to tackle almost any problem. Everybody . . . [wants] to try new things. The farmers are running ahead of the experts now. . . . Less labor, more speed, high efficiency . . . what interests me and keeps me excited all the time is watching how one little improvement in farming leads to another. Link 'em all together and the results are revolutionary. Makes me feel we haven't begun yet.

One of the farmers "running ahead of the experts" was a young man from California, who walked into the Garst & Thomas office in the fall of 1946. This young man wanted to buy 5,000 tons of corn cobs for use as cattle feed at his California lot. However, he did not buy the cobs because he found out that the freight rates made them too expensive. But Bob Garst suddenly realized that "cobs to cows" might be something for him to try. The corn cob pile at the Coon Rapids plant grew to as much as 15,000 tons in a single season, and the disposal of it was becoming a real problem. In the early days cobs were hauled away by home owners who used them in their heating and cooking stoves, but by the mid

1940's no one wanted cobs because they had all converted to electricity, L.P. gas, or fuel oil.*

Garst was even compounding his cob problem in 1948 when he started picking all of his corn earlier and shelling it immediately so that it could be artificially dried for storage. As the traditional corn crib disappeared, the cob pile became larger in the fall. At once he realized that his corn cob problem could be a blessing. He started searching for information at all the agricultural colleges, where he was a well-known figure.

During his search for a solution to the cob problem Garst observed something on his farm that opened his eyes to more questions. To protect their seed corn acres from weed infestation from neighboring fields and to comply with government programs, Garst and Thomas adopted the practice of planting brome grass strips completely around their corn fields. The size of the strip was such that it totaled the acres necessary to be diverted from corn production. Government programs permitted diverted acres to be used after September first of each year, so once the corn was harvested cattle could be turned into the fields to graze both the brome and the stalks. When corn was fertilized at different rates Garst observed that when the cattle were turned into the fields to graze they concentrated on the corn that had received 200 additional pounds of ammonium nitrate per acre. The cattle stayed on that heavier fertilized strip until it was entirely gleaned before they went to the other stalks or even the brome. After a lot of thinking and questioning he reached the conclusion that the cattle were after the stalks with the highest protein content—those with the heavier nitrogen application.

In the meantime, the experiments of Paul Gerlaugh at the Ohio Agricultural Experiment Station became known to Garst, as were other ideas he tracked down about cellulose feeds.** Using Gerlaugh's ideas, Garst started his next crusade with another experiment. Gerlaugh's experiments had indicated that "when properly fed corncobs were 2/3ds as valuable pound for pound as corn, up to 1/3d of the feed." In the fall of 1946 Garst purchased feeder

---

*There are 14 pounds of cobs in each bushel of corn, and by the 1940's about 400,000 bushels of seed was produced, which caused a cob pile of over 3,000 tons. By the 1970's over 15,000 tons of cobs were expelled each year.

**Ironically, a standard encyclopedia had considerable information on some of the questions Garst had asked himself.

cattle and divided them into five lots with different rations in each.*

The experiment was successful, but more had to be done to make cob feeding practical. As soon as the very muddy feedlot was dried out and cleaned in the spring of 1947 more cattle were purchased. These cattle were given the basic ration of cobs and protein supplement. They gained 1¾ pounds a day at a feed cost of 16¢ a pound and were graded "choice" by the packer. The protein supplement contained one-third urea. After the second experiment Garst became bold and purchased 500 head of steers just at a time when many cattlemen were refusing to buy cattle because of skyrocketing corn prices. These cattle once again received the full corn cob ration along with Vitamin A concentrate, and 5 pounds of protein supplement, plus minerals. The third try and the first on a large scale was even more of a success from a feeding standpoint, and a bonanza on the profit side, because the cattle sold in excess of $25 per hundred.**

Realizing that he had a sound program with the corn cob ration, Garst experimented with new ways to reduce feed cost still further. In 1949 he turned from using vegetable protein to a synthetic protein, urea, which was first used as a source of animal protein about 1943. Garst mixed 11 percent urea with 89 percent molasses and greatly reduced his protein cost with no difference in performance of the cattle.

While experimenting with urea and corn cobs, he tried using calves, yearlings, and two-year-olds to see if all age groups reacted the same way to the ration. He came to the conclusion that it was most efficient to start with a heavy cob ration and gradually change to a higher amount of corn as the cattle came closer to

---

*Lot 1: Straight shell corn plus a pound of protein supplement.
Lot 2: ⅓ cobs ⅔ shell corn plus 2 pounds of protein supplement.
Lot 3: ½ cobs ½ shell corn plus 3 pounds of protein supplement.
Lot 4: ⅔ cobs ⅓ shell corn plus 5 pounds of protein supplement.
Lot 5: Straight cobs plus 5 pounds of protein supplement.

All lots received 2 pounds of hay per head per day. Lot 5, using corn cobs as the bulk of the ration, "gained ⅔ds as rapidly as Lot 1 at ⅔ds the cost per pound of gain. Cobs were worth ⅔ds as much as shelled corn up to all the animal's carbohydrate intake—if properly supplemented."

**This was a new all-time high for beef cattle prices and was nearly double the high for 1919. It was not until 1945 that fat cattle prices exceeded those of the previous high of 1919. The short corn crop of 1947 caused the fat cattle prices of 1948 to be nearly double the 1919 high.

Commenting on the successful feeding program and sale of those 500 steers Garst said, "Cobs and protein meal were horribly successful with a feed cost of just over half that of a normal ration.

being finished. At the same time, he drastically reduced the amount of hay in the ration.* He reasoned that if all ages of cattle could do well on cobs and urea it might be profitable to have a brood cow herd to utilize even more of the by-products of a continuous corn-growing operation. Such an enterprise could work into the existing farm programs.

"Visions of the drought of 1934 when feeder cattle prices fell to $.03 a pound in Omaha," Garst reminisced, "made me a conservative in the cattle business. I realized the cow business would be no good in Iowa without a steady supply of low-cost roughage. We had it [corn stover] then, but didn't know how to use it other than as stalks in the field to be grazed and some shreddings."

His experiments with feeder cattle from 1946 on finally convinced him that brood cow herds could be inexpensively produced in the Corn Belt and raising beef calves would be a profitable business. The cow business would have to be sustained with by-products of corn growing and on land that was not suitable for corn. The chemistry of urea had made corn cobs, corn stalks, straw, and chaff all profitable cattle feeds. With all the evidence on hand, Garst decided to go into the beef brood cow business. Before its materialization, however, he was discouraged by many who said it could not be profitable in Iowa. His greatest opposition came from bankers because most farmers had treated their beef brood cow business as a secondary enterprise and had not done well with it.

Roswell Garst had heard the bankers say no before and in spite of them he had always found a way to do what he wanted. He devised a method of raising a brood cow herd in a roundabout way. Because he was an experienced and successful cattle feeder and could easily get financed for that enterprise, Garst decided on a program of using the bankers' money to develop a cow herd. He did this through what he called the "feedlot approach." In the fall, when they were looking for feedlot replacements, Garst buyers purchased the entire production of steers and heifers from some of the top ranch operations. These cattle were all individually weighed and identified when they entered the feedlot at Coon Rapids. The cattle were weighed 120 days later and divided into three separate groups according to their average daily gain. The

---

*Several times articles have indicated that Garst used no hay in his feeding program. Garst frequently gives the impression that hay was not used, but 55,000 bales of hay were produced in 1971, much of which came as excess production from his pastures.

top one-third of the heifers were taken off the fattening ration and put on pasture for summer breeding. The remaining heifers and all the steers were sufficient to repay the banker and most cash expenses. Once the top one-third of the heifers were bred and became an established part of the farm inventory the banker was willing to loan money on them. There were tax advantages, too.

In the early 1950's Garst felt that feeder cattle prices were out of line. He decided to rapidly increase his beef cow herd. The feed-lot approach was used along with waste products to carry the cows. Roswell Garst's ideas were not going unheeded by others as he recalled, "During the 1952 to 1954 drought in Kansas we had droves of farmers up here every Sunday to ask about feeding corn cobs to save their cattle. It was a nice feeling to know that the experimenting we had done was of real benefit to them."

By 1958 the cow herd had expanded to 575 head, and by 1972 it reached 4,000. This meant that Iowa, with its relatively high-priced land and a reputation as a corn-growing state, had a beef brood cow herd that was larger than the majority of the herds in the states where ranching was the chief enterprise. The Garst cattle-feeding program was now fully integrated because no longer did he have to rely on purchasing ranch-produced calves for a feeder supply. Each year a greater portion of the total feedlot calves were raised on the Garst farm, so that by 1972 very few had to be purchased.*

The total impact of what Garst was doing with corn cobs for the nation's food supply would eventually be significant. But like his selling of hybrid seed corn and nitrogen fertilizer, the message of the food value of cellulose and urea had to be repeated often before others adopted the idea. The *Farm Journal* and *Successful Farming* were the first magazines to write about Garst's cellulose experiments, on which he had been working for over a decade. At that time Garst said: "I am the first to admit that many of my experiments have been failures, but those that have succeeded have paid well. . . . I always considered my innovations sound. I really proved them before they were completely adopted by us and rec-ommended to others." By then (1960) there were 3,000 cattle in his lots being fed on cobs and urea. At that time synthetic urea cost 2¢ a pound versus 10¢ a pound for natural proteins.

---

*To get a top-quality herd the cows were artificially bred because it was the cheapest way to get the best breeding available. Besides, the task was made rela-tively simple because the cows were in smaller areas, where they could be closely managed.

He was called "a creative and venturesome thinker—an economic daredevil!" He was critical of agricultural colleges, and declared that "the professors teach stuff that's really only the history of agriculture! ..." One of his articles, entitled, "Bob Garst's Latest Crusade," typified the way he did things: "Garst is now promoting an idea that makes sense, as do most of his proposals for more efficient farming. He's pushing the idea of feeding part of the corn crop we now often waste—the cobs and the stalks—to beef cows here in the Midwest."

As he succeeded with the cob-feeding program and as he received more and more attention for his innovation, he became bolder in his crusade to promote the idea. At a gathering of 400 farmers on his farm in early September, 1959, he reminded them that in the past 12 years he had fed 25,000 head of cattle on cobs, molasses, and urea. "You people think my ideas about corn are ridiculous, but the fact of the matter is you are ridiculous by wasting more money and energy on hay. If you were smart, you'd be out fishing while your cattle thrive on corn cobs. . . , If you don't make and put up hay you don't feel like farmers."

Those words sounded like a battle cry, and they were meant to be one. For Roswell Garst was now ready to wage an all-out war against legumes, particularly alfalfa. Garst was so outspoken in his new crusade that some farm editors who were sold on legumes hesitated to write about things that took place on his farm. It was his opinion that "legumes were a part of the history of agriculture but today [1972] are not only unnecessary but uneconomic." He argued nitrogen could be bought for one-fifth of the cost of raising it; and therefore it was no longer necessary to use legumes in the rotation, and continuous corn and beans were much more profitable.

As early as 1955 he lectured to a group of agricultural students at Iowa State University: "I feel sorry for you because you are learning things that aren't so. Your profs [there were several in the room] still preach clover in the rotation for proper land use. I tell you that clover is obsolete, that feeding hay is obsolete, and that the best land use is no rotation at all." He added a parting thought that his sons had degrees from liberal arts colleges and were "unspoiled by agricultural schooling."

As critical as he appeared to be of some of the teachings of the professors of agriculture, Garst was known to many of them as their most profound student and strongest supporter. He was the first to acknowledge Wise Burroughs and S. A. Ewing of Iowa

State for their pioneering on corn stalk rations with beef cows, just as he acknowledged Paul Gerlaugh's work with the use of corn cobs. The entire management staff at the Garst farms stated during an interview that at their farm "corn was by far the most efficient crop and because of it they could produce beef from cows through the feedlot more cheaply than at any other spot in the world. Besides," they argued, "with continuous corn they could greatly reduce their machinery needs." And Roswell added, "The first people who quit putting up hay . . . will have the best profit."

In the 1950's Roswell Garst forecast that the Midwest would have to increase greatly its beef brood cow numbers to supply the growing world demand for beef. He even had the formula for the average Midwest farmer to do the job. The first step was for continued progress toward increased corn yields, which were necessary to feed more feeder cattle. Next he advocated continuous corn on the best land and a continuous grass crop on the less suitable land. All of the corn not used as silage would have the stalks chopped after the corn was combined. The Garsts followed exactly his formula, and after 1958 they increased from two single-row field choppers to four double-row choppers. At the same time they converted 50 percent of their farm to permanent grass pastures. Their 1968 records proved that 100 pounds of nitrogen per acre on good pastures enabled them to increase their yield of beef on that acre by 100 pounds for a return of 250 percent on their money. Their objective was to raise beef yields at least another 50 percent on already highly productive pastures.

In 1968 David Garst complained, "Farmers can no longer afford the luxury of unfertilized and unimproved pastures, or the use of hay to winter their cows." He added, "Pastureland is one of the greatest untapped resources of agriculture." By that date experience had taught the Garsts that they could support more than one cow and calf per acre of pasture per grazing season. Corn cobs and stalks were used during the short season when cattle could not actually graze on the land. During the grazing season of various grasses as many as three cows and calves were placed on each acre. It was Garst's opinion based on records of his operation that an Iowa farmer with equally good management could earn as much net income on one-half section of rented pasture land as he could raising 150 bushel an acre corn on rented corn land. His labor requirements would favor the corn slightly, but they would be more than offset by the increased investment in corn equipment.

Over the years Garst had developed a rather unique formula of cost that remained constant. Production records proved that for each 10¢ cost of raising a bushel of corn it cost $10 to keep a brood cow and raise a calf, using a ration of corn by-products. The Garsts maintained that "with high fixed cost of land and machinery—we can't cut cost by reducing expenditures. We must raise yields and we must feed our livestock more efficiently."

Bob Garst pointed to the research at Iowa State, where for several years they had been pasturing 10 beef brood cows and 10 beef brood heifers at the rate of one head for two acres of corn stalks for a 112-day period, with no supplement other than Vitamin A. The cash cost was 80¢ per head. He added, "It is the simplest way to double crop." Long-term records of the conversion to pasture and the use of corn by-products on the Garst farm showed an increase in total farm income of 15 to 20 percent per acre, and at the same time production costs were reduced by an equal amount. This meant that the income per acre for the total farm was increased by an average of 35 percent. In addition, there was the benefit of a more complete utilization of year-round labor.

In 1950 Garst predicted that cellulose and urea feeding to ruminants plus increased use of fertilizer and power machinery on the land would be the basis for the increasing agricultural yields that were necessary to meet growing world demands. "Farmers," he said, "could best improve their net income by spending more money on seeds, chemicals, fertilizers, and power to increase their yields because these were the only variable costs they had." By the early 1960's, the Garsts had their corn production program, which included chopping to save the stalks for stover, down to five trips over the field. It was soon apparent that even though Garst liked being "ahead of the crowd" there were other farmers not far behind. Iowa farmers increased their beef brood cow numbers from 571,000 head in 1949 to 1.4 million in 1972. Farmers had quickly made use of cornfield by-products to support the new enterprise.

Garst had repeated many times that now there were three avenues open to the Iowa cattle feeders: "quit feeding cattle, pay over-increasing prices for feeder cattle, or raise their own feeder cattle by taking the stalks and getting 100 percent of the value of the corn crop."* Rising land cost made double cropping attractive.

---

*Roswell Garst predicted that southern commercial feedlot owners would probably attempt to become involved in developing beef brood cow enterprises to supply at least 40 percent of their needs. This would stabilize their cost of feeder cattle and reduce the overall risk associated with cattle feeding.

Again farmers, including the Garsts, needed a new machine—a combination corn combine and ensilage cutter that could shell the corn for fattening cattle and chop the stalks for the brood cows in one trip through the field. Iowa State anticipated this need and started research on various machines, but up to 1974 nothing had been developed for the commercial market.

Beef consumption had grown nationally from 9.5 billion pounds in 1950 to 23.2 billion pounds in 1970. Garst predicted that America would need 12 million more beef brood cows and 500 million more bushels of corn to meet the beef requirements of 30 billion pounds by 1980. "This," he said, "could be done because meeting a challenge is what farmers enjoy best." And he added: "Alfalfa hay is as obsolete as the walking plow." But when asked what should be done with acres not suitable for regular cultivation and being used for hay he answered: "Put it in pasture. And keep it in pasture. I mean improved pasture. For example, if Iowa farmers would quit putting up most of the 2.5 million acres of hay which they sweat over each summer, and pasture those acres which have been used for hay . . . we could raise all the feeder cattle we need in our state."

Roswell Garst had much evidence to support what many felt was a radical stand on cellulose and cow feeding. He liked to show his visitors a 1952 letter from W. M. Beeson, Professor of Animal Science at Purdue University, which said, "To me the day of crop rotation is over because, through new technology, we can raise corn continuously on the same land. . . . For example, we can produce 2,000 pounds of beef per acre from corn silage and only 500 to 600 pounds of beef per acre from the best legume hay crops. . . ."

Garst noted that in 1972 the 14 leading corn states produced only 5.7 million acres of corn silage, while 17.2 million acres of hay were being harvested. He regretted this and added, "The explanation is that for nearly a century the agricultural colleges, experiment stations, and all the farm magazines rightfully encouraged rotations including legumes. With the advent of nitrogen, legumes were not only no longer necessary, but too costly." He pointed out that the use of urea as protein for ruminant feeds expanded from 4,000 tons in 1950 to 650,000 tons in 1972. That also made "legumes like horses . . . uneconomic."

Jonathan Garst kept encouraging his brother to increase the yields on good land and thereby make lower-grade land available for conservation purposes. Roswell Garst stated in 1948 that his newest experiment was "important because if you can feed cattle on cobs which are mostly cellulose you can probably use straw and other types of cellulose . . . but it opens up simply horrible possibilities for cheaply increasing the meat supply. . . ." His prophecy was correct. In 1973 Robert de Baca, Professor of Animal Breeding at Iowa State University, was hired to take direction of the 4,000-head cow herd of the Garst Company. Under de Baca the already well-bred herd of many breeds will be used in experiment for greater production per animal. Cheaper beef will be the end result.

Louis M. Thompson, corn breeder and Associate Dean of Iowa State University, substantiates many of Roswell Garst's theories. Thompson, in an in-depth study of world food needs, concluded that in 1965 there was less than two months' food reserve in the world. That situation was paralleled again in 1972, and only the large reserves of the United States and Canada prevented a real food shortage throughout the world. Thompson predicted, "The potential in increased demand for U.S. agricultural products is very great. . . . If American food is to be competitive in the world market, it must be priced at levels which reflect our relative advantage in producing agricultural products."

Supporting Garst's theories about producing cheap beef, Thompson added,

We must make greater use of corn stalks for feed for ruminant animals. Because of the competition between man and animals for grain, hog population will probably decrease rather than increase over present numbers. . . . Because of the dependence of hogs on soybean protein and grain that might be consumed directly by man, pork will become a luxury.

It appears now that at age 76 Roswell Garst had successfully won another crusade to keep American agriculture ahead of the crowd, and he was still a leader.[4]

The Philosophy of an Innovator

Roswell Garst can be best described as an impetuous innovator. According to Viv Bell, who farmed with Garst for 20 years: "He was really great. I could spend money to improve things and never asked him. . . . He was always pushing me to try new things. If I did and it didn't work he would tell me to try something else. He wanted his renters to be progressive and we could never move fast enough."

The *Christian Science Monitor* in one of its reports quoted Garst on his frustration about the unwillingness of people to change: "They [people] have to see, see, see, see about ten times. Come to think of it, I've never been able to find out where extreme conservatism leaves off and stupidity beings." Then he added a comment about a pending visit with Soviet Premier Nikita Khrushchev: "Well, I think Khrushchev is a reasonably progressive guy. He has to see things only about twice at the most—and he's got it." Both Jonathan and Roswell have been disturbed about the slowness of man to better himself. In 1963 Jonathan wrote:

Technologically, man has solved the problems of providing food for all the world. The economic, social, political, and cultural differences which impede the flow of this advanced technology from one country to another are still with us. . . .

Our American farmers are no different from other people; they have to have opportunity knock at their own door. Even peasant farmers in remote areas know the world is galloping ahead. They will not change readily.

John Strohm, prominent American agricultural writer, wrote in 1959:

For years I've known yeasty Bob Garst, listened to his blunt and colorful talk, and observed his unorthodox methods, which are making a real impact on agriculture around the world. . . . [Garst says of himself] "I'm no do gooder—I just get excited about feeding hungry people because I know it can be done." He recognizes that food is the world's No. 1 problem and is fully satisfied that with the know how available the world could have food unlimited.

In a 1958 issue of *Fortune Magazine* Bob Garst was described as "a devoted evangelist in his desire to create universal plenty, and to put meat on every table in the world. . . . He is an innovator and a super-salesman." In 1958 his crops averaged 108 bushels of corn per acre on several thousand acres in contrast to the Iowa average of 66 bushels. The *Fortune* writer noted that Garst was criticized by many as being a disciple of increased production regardless of other costs. Garst was quick to defend himself at that point by saying that cheap and abundant food was the solution to the world's greatest problem. To him all other problems were secondary.

Speaking of his own experience at the age of 76, Garst commented:

I have been a conservative for I have tried experiments in the back corner of the farm so if they didn't work I didn't repeat them. We have made lots of blunders but over the years experimenting has been profitable and I have had the fun of being ahead of the crowd. In spite of our alertness to new things I cry at the opportunities I have missed. There is always challenge particularly in areas of poor or undesirable land which no one else could make work. Even though we are ahead of others there are still infinite opportunities for increased production and growth right on this farm.

He could cite his records to show that while it took him 30 minutes to produce a bushel of corn in 1930, he was able to do it in 5 minutes in 1950, 3 minutes in 1960, and only 47 seconds in 1972.* Even though authorities such as John Strohm said labor productivity on his farm was five times the national average, Garst was still predicting greater efficiency. Many adaptations to machines were made on the farm, for Garst firmly maintained that "only by increasing output per man can we raise the farmer's standard of living." It was the consensus of the entire Garst managerial staff that with their level of technology and skilled manpower they were "infinitely superior" to any of the world's other food producers. They were probably correct, for their corn yields were 3 times the average of the rest of the world and their labor efficiency was 10 to 1 over much of the rest of the world. Productive cost in 1972 was just over 90¢ a bushel, but the long-range objective based on a 1972 dollar was 60¢ a bushel.

It was a strong belief of the entire managerial staff of the Garst farm that the best way to improve profits was through reduction of fixed costs, which make 75 percent of the cost of farming. Yields then have to be increased through maximum use of scientific and mechanical input of the highest-yielding crop on any given piece of land. The same applied to animal production. This, the staff felt, was the strongest consistent management factor of this farm. The intense production of corn and grass on various fields of the farm is clear proof that the Garst family used the land for the crop for which it was best suited to maximize profits per acre.

The innovative spirit of the dominant figure of the Garst farms was infectious. Staff and employees knew the uniqueness of the farm because of its high innovative goals. Employees had a sense of pride knowing that they were working for one of the recognized leaders of agriculture. Because of that, good-quality labor was and is available to the farm, and a high level of labor productivity is maintained with one man for each 500 acres of row crop and one

---

*This reduction of 38 to 1 in the time to produce a bushel of corn on the Garst farm is probably faster than the national average. The 1930 national average production time to produce a bushel of corn was 70 minutes, more than two and three-tenths times what it took Garst.

man for each 500 cow-calf-feeder unit. The management feels that even with a large labor force, because of tight management the farm does not have many of the "loose ends" that are often found on many of the average one-man farm operations.

Even though the Garst farm is large, members of the staff feel that farmers are too often inclined to expand by buying land rather than intensifying their operation. Historically this has been easier than making capital expansion within an existing farm. Garsts have proven to themselves that vertical expansion has almost always brought a quicker return on their investment. It is the contention of David Garst that the size of a farm should be computed on the basis of capital investment and not acres. By intensifying his capital and labor inputs the farmer is likely to have the quickest and best labor returns for himself. This, of course, tends to imply a combination crop and livestock enterprise. It also discounts the speculative long-range benefits of land appreciation that have been the only profits many farmers ever realize.

This does not mean to imply that the Garsts were against larger-sized farms, for they were quick to point out that the American farmers had been setting a definite trend in that direction. That trend was made possible because farmers had proven to themselves that "big farms, within limits, are the most efficient. . . ." Roswell Garst even expressed the belief that someday the large-scale, well managed farms of Russia can and probably will become as efficient as his farms. But even when well capitalized, the Russian farms could probably not compete with "the incentive of private ownership" as provided by the American system. It was Garst's belief, and a well founded one, that the natural trend of the American farmer toward larger, more efficient units was a good indication that the Russians could become much more efficient on their large farms.

In 1955 agricultural journalist John Strohm acquainted V. V. Matskevich, Soviet Minister of Agriculture, with the exploits of Roswell Garst. This was during a period when the Russians were struggling desperately to increase food production. Shortly thereafter 10 Russians were visiting at Coon Rapids and, after being properly "dazzled" by Garst, invited him to the Soviet Union. Garst was accompanied by the first American farm group to visit Russia. In the next four years Garst visited the Soviet Union and eastern Europe three times to give advice on corn growing, to sell hybrid seed corn, and to help the Russians get American farm machinery. Roswell and Elizabeth Garst dined with Premier and

Mrs. Khrushchev and their daughters, Minister of Agriculture V. V. Matskevich, and Minister of Commerce, Mr. Mikoyan, at a Black Sea resort.

After the initial visit by the Russians in 1955, other Russians of various levels visited at Coon Rapids. Khrushchev gave orders to his farm planners to "follow Garst's ideas of corn and livestock production." He added, "We must send some people to Mr. Garst in the U.S.A. We have had very good relations with him. We can use his methods for fattening chickens in our own country." Gitalow, a tractor brigade leader, said, "This year [1959] I spent three months at Garsts where I studied mechanization in corn cultivation on American farms." In a comment on Gitalow's visit, Khrushchev remarked, "Look and see how they [silage choppers] worked for Mr. Garst. Even though he is a capitalist he does not act badly toward us and we must be respectful toward him. . . . Comrade Gitalow did go there to work and he learned a lot from Mr. Garst. We are indebted to him. We Communists are always reproached for allegedly just criticizing the capitalists. But, as you can see, we thank a capitalist farmer for useful information."

While in Russia, the Garsts extended an invitation to the Khrushchevs to visit Iowa. In 1959 President Eisenhower announced that "Khrushchev's stopover in Iowa will be the only one in which he will directly come in contact with rural America." An editorial in the *Coon Rapids Enterprise* proudly stated: "Khrushchev . . . considers Garst the best authority on farming in the United States. . . ." John Strohm reported:

When reports came out that Khrushchev was going to Coon Rapids, flabbergasted diplomats and journalists had to scurry around to find out what the Soviet Premier well knew—that this is the stamping grounds of shrewd, suspender wearing Roswell "Bob" Garst, World's No. 1 corn farmer. . . . Thousands of U.S. farmers visit Garst . . . annually, to see "what's new." They get the latest scientific gospel on everything from fertilizer to feeding, and hear a homespun lecture on money-making methods . . . and it's just not U.S. farmers who are interested. The Garst guest book reads like a diplomatic Who's Who from Latin America, Europe, and Asia.

"Bob Garst is ahead of most researchers . . . and Khrushchev is smart enough to know it," said Ray Anderson of the *Farm Journal*, " . . . if you could visit Bob Garst . . . as I have done several times in the past twenty-seven years, you'd have a pretty good idea why [Khrushchev] wanted to stop there."

The *New York Times* reported that Garst had visited several eastern European nations and that "he has done more to sow seeds of friendship for the United States in Rumania than millions spent in propaganda. . . ." The article continued that Garst believed that the best way to win friends was to see to it that everyone had a full stomach.

What the nation was really curious about was why Khrushchev would visit a corn farm in Iowa when there was so much more to see in America. The Russians wanted to learn about American corn production and its application to Russian farming. Garst in an outspoken manner informed the Americans about the Russians' desire through an almost constant flow of articles in newspapers and farm magazines. Each successive news release from the world's largest seed corn plant at Coon Rapids was a little more flowery than the previous one.*

On August 13, 1959, Garst said of Khrushchev's visit to his farm: "He respects me as an American farmer—he knows I know corn." Shortly after, he added, "We'll only be doing with [Khrushchev] what we do with thousands of farmers every fall—take a conducted tour of our operations." Garst refused to tell the press how large his farm was, but all they had to do was to look at some old issues of farm magazines and they could find the exact acreage. As he was quoted: "Just say it's a large farm operation." Prior to Khrushchev's visit the *Coon Rapids Enterprise* noted: "Garst said, 'Khrushchev is crazy to see corn cobs fed to cattle. . . . The Russians can't afford to be as wasteful as we are.'"

In an open letter to the community carried in the *Coon Rapids Enterprise*, Garst gave the background of his connections with the Russian leaders, asked for cooperation from the local citizens to make the visit a pleasant one, and once again reminded everyone of the purpose of the visit: "Khrushchev was interested in seeing mechanization of agriculture, effects of fertilization, and cattle eating corn cobs."

Garst never failed to get in a word for what might be called his philosophy, or, as some called it, his sales pitch. Naturally he took credit for launching Russia on its corn program, just as Tom Campbell took credit for mechanized large-scale wheat farms. Khrushchev wanted to learn how to grow continuous corn. In an interview Garst related:

. . . rotations are as obsolete as dinosaurs. So now he wants to know

---

*The big Garst & Thomas seed corn plant was getting publicity it could never afford to buy, and Bob Garst was not one to let such an opportunity pass.

F. A. Pazandak on Big-4 tractor pulling 10-bottom J. I. Case plow, 1910. Note steering aid and light on extension at right front wheel. This tractor, capable of plowing 18 acres in 10-hour day, is shown on dust jacket.

October, 1974. Two tractors, shown on dust jacket. Ed Dullea, Jr., with a 300-H.P. four-wheel-drive Steiger pulling a 42-foot field cultivator with liquid fertilizer attachments—a rig capable of tilling and fertilizing 250 acres in less than 10 hours. John W. Swenson, with 20 H.P. 1919 Fordson with two-bottom 14-inch plow capable of plowing 9 acres in 10 hours.

First four-wheel-drive Steiger, made in a cattle barn at Red Lake Falls, Minnesota in 1958, by Maurice and Douglas Steiger with encouragement from their father, John Steiger. Picture taken at Bonanzaville, U.S.A., Fargo, North Dakota, a pioneer farm village.

Roy Robinson on first Model D-100 four-wheel-drive Versatile, that retailed for $9,600. This tractor, still in use in 1976, popularized the four-wheel-drive concept.

A 400-H.P. Cummins Diesel in a Michigan four-wheel-drive tractor that weighs 30 tons with the blade; a favorite of the employees of Ward Whitman farm, Robinson, North Dakota. Used in rush periods for field work with three other four-wheel drives.

Two Minneapolis-Moline U B's in tandem on Donald and Kenneth Wiese farm, Comfrey, Minnesota. May, 1964. These four-plow tractors in tandem could plow 3 acres an hour or disc 10 acres. Tandem hitch saved one man, plus giving greater fuel efficiency. (Photo, The Farmer)

Seven 30-60 Aultman-Taylors at Campbell Farming Corp., Hardin, Montana. Total drawbar H.P. 210. (Photo, W.M.S.T.)

Five Versatile 900's representing a combined 1,500 engine H.P. and a retail cost of $250,000. Purchased in 1975 by Orten Brodshaug, Horace, North Dakota. Each tractor has eight tires 20.8 inches in width.

T. C. Ross plowing, 1887, with five oxen on a Buford 12-inch plow, Steele County, North Dakota. (Photo, SHSND)

Oliver Haugen's father breaking sod at rate of 1 acre per day, Ryder, North Dakota, 1907.

Walking plow capable of plowing ¾ to 1½ acres a day. Developed in 1830's and used as chief plowing implement until riding plows were introduced in 1880's. (Photo, MHS)

Standard three-wheel three-horse B & L sulky plow. The Moline Plow Co. produced its first successful model in 1884. This one-bottom plow could plow up to 1½ to 2½ acres per day. (Photo, MHS)

Five horses on a 16-inch walking, breaking plow in Traill County, North Dakota, c. 1911. Could do 1⁶/₁₀ acres per day. Note nose bags for fly protection. (Photo, June Tweten)

Steam powered plowing on Pazandak brothers' farm, Fullerton, North Dakota, 1910. Note steam pistons to lift plow out of ground. Tractor is a 4-cylinder 6- by 8-inch, 650 rpm, 30-H.P., Big-4.

Howard Nelson, age 17, Casselton, North Dakota, 1935. General-purpose John Deere, pulling a three 14-inch bottom plow at 3 mph, is capable of plowing 11 acres in eight hours. Tractor sold for about $1,100, and plow for about $200. (Photo, R. D. Offutt, Inc., Casselton, North Dakota)

Early model Steiger 3300 in 1963 on Gregory Weiland farm, Euclid, Minnesota, pulling three seven-bottom 16-inch plows, is capable of plowing 160 acres in 12 hours on 160 gallons of diesel. This is 320 times as fast as a one-bottom ox plow could do the same job.

A 10-foot grain drill could plant 12 acres a day, Traill County, North Dakota, 1905. (Photo, June Tweten)

A 102-foot grain drill used on Kolstad farm and ranch, Chester, Montana, capable of seeding over 500 acres per day. Hitch was made in 1952 and has been used ever since. First pulled with I.H.C. TD 21 crawler; now powered by John Deere four-wheel-drive tractor. One man rides drill. Two trucks haul fertilizer and seed to load drill.

Cradle-scythe operation—a standard from about 1790 until replaced by reaper or binder. Some cradles were known to have been in use in the United States as recently as 1950's. Observed by Oswald Daellenbach, longtime Minnesota Extension Agent. (Photo, MHS)

A 6-foot binder could do 6-10 acres a day. Mayville, North Dakota area, c. 1880. (Photo, George Hilstad)

Big-4 pulling five 8-foot Deering binders, Hansman hitches. Twelve shockers used to keep up with binders. Horse binder used to open fields and finish final rounds. Pazandak brothers' farm, Fullerton, North Dakota, 1918.

I.H.C. 14-foot combine pulled by I.H.C. 15-30 tractor in 1924. Combine did not have grain tank so wagon had to be pulled to catch grain as it was threshed. A two-man crew was required to operate this rig which could do 20 to 25 acres a day in 15-bushel wheat. This machine plugged quite easily. Operated by New Era Tractor Farms. (Photo, E. M. Young)

Holt 120 crawler pulling Holt combine and grain wagon, 1924, Campbell farms, Hardin, Montana. (Photo, J. R. Taylor)

August 8, 1927, three new 20-foot steel combines pulled by I.H.C. 15-30 tractors. These outfits could direct cut and thresh 50 to 60 acres each per day. Owned by New Era Tractor Farms, as Young brothers called their operation. Note wooden cab on tractor. (Photo, E. M. Young)

Holt 24-foot wooden, self-propelled combine that ran on single, Caterpillar-type track and two wheels, one wheel in front and the other on outer end of platform (extreme left). Photo shows driver, platform operator, and thresherman. Campbell Farming Corporation, Hardin, Montana, 1927. (Photo, Rex Wemple)

Ted A. Dilse, Scranton, North Dakota, operating Cletrac crawler pulling first 20-foot model Holt combine, 1927.

A 3,250-acre grain field with 17 combines direct harvesting on Campbell farms, Hardin, Montana, 1928. Pulled by 18-36 I.H.C. tractors. (Photo, J. R. Taylor)

First self-propelled, 10-foot-cut swather made by Julius Ommodt, New Ulm, Minnesota, early 1940's. Owatonna Manufacturing purchased rights on Ommodt's patent, 1950.

Dan Dullea, Georgetown, Minnesota, with a 30-foot Versatile swather taking 17 rows of soybeans traveling 4 mph for an average of 120 acres per eight-hour day. In wheat, it can easily swath 180 acres in a "long afternoon." Approximate price, $15,000.

Seventeen combines and grain trucks that harvested over 20,000 bushels a day on Jarrett farms and ranches, Britton, South Dakota, September, 1961. Most of the combines belonged to custom operators. Left to right, Don, Ray, and Ron Jarrett.

Holt 120 pulling nine wagons with 200 bushels of wheat each, loaded for 40-mile trip to Hardin, Montana, Campbell farms. (Photo, J. R. Taylor)

Last year that F. A. Pazandak used a threshing machine. Wagon train delivered grain direct to elevator. Twin City 20-35 tractor, late 1930's.

Ox and horse provided power for farmer's "pickup truck" at Fergus Falls, Minnesota, c. 1890, and were capable of hauling 1,000 pounds at 3 mph. (Photo, Oxley)

A. H. Oilinge, who left Britton, South Dakota, 1912, to homestead in Saskatchewan, reported that he made $55 and board, which included time and a half for overtime—and that he was "getting lots of overtime." Photo taken at Harris, Saskatchewan depot, April 12, 1912. (Photo, Ray Jarrett)

A 1909 International pickup owned by O. B. Garnaas, Sheyenne, North Dakota. Note rope chains on rear wheels. (Photo, T. B. Garnaas)

A 1912 Avery truck hauling hogs to market, 10° below zero, February 10, 1914. Rear tires were wooden pegs driven into cast iron rims. Note chain drive and single headlight. Owned by August Loffelmacher, Fairfax, Minnesota. (Photo, WMSTR)

A 1917 Dort with bucket seats and factory-built trailer used to haul fuel to tractors in field on Young brothers' farm. (Photo, E. M. Young)

Frank Kiene, Northcote, Minnesota, farm, 1922. Model T Ford with 56 bushels (3,360 pounds) of wheat. (Photo, Charles Lysfjord)

A 1925 Mack truck with trailer used on Campbell farms, Hardin, Montana. Driver and owner was Grady Woodward. Rig could haul 200 bushels on truck and 100 bushels in trailer. (Photo, Rex Wemple)

Nine semitrailer loads of lambs bound for West Fargo market from Jarrett farms and ranches, January 17, 1955. Shipment contained 10 more semiloads two days later. Total consignment was 3,400 lambs, largest sheep shipment in history of the then 20-year-old West Fargo market. At this time, Jarretts had 6,000 ewes as the basis of their flock.

Scale house and office of T-Bone Feeders, Shepherd, Montana. Company has several of these trucks to haul grain from farmers to feedlot storage.

One of 45 semitrailer units owned by Vern Hagen Farms, East Grand Forks, Minnesota, 1975. Hauling farm produce from Hagen's own farm is primary function of this truck fleet, but men are kept busy year round hauling produce for other farms.

Three planes dusting potatoes on Ole Flaat's Red Wing farm near Grand Forks, North Dakota, 1947. Each plane did 50 acres per hour dusting on 160-acre fields. After spraying commenced in 1948, each plane could do 100 acres per hour.

Cessna 180, one of two planes owned by Freddie and Marlys Mutschler, Wimbledon, North Dakota. This plane is used only for checking fields on more than 30 sections of land. The other plane is a Cessna 414, used for long distance flying only. Marlys flies to fields when extra help or parts are needed.

Eddie Velo and his "dream" of a better turkey barn cleaner. Machine was built for Velo by Cyril and Louis Keller, blacksmiths at Rothsay, Minnesota, 1954, at a cost in labor and parts of $1,400. The Bobcat emblem on this three-wheeled, two-wheel-drive machine gave name to the eventual four-wheel-drive Melroe Bobcat, which has eliminated the scoop shovel in many farm jobs; especially in potato storage, dairy barns, poultry barns, and fertilizer warehouses.

Two-row Iron Age potato planter, spring, 1940, Kiene farms, Kennedy, Minnesota. Left, Glen Bowman and right, Frank Kiene. Despite fact that Mr. Kiene was very active in physical operation of his farms, he always wore a suit, whether sorting cattle or hogs or repairing machinery.

German-Russian homesteaders, Mr. and Mrs. Carl Hoffman, Parshall, North Dakota, 1912. (Photo, Martin Kreft)

Milt and Carol Hertz at their home in Mott, North Dakota. After farming for 10 years, the couple built this house. Hertz was one of the nation's Outstanding Young Farmers in 1969. Both are college graduates and were school teachers before they started farming. An internal radio system based in house and operated by Carol keeps in contact with trucks and tractors in fields 35 miles on each side of Mott.

Bert Hanson, Vernon Center, Minnesota, January, 1962, as he made a testimonial on the John Deere No. 12 chopper for *The Furrow* magazine.

Ronald D. Offutt, Jr., Moorhead, Minnesota, assumed management of his father's farm in 1962 when his father was hospitalized with a heart condition. After graduating from college in 1964, Offutt went into farming full time and in 1975 owned and operated nearly a township of land.

Roswell "Bob" Garst on stalklage pile, 1975. Tractor blade levels the load while tractor packs to seal out air to reduce spoilage. Stalklage is corn residue after corn grain has been combined. (Photo, Dick Seim, *Farm Journal*)

Campbell farms' top staff, c. 1922, left to right, J. R. "Punk" Taylor, foreman and later general manager; Tom Hart; and Thomas D. Campbell. (Photo, J. R. Taylor)

Thomas D. Campbell in standard dress at headquarters in Hardin, Montana, age 42, 1924. (Photo, J. R. Taylor)

Loading beets, 1930, Model A Ford, left to right, Paul Luebbe, Raymond Schieldt, and Ray Koester, Wilkin County, Minnesota. (Photo, Art Overby)

A 1935 International truck hauling its first load of 6 tons of beets from field of Walter J. Ross, one of pioneer sugar beet growers of Red River Valley. Prior to 1942, beets had to be hand loaded as shown. Workers received 25¢ a ton for loading and shoveled from 10 to as high as 20 tons per day.

Cultivating sugar beets using three four-row converted horse cultivators pulled by I.H.C. F-20 tractor on Walter Ross farm, Fisher, Minnesota, 1937. Using horses, Ross could cultivate 12 acres per cultivator; by putting three cultivators behind tractor, four men could do 50 to 54 acres in 8 hours.

A six-row sugar beet planter with fertilizer boxes, Peterson truck farms, Moorhead, Minnesota, 1939. By using 20-inch row spacing, 15 to 20 acres could be planted each day, depending on conditions.

Some of first irrigated sugar beets in Red River Valley. Charles Peterson farms, Moorhead, Minnesota, 1934. In 1939, Hank Peterson got a 30-ton crop of sugar beets when average dryland yields were less than 7½ tons.

First known sugar beet loader in Red River Valley on Walter Ross farm, Fisher, Minnesota, 1939. Machine could load 1 ton a minute in contrast to 10 to 20 tons that a man could load by hand in a long day. Ross believed that loader came just in time to prevent extinction of sugar industry in Valley. Beets were lifted by machine but still had to be hand topped and rowed for loader.

One of two sugar beet harvesters in Red River Valley, 1945. The other was on Jay Wilder's farm. Machine topped, lifted from the ground, and loaded beets, taking but one row at a time, and could do about 3 acres a day. Each beet had to be handled by one of men to clean off dirt lumps or part of tip. Walter Ross farm.

Mrs. Andrew Benson homestead, Chinook, Montana, 1914, first year. Left to right, George, Harvey, Mrs. Benson, Nora, and Lillian Benson. (Photo, Mrs. Harvey Benson)

Mrs. Andrew Benson homestead shack, Chinook, Montana, second year, 1915, after application of tar paper and new addition and porch on right. (Photo, Mrs. Harvey Benson)

Farm home of Don and Jeannine Jarrett, Britton, South Dakota, September, 1975. Jarrett is in partnership with his father, Ray, and his brother, Ron, in Jarrett Farms and Ranches. Total electric house. Note tower for business radio system used by farm.

A camp on Campbell farms, Hardin, Montana, early 1920's. Dining halls, bunk houses, supply house, and shop. Wagons in background. (Photo, J. R. Taylor)

Dugout barn on Mrs. Andrew Benson homestead, Chinook, Montana, 1914. (Photo, Mrs. Harvey Benson)

Interior of Baxter barn, one of many turkey confinement units operated by Earl Olson of Willmar, Minnesota, probably the nation's largest private turkey grower. This building has 10,000 to 33,000 turkeys which are housed in this manner for 8 to 21 weeks.

Office-window view of some of the 40,000 cattle in T-Bone Feeder's 126 pens at Shepherd, Montana. The company has two other lots of 5,000 and 10,000 capacity. Lot is lit at night for cattle comfort and safety with 135 night lights. Trucks are used to fill fence-line feed bunks. About ¼ of the lot is visible on this photo. (Photo, O. J. Deveraux)

A 60- by 208-foot confined slatted floor cattle feeding barn on Bill Noy, Jr., farm, Vernon Center, Minnesota. Building was erected by Bert Hanson in 1970, and was paid for completely by first 1,200 cattle to be fed in it. One-time capacity is 600 head. The four oxygen-limiting feed-storage structures are 17 by 50, 20 by 50, 20 by 60, and 25 by 80 feet. This was the site of the world-plowing contest held in 1971.

Potato cutting, Peterson truck farms, Moorhead, Minnesota. Prior to mechanical cutters in 1950, it took eight people to keep up with four-row planter. With mechanical cutter, two men working part time could do same job. Men could cut about 400 pounds per hour with others handling sacks.

Cultivating potatoes with one-row horse cultivators on Joe Thompson farm, west of Grafton, North Dakota. Discs are used to throw dirt on potato row. About 4 or 5 acres a day could be worked by one team, spring, 1927. Joe Thompson, pioneer potato grower, and his daughter, Bernice, who later became Mrs. Bill Hall.

Two-row, three-horse potato planter, Grant farm, Glyndon, Minnesota, 1930's, could plant 10 acres a day.

Five one-row, horse-drawn potato planters, early 1920's, on Northcote, Minnesota, farm of Kiene Farms, Inc. Each rig could do about 7 acres in 10 hours (35 acres for five crews). (Photo, Charles Lysfjord)

Ray Jarrett using double-row potato planter on his father's farm near Britton, South Dakota, May, 1923.

Rear view, five-potato planters, Kiene farms, Kennedy, Minnesota, 1944. (Photo, Charles Lysfjord)

Two two-row potato planters made into one planter. Believed to be one of first four-row planters in Red River Valley. Could plant two acres an hour. Hank Peterson, Moorhead, Minnesota, 1936.

May 22, 1948, D-4 Caterpillar diesel pulling factory-made Iron Age four-row potato planter. Potatoes were planted on 165 acres in 6 days of 12 hours each. Each hopper held 300 pounds of potatoes and each fertilizer box held 200 pounds. Peterson truck farms, Moorhead, Minnesota.

Bottom-unloading potato truck powered by gas motor. Elevator used to load up to 3 tons of potato seed into a six-row planter. Fertilizer is loaded in the same manner using an auger. Driscoll brothers, East Grand Forks, Minnesota.

Field-potato loader capable of holding enough potato seed and fertilizer to fill six-row planter instantaneously by gravity. Planter covers are opened, and planter is backed under. Developed by Ryan brothers, East Grand Forks, Minnesota, 1966, Arden Burbidge farm, Park River, North Dakota.

Potato harvest on Henry Schroeder farm near Sabin, Minnesota, 1915.

Two-row potato digger powered by F-20 I.H.C. tractor on Joe Thompson farm near Grafton, North Dakota, 1934. Machine dug potatoes and laid them in row for hand picking.

Hand picking potatoes on Joe Thompson farm, Grafton, North Dakota, c. 1930. Pickers varied from 60 to 120 bushels per day and over the years were paid 3¢ to 15¢ a bushel.

Part of potato-loading and picking crew on Charles Peterson farm, Moorhead, Minnesota. Each picker picked 50 to 100 bushels potatoes into sacks in field and then sacks of potatoes were loaded and hauled to storage. This crew could harvest a maximum of 400 bushels (24,000 pounds) in one day.

Ray Jarrett hauling potatoes to Britton, South Dakota for his father in 1921.

Grading potatoes as they are hauled in from field, Jarrett farm, Britton, South Dakota, October, 1923. Ray Jarrett in light shirt and hat.

Loading sacks of potatoes that have been filled by hand pickers. This was common practice of harvest until mechanical harvester became available on large scale in 1952. L. E. Tibert farm near Voss, North Dakota, 1948. (Photo, J. Budd Tibert)

Sack unloading of potatoes, J. G. Hall & Sons, Hoople, North Dakota, 1950, before advent of mechanical harvesters and unloading boxes.

Ole Flaat and J. W. Lambie built this elevator in 1940's to reduce labor. As truck drove through field, men emptied sacks of potatoes onto elevator. This helped knock off more dirt and saved number of sacks required in field. Hoist on truck enabled driver to dump load into elevator at potato house. Flaat farm, East Grand Forks, Minnesota.

Ole Flaat and J. W. Lambie developed an experimental harvester in 1950 that they attached to a standard two-row digger. This was one of first efforts to eliminate hand picking and sacking in field since a few experimental models in 1920's. Ole Flaat, East Grand Forks, Minnesota.

Harvesting potatoes with five new harvesters requiring five workers, plus tractor driver, for each unit. Pickup contains fuel, repairs, and mechanic for instant repair work in field. This is a 960-acre field with 1½-mile rows. Corn planted every 15 rods to reduce wind erosion. J. G. Hall & Sons, Hoople, North Dakota, 1953.

Three-man crew can harvest a 40,000-pound load in 40 minutes. Vern Hagen farms, East Grand Forks and Crookston, Minnesota, 1975. (Photo, John Sundvor)

Northern Potato Co. Warehouse, East Grand Forks, Minnesota, one million hundredweight capacity. Believed to be largest single potato storage building in Red River Valley, where ⅛ of the nation's potatoes are produced.

Bin pilers putting potatoes in storage, Driscoll brothers, East Grand Forks, Minnesota. Skid steer Melroe loaders haul out dirt and damaged potatoes that are sorted from field truck.

Loading conveyor putting bulk potatoes into refrigerated railroad car. Note operator sitting at car entrance. He controls conveyor by hydraulic levers. A 60,000-pound carload takes less than 1 hour (1,000 pounds a minute). Driscoll brothers.

First semitrailer outfits owned by Campbell farms, Hardin, Montana, 1935 Chevrolet trucks which could haul 250 bushels of wheat at very top speed of 30 mph on 40-mile company-built road to rail siding. (Photo, Johnny Owens)

Nine Aultman Taylor 30-60 tractors pulling Holt combines, one experimental Rumely tractor and combine, and one Hart-Parr tractor pulling experimental Nichols and Shepard combine. Campbell farms, Hardin, Montana. There were actually 13 units in the field when this photo was taken in 1929. (Photo, Johnny Owens)

Very bad smut conditions added to normal grease and dirt, 1928. Left to right, Tom Haines, foreman; Johnny Owens, mechanic; Fred Hodges; Thomas D. Campbell, owner; Dan Maddox, attorney; unknown; unknown; Jim Miller; and Dick Melke. Extreme right, Frank Buckner, still with CFC in 1971. Campbell farms, Hardin, Montana. (Photo, Johnny Owens)

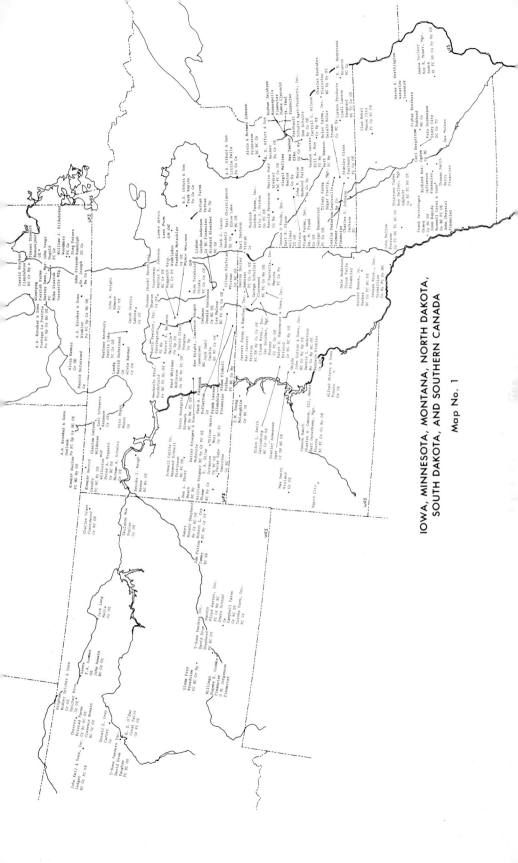

IOWA, MINNESOTA, MONTANA, NORTH DAKOTA,
SOUTH DAKOTA, AND SOUTHERN CANADA

Map No. 1

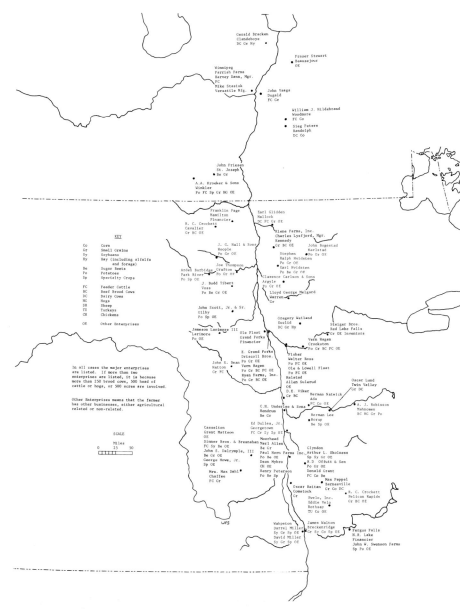

THE RED RIVER VALLEY

Containing parts of Manitoba, Minnesota, and North Dakota

Map No. 2

Note: Symbols for all maps are on this page.

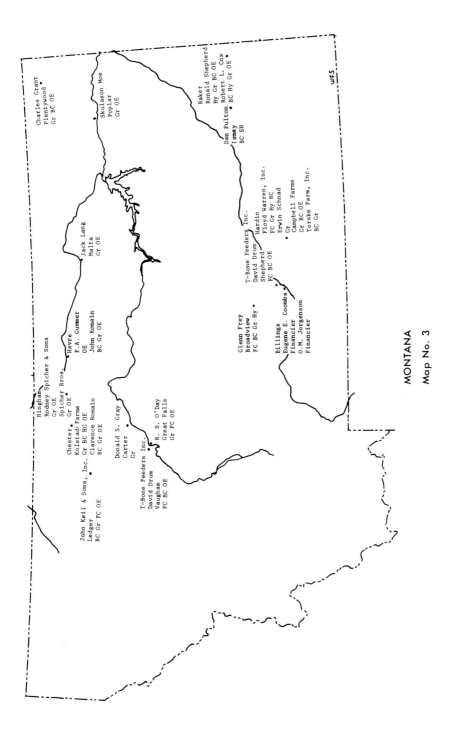

MONTANA

Map No. 3

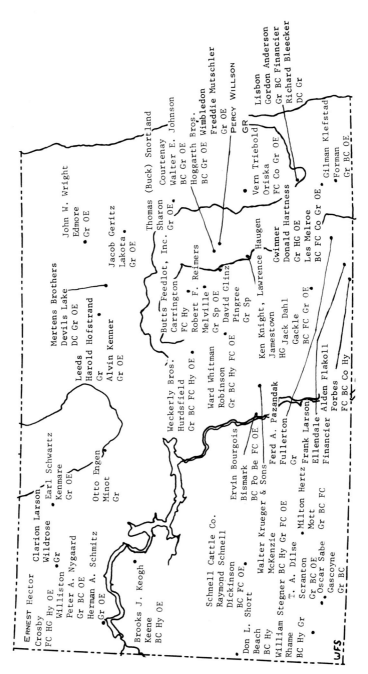

NORTH DAKOTA

Map No. 4

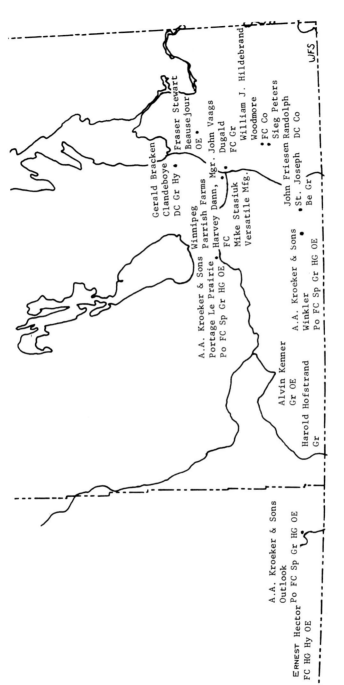

SOUTHERN CANADA

Map No. 5

A.A. Kroeker & Sons
Outlook Po FC Sp Gr HG OE

ERNEST Hector Po FC Sp Gr HG OE
FC HG Hy OE

Alvin Kenner
Gr OE

Harold Hofstrand
Gr

A.A. Kroeker & Sons
Portage Le Prairie
Po FC Sp Gr HG OE

Winnipeg
Parrish Farms
Harvey Dann, Mgr.
FC
Mike Stasiuk
Versatile Mfg.

A.A. Kroeker & Sons
Winkler
Po FC Sp Gr HG OE

Gerald Bracken
Clandeboye
DC Gr Hy

Fraser Stewart
Beausejour
OE

John Vaags
Dugald
FC Gr

William J. Hildebrand
Woodmore
FC Co

Sieg Peters

John Friesen Randolph
St. Joseph   DC Co
Be Gr

WFS

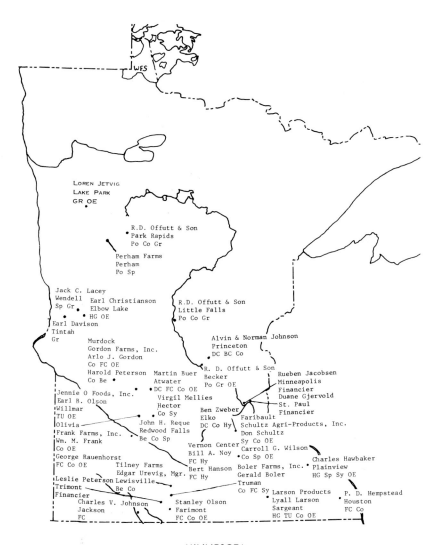

LOREN JETVIG
LAKE PARK
GR OE
•

R.D. Offutt & Son
• Park Rapids
Po Co Gr

Perham Farms
Perham
Po Sp

Jack C. Lacey
Wendell   Earl Christianson
Sp Gr •   Elbow Lake
•   • HG OE
Earl Davison
Tintah
Gr   Murdock
          Gordon Farms, Inc.
          Arlo J. Gordon
          Co FC OE
          Harold Peterson Martin Buer
          Co Be •   Atwater
Jennie O Foods, Inc.   • DC FC Co OE
Earl B. Olson   Virgil Mellies
Willmar   Hector
TU OE   • Co Sy
Olivia   John H. Reque
Frank Farms, Inc.   Redwood Falls
Wm. M. Frank   Be Co Sp
Co OE
George Rauenhorst
FC Co OE   Tilney Farms
                  Edgar Urevig, Mgr.
Leslie Peterson Lewisville
Trimont   Be Co
Financier
          Charles V. Johnson
          Jackson
          FC

R.D. Offutt & Son
Little Falls
Po Co Gr

Alvin & Norman Johnson
Princeton
DC BC Co

R. D. Offutt & Son
Becker
Po Gr OE

Ben Zweber
Elko
DC Co Hy

Vernon Center
Bill A. Noy
FC Hy
Bert Hanson Boler Farms, Inc.
FC Hy   Gerald Boler
             Truman
             Co FC Sy
Stanley Olson
• Farimont
FC Co OE

Rueben Jacobsen
Minneapolis
Financier
Duane Gjervold
St. Paul
Financier

Faribault
Schultz Agri-Products, Inc.
Don Schultz
Sy Co OE
Carroll G. Wilson
• Co Sp OE
                Charles Hawbaker
             • Plainview
             HG Sp Sy OE

Larson Products
Lyall Larson   P. D. Hempstead
Sargeant   • Houston
HG TU Co OE   FC Co

MINNESOTA

Map No. 6

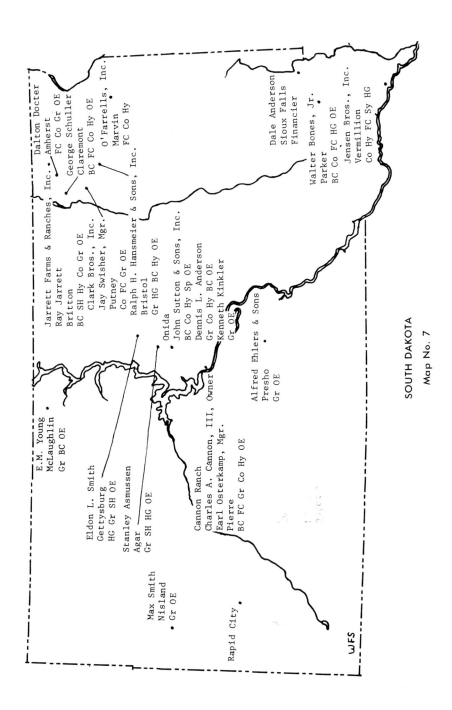

E.M. Young
McLaughlin
Gr BC OE

Dalton Docter
Jarrett Farms & Ranches, Inc. Amherst
Ray Jarrett   FC Co Gr OE
Britton   George Schuller
BC SH Hy Co Gr OE   Claremont
         Clark Bros., Inc.   BC FC Co Hy OE
         Jay Swisher, Mgr.   O'Farrells, Inc.
         Putney   Marvin
         Co FC Gr OE   FC Co Hy
Ralph H. Hansmeier & Sons, Inc.
Bristol
Gr HG BC Hy OE

Eldon L. Smith
Gettysburg
HG Gr SH OE

Stanley Asmussen
Agar
Gr SH HG OE

Max Smith
Nisland
Gr OE

Cannon Ranch
Charles A. Cannon, III, Owner
Earl Osterkamp, Mgr.
Pierre
BC FC Gr Co Hy OE

Rapid City

Onida
John Sutton & Sons, Inc.
BC Co Hy Sp OE
Dennis L. Anderson
Gr Co Hy BC OE
Kenneth Kinkler
Gr OE

Alfred Ehlers & Sons
Presho
Gr OE

Dale Anderson
Sioux Falls
Financier

Walter Bones, Jr.
Parker
BC Co FC HG OE

Jensen Bros., Inc.
Vermillion
Co Hy FC Sy HG

SOUTH DAKOTA
Map No. 7

WFS

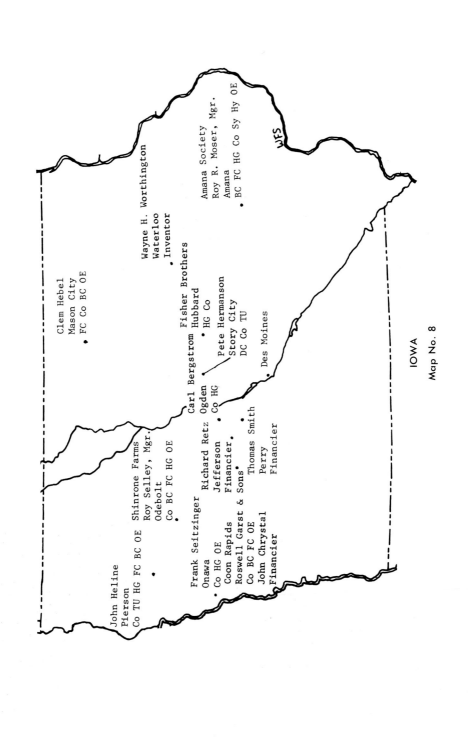

IOWA

Map No. 8

about cellulose and urea . . . large-scale machinery to reduce labor . . . the effect of heavy application of commercial fertilizer to corn . . . why 12 percent of the American nation can produce the world's finest diet when it takes 50 percent in U.S.S.R. and 85 percent in India. . . . Why it takes two hours of labor to produce a bushel of corn in Russian when Garst can do it in three minutes.

"Labor costs are high in Russia because yields are low." Garst agreed that the Russian collectives could become more efficient because it was clear that the large American farms were more efficient. Referring to the small 10-acre plots of eastern Europe he added, "There's no way you can mechanize a thing like that at all. They're individually owned . . . but the only guarantee the guy has [who farms it] . . . is to starve to death."

Asked if he approved of the political tactics of Russian leaders, he answered that in light of what Stalin and Khrushchev had done to improve living conditions in Russia their ruthlessness in earlier times was justified to prevent many more from starving to death.

In 1962, after the Russians had to rely on the world market for food, American journalists again visited Garst because he was the "American regarded as closest to the Russian farm problem." Garst's explanation of the Russian failure in crop production was as follows: "In 1955, Russian agriculture was at about the same stage that American agriculture had reached by 1930. . . ." In his opinion the Russians were unable to capitalize agriculture prior to 1955 and he added, "I don't think that the Russians really believe that here in America the investment per worker in agriculture is $30,000., while the investment per worker in industry is only $15,000." He felt that if the Russians would invest more capital in agriculture its share of that nation's labor force could be reduced from 45 percent in 1962 to 20 percent by 1970.

Being reminded that a farmer of their community was "one of the world's best known corn breeders and producers . . . [who was] known about as well abroad as in America . . . and one of the master farmers of this planet" was sometimes not too well received in the little community of Coon Rapids. Many saw Roswell Garst's interest in helping the underfed Communist nations as a sign of affection for their political system. This could not have been further from the truth, for Bob Garst was a true disciple of the free-enterprise system. Others were critical because they felt that Garst was motivated only to sell hybrid corn and sorghum seed. The announcement in national farm magazines in the month following Khrushchev's visit that 200,000 bushels of butts and tips of hybrid seed had been sold to Russia appeared to affirm the suspicions of those people.*

Garst was the first to admit that he lost friends because of Khrushchev's visit, but he had two rather solid defenses for himself: "Khrushchev was interested in finding out how to produce more food with less labor, and I showed him how to do it." Another defense was that many of those who criticized him in 1959 were the first in line to sell wheat and corn when the Russians had to buy those commodities in 1972 and 1973. The editor of one of the nation's largest farm magazines, after visiting Garst's farm in 1973, wrote: "I view with mixed emotions Roswell Garst's sorghum seed negotiations with the Russians. It's humanitarian to share our know-how with them to increase their food supply. But, from a selfish standpoint, if the Russians grow more of their own feed grains, it follows that we'll sell them less of ours. That's not good for American farmers."

To such arguments Garst replied that because of climatic differences between the United States and Russia he doubted that the Russians would ever be able "to catch up to us," and as they became more accustomed to a better diet they would want to buy more than ever from America.

Senator Hubert Humphrey substantiates Garst's opinion. Senator Humphrey visited Russia after the large Russian wheat purchase of 1972 and met with Kosygin, the Chairman of the Council of Ministers, and with the Minister of Foreign Trade and the Minister of Agriculture. From that visit Humphrey learned that:

[the Russian wheat purchase was] carried out under the specific instructions of the Minister of Foreign trade and Chairman Kosygin . . . [as] a part of the Brezhnev policy to sustain consumer goods. It was this policy which necessitated the purchase of wheat.

In April 1971 Brezhnev . . . assured the Party and, of course, the Soviet people that the high priorities of the Soviet government . . . would be in the consumer area, including . . . above all, food. When the Russian crop went bad in the summer of 1972, Mr. Brezhnev had to make a choice as to whether or not the government was going to ask the Russian people to tighten their belts or whether or not the Russian government was going to spend its hard currency to purchase . . . American wheat. They decided to purchase the wheat. I'm sure they did so because they knew that to fail to keep a commitment relating to the consumer needs might very well pre-

---

*Those 200,000 bushels of hybrid seed were the parent stock of the hybrid corn operations in Russia. Unfortunately for the Russians a very large portion of the total acreage of corn seeded from that stock froze before it matured. This diminished the Russian hopes of ever being able to produce enough corn for their needs.

cipitate a political crisis in the country. The Russian leaders were very mindful . . . [of the Polish revolution of 1970] . . . due to low wages and the lack of consumer goods, particularly food. Apparently the Russian leaders didn't want that to happen in the Soviet Union so they made a political decision to purchase food.

Senator Humphrey added that not only did the government want to "satisfy the consumer needs," but it also desired to "preserve the livestock."

After Khrushchev's visit of 1959 the Apple Valley Farm was the site of many more visits by foreigners; and its owner, to his delight, was called upon to expound his philosophy more than ever. Within a month he addressed United Nations' delegates who wanted to learn how to increase food production in their respective countries. That was followed in the same month by an appearance at the International Trade Council in Chicago.

The six key staff members of the Garst farms are well informed about the nature and future of agriculture. Dr. Robert de Baca, who is now in charge of the breeding program of the 4,000 cows, says, "Crossbreeding's time has come . . . [and] we will be the generation that really puts it to work." Dr. de Baca, long associated with Iowa State University and one of the nation's leading animal scientists, is head of the Garst crusade to crossbreed animals to create a meat animal superior to that of any of the existing breeds. Dr. de Baca said, "We have no perfect breed, but crossbreeding makes it possible to bring together the desirable traits of several breeds." Those who have seen the Garst crossbred cow herd know that it is already well ahead of much of the nation in this new venture for better agriculture. It is the opinion of the Garsts that on a large-scale cattle crossbreeding program in the Midwest only Ted Jennings of Highmore, South Dakota, is ahead of them.

Dr. de Baca points to the evidence from 14 experiment stations with over 5,000 cattle that crossbreeding brings substantial economic gains over straight breeds. He adds, "Through systematic crossbreeding among British breeds up to 25 percent more output per cow exposed can be expected compared to straight breeding. . . ." With 4,000 cows and a top-notch scientist to conduct his program, Roswell Garst once again will be "having fun keeping ahead of the crowd," for his lead is substantial.

Stephen Garst, who is responsible for much of the livestock operation, feels that "the progressive feeder and cowman has a great opportunity to determine the future direction of the beef industry. This should improve our market and help the packing industry by giving them a more uniform animal for economy of butchering—which means a better product for the consumer."

The Garst farm staff are highly oriented toward the production of beef. Their research indicated that since 1930 the American consumer had spent about 2.5 percent of his personal disposable income each year for beef, regardless of price. They reasoned that if they could produce more beef at a lower cost they could increase total sales. With the use of the total corn plant, urea, crossbreeding, and intense scientific and mechanical inputs to the land, the Garsts know that production costs can be reduced. They would like to produce a "beef carcass that the consumer wants and [that] satisfies the packaged beef concept of the packer—but [in 1973] neither the consumer, packer, or feeder know exactly what kind of an animal will fit that demand."

Few farmers are as conscious of agriculture's total marketing problems as are the staff of the Garst farm. That fact was recognized when in 1971 the American Meat Institute bestowed its annual award on Roswell Garst "as the man who has made the most contribution to the meat industry" that year.

In their planting practices the Garsts looked to the acra-plant system of proper placing of the seed to be the next major breakthrough in corn production. David Garst was making acra-plant his main project in conjunction with Don Williams. With better planting techniques, heavier fertilizer application, and intense use of herbicides and pesticides, yields will improve enough to justify funds on upgrading land through better drainage systems where needed. Irrigation is an obvious part of the future for the crop producer. Such capital inputs indicate that the minimum capital investment on a one-man farm must be at least $150,000, but there must also be at least $250,000 at work sometime during the year. The Garsts believe that to make a larger farm succeed, the capital investment must be $200,000 to $250,000 per worker (1973 dollars). This is considerably higher than the average investment for American industry, but it is necessary if the agricultural laborer is to have a just wage.

American agriculture has already pointed the way to achieve this, according to the Garst farm staff. They point out that with only 3 percent of the labor force involved in farming, over one-fourth of all exports from the world's greatest industrial nation are agricultural goods. David Garst used the well proven example that if labor is worth $20 a day, a man using a pitchfork and shovel cost 10¢ a pound of gain per animal, but with automated equipment feeding 10 times as many cattle the labor cost per pound of gain is 1¢.

The Garst philosophy about the future of agriculture is best summed up by his son David:

It is a paradox that despite our trememdous wealth, few farmers make a fair return for labor and investment. It is a paradox that despite our astonishing efficiency, few farmers have even scratched the surface of proven technology. It is a paradox that despite our almost complete freedom of action, farmers have failed to take advantage of the opportunity before them.

It is the changing nature of the agricultural revolution that holds the promise of farm prosperity.

To prove his point Garst used the example of an Iowa farm neighbor whose records established that he was raising a ton of beef per acre. Four workers were producing the crop and fattening the cattle to provide the meat supply for more than 5,000 average American consumers. "This man," David Garst says, "freely admits that he still is a long way from optimum efficiency." No one on the Garst Apple Valley Farm at Coon Rapids, Iowa, will disagree with this neighbor even though few American farmers had matched those 1965 production figures nearly a decade later.

Recycling of animal excreta ranks high on the priority list at the Garst farms. Roswell observed in the 1950's that it was the urine more than the solids that seemed to give the greatest benefit to the following year's crop. As an experienced corn-beef-hog farmer he long ago learned that hogs did very well following beef animals. Iowa beef-hog farmers had a favorite saying: "Fortunate indeed is the pig that follows the steer." According to Bob Garst: "Nature has a lesson for us in that practice—manure does have great feed value."

The overriding mission of Roswell Garst is to see to it that the world has ample food. Bob Garst's brother Jonathan maintains that "the well-fed nations of the earth have their growth rate under control. . . . More food, fewer babies is strictly a human affair." From this, Roswell argues that the more rapidly we increase the food supply of the world the sooner we will reach the point when the birth rate will decline and the world's number one problem of feeding itself will be under control. Garst points out that the well-fed man requires 3,000 calories per day and a starvation diet is 1,500 calories. Once a man achieves the 3,000-calorie level, he becomes interested in the luxuries of life that finally have come within his reach, and he limits the size of his family so that he can

Beyond the Furrow

enjoy those luxuries. In countries where the daily 3,000 calories are available, the percent of income spent for food is drastically lower than in the countries where the food supply is marginal. This evidence is overpowering, and there is nothing the crusader Garst would more like to do than show the world how it can be done. Whether he will live to see the world convinced of this theory is unknown to us. Even if he does not reach the highest achievement, history will someday record that Roswell Garst "influenced agricultural people as much as anybody."[5]

NOTES

[1]Tom Patrick, "Encounter with Garst: Challenger of Tradition," *Big Farmer*, April, 1972, Vol. XLIV, No. 4; Letter of David Garst, September 30, 1971; Interview of Roswell Garst, Coon Rapids, Iowa, June 14, 1972; Interview of Viv Bell, Glidden, Iowa, June 15, 1972. Mr. Bell rented land from Garst for 20 years; *Coon Rapids Enterprise*, May 16, 1930, June 20, 1930, February 2, 1931, July 18, 1930, February 13, 1931; April 6, 1934, June 5, 1931, June 12, 1931; December 4, 1931, December 25, 1931, January 30, 1931; Jonathan Garst, *No Need for Hunger*, Random House, New York, 1963, pp. 13-16; Roswell Garst, "Forty Exciting Years, 1930-1970, A History of Garst and Thomas Hybrid Corn Company," privately published, 1970; Interview of Charles and Bertha Thomas, Coon Rapids, Iowa, June 15, 1972, Charles Thomas was a partner of Garst for 42 years. He was born in 1894, only four years older than Garst.

[2]R. Garst, "Forty Exciting Years, 1930-1970, A History of Garst and Thomas Hybrid Corn Company"; David Garst, "Agricultural Progress and Opportunities in the 1970's," a speech delivered to the N.A.A.M.A., October 1, 1969; Charles Thomas Interview; *Coon Rapids Enterprise*, August 27, 1959, October 23, 1931, March 8, 1940, March 15, 1940, October 12, 1934, October 19, 1934, August 8, 1930, August 15, 1930, October 10, 1930, November 21, 1930, February 2, 1934, July 13, 1934, August 17, 1934, August 24, 1934, October 24, 1934, February 11, 1938, March 18, 1938, May 27, 1938, September 16, 1938, September 23, 1938, October 7, 1938, October 21, 1938, November 4, 1938, December 2, 1938; John Strohm, "A Farmer with a Message—and a Method," *Reader's Digest*, May, 1960, p. 149; Ray Anderson, "The Farm—and Farmer—Mr. K. Chose to Visit," *Farm Journal*, October, 1959, LXXXIII, No. 10, p. 39; Orville L. Freeman, "Malthus, Marx, and the North American Breadbasket," *Agricultural Policy in an Affluent Society*, ed. Vernon W. Ruttan, Arley D. Waldo, and James P. Houck (W. W. Norton, New York, 1969), pp. 289-290; David Garst, "Address to West Coast Farm Institute," delivered Spokane, Washington, February 11, 1963; J. Garst, *No Need for Hunger*, pp. 16, 18, 22, 35; Roswell Garst, "The Agricultural Prospect Before Us," A manuscript written in 1950 and published in *The Farm Quarterly*, Spring, 1951; Everett M. Rogers, *Diffusion of Innovation* (The Free Press, New York, 1962), pp. 85, 113; George M. Beal and Joe M. Bohlen, "The Diffusion Process," Special Report No. 18, Cooperative Extension Service, Iowa State University of Science and Technology (Ames, November, 1962); Bryce Ryan and Neal Gross, "Acceptance and Diffusion of Hybrid Corn Seed in Two Iowa Communities," Agricultural Experiment Station Iowa State College Research Bulletin 372 (Ames, January, 1950), pp. 671-672; "Corn: Acreage, Yield, and Production by States 1866 to 1943," U.S.D.A., 1954, p. 2, 27; *Big Farmer*, LXIV, No. 4; Roswell Garst Interview; Roswell Garst, "Cows and Calves Can Add Profits," a manuscript based on a study of the Garst corn roughage feeding experiments; *Land of Plenty*, p. 58; Interview of Bert A. Hanson, Vernon

Center, Minnesota, January 11, 1971; William J. Peterson, "Iowa Weather Statistics—1873-1968," Of Time and Weather, The Palimpest, Vol. L, No. 1, January, 1969, p. 63; "Iowa—All Corn—Acreage, Average Yield per acre, and Production," Iowa Crop and Livestock Reporting Service; Iowa Annual Farm Census 1971, Bulletin 92-AG, Iowa Department of Agriculture and U.S.D.A., pp. 12-13.

³Coon Rapids Enterprise, December 26, 1930, October 4, 1940, October 18, 1940, January 9, 1942, March 13, 1942, April 3, 1942; J. Garst, No Need . . ., pp. 26, 28, 29; R. Garst, "Forty Exciting Years . . ."; "Iowa—All Corn—Acreage, Average Yield per Acre, and Production"; Roswell Garst Interview; Strohm, Reader's Digest, May 1960, p. 150; Roswell Garst, "Cows, Calves . . . Profits;" Average Prices Received by Iowa Farmers—Corn, 1908-1969," Iowa Crop and Livestock Reporting Service; Combined Interview of Roswell, David, and Stephen Garst, John Chrystal, and Bob Henah, Coon Rapids, Iowa, June 13, 1972; John Dos Passos, "Revolution on the Farm," Life, XXV, No. 8, August 23, 1948, pp. 95-98; Minnesota Crop Reporter Bulletin No. 503A, November 1972, Minnesota and U.S.D.A. Cooperating; Big Farmer, XLIV, No. 4; R. Garst, "Agricultural Prospects . . ."; Farm Journal, LXXXIII, No. 10, p. 86; Roswell Garst, "Bob Garst's Latest Crusade," Successful Farming, July 1962, p. 70; Letter from David Garst, January 5, 1972; D. Garst, "Agricultural Progress . . ."; David Garst, "Cows and Calves a Companion Crop for Feed Grains," A speech delivered to the Iowa-Kansas Fertilizer and Agricultural Chemical Association, January, 1969.

⁴Life, XXV, No. 8, pp. 101, 103; Letter of Roswell Garst to Lane Palmer, editor of the Farm Journal, May 27, 1972, p. 70; Roswell Garst Interview; Successful Farming, July, 1962; "The Use of Corn Cobs, Corn Stalks, and Grain Sorghum Stubble for Cattle Feed," Garst and Thomas Bulletin No. 5; "Average Prices Received by Iowa Farmers—Beef Cattle," Iowa Crop and Livestock Reporting Service; Successful Farming, July, 1962; David Garst, "West Coast Farm Institute, 1963; David Garst, "Cows and Calves . . ."; Reader's Digest, May, 1960; Farm Journal, LXXXIII, p. 86; Coon Rapids Enterprise, September 17, 1959; Roswell Garst, "Cows and Calves Can Add Profits"; Interview of Stephen Garst, Coon Rapids, Iowa, June 15, 1972; Roswell Garst, "The Agricultural Prospect . . ."; Iowa Annual Farm Census Bulletin 92-AC, p. 29; Roswell Garst, "How to Have Cow Feed That Costs Less Than Nothing," A script prepared March 28, 1972, stating Garst's full position on the development of beef brood cow enterprises; Garst and Thomas Bulletin No. 5, "The Use of Corn Cobs . . ."; Roswell Garst, "Be Patient, Beef Consumers," The New York Times, May 9, 1972; Roswell Garst, "The Growing Demand for Beef Means We Need More Beef Brood Cows," A manuscript written January, 1972; Al Oppedal, "Legume Hays Are Wasted Energy," Feedlot Management, Vol. XIV, No. 9, September, 1972, pp. 60-62, 74; Roswell Garst, "Growing Demand for Beef Means We Need More Beef Cows," Big Farmer, Vol. XLIV, No. 5, August 1972; Beef, July, 1973, news notes in Beef Business; Louis M. Thompson, "Iowa Agriculture, World Food Needs and Educational Response," Center for Agricultural and Economic Development, Iowa State University, AAC, 156 (Ames, 1965), pp. 14-15; Louis M. Thompson, "World Food Situation in 1973," Iowa State University, November, 1973.

⁵Viv Bell Interview; Coon Rapids Enterprise, September 17, 1959, including a quote from Christian Science Monitor, September 10, 1959; Reader's Digest, May, 1960, pp. 148-152; J. Garst, No Need for Hunger, pp. 5, 6, 11, 100, 109; Editorial, "How Tall the Corn?" Fortune, LX, No. 5 (November, 1959), pp. 118, 120; R. Garst Interview; Combined staff interview; R. Garst, "The Agricultural Prospect. . . ."; Interview of David Garst, Coon Rapids, Iowa, June 12, 1972; Interview of Bob Henah, Coon Rapids, Iowa, General farm manager, June 14, 1972; Coon Rapids Enterprise, September 10, 1959, April 3, 1959, February 27, 1959, September 3, 1959, August 13, 1959, August 27, 1959, October 15, 1959, November 26, 1959; "The Real Story of Khrushchev's Farm Failure," U.S. News & World Report, LII, No. 14 (April 2, 1962), pp. 78-80; Letter from David Garst, January 5, 1972; L. Soth,

"The Iowa 'K' Will See," *New York Times Magazine*, September 6, 1959, pp. 6-7; *Farm Journal*, LXXXIII, p. 39; "What Russia's No. 1 Man Will Learn on an Iowa Farm," *U.S. News & World Report*, XLVII, No. 11 (September 14, 1959), pp. 52-55; "From the University of Experience," *Better Crops with Plant Food*, LV, No. 4 (Winter, 1971-72), pp. 6-7; Senator Hubert H. Humphrey letter, August 4, 1973, to the author; Dick Hanson, "Editorial," *Successful Farming*, LXXI, No. 1 (January, 1973), p. 4; Guy Price, "Crossbreeding by Computer," *Big Farmer*, XLIV, No. 2 (February, 1974), Cattle Guide; Stephen Garst Interview, David Garst, "Agricultural Progress . . ."; David Garst, "Agriculture's Opportunity Knocks!," A manuscript prepared January, 1966, for delivery to various Midwest farm groups.

CHAPTER VIII

Bert Hanson—Disciple of Alfalfa

A Progressive Home

IN 1910 Honnace A. Hanson left home and went to Nicollet County, Minnesota, where he purchased a 400-acre farm for $80 an acre. The following year he started a herd of registered Short-horn cattle, which later became the property of his son, Bert. In 1917 Honnace A. Hanson purchased a 12-20 Heider tractor with a cab and three-bottom mounted plow for $1,720. He also acquired a herd of registered hogs and a stable of outstanding Percheron horses.*

This was the farm setting in which H. A. Hanson's son, Bert, was reared. After finishing the eighth grade the young Hanson, energetic and aggressive, decided to farm with his father. Bert's mother died when he was 20 years old. In 1926, at the age of 25, Bert married a local farm girl, Irene Anderson, and rented his father's 400 acres for $7 an acre—cash. He purchased livestock and machinery from his father on a contract.

But Bert Hanson was not satisfied just to start with his father's machinery and fancy horses. Even though he was a horse fancier he felt that tractors were the way to farm and decided to buy a Model D John Deere (15-27) and a three-bottom plow for $1,000. This tractor could plow at four miles per hour and could work about 14 acres in 10 hours in contrast to 5½ acres with the two-bottom horse plow. He also purchased a new John Deere two-row corn planter with fertilizer attachments for about $100. This was reputed to be the first John Deere corn planter with fertilizer

---

*In 1910 the average farm in the United States was 138 acres. There were few 400-acre farms and even fewer herds of registered cattle and hogs. Tractors with cabs and mounted plows were rare indeed.

boxes sold in Minnesota.* Cooperating with some fellow Farm
Bureau members, Hanson secured 3 tons of fertilizer on the first
carload of commercial fertilizer ever shipped into Mankato, Minnesota. Deere and Webber were very excited about this new
young farmer and his purchase of their latest equipment. To learn
more about him they sent a reporter and a photographer to Hanson's farm to take pictures and get information for an article about
him for the company magazine, *The Furrow*.

From 1926 through 1932 Bert and Irene Hanson operated the
home farm, and because they were paying cash rent they experienced some operating losses. To make up for the loss Irene made
butter which sold for 12¢ a pound. At that time corn was 9¢ a
bushel, and a 900-pound sow brought 3¢ a pound. "I was well
aware of the problems of farming by then, but I was determined to
get on with a farm of my own," said Hanson later.

The young couple soon became known in Nicollet County as
successful hog breeders. By breeding sows as quickly as possible
after farrowing they were able to average nearly three litters per
year from each sow.** At this time *The Breeders Gazette* carried
an article about the young Minnesota farm couple's hog-raising
enterprise. In reference to the article, the Hansons received many
critical letters and comments from visitors and from farmers who
said it was impossible to do what they were doing. But when a
hog breeder from Germany visited them and mentioned that this
had been done for years in Germany on his farm, they realized
that they were on the right track.

Fertilizing corn, using the latest equipment, and experimenting
in hog breeding did not consume all of Bert's time. He loved
competition and was looking for a new challenge every day. He
liked to fish, hunt, travel, and do anything that took endurance. At
age 72 Hanson recalled the activities of his earlier days: "I always
liked being in first place. I don't know when I have been second
in anything that I have tried." It was his competitive attitude that
prompted some of his neighbors to challenge him to enter the
local corn-husking contests. Those spirited husking bees were just
the thing to bring out his strongest impulse. As he mentioned in
later years, "I was challenged to husk corn in the 1925 state con-

---

*Farm Equipment Institute recorded that 1929 was the first year that band fer-
tilizer attachments were available commercially. But generally companies have
equipment out on test at various farms a couple of years before they are put into
full production.
**The gestation period for hogs is 112 to 115 days.

test. I had never given it a thought but I entered." That was the beginning, for in 1926 Hanson was second in the state and eighth in the national. In 1927 he was the Minnesota Corn Husking Champion. In the national contest held at Winnebago, Minnesota, he placed second. In 1931 Bert Hanson was Minnesota Husking Champion again, making him the state's first two-time winner. He did not do well at the national contest held at Grundy Center, Iowa, "because it was the one day in my life that I have been sick," he recalled.* There was some satisfaction from that contest because Bert visited with Henry A. Wallace, later the U. S. Secretary of Agriculture. Wallace informed him that he had upset all the statistics about what constituted the ideal corn husker. According to Wallace the ideal husker was to be over 6 feet, quite slim, and less than 25 years of age. At the peak of his competition Hanson was 30 years old, just 6 feet, and 260 pounds, all bones and muscle. Hanson's best record as a husker was 29 bushels in 80 minutes. In connection with his activities in corn husking Hanson became exposed to many different corn-growing techniques that he applied to his farming operation in order to improve the production. This was more than ample reward for the time he lost at all the husking events.

His father's herd of registered polled Shorthorns which Bert had purchased in 1926 numbered 51 head and was made up of calves, yearlings, and mature stock. Bert paid his father $5,100 for the entire herd, which was a true market value for that period.** This record herd later became one of the greatest sources of satisfaction for Bert and Irene. In 1928 the herd was shown at 28 livestock expositions, and although time and cost factors were involved in participating in these shows, the Hansons were quick to see the value in being exposed to outstanding farmers of the nation. This was particularly true in the case of Bert because he was a natural conversationalist who could talk as well as listen.

In his first years of farming, Hanson employed John Morris as his herdsman, because "Morris liked showing cattle better than eating and really knew how to win the ribbons." In the 10 years that Morris worked for Hanson they won nearly one-third of the International Championships in the polled Shorthorn breed besides winning every premier award at the Minnesota State Fair in

---

*Mrs. Hanson verified that from the day of their marriage in 1926 to 1973 he had been sick only the one day of the 1931 national husking contest.
**Hanson still had the cancelled note in 1974 to show as proof of his purchase.

every class in which they were eligible to compete. Along with Morris, Bert Hanson became thoroughly convinced of the value of feeding ground ear corn to "keep the bloom on cattle," a practice that he never discontinued in his farming career. From 1926 through 1935 the Hanson herd won 126 ribbons at the International Livestock Exposition.

"Shadybrook Monarch," the best bull of the Hanson herd, was grand champion at the International in 1931. In all, he took grand champion ribbons in 69 shows. Another record held by one of the Hanson bulls is for the heaviest bull for age ever shown in the Shorthorn range trials. His commercial cattle also did well, with a record dressing of 68 percent cold at the packer; and up to 1973 he had the highest-priced load of mixed steers and heifers sold at the South St. Paul Union Stockyards. In 1947 three Hanson bulls were sold to Mr. Bianchi of Guatemala. Through this sale Hanson became associated in what is reputed to be the first lot of cattle ever assembled for air shipment out of Minnesota and the first air shipment of registered Shorthorns to a foreign country. The 31 cattle loaded under the supervision of Hanson were the heaviest livestock shipment to that date by American Airlines. The flight took place on March 20, 1947, in a converted army C-54.

The original herd, although a costly venture to Hanson in 1926, was eventually built into 155 producing cows. It provided several functions for the Hanson Farm, for in addition to the pride of owning a fine herd and the opportunities to exhibit it, there was the chance to experiment with his own cattle in the feedlot, as contrasted to purchased animals. Hanson's banker, Clarence Banks, related that the valuable cow herd provided an "excellent financial lever which Bert knew how to use to the hilt." "This," Mr. Banks added, "was a strong factor in the total success of Shadybrook Farms."[1]

Shadybrook Farm

In 1931 the Hansons started looking for more land, and when land owners found them to be prospective buyers they were beseeched with offers. Hanson said:

The Federal Land Bank had between 1,300 and 1,400 farms available in Iowa, Minnesota, and North Dakota. I think I must have looked at a thousand of them—I always did like to travel—before I decided on the one we wanted. . . .

I think we took the place with the worst buildings and the most run down land, but it had many positive points. It had Clarion Webster soil and in my opinion that has got to be the best in the world—there are only about 19 million acres in Minnesota and 13 million acres in Iowa—you know! At the time we bought the farm it was reported to be so poor that if a jackrabbit wanted to cross it, he had to carry a lunch bucket or he would starve to death.

Clarence Banks, who financed the Hansons for over 40 years, remembered: "When Bert bought that farm everyone gave him six months because it was one of the most rundown in the area but today (1973) it is one of the best."

This farm contained 252 acres and was located one mile north of Vernon Center, Minnesota. Hanson recalled:

We were excited because it had so many good features, a paved U. S. highway ran right by the place, it had electricity so we could have lights, pump water, and run the washing machine. These were really big labor savers which were important because in 1932 we ran the 400 acres in Nicollet besides this farm.

To which Mrs. Hanson added:

After one week's honeymoon I started to cook for hired men; they were good reliable men but it took so many people to do the job in the twenties and thirties. In 1932 on less than 700 acres we had five year round hired men. Today [1972] we have a neighbor who farms 320 acres of continuous corn and he puts in less than 750 hours a year because it takes him less than a minute to produce a bushel of corn. He could do much more but he is satisfied.

With a smile, Bert added,

I think if I had not married when I did I would have become a hobo because I had such a desire to see the world. . . . But we learned something in 1932 that became a part of our continued operation. We learned that volume was important. With enough volume you need the manpower around to do the job so every once in a while you can take time off to do the things you like—fish, hunt, and travel. See this picture of sixteen deer? That was taken in 1932 when some of my friends and I took off to hunt while the work was taken care of by the men. How could I have done that without some volume?

Shadybrook Farm had been purchased by the Wilber family in the 1860's for $5 an acre, but when the Hansons bought it in 1932 it carried a $19,200 mortgage ($78 an acre). The Federal Land Bank sold it to Bert and Irene Hanson for $17,000, on August 18, 1932. As a down payment the Hansons borrowed $1,000 on a personal note at 4.5 percent. It took eight and one-half years to pay back that $1,000 in addition to the $250 semi-annual payments required by the Federal Land Bank. The Hansons recalled, "We always met the payments because from 1933 on we have never had a year in which the farm had an operating loss. . . . We didn't dare to have a loss because when we got going our liabilities were more than our assets. But in 1972 I sold a truck load of fat steers for more than all of our debts in 1933."

With two farms about 20 miles apart Bert Hanson was forced to buy a truck to transport equipment and crops between them. He soon found himself using the truck to an advantage in marketing his production. He realized that he needed year-round men, and with the use of the truck he could keep them busy in the "off season." However, because of the business of marketing, Hanson personally had to spend much time on the road throughout the 1930's.

In 1933, the first full year on his newly purchased farm, Hanson planted all corn because that was the traditional high-income crop for Blue Earth County. The weather was good, and he averaged 80 bushels an acre. After receiving 9¢ a bushel for corn in 1932, he was prepared for the worst; but the 26¢ a bushel price for corn in the fall of 1933 looked good, so he sold 5,200 bushels to "pay up his bills." He held back on the rest of the crop and in 1934 was able to seal it under the government program for 45¢ a bushel. That was his best break, for 1934 was the "toughest year" the Hansons had. A 56-day midsummer drought reduced corn yields to 28 and oats to 45 bushels per acre.

Remembering those years of stress he said:

After 1934 I had a steady climb in progress—we passed our low point in price in 1932 and yield in 1934. Besides it was 1934 that I bought a new Chevrolet ton and one-half truck and scouted the country to try to find a better market for my crop. I made a deal with the Farmers Feed Store at Hinckley, Minnesota. They would take my corn if I could deliver four loads a week. . . . By leaving by 2 a.m. I could haul the corn to Hinckley, take a load of oats to Minneapolis, and haul five tons of flour to Mankato in time to get unloaded by 5 p.m. It was a 350 mile round trip and I made four a week. The men gassed and loaded the truck each night after I got home so it was all set to go the next morning. I could be gone lots because Irene knew how to take care of the place. Besides our head man had been with us since 1926 and knew how we wanted things done.* Of course I was home a couple hours each evening and over the weekends.

By working this schedule Hanson grossed $46 per 350-mile trip, or $184 a week. He actually received 15¢ a bushel more for his corn, excluding his labor, because the load of oats and flour on

---

*The employee referred to was Harold Parsons, who started with the Hansons in 1926 and remained with them until he retired in 1955.

backhaul paid his cash expenses.* That year Bert Hanson traveled 48,000 miles and received about $3,600 extra income either for his corn or labor "which ever way you figure it." Hired help at that time cost $30 a month plus room, board, and washing, so for a $720 cash outlay Hanson could hire two year-round employees to do his work while he was on the road and still had a good profit for his time. The tensions of working with hired labor fell mostly on Mrs. Hanson, but she was able to handle this responsibility. Besides, Bert and the men got along better if he let them work at their own pace.

The big bonus from trucking his farm produce directly to the consumer came in 1936. As is so often true in agriculture, "one man's loss is another's gain," and what happened in 1936 proved this point to Bert Hanson. Because of the very dry season, grain was in short supply in the Hinckley area. Hanson bought back his 1933 corn that he had sealed in 1934 for 45¢ a bushel and sold it to the Farmers Feed Store for $1.56. He netted over $5,000 additional income on his 1933 crop in this manner. "At that point," Hanson emphasized, "I started buying land so I could get volume. It was easy to do because there was lots of land being dumped on the market. I added ten additional pieces of land to the original farm until I stopped buying in the 1950's."[2]

Alfalfa Takes Over

As indicated, Bert Hanson reached a turning point in 1936 by reaping a bonanza income in the sale of his 1933 corn crop. But equally important for his long-range objectives was an adversity that struck in 1932. That year the Hansons discovered that their newly purchased Shadybrook Farm was so infected by diseases that they were forced to discontinue their registered hog operation at once. This was a blow to the young couple, because they had wanted to remain diversified and their hog enterprise resulted in a good cash flow which is so essential to beginning farmers. Little did they realize that being forced to quit hogs would eventually lead to an enterprise in a field where Bert Hanson would make his fortune and establish his reputation. It was just another challenge to overcome and the versatile Hansons were not stopped by it.

As Hanson's banker, Clarence Banks, recalled:

---

*Hanson had an excellent set of records and pointed out that his 1934 truck had cost him $.01728 a mile to operate, exclusive of gas, oil, and tires, but including replacement with a 1935 truck.

I knew Roswell Garst when he started farming and I have often thought of the similarities of the two men, [Hanson and Garst] both wanted to be first and both liked experimenting with new things. Bert was always experimenting but he did it on a small scale so if something went wrong there weren't many dollars involved. I knew Bert from the start here in 1932 and it was obvious then that if anyone would make it he would. He was a good cattleman and a top notch manager. There were many others in the community no worse off than he who had the same opportunities he did, but only a very few took advantage of them.

Mrs. Myrtle Harkins, local newspaper editor, said:

I remember Bert and Irene from the start; they were a terrific team and knew how to get things done. Merton, my husband, frequently went out to work at Bert's nights after he closed the press for the day because they lifted each other's morale. Even in the depths of the depression Bert and Irene were always looking ahead for new things to do. They were really hardup but they never gave it a second thought.

Bert Hanson's father had liked raising alfalfa but had some reservations about that crop. Bert recalled: "My dad always cussed alfalfa because it interfered with corn cultivation and that was more important." But alfalfa was important in the rotation and was needed on a farm with a cow herd of 51 head. When Bert and Irene Hanson had to quit the hog business they decided to increase the cow herd as rapidly as they could afford to. Because most of Shadybrook Farm could be tilled it proved economical to bring feed to the cows rather than graze the land.

The Hansons reasoned that if they were going to raise alfalfa it would have to be in a large enough volume to justify a crew and the best possible equipment to do the job. The concept of volume once again proved to be a proper approach. Therefore, after 1933, alfalfa became the major crop in acres and income on Shadybrook Farm. So, in the heart of some of the nation's best corn land, corn took second place to alfalfa. Hanson, pointing to his records from 1933 on, proved that "alfalfa has consistently made us more money per acre than corn besides being a much more trouble-free ration for my cattle." He pointed specifically to one year in the early 1940's when he sold 5 tons of alfalfa per acre at $33 a ton and his 100-bushel corn crop sold at 80¢ a bushel. Then he referred to Morrison's *Feeds and Feeding* to prove his contention about alfalfa. As an added bonus he felt there was much less risk in alfalfa than corn, not to mention the fertility benefits so necessary in the days prior to the mass production of nitrogen.

To be in the commercial alfalfa business on the scale that Hanson desired forced him to buy a semitrailer truck. Once again he gloated about being a first—this time he was the first farmer to ever purchase a Chevrolet truck and trailer unit from a large dealer at Mankato. A 1½-ton truck was not a profitable unit to haul a bulky feed like hay. He soon discovered that if he could guarantee a steady supply of quality hay he could sell more than he could produce. His first big customer was the South St. Paul Union Stockyards. Eventually he delivered hay as far south as San Antonio, Texas, as far west as Baker, Montana, and as far east as Baltimore, Maryland. He expanded his alfalfa acreage as rapidly as he could buy land.

Hay-making technology changed tremendously during the 40 years that Bert Hanson was in the business. When he first started cutting alfalfa in 1926, a two-horse, 5-foot mower, a side-delivery rake, horse-drawn hay loaders, and slings were used. He could harvest a hay crop at the rate of 1 ton for every four man-hours. It took three full-time men to operate his 400 acres. After he became a commercial hay producer most of the hay was put up directly into bales in the field. A John Deere stationary wire-tie baler driven by a belt from an F-20 tractor was placed in the middle of a field. A man with a two-horse, 10-foot dump rake pulled the hay up to the baler. Two men pitched the hay onto the platform where another man using a pitchfork fed the baler. A fifth man tied the wire and stacked the bales. In that manner about 40 acres yielding 80 tons of hay per cutting could be baled in two good working days. About five baler settings were required in a 40-acre field. With this mechanized crew it now required only about 1 hour and 20 minutes per ton up to the point of loading it in the field.*

In 1940 Hanson purchased a Case wire-tie pickup baler that was driven by its own engine and was pulled down the field by tractor power. The major problem here was to have enough wagons and men to keep the baler going steady. By this time the 5-foot horse-drawn mowers were discarded for 7-foot tractor-drawn implements, and side-delivery rakes were converted to be pulled by tractors. Because he had the equipment and the labor force, Hanson was flooded with requests for custom hay making. For many years a crew of 15 men was kept busy putting up the three crops on a thousand acres of alfalfa land. Handling 100-pound, wire-tied bales in the heat of summer was a strenuous and distasteful job.

---

*In addition to the time required per ton, the back-breaking efforts of this crew have to be kept in proper perspective. It should also be noted that handling hay in this manner caused a high percent loss of the leaves, which are the high-protein portion of alfalfa. This was about six times faster than the average farmer could put up hay in the 1930's.

Once again Hanson was first. It was 1937 when he purchased the first twine-tie baler sold west of the Mississippi River by New Holland. This baler had a 15 H.P. mounted Wisconsin engine. The lighter twine-tied bales were a real relief for the labor force, and therefore Hanson adopted that method.

When the 15-man crew finished making hay, many remained with him for corn picking. The corn-picking record on Shadybrook Farm was achieved by a nine-man crew averaging 105 bushels a day throughout harvest. The corn averaged 85 bushels, the weather was favorable, and they picked just over 11 acres a day. These men were paid 2¢ a bushel and room and board, but they had to furnish a wagon and a team. In the winter it cost Hanson another 15¢ per bushel to shell the corn and haul it to market.*

The commercial hay-making business grew until 1951. That year Hanson was the largest hay producer in Blue Earth County, Minnesota, and produced over 5 percent of the total alfalfa crop on 2 percent of the land seeded to alfalfa. His weighed average yield was consistently more than double the county average, according to official figures. That year Hanson sold 2,000 tons of alfalfa at $25.50 a ton from 400 acres and had enough additional to feed 115 cows and about 100 feeder cattle.

Because he had a beef brood cow herd, it was only natural that Hanson should decide to get into beef feeding. This became even more essential since he felt he could not produce hogs to his satisfaction. Each year calves that did not meet his standards of quality to merit being used in his registered herd were finished into fat cattle on his place. Because his truck was hauling hay to South St. Paul it was a simple matter to watch for good buys at the Union Stockyards. The drought of 1936 which forced many farmers to market a large number of cattle provided "good buys" for anyone who had feed available. This presented an opportunity for Hanson, and he got started with a commercial cattle-feeding venture that eventually evolved into his major farm enterprise. Alfalfa and ground ear corn were the basic ingredients for his cattle ration and remained the sole ration until his retirement in 1973. Bert Han-

---

*By contrast, in one of his last years of farming, an eight-row combine was used to harvest 9,000 bushels in 10 hours. An extra man using two trucks was kept busy hauling it to storage. The cost per bushel with almost no strenuous effort by the men was 43¢ delivered to the elevator. The total earned by that custom combiner and one helper was $387 per day. This was in contrast to 35¢ per bushel, or $18.90 a day, for nine hand pickers with 10 horses and five wagons. The highest record Hanson could recall for an 80-minute husking contest was equal to 247 bushels in 10 hours.

son's pioneer feeding innovations with those ingredients gave him national recognition.

When prices started to rise in 1942 because of wartime demands, farmers were encouraged to increase production. However, they were afraid of losing their labor force to the military or war industries and at the same time faced a shortage of machinery. There were many complaints. Hanson appeared before a county rationing board and said, "One man with a tractor could do as much as four men and eight horses—besides the horses ate the crops of nearly thirty-five acres of land. I surveyed my township and it is my opinion that the farmers there could maintain full production with one-half the machinery and labor they possess."

This did not make Hanson popular, but the facts proved him to be correct. When labor became less abundant Hanson, who always preferred large equipment, was called on to do custom work. One year his four-row corn planter was used to plant 4,000 acres of row crops (corn and soybeans). He recalled, "That planter was kept running every hour of the day that it was fit to work." In this manner he proved his point to the neighborhood.

During World War II Hanson decided to quit trucking and custom work, but he had found a new project. In 1942 he had purchased a Buckeye No. 1 trencher for $4,227. He was not first this time, for there had been one earlier mechanical machine in the area; but most tiling was still being done by hand. Two men could put in 150 feet of 10-inch tile a day. With the wartime labor shortage tiling stopped, but farmers were more anxious than ever to tile because they had money. The rationing board would not give consent for the purchase of a machine unless it was proven that a minimum number of acres would be drained. Hanson was quick to get the signatures for much more acreage than the rationing board required, and he was able to buy the machine.

Using four men the machine could lay 3,000 feet of 10-inch tile per day, 10 times faster per man than it could be done by hand and with much less effort.* Shadybrook Farm had some tile installed as early as 1905, and Hanson continued tiling by hand whenever he had spare time and funds. Some tile on his farm was 18 inches in diameter and 21 feet deep. Once he purchased the machine, he completed the necessary tiling on his farm, ending up

---

*Hanson added that in 1973 a four-man machine using a laser beam to guide the depth could lay 10,500 feet of plastic tile in an eight-hour day. That is 33 times farther per man than could be done with a spade. The cost of operating Hanson's machine was less than 4¢ per foot.

with a total of 297,000 feet on 750 acres. Then he tiled nearly two million feet for farmers in the area before he disposed of the machine.

"Tile on land that needs drainage pays better than anything else you can do with it," says Hanson, "but sometimes there are other problems—sharp steep hilltops that are miserable to work and seldom produce a crop." Roswell Garst moved dirt to fill potholes and Bert Hanson cut down hilltops to get better production for the same reason. There were six sharp high spots on his farm that he wanted to get rid of. He hired a large commercial dirt mover to push aside the topsoil from his hilltops and his low spots. Then from each hilltop 6,000 to 8,000 cubic yards of clay were moved to the low spots and the topsoil replaced on both the new filled low spot and the hills. In all cases the newly-formed areas produced better than they had done previously, and poor field conditions were eliminated. The cost amounted to over $1,000 per hill, but Hanson indicated the project had paid for itself many times.[3]

Removing the Ache from Haying

"Some of my concepts about farming changed drastically after World War II," recalled Bert Hanson. "The labor shortage made good labor more difficult to secure, but at the same time some of the new automated equipment gave us an alternative which really was more profitable than sticking with labor."

Hanson had always been an advocate of volume operation, using the most advanced equipment; and in keeping with his habit he was very outspoken about advising farmers to take the same direction. Because he was a strong man both mentally and physically, he occasionally had difficulties with his labor force. Few men appreciated going at a pace equal to Hanson's. His disposition also caused him to have labor problems. Once he remarked to a life-long friend after one of his veteran employees had left, "It's sort of like driving over rough sod for a short distance before the going gets smooth again." There were those employees, however, who stayed with the Hansons for long periods because they liked working for an innovator who had good equipment. Besides, during much of his farming career Bert Hanson spent more time off the farm than at home because he was so frequently involved in outside activities. Mrs. Hanson, in contrast, had an excellent sense of humor and direction with the labor force, besides being an outstanding cook, which appealed to the men.

Labor problems eventually led Hanson to discontinue the commercial alfalfa business. Yet by the same twist it made him an even more dedicated proponent of alfalfa. Just at the time when he was considering changing his mode of farming a salesman stopped by to "sell him" on the merits of a Harvestore oxygen-controlled storage unit.* Of that event Hanson said:

I had never paid one bit of attention to Harvestore, but since 1926 I had complained about spoilage in my bunker silos and I just hated working with frozen silage in my tower silos. The labor requirements were horrible. After the war [World War II] I could see there was little future in baling hay. When I had to pay seven high school boys $13 each to load 1,700 bales of alfalfa I decided that's got to be the end of bales. That was seven times as much as I had to pay per bale before the war and they complained about doing it worse than ever. But, I loved alfalfa and when I saw that bottom unloader with sealed storage plus automated handling, it appealed to me. I can say now [1973] that was the greatest stroke of business I ever did when I bought that first Harvestore. Besides, in twenty-two years of cattle feeding since, I have never had a year which has shown a loss. Of course, I have always fed calves and have always sold when they are ready.

Hanson's banker, Clarence Banks, who was at first quite skeptical about Hanson's new venture, had the following comments:

When Bert wanted to buy that first Harvestore I really questioned him. He was a good cattleman and was doing fully as well as any of our customers. But I went along with him and after he showed me his records on feeding just one lot of cattle on haylage from a Harvestore I had no choice but to go along. His feed costs dropped sharply. But I still doubted.

The initial purpose of going to Harvestore feeding was to automate the feeding of his registered brood cow herd. "Besides," Hanson said, "feeder cattle were too high so I wanted to raise my own calves to more completely use all of the crop we were producing."**

Automated alfalfa production and feeding gave benefits to Bert Hanson which he had never anticipated. Fortunately, he expanded in a good year and so was off to an excellent start. As the Hansons recall it:

---

*Harvestore is the trade name of glass-lined steel silage, haylage, and grain storage units. This structure has an air breather system which controls the oxygen that can enter and come in contact with the stored feed. A bottom unloader system makes it possible to have a basically air-tight structure. For those unfamiliar with Harvestores, these structures are commonly referred to as the "blue silos."

**At this time, 1951, Roswell Garst had the same thoughts about building his cow herd. The basic difference in the thinking of the two men was in the type of feeding ration.

The first Harvestore in 1951 cost us $5,100 and everybody said that we would be broke before we ever emptied it. Once we could see how it was going to work we decided to build a house, so we spent $40,000 for a new house. The next thing we heard was that we would be bankrupt before we would get to live in it. But what the neighbors didn't know was that the cattle paid for the Harvestore and then the house before we moved into it. There was no magic to it—just that we had terrifically low feed cost and there was a nice fat cattle market.

Bert Hanson loved to talk about what he could do on his farm. He was direct and outspoken, which sometimes caused resentment in his local community, but at the same time made him known nationally as a speaker in farm circles. It all happened quite accidentally. The Hansons had the third Harvestore built in Minnesota when only a few hundred were in use nationally. As with many new products, flaws in the mechanical aspects occur, once in use. Hanson had his share of trouble with the bottom unloader, which was very frustrating because there was no way to get feed out of a sealed structure if the unloader malfunctioned.

Hanson was loud and persistent in his complaints to the company about the defective unloader. To appease him he was given a new improved unloader at no cost. In 1953 he was invited to participate at a Harvestore Dealers meeting in Wisconsin because he had become one of the most outspoken critics of poor service offered by the dealers. Obviously the company did this to show the dealers an irate customer at his worst, for Hanson was the kind of man who could make his point stick. Hanson said of that invitation, "I was all set to really raise h— with the dealers like I had with the company. I did, but I had to tell them the good side too." Even though he was a severe critic he also proved to be the greatest booster of Harvestore, for he was a precise record keeper and he could prove some outstanding results derived from it. That, in addition to the great benefits of automated convenience, made Hanson a firm disciple of what to many people was simply "an overpriced silo."

The loss of two large stacks of baled alfalfa to spontaneous combustion gave Hanson even another incentive to buy his second Harvestore in late 1953 and still another in 1955. By 1972 he had nine of them, representing an investment of nearly $200,000. With the second Harvestore, Hanson, who was a long-time advocate of ground ear corn and alfalfa, simply converted from baled alfalfa to alfalfa haylage. His records tell the story of what could be done. An extremely simple total ration of trace minerals, salt, di-calcium phosphate, haylage, and limited ground ear corn did an outstanding job. He used the total ration concept because he was convinced that most animals did not balance their own diets if given the traditional free-choice ration.

Allowing himself $100 an acre for each acre of crop that went into cattle feed, in 1955 Hanson was able to produce beef at a feed cost of 8.5¢ per pound, from 450 to 1,050 pounds when they were marketed.* That same ration was used on a mature 900-pound animal for a gain of 605 pounds in 99 days. In a third experiment 11 210-pound calves were culled out of 86 head that averaged about 400 pounds. Those 11 culls were isolated and nursed back to health and after 180 days on full feed gained 1.96 pounds per day at a feed cost of 6.8¢ per pound.

After such convincing feedlot experiments and after establishing the fact that he could feed a cow for one year and her calf from birth to slaughter on 1½ acres of alfalfa and one-sixth of an acre of corn, Bert Hanson knew he was on the right track. His cow herd was limited to an exercise pasture and put on a haylage ration that was restricted to all a cow could eat in 20 minutes, twice a day. The calves were weaned early because they grew more economically on haylage than they did by milking their mothers. From 1951 until he retired in 1973 this simple but carefully designed program produced 1,200 to 1,400 pounds of beef per acre per year.

After his experimental results became known, the promoter of alfalfa was frequently asked to talk about what he was doing on his rolling southern Minnesota farm. Before he appeared at a meeting of a group of farmers in Illinois he decided that he needed a motto. It came out K.I.S.S. which stood for "keep it simple, stupid," terminology much in keeping with Hansonian philosophy, even though it might have provoked some who heard him use it. The motto reflected his attitude toward the exotic rations and so-called "hot rations" advocated by many in the beef-feeding business. Hanson reasoned that with haylage and high-moisture corn plus force-fed minerals, the animal had a total ration. Besides, the ration was virtually free of any digestion-causing trouble. During the next 20 years, while his Blue Earth County farm neighbors reduced their alfalfa acreage to half, Bert Hanson kept his acreage high and steadily increased his yields.

---

*Hanson's total actual production cost, including the cost of land, ranged from $76.76 to $79.32 per acre in 1972. Over the 20 years as land charges rose they were offset by increased yields and larger equipment, which in every case proved more economical. Hanson's records had the cost of every operation pinpointed. Allowing himself $100 an acre for each acre of production used in the feedlot, he averaged over $20 per acre per year net return to his land, including interest.

During the 1950's, 1960's, and early 1970's he enjoyed disturbing people with dramatic but well-documented statements as he spoke to nearly 2,000 different audiences in virtually every state of the nation and parts of Canada. He gave nearly 900 talks to farmer-feeder groups sponsored by A. O. Smith. A provoking speaker with a "down on the farm" philosophy, Hanson was able to motivate his audiences into thinking, even though his listeners did not always agree with him. While Roswell Garst crusaded for corn, Bert Hanson crusaded for alfalfa. Hanson would appear before farm groups and address them in the following manner: "Your own Nicollet County isn't producing 50 percent of the food products it could, right now, with the technical knowledge of agriculture we already have." Then he proved his point by showing that with one full-time employee he was able to crop 820 acres, feed 600 cattle, and care for his brood cow herd:

We had to hire twenty days' labor this year because I personally was able to put in only 210 hours tractor time [at age 70] because I'm only home about two days a week. We are overpowered on our farm but power is cheap labor. To justify another man we would have to add 350 acres of land and increase our feed lot by 50 percent. . . . In 1940, it took fifteen men to do the same number of acres of alfalfa at five tons per acre that one full-time and one part-time man can do with seven and one-half ton yields in 1970. And they don't have to sweat today because haylage takes the ache out of alfalfa. With our automated system we do less than one-third the work required with corn. . . . We could run our township with one-half the machinery and labor force that is being used at present [1971]. We have got to consolidate land into economic units. I just came from England and they are consolidating their farms to gain more of the economics of production.*

He delighted in tackling head-on those who had opinions contrary to his. Speaking to the National Farm Equipment Manufacturers Association he said: "I know what Bob Garst told you . . . about the futility of feeding forages and how hay tools were obsolete. I don't agree with him. I am a successful beef cattle finisher . . . and I know that you can't change a cow from a forage eater without sacrificing some efficiency." His speech then proceeded to prove his thesis based on his records.

On two occasions Bert Hanson employed commercial dining facilities to provide steak dinners to agri-business people who de-

---

*Hanson probably was overlooking the fact that he had automated his own corn production, so it took less time than formerly. However, on the basis of feed value per acre produced on his farm, alfalfa took just over one-half as much time per acre as did corn. So there was much validity in his comment.

bated the merits of haylage-fed animals as compared to corn-fed animals. At one of these dinners the president of George A. Hormel & Company was a guest and lost a bet to Hanson. The point Hanson was trying to make was that the packers discriminated against haylage-fed beef but had no basis for their judgment. His success in marketing his cattle once the packers overcame their bias is proof enough that his opinion was well founded.

The similarity of Roswell Garst and Bert Hanson relative to their respective corn-alfalfa crusades has an ironic touch. Garst always praised corn and tried to tear down the image of hay, yet he was a successful hay producer even though he seldom discussed the issue. Hanson always talked the value of haylage and maintained that the only reason he used corn in his ration was to take advantage of the excessive protein in a straight haylage ration. However, Hanson was an extremely successful corn producer, averaging 140 bushels per acre from 1957 through 1972.

The ever-restless Hanson was not content just to be speaking, and raising cows, feeders, and alfalfa. He was constantly looking for more innovations. Others were aware of this, and Shadybrook Farms sometimes resembled a small-scale state experiment farm. In 1942 Hanson started conducting experiments in conjunction with chemical and fertilizer companies. That year he raised his first sweet corn and performed some fertilizer experiments. Since 1942 he has had one 80-acre field in continuous sweet corn, constantly increasing yields because he applied a high rate of commercial fertilizer.

Eight well-known alfalfa and seed corn companies have had experimental plots on his land most of the time since the mid-1950's. Two chemical companies operated experimental plots with Hanson for his last 17 years of farming. In addition there was another unusual aspect of his fertilizer testing program. Much of his land received very little commercially processed fertilizer during the last 15 years Hanson farmed. He had always been an advocate of fertilizer and was a pioneer user in his area, but his experimental plots in late years showed little need for it on most of his land. The answer obviously is that much of his farm received up to 15 tons of manure each year from his livestock operation. Measured average corn yield over his last 15 years was 140 bushels an acre, and alfalfa was 7½ tons dry yield per acre. The highest yield attained in his test plots was 186 bushels of 14-percent-moisture corn, but most years the test plots at various levels of fertilizer did not surpass his average field yields.

Hanson did have a long-range fertility program, which gives some understanding of why his farm was a consistently high producer. He always had more than one head of livestock per acre on the farm at all times. Besides that, every five years 500 pounds of rock phosphate were applied to his land, and every five years three to four tons of lime. Hanson had figures to prove that those applications cost 77¢ per acre per cutting of alfalfa and 22¢ per bushel of corn produced. The rest of his fertilizer cost was the cost of hauling the manure from the feedlots and the confinement pit.

Because of his strong interest in reducing labor requirements, it was no problem to get Hanson to agree to experiment with new labor-saving machinery. His major experimental work was on corn and hay harvesting equipment, most of it in the form of field choppers, swathers, forage wagons, blowers, and feeding equipment. His most extensive experimental work was with John Deere, Owatonna Manufacturing Company, and A. O. Smith. These companies and the chemical companies credit Hanson with being one of the best salesmen for their products. The "selling" was done when the people visited him to see how things worked by being used on Shadybrook Farm. If the product was good Hanson was quick to recognize it and could explain why it was good, and most visitors were quick to see that Hanson knew his farming.

Gradually Shadybrook became one of the most visited farms in Minnesota. The guest book revealed the names of several thousand people from distant places who came each year. In 1971, the year before the world's plowing contest took place on the farm, there were over 7,700 visitors at Shadybrook. Of that number 1,500 were served meals by Mrs. Hanson. Hanson, explaining his experimental programs, said:

I did not receive special rates or discounts as many people think on most of the experimental work. Any concession gained seldom offset the additional cost and time involved because of the experiments. But we have met lots of people from all over the world, have had lots of fun doing the job besides getting the jump on everybody else when something really clicked and that is where the profit is.

Bert Hanson was also invited to appear before many groups to speak on the results of his experimental work. For a man who learned from travel and people, the exposure brought more satisfaction than the honorarium involved. For 20 years he was on the Minneapolis Chamber of Commerce speakers' panel; and because of his ability to portray a fresh image of agriculture to those not on the farm, he was widely sought as a speaker throughout the Midwest.

In 1952 Hanson confined his 155-cow herd to a limited area and put them entirely on haylage. From that day on none of his cows ever received any grain. In 1973 he had 20-year-old cows weighing 1,600 pounds and still producing calves. His senior cow had produced 21 calves in as many years. Hanson was convinced that confinement and haylage easily increased a cow's ability to produce five extra calves in her lifetime. "On $300 an acre land confinement was tremendous from an economy standpoint, and today on thousand dollar land it is necessary," said Hanson. "The only problem is that the cows really should get a little more exercise." His records and experience were proof of his contention that cows were forage animals and fared the best on that type of ration. Hanson was so persistent about using a heavy forage ration that he resigned from Performance Registry International, giving as his reason: "After seven years I got disgusted because all they talked about was feeding cattle on a hot ration for 150 days and I insisted that it is lifetime gains that's important."

After feeding cattle for many years with traditional sheds and bunk feeders, Bert Hanson decided it was time to completely automate his feeding operation. In 1971 he built a $100,000 cold confinement feeding barn. This 60- by 208-foot slatted floor building could house 500 to 600 cattle, depending on their size. It had a fully automated shuttle feeder that was one of the first half-dozen installed by the manufacturer. The expansion was perfectly timed, "lucky for me" said Hanson, and the ultra-modern structure was paid for by the first 1,200 head of cattle to be fed out in it. There were two factors that enabled Hanson to realize such a rapid return on his investment: the increased feeding efficiency in the new structure kept feed cost low; and the market rose sharply, enabling him to sell fat cattle at almost as much per pound as he paid for heavy calves. Hanson said of the experience: "It was a lot like 1951 and 1952 when everything worked in your favor, but those years don't come along very often."

The confined feeding building supplied with feed from Harvestores that were push-button-operated was just another step in the direction of using capital to substitute for labor. With all cattle fed by a shuttle feeder Hanson could feed a thousand cattle three times a day, requiring only 25 minutes for each feeding. Labor cost per pound of gain based on 2 pounds per head daily gain and $3.50 an hour for labor would be .00219¢ in contrast to about 10¢ a pound labor cost if the hand-, shovel-, and basket-feeding system were used. Using tractor-powered manure pumps and a power-operated 3,100-gallon-tank manure spreader, one man was able to haul a year's supply of manure from 500 cattle in 16 days' time. His entire job consisted of running the tractors and turning valves—a far cry from the day when all manure hauling was done by pitchfork.*

In 1972 the value of good farm land in the Vernon Center area was $700 per acre, but on Hanson's fully automated farm his total investment, including the land, was $2,000 an acre. "That $2,000 an acre farm pays because we get an efficient job done when it is supposed to be done and the capacity of the man with fully automated equipment is almost hard for me to believe when I think of how I first made hay and fed cows with the pitchfork as my main tool. If a farmer wants to grow he has got to think of automation—too many just think that growth means more acres and more tractors."**

Hanson, one year-round man, and a third man who was employed for 20 days, ran the 830-acre farm, maintained 100 registered brood cows, and fed an average of 750 cattle a year. There were six tractors totalling 420 H.P. on the farm besides the fully automated feedlot which operated with 15 electric motors. There were 400 acres in alfalfa and 150 acres in corn, 80 acres in sweet corn, 40 acres in peas. "Besides," said Hanson, "there was a hundred acres in government diversion which made a good place to spread manure—that's about the best benefit of the farm program." Enough beef was raised to supply the average annual consumption of 2,932 people, enough sweet corn was raised to fill 48,000 cans, or the average annual requirement of 1,800 people and 100,000 pounds of peas adequate to provide for 3,400 people per year.*** In addition, as much as 8,000 bushels of corn were sold from the farm in a single year. Referring to his annual work year Hanson said, "We could do this amount of work even with

---

*The first manure spreaders with endless aprons were not commercially produced prior to 1877, and the first tractor-powered loaders came on the market in 1938; but it was not until after World War II that power manure-handling equipment became commonplace. It was about 1937 that automatic barn cleaners were commercially produced.

**Hanson had a 10-year average death loss in his feedlot of less than 1 percent. For the first 1,200 cattle through that structure a "needle" was used only seven times. He attributes his success to the balanced total ration concept based on haylage.

***The cost of the sweet corn averaged 3¢ per can as paid to Hanson, but it was sold by the retailer for 24¢ a can.

me gone at least three days out of each week and my year round employee was working about 55 percent of his time during many months of the year. The only reason we never really wanted to do anymore is that it just created bigger income tax problems."[4]

They Threw Away the Mold

The ability to convey his ideas to audiences throughout the nation was based primarily on Bert Hanson's natural talent of shocking people without making them too provoked in the process. Even those who didn't agree with him frequently made the remark, "After Bert Hanson was created they threw away the mold." Bert Hanson, like Roswell Garst, was a great crusader who gained the respect of many. Hanson was not a highly polished speaker, but that did not detract from his message. In his home community of Vernon Center many of the local people heard him speak publicly for the first time when he gave travelogs on his trip to Russia.

Hanson obviously was a strong disciple of the American free-enterprise system and liked to compare it to the government-controlled economy of Russia. To emphasize the contrast Hanson frequently referred to "the tremenjous [sic] opportunities we have in America in contrast to the Russians." For the first time many local people who had long avoided Hanson realized that he was not the highly-educated man they had always assumed him to be. "For the first time many saw him as he really was," said the local editor, "a highly motivated self-made man. They seemed to appreciate him more after that."

Bert Hanson's philosophy toward life was very simple. In his own words:

There are only three things in life that are positive; death, taxes, and change. Change is the only one you can do anything about so you better do it. If a farmer wants to keep up with the world he must grow at the same rate as the population and national debt combined. If he wants to get ahead he must grow at a rate faster than that. The man who is willing to work and can manage probably has a greater opportunity in farming today [1971] than anytime in history. I have seen a good many opportunities go by the wayside. The lure of attractive wages and short hours in the city still beckons strong in contrast to long hours plus the tension of managing your own farm. We have farmed in six different decades and have gone through the ups and downs but we don't think we could have done as well anyplace else as we have in farming. . . . When I started farming with horsepowered equipment I was already ahead of where some of the world's farmers are today [1971]. In the world plowing contest in 1971 in England I met a man who had never worn shoes or a dress suit until he came to England that week. And this man's equipment was way ahead of the great majority of the farmers in his country. . . . I think of the breaks I've had—for over forty years I've had the same banker. He likes cattle—cattle loans make up over 40 percent of the total loans of his bank and he's always worked with me. Sometimes it was pretty tight, but I consider myself a master of high finance. We have always borrowed lots of money. I have never been refused by a financial agency, even though I remember begging awfully hard at times. But I never went anyplace else for money so Clarence [the local banker] knew he could depend on me. The feedlot, cattle, Harvestores, and the $100,000 confinement barn were all financed by that little bank.*

Contrary to the pattern of many farmers, Bert Hanson's success was not dependent upon the appreciation of land. He bought little land after World War II; instead he concentrated on developing the land he owned. Heavy capital investments such as tiling, Harvestores, two new houses, and confined feeding facilities do not appreciate in value as land does. They do, however, generally bring a faster and higher return to capital, and Hanson preferred to take that direction.

That does not mean to imply that the Hansons were not interested in more land. At the age of 57 Hanson was contemplating buying 12,400 acres in central Minnesota to use as a ranch on which to develop a cow herd. The calves from that ranch would have supplied Hanson's feedlot and given him a totally integrated livestock operation. The death of their son prevented the Hansons from making the purchase, for much of the motivation to expand was then lost. Myrtle Harkins, the local editor, reported in an interview:

The Hansons were not money minded people, but they never stopped striving from being number one. They just liked to prove that they could master anything they tried. Bert really liked being a leader and Irene was always strong encouragement. I have seen him under tremendous pressure and it was obvious that he had a powerful character, but at the same time a man with a big heart. His strong personality and confidence came through, but those also were the very things that provoked many in the neighborhood.

Many incidents could be cited to show strength of character and courage. Speaking before the Nebraska Society of Farm Managers and Rural Appraisers, Hanson said: "Farmers and ranchers are

---

*Having 40 percent of all the bank's loans in cattle indicates great faith in the cattle-feeding business. Very few banks would concentrate their total borrowings so much in one line unless their experience had been outstanding. The State Bank of Vernon Center could by law only make direct loans of about $65,000 from its own funds to any one individual, so it had to rely on correspondent banks for many of the larger Hanson loans.

only about 40 percent efficient and because of that they really are paying about 25 percent interest rates. If you don't change, you will soon be out of business. We've got to farm cheaper, use more labor-saving machinery, and we've got to change our system of farming. The world of substitutes will drive us out of business, if we cannot learn to keep up by innovation."

Such talk is not always popular with many rural people, who already think that they are doing a good job of low-cost production.

The extremely aggressive nature of the man surfaced during some of the more emotional periods of farm organization activity. Hanson, although active in many organizations, was a completely free man and wanted very much to remain an independent operator. Having great self-confidence, he felt that he could face any problem of farming. This attitude brought him into direct open conflict with farm organizations twice in his farming career. In the 1930's when the Farmer's Holiday Association attempted to block the roads to prevent farmers from marketing their products, Bert Hanson decided that they would not stop him. He armed his men with clubs and pitchforks. When his truck, loaded with grain, was ready to go, his men went on ahead to clear the way. Hanson led the pack, and there was no physical violence. That stopped the local agitators.

In the 1960's when the National Farmers' Organization was attempting to force a withholding action, Hanson let it be known that he was not joining anyone and that when his cattle were ready to be marketed they would be shipped. Several cattle feedlots in his area were opened at the time, and cattle were allowed to escape, causing considerable economic loss. Rather than be intimidated, the 61-year-old Hanson asked to be sworn in as township constable. He picked his own deputies, "men who wouldn't back down if there happened to be a little physical violence," he said, "and we patrolled the township roads in our own cars at our personal expense. That was the end of trouble in our township." Other people in the area were less brave than Hanson and joined out of fear for their wives and children.

When asked what troubled him most about his way of farming, Hanson remarked that he was concerned because "new varieties of alfalfa have not given any significant boosts in yields." In the 1930's Bert Hanson sold about 5 tons of alfalfa per acre. In the 1970's his yields were 7½ tons per acre, but Hanson thought they should be more. Minnesota agricultural statistics credit Blue Earth

County as being the county with the highest average yields in alfalfa production. For many years Blue Earth has been one of the consistent leaders in top average alfalfa yields. Bert Hanson farmed in about the middle of the county. It may never be proven that the example of Bert Hanson made Blue Earth County the champion alfalfa county of Minnesota, but the fact has to be more than a coincidence. In any case Hanson's yields over a span of 40 years have been nearly double, or more than double, the county average. He insists that manure and alfalfa are the major basis of soil fertility; however, he does not overlook the richness of Clarion-Webster soil or the huge doses of phosphate and lime he regularly applied.

The Hansons liked to be leaders in other aspects of farming besides Shorthorn breeding, cropping alfalfa, and automated feeding. Bert Hanson achieved a number-one distinction again in his life when he became charter member number one among the Flying Farmers of Minnesota. It had been his ambition to own his own plane by his fortieth birthday, but World War II prevented that. However, in 1945 he became the first farmer in Blue Earth County to buy a new airplane—it was the thirteenth Ercoupe sold after the war.

At St. Paul in 1945 the Minnesota Flying Farmers Association was formed. Hanson was one of the Minnesota Flying Farmers to attend the organizing session of the National Flying Farmers at Stillwater, Oklahoma. He was elected the first secretary-treasurer of the national organization in 1945 and held the post for three years. He predicted that farmers would buy 75 percent of the light planes in the United States because the plane was a production tool for them. Not long after, he purchased the first Flying Station Wagon by Stinson. In all, he owned four new planes. In 1952 Hanson was elected vice-president of the National Flying Farmers and spent a great deal of his time on that job. Of his experience with the Flying Farmers Hanson said, "It took lots of time and a pretty penny out of the pocket, but the exposure was priceless— among other delightful experiences there was dinner with the President of Mexico." By 1952 he had logged 800 hours of pilot time, using the plane to fly to many of his speaking engagements in smaller agricultural communities that lacked good commercial transportation.

Few achievements gave Bert and Irene Hanson more satisfaction than his being elected Outstanding Conservationist for Southwestern Minnesota in 1948. Shortly after, he became the first president of the Southern Minnesota Conservation Association. That distinction earned him the position as the first full-time Minnesota farmer to sit on the Minneapolis Chamber of Commerce Agricultural Committee, and in 1968 he was presented a distinguished service award for his leadership. Hanson said, "There were lots of honors, but being a charter member of the Cowboy Hall of Fame is as unique as any I have—it gives me great pride." The alfalfa crusader and innovator had a very real human side to him. Like so many of his kind he was recognized by those far beyond the boundaries of his home community, but avoided by those nearest him.

From September 11 to 17, 1972, much of the agricultural world focused its attention on 1,400 acres of land owned by 14 farmers at Vernon Center, Minnesota. The occasion was Farmfest U.S.A. and the World Ploughing Contest.* The Hansons donated their entire farm to "this world series of agriculture in the interest of good neighborliness to the world." Thirty seed corn companies had 600 acres planted to corn, and six companies had 21 varieties of soybeans on another 130 acres; the remainder of the 1,400 acres was used for demonstrations, competitive events, and parking. Over 500,000 people attended the highly successful affair, that, unlike most similar exhibitions, netted the sponsors a profit. Even the State of Minnesota was repaid its financial contribution of $50,000.

With most countries of the world competing for this significant event it is a distinct honor for a nation to be given "its turn" and even more of an honor for the farmers whose land it is held on. Such an event was a hardship for all of the farm owners, but Bert Hanson worked long and hard to see that the world ploughing contest was a success.

This was also a fitting climax to his lifetime of progressive farming. Many of his neighbors who had long been critical of "the crusader" in their community saw that in the agricultural world Bert Hanson was an acknowledged leader. Their attitude toward him mellowed.

After the big event was over Irene Hanson was again free to give a helping hand to anyone in the community, and Bert was free to use his white Cadillac to hunt jack rabbits and trap gophers. Or they could be one of 16 farm couples invited to the White House during President Nixon's Salute to Agriculture in

---

*Hanson preferred that the more internationalized word "ploughing" rather than the American word "plowing" be used.

1972. To pitch in and help on all the projects around Vernon Center was a big achievement, for "next to the bank he has been the most generous benefactor to the community with his wealth and talent." Or Bert could travel to South America to give some know-how to the farmers on that continent.

True to his philosophy and his character he wrote to a friend who was copying Hanson's system: "You won't need a prayer if you follow my advice all the way for it won't fail . . . as long as you keep people eating beef . . . feed calves . . . sell them when they are ready to go choice and buy replacements. Never hold for a higher market. If it looks bad sell and buy on the same day."[5]

Successor to a Crusader

The Bert Hansons had three daughters and a son. When their son lost his life in the late 1950's much of their desire to really "step out and farm" vanished. It was only the energetic nature of the Hansons that motivated them to continue after that. They hoped that one of the sons-in-law would be a farmer. Fate decided otherwise, so Shadybrook Farms, like many other proud homesteads, was put up for sale. The alfalfa and Harvestore farm that had been visited by thousands during the decades from the 1940's to the 1970's was the dream of many young farmers. However, only an agressive farmer with ardent drive and ambition could continue Bert and Irene's tradition.

Bill A. Noy, Jr., grew up on a 1,040-acre grain and livestock farm not far from the Bert Hanson's. This is Bill's version of how he got into farming: "At age eighteen Dad gave me the responsibility of getting crops in and out. He gave me guidance. To this day [age 45] I go to him for suggestions because I like his views. Dad has often opposed my ideas, but when I make my decision he generally will end up agreeing with me."

From those comments it is obvious that Bill Noy, Jr., came from a family who liked farming and were successful at it. In 1951, his first year of farming on his own, Bill rented 500 acres and had two men working for him. "I was share renting so I didn't need much capital and this got me rolling. I was able to use some of Dad's machinery; this helped too. I was twenty-three then."

In 1956, after five years of renting, Bill had enough cash to pay down on a 240-acre farm, which he bought on contract. Having farmed independently for a few years he was fully confident of his financial base. Of those early years he remarked, "I was satisfied from the start that farming paid and now after twenty-four years I can say I am pleased with the steady progress I have made."

From 1951 through 1972, only 1963, a disastrous year for cattle feeders, ended in a reversal for Bill. Because he was a good manager and was accustomed to progress, he withstood the shock of financial loss. But it made him reassess his operation, and, as is often the case, he found an opportunity to profit from his adversity:

This was a serious loss and I knew I had to change my methods of feeding. We had been using a concentrate ration based mostly on corn, completely ignoring feed cost, counting on the breaks in buy and sell to make it for us. I had visited with Bert Hanson often over the years about his ideas on using haylage to feed cattle, but the investment always concerned me. After getting pinched in a down market I decided that the cost of production had to be reduced. Bert could prove to me that he was not losing money on cattle during that period when all the feeders were being whipped. It looked like Bert had been right in his ideas and the Harvestore and haylage were the answer; however, I was skeptical so I only bought one for part of my storage needs, hoping I could fit it into my overall program.

Bill Noy put up his first Harvestore in the spring of 1963 and he intensified his record keeping on his feed cost:

We even hauled cattle to town every twenty-eight days to weigh them. After the fourth weighing I had coffee with Clarence Banks, the banker, and we proved we were feeding cattle up to 600 pounds for seven cents feed cost. Clarence looked at me and said, "This is what Bert Hanson has been trying to tell us since 1952." For the first time in all those years it was clear that what Bert had been doing all the time was right.

The banker did not hesitate to finance the purchase of a second Harvestore. Noy recalled:

I had not grown much alfalfa prior to that time so it was a new experience for me. Bert was an eager, willing teacher and after a few bad experiences it has become my best crop on a net return per acre basis. I was traditionally convinced that corn planted for the government program was the way to go. Once I got into haylage the records changed my mind. After that second Harvestore I hit a plateau of 320 acres and 320 cattle besides some beans and I knew we had all we needed for a living. Financing was no longer a problem.

Alfalfa and automation prompted Bill Noy to think ahead. Economically he was perfectly content with his half section, but he began to visualize new plateaus of a challenging growth. "Besides," his wife said, "he made me nervous being around the house because automation gave him so much free time." Noy became bored until he realized that if he expanded his farm he could afford better equipment and could get the job done even more easily, for he had a personal problem so common to many farmers. Here is how he explained that: "I personally got great satisfaction out of manual labor and hated to go push button, but since then I have learned that it is just as much fun to watch others do the work as to do it myself, even though I personally prefer to work alone."*

Once Noy overcame his reluctance to mechanize farming he watched for new opportunities to expand his operation. His type of farming would no longer be tied down to manual labor but would require lots of time for thinking and planning. His own 240-acre farm grew by his renting additional acres whenever more land became available. By 1971 there were five Harvestores on the Noy farm, and plans were being laid for five more in preparation for the day when the son would finish school.

One noon at the dinner table a "thunder clap" hit the Noy home. A salesman dropped in and informed them that Bert Hanson's Shadybrook Farm was for sale. Bill Noy had often dreamed that some day he would like to own that farm, but he never thought it could happen. Shirley Noy's reaction was that it was purely rumor and she did not give it a second thought. She remarked, "I said 'it would take over one-half million to buy Bert out.' But I could see that Bill was figuring how to buy that place while he was eating his dinner."

Within less than two hours from the time he first heard the news, Bill Noy was at the Hanson doorstep. Bill said:

This was completely unexpected, but I felt it was the chance of a lifetime. When Bert said the report was true it took me less than thirty seconds to say I'll buy it because I had figured the economics of it while eating dinner. In one stroke I doubled the size of my land operation from 700 to 1,400 acres besides adding nine Harvestores and a confinement barn for nearly 1,000 head. Plus, I added real quality to my existing operation. Bert agreed to stick around to guide me. Bert and I sat down to talk terms and it came to me I was putting myself on the limb for a million dollar deal.

Shirley Noy stayed at home while her husband was buying the Hanson farm. She said, "I didn't know if I should pray for or against the purchase of the Hanson farm. There were other farms

---

*Bill Noy, Jr., like nearly all the farmers interviewed for this work, were hard drivers who had the capacity and the desire to out-work most individuals. Physical work and mental tension were a challenge to them.

available for less money, but Shadybrook was the best buy because of the quality and setup. We could make it pay at once because it was in top condition."

After making the agreement with Hanson, Bill Noy had to see Clarence Banks who had financed both farmers over the years. "The immediate reaction of the banker was positive," said Noy, "because he knew Shadybrook's production records and he knew my profit history." The banker said, "I feel more comfortable with Bill Noy and that big loan than I do with a lot of my $6,000 and $10,000 loans where there is not the management or the necessary volume to make a farm pay." Securing the money for a million-dollar farm was not a great obstacle, once the decision had been made to buy.

When Noy was asked to give the reasons why he decided to more than double his operation in one purchase while he admittedly was already doing well on his present farm, he was unable to give an immediate answer. He could only say, "I had always dreamed about it; I could give a thousand reasons why but I'm not sure. Maybe it's because my boy wants to farm. There certainly was no monetary desire—maybe I just wanted to own it out of desire." His wife, Shirley, was more precise: "Bill is special; he likes to do things. He is always looking for a new challenge which seems to be out of reach." To which Bill added, "This just sort of fell into my plans and I cannot really explain it. I am not afraid of the risk other than how the setback could disturb the family."

There were five other solid offers for Shadybrook Farm, all of whom had the money to make the deal, according to Hanson. "These were farmers from the area. And it was not advertised— just word of mouth. But we liked Bill and Shirley and we were hoping they would ask to buy it. Bill really adopted my ideas and we sort of looked to him as a star pupil."*

Bill and Shirley Noy, owners of 940 acres of land and 14 Harvestores** besides renting another 460 acres, were asked after the purchase of Shadybrook Farm what their biggest problem would be in the future. Neither could think of a major problem. Bill's only comment was:

I am going to make it a single unit and keep the whole thing rolling. As

---

*The banker verified that there were five solid and capable potential buyers for Shadybrook.

**With 14 Harvestores Bill Noy possesses as many of those structures as any other private owner in Minnesota. Art Hyde of Wilder Farms also possesses 14 Harvestores.

it looks now it should be easy. Maybe I should have moved faster. I have the equipment to do the job but I would like to buy a four-wheel-drive tractor because I prefer the insurance of being overpowered. I guess my only concern is what's going to happen to the farmers who can't keep up?

That was the parting thought of an outstanding farmer, who had good training on a small family farm and who later made a single purchase that was beyond the vision of most of the farmers around him. Bill Noy had strong character traits that made him a real competitor—youth, management experience, and success. Besides, he had learned at home to think positively about farming.[6]

NOTES

[1]Interview of Mr. and Mrs. Bert A. Hanson, Vernon Center, Minnesota, February 25-March 2, 1973; *Land of Plenty*, p. 58; Interview of Bert A. Hanson, at Moorhead, Minnesota, January 11, 1971; "World Ploughing Contest," *Minnesota Valley Breeders Association Bulletin VII*, No. 7 (August, 1972); Interview of Clarence M. Banks, President of State Bank of Vernon Center, Vernon Center, Minnesota, March 1, 1973; "The Hanson Story," *Journal of National Polled Cattle Club*, April 1, 1953; "He's Always First," *The American Magazine*, CLV, No. 1 (January, 1953); *Kansas City Daily Drovers Telegram*, July 25, 1947.

[2]Mr. and Mrs. Hanson interview, 1973; Bert Hanson interview, 1971.

[3]Mr. and Mrs. Hanson interview, 1973; Bert Hanson interview, 1971; Clarence Banks interview; Interview of Mrs. Myrtle Harkins, Vernon Center, Minnesota, March 1, 1973. Merton Harkins was editor and owner of the *Vernon Center News* until his death, at which time Mrs. Harkins assumed the position. The Harkins and the Hanson families were closely associated on a social basis; "Hay, Alfalfa, and Alfalfa Mixtures, Acreage, Yield and Production, by Counties," *Minnesota Agricultural Statistics*, 1950, 1954, and 1973.

[4]Mr. and Mrs. Hanson interview, 1973; Bert Hanson interview, 1971; Clarence Banks interview; *St. Peter Herald*, June 24, 1954; Mrs. Myrtle Harkins interview; "Haylage Takes the Ache out of Alfalfa," *Pioneer Corn Profits*, Summer, 1967, Pioneer Hi Bred Corn Company; "Farmfest, U.S.A.," *Hormel Farmer* XXXVI, No. 10 (June 15, 1972); "Cows Can't Be Changed from Forage Eaters," *Implement and Tractor* (December 1, 1962); *Owatonna (Minnesota) People's Press*, July 18, 1972; *Land of Plenty*, p. 59.

[5]Mrs. Myrtle Harkins interview; Mr. and Mrs. Hanson interview; Bert Hanson interview; Clarence Banks interview; *Beef Producer*, Broken Bow, Nebraska, June 22, 1965; Interview of Bill A. and Shirley Noy, Jr., Vernon Center, Minnesota, February 28, 1973; *Minnesota Agricultural Statistics—1973*, pp. 34-35; *The American Magazine*; *Minneapolis Sunday Tribune*, August 25, 1947; *Journal of National Polled Cattle*; *Minnesota Valley Breeders Journal*, August, 1972; "Farmfest Site— 1400 Acres Owned by 14 Farmers," *The Farmer*, XC, No. 17 (September 2, 1972), p. 67; "Farm Innovator Received Award," *Minneapolis Star*, June 12, 1968; *Washington Post*; *The Hormel Farmer*, XXVI, No. 12, (August 15, 1962); Bert Hanson letter, February 21, 1974.

[6]Mr. and Mrs. Noy interview; Mr. and Mrs. Hanson interview; Clarence Banks interview.

CHAPTER IX

Red River Potatoes

THE RED River Valley of Manitoba, Minnesota, and North Dakota is commonly referred to as the garden spot of the Northwest. The treeless, fertile, stoneless prairie along the Red River, though not always attractive to the pioneer farmer, has become one of America's choicest farming regions. The flat valley permits the use of large equipment, enabling efficient production. This is particularly important where crops requiring intensive use of labor, such as potatoes, are grown. In this northern region long days during the growing season and cool nights provide an excellent climate for potatoes.

During the 1870's and 1880's the farmer's efforts were almost totally concentrated on subsistence farming plus the production of wheat for his cash income. Wheat had long served as the traditional frontier crop because it was not readily perishable and it generally had a relatively high value per unit of weight. This enabled the farmer to bear the cost of long hauls to market. Besides, wheat allowed the pioneer to use his most abundant commodity—land. But wheat had a disadvantage; it did not give the pioneer farmer any chance to cultivate against weeds.

In addition to wheat the Valley farmer needed a row crop that he could cultivate for weeds during the growing season. Corn, a traditional crop for that purpose, was ideal in rotation with wheat, especially if the farmer had livestock. However, unless corn was needed for livestock it was not so profitable as wheat in the northern latitudes of the Red River Valley. Because livestock production on any large scale was never popular in the Valley, some other row crop had to be found.

The Pioneers

From the beginning of the white man's settlement in the Red River Valley it was known that vegetables of all kinds did very well in the rich mellow soil and even did well in the heavier clay

311

soils that bordered the Red River. Alexander Henry, an early explorer, grew potatoes and many other vegetables in the area around Pembina (North Dakota) between 1800 and 1808. The ability to produce different food crops encouraged Lord Selkirk and the Hudson's Bay Company in the first two decades of the nineteenth century to bring permanent farm settlers to Canada. Conversely it was the 1813 crop failure—including potatoes— which was caused in part by a heavy weed infestation, that nearly put an end to the settlement of Fort Garry, now called Winnipeg.

R. M. Probstfield came to the Red River Valley in 1859, settled north of Moorhead in 1868, and immediately started raising garden crops. Probstfield likely was the first sizeable, long-time truck gardener in the Valley. He filled a great void for the homesteaders, who often lacked potatoes and other vegetables.* With the exception of Clay County, Minnesota (Moorhead), no county in the Valley had any commercial potato acreage prior to 1890. The influence of the railroad and the growth of Fargo-Moorhead provided a sizeable market and probably was the cause of Clay County's leadership in potato production.

In 1890, 12 Dakota and Minnesota counties and six counties of Manitoba in the Valley had 17,307 acres of potatoes. This was far less than 1 percent of all the land being tilled, and such a small potato production certainly was not large enough for export. It was only in the 1890's that commercial potato production developed and Clay County rose to an early leadership in that agricultural venture. Henry Schroeder, a German immigrant, arrived in Clay County in 1878 and in 1893 started commercial potato seed production. Schroeder is generally credited as the father of the potato industry in the Valley. George W. Bilsborrow, who arrived in Wilkin County in 1894, is sometimes recognized as the person who introduced certified seed production into the area. Bilsborrow, however, never attained the volume of production or recognition in the potato industry that Schroeder did.**

---

*The Probstfield family remained in the truck-garden business through three generations, from 1869 to about 1964. The farm was still being rented for that purpose in 1974. Settlers along the Buffalo River east of Probstfield's in 1870 had to go 20 miles to Jimmy Rice's farm near Comstock for their winter's supply of potatoes.

**Henry Schroeder had financial difficulties late in his career and lost much of his land, but the name is still well remembered in the Valley. A grandson, Henry E. Schroeder, lives on the original Schroeder homestead and is a leader in sugar beet production. On another farm only a few miles distant three generations of the Schroeder family were still living in 1974. A son, Ernest Schroeder, was retired, but Robert Schroeder, a grandson, and Stephen Schroeder, a great grandson of the original Henry Schroeder, were active in large-scale potato production.

Henry Schroeder influenced fellow farmers Ewald Wiedemann, Edward Grover, and Eugene Grant to grow potatoes, giving Clay County four sizeable growers from the beginning. These men all farmed on mellow dark beardon soil along the Buffalo River, which was fertile and ideal for potato production in the days of hand harvesting. The industry grew rapidly under the leadership of the energetic Henry Schroeder, who exported 8 carloads of seed from Clay County in 1894, and 60 carloads in 1895. In 1899 he produced enough potatoes from 260 acres to fill 85 railroad cars, and by 1915 his production had swelled to 86,000 bushels from about 700 acres.

Turn-of-the-century farmers were searching for another crop because wheat monoculture was having a depressing effect on yields. Potatoes, which could stand frequent deep cultivation, not only helped to clean out the weeds but revived soil fertility. Area experiments clearly demonstrated the economic value of rotation. In addition, 4-H potato-growing contests had made farmers aware of the yield potential of the crop.

In 1902 C. W. Mundstock netted $135 an acre from four acres of potatoes. That contrasted to a gross of $15 an acre from wheat. The Wilkin County development people saw to it that everyone learned of Mundstock's success by giving it great play in the county's promotional brochure.

Charles Peterson migrated from Sweden in 1888. He worked as a hired man in the Red River Valley for a few years until he had enough money to homestead at Garrison, North Dakota. Peterson liked truck gardening, and he liked the fertile flat prairie land along the Red River. As soon as he had built up some capital he sold his Dakota farm and returned to Moorhead. In 1904 he planted 20 acres of potatoes on the land which at present is a part of the Moorhead State College campus. Peterson also planted onions and other truck-garden crops besides potatoes and concentrated on the expanding Fargo-Moorhead retail market. He soon surpassed Probstfield in total production and became a leading supplier of truck-garden crops in the southern Valley from 1904 on.

While Peterson was enhancing the reputation of Red River Valley potatoes in the upper part of the Valley, Nels Folson of Hoople, North Dakota, started potato production in the lower Valley. Folson is credited with shipping the first carload of table stock potatoes from North Dakota. The seed for those potatoes came from Henry Schroeder, the potato king of Clay County,

Minnesota. His descendants, like Peterson's, have continued to be leading potato producers and agricultural innovators in the area.

Folson, who started in 1905, became the first large-scale potato grower in Walsh County. Walsh County took the early lead and has generally been North Dakota's leading potato-producing county. In 1908 Folson built a 100,000-bushel potato warehouse prior to expanding his holdings to 100 acres in 1909.

Folson was a neighbor of Tom Thompson, who farmed about 15 miles west of Grafton. In 1908, because of Folson's influence, Thompson planted 15 acres of potatoes. That was the beginning of a dynasty of potato families, for Thompson's son Joe was an early large-scale potato grower in Walsh County and for 40 years he would be an innovative force in the Valley with his potato ventures.* Lars and Hans Larson, the Halversons and Midgardens from Park River, L. E. Tibert from Voss, along with a few farmers in the Crystal area, represented most of the earliest potato growers on the Dakota side of the Red River.

East of the Red River at Kennedy, Minnesota, Frank Kiene was pioneering in large-scale potato growing. Farther south Alfred Hvidsten at Stephen was operating on such a large scale by 1916 that he felt the necessity of buying a mechanical harvester. This machine, a Cudigan-Gear harvester drawn by six horses and ground driven, cost $200. It was operated for four years and was probably the object of more photographers than any other aspect of potato farming in the Valley during the World War I era.

Potato production had tremendous profit potential, and many farmers jumped into the business without also being aware of its hazards. However, management became crucial as labor, disease, storage, and marketing were all more complicated with potatoes than in traditional grain farming.

During the decade after 1910 the potato acreage in the Valley expanded rapidly. Potatoes had proven themselves as profitable, and farmers were seeking new crops. The seed departments and extension service of the governmental units encouraged potato raising. World War I brought an unprecedented demand for potatoes, causing prices to increase five- to eight-fold during those years. But potatoes were perishable, and if the supply was just a

---

*Joe Thompson's operation eventually became part of the Hall Brothers potato firm, one of the largest in the Red River Valley since the 1940's. Thompson's daughter, Bernice, married William Hall. Chester and A. B. Thompson, also sons of Tom Thompson, were large early growers of potatoes.

bit larger than the demand the prices dropped sharply. Many of the early growers in the Park River and Hoople area in North Dakota were caught in such a price squeeze with the 1917 crop and were unable to sell their surplus potatoes for enough money to pay the freight, even though potatoes averaged $1.35 per hundred for the crop year. Hans and Lars Larson were two of those who were caught on the down market. Ironically production was still higher in 1918, but because of world demand prices remained high. After a sharp break in the price in 1921 and 1922, potato income remained relatively good by contrast to other crops.

The biggest change in potato profits came with the increased cost of production. Potatoes are a labor-intense crop, and wages rose sharply during the period of wartime shortages. When the war was over, potato prices dropped, but wages did not. However, because they were a good crop in the rotation, farmers were willing to stay with them. Grain yields generally improved if grain followed potatoes in the rotation.

In the days of limited scientific farming the farmer was almost helpless in combatting potato diseases. Nels Folson, a leader among the farmers in investigating the disease problem and also in studying mechanization of potato production, visited with many growers in the area west of Minneapolis and learned that blackleg and scab were two of the major disease problems. Because of its northern location Walsh County, North Dakota, was a superior area for growing seed potatoes. Therefore, it was essential that they produce a disease-free product. Dr. Henry L. Bolley, plant breeder at the North Dakota Agricultural Experiment Station, had developed a formaldehyde solution for treating potato seed that greatly reduced the danger of scab. Unfortunately, farmers often avoided using formaldehyde until they had lost a crop to scab.

When blackleg disease was identified in North Dakota, Dr. Bolley called a meeting of lower Red River Valley growers at Grand Forks to explain the problem and how to control the disease. Farmers were asked to bring 12 tubers each to be judged and analyzed. Very few farmers attended the meeting, and fewer still brought potatoes to be judged. Of the eight samples exhibited, Bolley found six unfit for seed. Only Halversons from Forest River and Midgardens in the Park River area had exhibits that were eligible for resale as seed. As a result the Midgardens became the first growers to ship a carload of state certified seed potatoes out of Nash, North Dakota. Midgardens had North Dakota Certification Number Six, placing them among the pioneers in certified seed production. The potatoes were Early Ohios, long a popular variety in the Valley.

To be eligible for state certification a farmer had to quarter his potato seed and submit for inspection a quarter part of several of the potatoes that he was using for seed.* The state indexed the seed and planted them in their greenhouses, where they were carefully watched for signs of disease. At the same time, the farmer had to submit to state inspectors, who looked for diseased tubers in the fields where corresponding seed was planted. To avoid blackleg, farmers had to walk through the fields and destroy the infected plants.** This was a labor-consuming and costly process, but it was and still is the only way to avoid an outbreak of blackleg.

Producing certified seed required more careful management than many farmers cared to provide for potato production. As more potatoes were produced in the Valley it became increasingly difficult to raise disease-free seed. To reduce the need for intense roguing it also became necessary to have foundation seed raised in more isolated areas. Folson, A. B. Thompson, L. E. Tibert, and the Midgardens, in cooperation with the State Seed Department, went to the western part of North Dakota in search of farmers who wanted to specialize in growing foundation seed. Many of the early growers in the West were trained through 4-H club activity. Big acreages were not needed, and children could do much of the work necessary in producing disease-free seed. Western production helped the Midgardens, for it was not until their third year of using western seed that they discovered their first spindle tuber. Next the Midgardens discovered leaf roll in some of their potatoes, forcing them to rerogue their fields several times. The first summer leafroll was noted, A. N. Midgarden was in the potato fields almost steadily and walked about 2,000 miles looking for diseased plants. He said his back muscles became very sore.[1]

From Hand to Machine Production

The wide open spaces and flat fertile fields of the Red River

---

*Potatoes used for seed were cut into quarters to stretch the seed supply. The only requirement was that each quarter have an eye in it to make it a bearing seed. Quartering was the quick way to cut potatoes, and generally each quarter contained an eye.

**Walking through the fields to glean out the infested plants is called roguing. This was very slow, difficult work, which required observing every plant as the roguer walked down the rows. It was also hard on back muscles and legs.

Valley have made it attractive to large-scale potato operations.* Large-scale potato farming started in the 1950's with new technological developments and a change in the marketing system. Generally the reason for this growth to rather large units was brought about by a simple matter of economy of scale. In a labor-intense enterprise some of the major costs can be best reduced by using big machinery. The big fields of the Valley were ideal for efficient use of the largest possible machines, enabling potato growers to substitute capital inputs for labor inputs.

Table 1, at the end of this chapter, illustrates the scope of operation of the farmers interviewed for this study. These potato farmers are well-known throughout the Valley and can be regarded as pace setters for the potato industry.**

Charles Peterson's first potato crop of 20 acres was planted in 1904 within the present-day city limits of Moorhead. Although horse-drawn one-row potato planters were produced in the 1880's, Peterson could not afford one in his first year of potato raising. So like most of the early growers, Peterson used a walking plow to open a furrow for hand planting of potatoes. The team and plow

*The average labor force on the highly mechanized potato farms of the Valley would astound the small-scale one-man grain or dairy farmer of other regions.

**Because of financial and time limitations a total survey of Valley potato growers was not possible. The names identified below are recognized leaders in Manitoba, Minnesota, and North Dakota. In no way should this be considered a list of average growers, nor does the study imply that these are typical units; however, for those who know the industry it is clear that the trends followed by these people are indicative of the potato business. When quotations are used in this chapter it will be from one of these growers. Sometimes the individual quoted will be cited by name; however, in many cases they were willing to make a statement to the writer, but they preferred not to be publicly identified as the author. Two of these farmers have discontinued growing potatoes since the time of interview and converted to enterprises more to their personal liking:

John H. Bogestad
Arden Burbidge
Clarence Carlson
David Carlson
Ron Carlson
John S. Dean
Keith Driscoll
Ray Driscoll
Ole A. Flaat, retired
O. Lowell Flaat
Cliff Hagen
Vernon G. Hagen
Bill Hall
Paul Horn, Jr.
Paul Horn, Sr.
Earl Hvidsten

Ralph Hvidsten
Don Kroeker
Walter Kroeker
Jameson Larimore III
Charles Lysfjord
A. N. Midgarden, retired
Ronald D. Offutt, Jr.
Ronald D. Offutt, Sr.
Henry R. Peterson
Gerald Ryan
Thomas Ryan
John W. Scott, Jr.
John W. Scott, Sr.
John W. Swenson
Joe Thompson, retired
J. Budd Tibert

could turn 1 to 2 acres of land each day. It took two people to hand-drop the seed in the furrow and another person to hand-cut the seed just prior to planting. After the potatoes were planted a fifth person used a Boss Harrow to go over the ground several times to cover the potatoes.*

Tom Thompson of Nash, North Dakota, planted his first 15 acres in 1908 in this manner. During the summer both Peterson and Thompson used a one-row horse cultivator and cultivated their potatoes five times. In addition, hoers went through the field to eliminate the weeds that the cultivators could not get, and they rogued the fields for diseased plants at the same time. When fall came Peterson's 20 acres of potatoes were dug by men with forks. It required at least four men to dig an acre a day. Others picked the potatoes and put them in sacks. Thompson used a horse-drawn wheelless potato digger that had two handles used by the operator to steer under the row of potatoes. This laid the potatoes on top of the ground for hand pickers.**

Thompson's potatoes were hauled by wagon to the railroad car, where the sacks were emptied directly into the car for bulk shipping. At the destination store employees came to the car and put potatoes in barrels or sacks, according to their needs. The Petersons delivered as many potatoes as they could directly to customers in Fargo and Moorhead. The rest were stored and sold during the winter season. The yields were from 30 to 40 hundredweight per acre, and the price from 30¢ to 60¢ a hundredweight. It took about 60 hours of labor to grow an acre of potatoes.*** Total machinery investment for Peterson was $16.90 for a 14-inch walking plow, $14.30 for a two-section 12-foot harrow, $34.80 for a single-row hand-lift riding cultivator, and $2.25 for a steel double tree plus the cost of a team of horses and harness. Thompson had the additional cost of approximately $40 for a potato digger.

From those primitive beginnings both the Petersons and the Thompsons, along with the other potato growers of the Red River Valley, adapted and invented new ways and means to reduce the

---

*Peterson had five acres of onions in 1904, which were also hand-seeded and hand-dug.

**The Thompsons had their pickers sort in the field, so that the "picked" potatoes could be loaded directly into a railroad car. One night during their harvest in 1908 they received 11 inches of snow. To keep the potatoes in the railroad car from freezing, lanterns borrowed from neighbors were placed in the car and kept burning.

***In a 1962 study of 82 Red River Valley potato growers the average production per acre was 145 hundredweight, and total labor time about 14 hours. The productivity of a 1962 laborer was 17.7 times that of a 1904 laborer.

work load and increase the yields. Charles Peterson understood the fertility potential of Valley soil, but he knew that improvements could be made to boost yields. Because he was close to Fargo-Moorhead he took advantage of the surplus manure that accumulated in the livery stables and private horse and cow barns of those communities. Peterson had two men with separate teams employed the year around hauling manure from those barns. He felt this was necessary to mellow the heavy Fargo Clay soil as well as providing fertilizer. With this practice Peterson was able to build yields to 90 hundredweight per acre.*

It wasn't long before most of the potato growers purchased the commercial one-row potato planter capable of planting about 5 acres a day. In 1917 a two-row potato planter became available. By 1916 most farmers were using the double-bottom riding plow that sold for $82.20. Following plowing, the land was disced with an 8-foot single disc that cost $38. After the potatoes were planted, the land was harrowed at least twice before being cultivated with a Boss Harrow. Each summer the potato crop was cultivated from three to five times. After a few years the potato weeder came into use, and that was used in place of the harrow and after each cultivation.

About the time of World War I more of the farmers were converting to the four-horse, one-row potato digger with a vine turner. This new digger cost $102.70 in 1916 and could dig 5 to 8 acres a day. With 40- to 45-hundredweight yield it required eight pickers. Pickers receiving 2¢ to 5¢ per bushel could pick from 50 to 100 bushels a day. Two loaders in the field loaded 80 70-pound bags on one of the two wagons used to haul potatoes to storage. Because the fields were generally soft, four horses were used on each wagon. At the potato storage the driver unloaded the sacks into a sack-type spout that was held closed by another man to reduce bruising. Loaders, haulers, and warehousemen were commonly paid $2 to $3 a day for the harvest season. In this manner 14 workers could harvest 560 bushels of potatoes a day, an average of 40 bushels per person per day.**

---

*Prior to World War I Peterson, as well as the Grants and Schroeders, tiled some of his land into the Buffalo and Red rivers. This was some of the very limited subsoil tiling done in the entire Valley, a generally impractical procedure because of the flatness of the region.
**A hundred 1-bushel bags per day was generally considered a good picker's quota; however, many pickers did not do much more than 50 or 60 bushels. The very best pickers could do up to 200 bushels per day. A bushel of potatoes weighs 60 pounds.

The Cudigan-Gear potato harvester purchased by Alfred Hvidsten in 1916 was the first serious attempt to get around the expensive labor requirements of the potato harvest. This machine, drawn by six horses, could dig one row of potatoes and elevate them directly into a wagon that was pulled alongside the harvester. A triple-box wagon was used, and four horses were required to pull it. A chain-type vine eliminator was used, which worked fairly well. The big problem was the inability to separate the clumps of dirt from the potatoes, and far too much dirt was hauled into storage. Once long-term storage was adopted, the use of the machine was discontinued. The Flaats had two of these machines that they used only a couple years before they too reverted back to hand picking.

Because of the tremendous labor requirements in the potato harvest, Indians, lumberjacks, housewives from nearby towns, and school children worked in the fields. It was a common practice throughout the Valley to dismiss school during the peak of potato harvest. This continued until World War II. With the adoption of mechanical harvesters in the early 1950's, the traditional parade of potato harvest brigades came to an end.

Once the potatoes were harvested they were commonly stored in trenches in the ground and covered with straw to prevent them from freezing. A sliding shed was moved over the trench when potatoes had to be sorted and sacked for delivery. The sorting crew consisted of one man who hand cranked the grader and watched for bruised and cut potatoes. A second man shoveled potatoes onto the grader, while a third man placed the sacks on the grader, sewed and weighed the full sacks, and piled them. A three-man crew could grade, sack, and load a railroad car in a 10-hour day. The first potato sacks contained 150 pounds and were called "backbreakers." About World War I the 120-pound sack was introduced and remained the standard for the industry until the late 1930's.

If certified seed potatoes were being shipped, each tag on every sack had to be hand signed. Seed potatoes were always shipped in sacks, but table-stock potatoes were frequently shipped in bulk. Grading and sacking were done in the usual manner for bulk potato shipments, but the sacks were not sewed shut. The filled but open sacks were elevated into the railroad car, and the potatoes dumped into the car. In this manner sacks could be reused.

If potatoes were sold directly from the field they had to be graded from a large bulk bin that was built above the ground and could hold nearly a carload. The men who hauled potatoes from the field dumped their sacked potatoes into this bin. Men in the underground trench opened the bin and let potatoes flow over the grader into the sacks. The graded sacked potatoes were then elevated into the rail car. This was an efficient method, but it could be used only during periods of above-freezing temperatures.

Again and again progressive potato farmers were looking for new ways to improve quality and increase yields. Because most growers had 20 to 40 acres of potatoes they were willing to spend extra money on them because any increase in yield could produce a sizeable increase in net profits. Charles Peterson was one of the first growers in the Valley to use a sweet clover plow down to improve soil conditions for his potatoes and other vegetables. Peterson was doing this in 1916 and along with Henry Schroeder was one of the leaders in the practice, even though personnel at the North Dakota Agricultural Experiment Station had been recommending a clover plow down fallow since about 1900.

Very little fertilizer was used for the potato crop in the Valley prior to 1916. Charles Peterson was convinced by agricultural research people to try some phosphate (0-20-0). Because there were no commercially built fertilizer spreaders available to Peterson, the phosphate was spread by men carrying pails and hand tossing it. Up to 250 pounds were spread per acre on several acres each year until the mid-1920's, when a commercial spreader became available.

A. N. Midgarden and Joe Thompson believe that next to James J. Hill they were the earliest users of commercial fertilizer in Walsh County, North Dakota. Thompson was of the opinion that commercial fertilizer, the tractor, and the mechanical harvester were the three greatest boons to the potato industry in his farming career from 1918 to 1960. Midgarden and Thompson thought it was 1920 when Hill introduced the use of phosphate around Nash with excellent results. Thompson saw an opportunity and became a dealer. He received orders for seven carloads of 0-20-0 for use in 1921. Midgarden used an old Dowigac grain shoe drill to spread 300 pounds of phosphate per acre.* They plugged every other spout and dropped the phosphate in 12-inch spacings and about 1 inch deep. The biggest trouble was keeping the fertilizer from caking and plugging the spouts. Midgarden said, "We had all

---

*Midgarden, Peterson, and Thompson all indicated that 0-20-0 was applied at the rate of 250 to 300 pounds an acre. This was rather high for early users.

kinds of trouble getting it spread because of the plugging but we realized it was worthwhile when we got more even maturity, better quality, and a large increase in yield that fall." The Midgardens shipped the first carload of certified seed potatoes out of Nash (1918) and were very concerned about quality as well as yield.*

Joe Thompson spread his fertilizer using a two-wheeled hand cart with a barrel and a fan on it, that had been purchased from Sears, Roebuck for spreading grasshopper bait. Thompson recalled, "We did this in the fall of 1920 and this was lucky because as the wheels bounced on the frozen clods it shook the cart enough to keep the fertilizer running freely. This way by luck we got good distribution. We tried some on barley following potatoes—where we put the phosphate we got about twenty bushels an acre more." Midgarden added, "We made cross patterns on some fields just to see if fertilizer really paid. In the grain field we could see the fertilizer pattern all year and in the silage corn field there was nearly a foot difference in height. We could see long before harvest that fertilizer would pay."

Midgarden, Peterson, and Thompson all agree that the high prices for potatoes during World War I was one of the reasons they were willing to try commercial fertilizer. Midgarden got $2.00 for 120-pound sacks of potatoes in 1918 and $2.25 in 1919. Thompson said 1918 was the year he started on his own and it was the best year he had in his total career. From a rented 320-acre farm he sold 8,000 bushels of wheat for a $2.62 average, and 6,000 bushels of potatoes for a $2.00 average. Thompson added, "Everything went right for a first year farmer; from a $33,000 gross income I had $20,000 left after all expenses. I paid cash rent of $4.50 an acre. That all changed because in March 1920 I paid $150 an acre for land but fortunately had the cash for most of it or I could have lost it in the early twenties."

The best year for Charles Peterson, who farmed from 1890 through the 1930's, was 1919. That year Peterson had 20 acres of onions that went 900 bushels per acre and netted him over $20,000. There were 40 acres of potatoes, most of which were re-

---

*This certified seed sold in 1918 was not tagged, according to Midgarden. Starting in 1919 every bag of certified seed sold by North Dakota growers had to bear a hand-signed tag. When the tagged bags got to Kansas City the broker had difficulty selling to farmer growers because they were not acquainted with the idea of certification.

tailed in the Fargo-Moorhead area, for over $3 a hundred. Henry Peterson grinned as he commented:

The twenties and thirties were not as difficult for Dad as for most farmers because from that date on he always had money on savings. He deserved it for when he came to the Valley in 1888 to work as a hired man he had less than a dollar to his name. But Dad knew soil and was always reading. He never complained about conditions, not even 1921, but he understood people and was a good marketer. In 1921 he owned 120 acres and rented 100 acres and hired up to twenty people during cropping season.

A newspaper article of 1929 credited Peterson with using a full carload of commercial fertilizer on his farm that year. Very few farms could boast of such a record at that date.

Ole A. Flaat, although not the first to use commercial fertilizer in the Valley, did much pioneering in its use. Flaat started farming on his own in 1918 on land rented from his father. Even though the senior Flaat considered himself very progressive and helped Ole become one of the earliest certified seed growers in Minnesota, he was reluctant to let his son use commercial fertilizer. After farming for seven years Ole Flaat decided in 1925, in spite of his father's advice, that it was time to experiment with commercial fertilizer. Flaat said:

There were lots of rumors around that adding nitrogen, phosphate, or potash to the soil would only throw it out of balance with other minerals that were still unknown. Then there was the typical story that floats in agriculture. The farmers out east or down south were having trouble with their soil after using commercial fertilizer. Even the experiment stations were telling us that in the Valley we could ignore nitrogen and potash because phosphate was the only one that paid. I listened and then got thinking. Fertilizer can't hurt, it can only help, so quietly I started using 0-20-0 on a limited basis in 1925. In 1926 I used quite a bit more. In 1927 I used 0-20-0 at about 175 pounds an acre on most of my potatoes with excellent results in quality and yields. I had to buy from the Agricultural Chemical Company in St. Louis because there were no local outlets. Some people were a little disturbed because they thought it was all wrong and it would be the ruination of Valley agriculture, but there were very few potato farmers who felt that way. Most of them were curious and had a wait and see attitude.

It took many of these farmers quite a while to really be convinced. In 1931 Flaat Farms hired Larry Brown, a college graduate, to work on a fertilizer program as his first management project. A few years later when Flaat employed Dr. A. C. Wardner, a chemist, to make a soil analysis and recommend fertilizers, many neighbors could not understand Flaat's reasoning. Those farmers were still either not using fertilizer or using only 0-20-0 at minimum rates. Brown and Wardner developed a progressive fertilization program for the Flaat Farms. By 1943 the Flaat Farms Supply Company had the first fertilizer-blending plant in North Dakota. Flaat was the chief investor, and Brown was the builder, manager, and salesman. This firm eventually grew into Agsco Company, Inc., with several branches throughout the Valley.

In 1938 the Marshall County agent reported that Alfred Hvidsten of Stephen had been experimenting with a fertilizer plot for potatoes. Hvidsten's conclusion was that a "mixed balanced fertilizer gave a distinct yield advantage."

It was not until the late 1930's that a noticeable average Valley yield increase took place because of the use of fertilizer. While World War I helped the adoption of fertilizer, it was the demand and higher prices of World War II that stimulated the average potato farmer to step up the use of fertilizer. Between 1919 and 1936, potato yields ranged from 3,500 to 5,900 pounds per acre. From 1937 to 1948, yields increased to over 9,000 pounds. Since 1949, Valley yields have consistently been over 10,000 pounds per acre. One county had an average one-year yield of 19,000 pounds per acre. By the 1960's farmers were expecting 15,000-pound yields to keep up with rising costs. The consumer was the chief benefactor of these yield increases because in spite of rising costs the price of potatoes has remained relatively stable.

Production ability changed slowly for the potato farmer during the 1920's. When Johann Hall started with potatoes at Hoople in 1928 to improve his farm income, he raised certified seed. Hall preferred to farm on a pay-as-you-go basis and would not spend money for potato equipment until he could pay cash. Because of this decision the first crop of potatoes was dug by fork and hauled out of the field by wheelbarrow. But Hall's potato income was encouraging, so he purchased equipment rapidly and by 1931 was using I.H.C. F-12 tractors.

Alfred Hvidsten's 1916 horse-drawn mechanical harvester was a pioneer in harvesting, but although it was used by a few other innovators, it did not prove successful under many conditions. The Flaats of Grand Forks had two horse-drawn mechanical potato harvesters prior to 1920 that worked fine on beardon loam soil; but on heavier soil too much dirt was left on the potatoes, and this created an excessive cleaning problem at the potato storage. There would have to be many improvements before hand harvesting could be eliminated. Arden Burbidge, long-time leader in the potato industry, said in 1970: "Malthus and coolie labor were a problem for the American farmer thirty years ago, but with mechanical power and the equipment to go with it we have become the world's most efficient producers—maybe to the point that we could hurt farmers of other nations if we went all out."

It appears that Valley potato farmers were quick to see the advantage of mechanical power in the 1920's for tilling purposes even though it would be nearly 30 years before that power could be applied to harvesting with full advantage. They did buy tractors and apply them to potato production as rapidly as they could. Joe Thompson bought a John Deere tractor in 1921 for $526. The Thompsons were having grasshopper problems, and they discovered if they plowed at night the grasshoppers were plowed under. A man walked ahead of the tractor with a lantern to show the way.

The first tractor was such a success for Thompson that in 1922 he purchased a Fordson. It became the popular tractor on their farm because it was easier to handle than the bigger tractor. There were many light jobs to be done on a diversified farm that were handier to do with the Fordson than with horses or bigger tractors. Thompson liked horses, but he was quick to see that tractors reduced labor requirements and freed land for cash crops.*

Joe Thompson had a year-round man when he started on 320 acres in 1918 and in the busy season needed as many as 15 additional workers. Even though he enjoyed working with people he soon realized the economics of tractor power to reduce the payroll. As soon as he purchased more land in 1923 he bought his third tractor, a John Deere D that pulled three bottoms. Only two years later he owned a fourth tractor, this time an I.H.C. Farmall. It had a power takeoff, could handle a loader, and could be used for row crops. "The men all liked that tractor," Thompson recalled.

Looking back at his growth in farming Thompson reflected:

My interest in farming was to get bigger. I was ambitious and it was fun. I had progressive neighbors around me for competition but it was the friendly kind. I really had little interest in machinery, especially tractors, but they were the secret to growth. It seems each model was a little better than the previous one so we just kept on getting the improved models. The combine I bought in 1927 for $700 was the next big labor and dollar

---

*On Tom Thompson's farm of 1,280 acres, 240 acres were in pasture, and 120 acres in hay land to feed the horse herd and a few cows. Joe Thompson had only 320 acres when he started farming. He had 260 acres in cash grain and potatoes and on the rest raised crops to feed the horses.

saver. The tractor and combine were big reasons for my growth in farming.

Ole Flaat, a contemporary of Thompson, would have liked to have purchased a tractor when he started in 1918, but he also was fond of horses, and his dad "held back." Ole's father was 58 when Ole purchased an I.H.C. Farmall in 1927. Ole realized its value at once. Because of heavy soil conditions the Flaats turned to track-type tractors in the early 1930's. By 1937 there were 12 tractors on the Flaat farm, including three Caterpillars. In the early 1940's there were 20 tractors "because expansion was economical with mechanical power."

Charles Peterson "permitted" his boys to buy a John Deere tractor in 1928. Once they owned one, their advantage became so obvious that in 1929 they purchased a Caterpillar for plowing and tillage on their heavy soil. Next they used the Caterpillar on the potato planter. Because it had so much surplus power they decided to put two double-row planters together to make a four-row.* Henry Peterson said, "We used two double row Iron Age planters that had the canvas hoppers."

With the availability of tractors to plow, till, cultivate, pull a two-row planter, and two-row diggers, the more alert potato farmers continued to expand their acreage. Quality improved, and income was generally better than from any other crop except sugar beets. It became common to see two or three tractor planters and diggers on the larger farms; and eventually some farms, like the Flaat's, had as many as 10 double-row tractor-powered diggers. Only one major bottleneck remained—the task of getting the potatoes from the field into storage.

Few examples in agriculture provide a more dramatic breakthrough in the technology of production than that of the mechanical potato harvester. The single-row, 1,095-pound, 7-foot potato digger with a vine turner, that cost $102.70 in 1916, changed little mechanically up to 1950. In 1950 the standard digger was 9 feet long and weighed 1,156 pounds at the price of $409.50. For $713.75 a more sturdy 2,097-pound, two-row, pull-type digger with tractor hitch could be purchased. All the digger did was lift the potatoes out of the ground, shake off some of the dirt, and let the potatoes fall into a row on top of the ground to be hand picked.

For the potato grower, like all other farmers, the 1930's were not

---

*Peterson did not know of another four-row potato planter when they made theirs, so probably it was the first one in the Red River Valley.

especially attractive years to be in agriculture. The Valley, with its ability to hold subsoil moisture, was in some respects a garden spot in contrast to the drier areas to the west. John W. Scott, Sr., who started farming in 1924, said that 1929 and 1936 were the two years that put him "on his feet." Scott recalled:

Except for sheep the only crops I could make money on during the twenties and thirties were beets and potatoes. It was kind of a break for me but in 1929 I just could not borrow money for anything until one day late in the summer the banker came out and saw that I had a good crop coming. He was still afraid of what happened in 1928 when the price of potatoes got so bad that many were not even harvested. In 1929 production was way down and prices were up.

This happened to me in 1935 too. There was a big crop just like in 1928 and the bottom dropped out of the market. So, in 1936 everybody decided to plant wheat. I went out and bought all the seed potatoes which nobody wanted to plant. We planted and planted—I think almost to the first of July. The year was dry, production was way down but did we get a price. From that day on my banker never gave me trouble and always loaned me the money I asked for. He put me on the board of directors of the bank and I was on until 1971.*

Halls, Flaats, Petersons, Thompsons, Scotts, and Tiberts all increased their farms to two or three sections of crop land each during the 1930's and expanded potato production to 300 and 400 acres. A four- or five-year rotation with potatoes was considered ideal to reduce the risk of disease. More potato acres would have been desirable, but the harvest problems in the Valley with its short fall were virtually insurmountable. The older progressive growers and the younger foresighted ones, like Vernon Hagen, the Hvidstens, Ryans, Horns, Driscolls, and Kroekers, all seemed to sense that improved conditions were imminent.

---

*According to John Scott, Sr.: In 1928 North Dakota potato production averaged 5,900 pounds per acre and the price averaged $.50 cwt. In 1929 production averaged 3,400 pounds and the average price was $2.20. In 1935 production reached the volume of 1928 and the price tumbled to $.63 cwt. In 1936 total acres planted were down and because of low moisture average yields were down too. The price reflected the short supply and rose to $1.98 cwt average for the year's crop. In Minnesota the 1929 and 1936 prices were triple the 1928 and 1935 prices. Joe Thompson remembered that those were good years for him too. He made a personal loan to his banker during the early 1930's. He remembered two days in 1936 when the hot winds just shriveled the grain and potato leaves. He said, "That caused potatoes to sell for $2.00 to $3.00 a hundred." Thompson purchased a farm from North West Bank Corporation in early 1929 under a contract that he would pay a cash price of $35.00 an acre plus interest. The contract specified that so many bags of potatoes were to be sold each year until the purchase price was fulfilled. The price was so good in the fall of 1929 that he was able to pay for the entire farm from his first crop.

World War II, which created an increased demand for potatoes and caused better prices, was also responsible for a labor shortage. This more than ever before put pressure on technological improvements. In the 1920's potato blight was combatted by using a one-horse, three-row potato duster. By the 1930's the standard duster in the Valley used two horses and had six rows that could do about 40 acres a day. Ole Flaat recalled that the duster was a modification of a machine used in the cotton fields. The dust was blown straight back from the hopper boxes and created a white cloud in the field. But it was better than the earlier liquid sprinkler. The Marshall County agent recorded that in 1944 airplanes were used for the first time to dust potatoes. The reason was that the fields were too wet to drive on. A plane could dust 35 to 40 acres per hour. By that time Ole Flaat used three airplanes on his 4,000-acre potato crop and did the entire job in 40 hours. The time needed for dusting was important because it took only a few days after the first signs of blight to greatly reduce the yields.

J. G. Hall, who started with potatoes in 1928, watched his farm grow as each son became old enough to help manage another expansion of his operation. By 1938 the Halls were using tractors and two-row planters to plant 20 acres a day. At that time the Petersons, who were using a track-type tractor and homemade four-row planter, could do about 35 acres a day. The fertilizer was broadcast on the field in bulk form. By the 1970's users of four-row planters could do 50 acres a day, and six-row planters could do 75 acres and put on fertilizer at the same time. Those tractors developed speeds up to 6 miles per hour, and loading of the potato and fertilizer boxes by hand had given way to mechanical loading. A six-row planter required at least a 125-H.P. tractor because of the heavy load of potatoes and fertilizer required to make a trip back and forth across the field.*

To speed up planting, Arden Burbidge and the Ryan Brothers have built large portable overhead hoppers that are filled while the planter is making a round. At the end of a round the planter is driven under the hopper and refilled instantly without tractor or planter operator leaving his position. Arden Burbidge has calculated that in using a six-row planter and traveling at 5 miles per

---

*A six-row planter, using 38-inch rows, plants 2⁴/10 acres on ½-mile fields in each round. Since it is desirable to make a 1-mile round without stopping, the fertilizer and seed load was 8,000 pounds. The weight of the planter and the planting action in soft soil add immensely to the power requirements. Arden Burbidge has calculated that each hour of seeding time with a six-row planter, using $2.50 a hundred seed cost, amounts to $522.

hour, 5⁸/₁₀ acres can be planted in one hour, if it takes 10 minutes to refill. Using the overhead hopper and refilling in three minutes' time, 8¹/₁₀ acres can be planted per hour. Over 27 acres more could be planted in a 12-hour day with a single six-row planter. Unfortunately, bridges and roads often limited them to the use of four-row equipment. In this manner maximum farm efficiency was being restricted by a non-farm obstacle.

As total acres planted per farm increased, the necessity for better harvesting methods became even greater. In 1945 Thorbald Hagen had planted 75 acres of potatoes before becoming seriously ill. That fall his son, Vern Hagen, had to quit ninth grade and take over the operation of his dad's 400 acres. A crew of 19 people was required for 10 to 11 days to harvest those 75 acres, which averaged about 130 bushels per acre.

The Hagen crew on a typical small-farm potato operation consisted of a tractor driver and one person on the two-row digger, an average of 13 pickers,* and a four-man crew on a truck for loading the sacks in the field and dumping them into storage. By the 1940's pickers were getting 10¢ a bushel, and haulers and equipment people were paid by the day. The crew averaged 51 bushels per day per person to harvest the 1945 crop.

The five Hall brothers, who had 1,000 acres of potatoes in the late 1940's, tried to schedule their harvest so that it could be completed in 23 work days.** The Halls hoped that their pickers could average 100 bushels each per day (in the 1930's they paid 3¢ a bushel, and by the early 1950's they paid up to 15¢ a bushel). Obviously, the crew required was determined by the average yield; but in their last years of hand harvest, the Halls required a crew of 150 to get the job done in their planned 23 days. About 90 people were required to pick the potatoes in the field and to put them in bags which were set in rows, making it easier for loading. A crew of 32, using eight trucks, loaded the potatoes and hauled them to storage. The Halls rented much of their land at this time, and because of that, quite long trips had to be made from the scattered fields to storage. Six tractors, each pulling two-row diggers, required another 12 men. In addition, two men were needed to keep

---

*The Hagen crew averaged only about 70 bushels a day. This was during World War II, and because of the intense competition for pickers in harvest season, housewives, school children, and even people in jail had to help. Vern Hagen felt that he had only a small operation and therefore could not get so good a class of laborers as the larger farmers in the area.

**The town of Hoople, which normally had just over 300 residents, would have as many as 3,000 transients in the area during potato harvest.

the equipment moving and supply sacks to the pickers. At least 12 men were needed in the warehouse to grade potatoes for immediate shipment and supervise the bins for proper storage. The average production during harvest was 55 bushels per person per day.

The most dramatic potato harvest scenes occurred on the Flaat farm near Grand Forks during World War II. It was almost a repetition of the drama of the bonanza wheat farm harvests of the late 1800's. Ole Flaat grew from 100 to 200 acres of potatoes each year during the 1920's. At that time the Flaat farm consisted of less than a section, but five full-time men were employed, and during the six-week potato harvest an additional 20 people were required. Once the Flaats started buying tractors and converting to larger equipment, their acreage increased regularly during the 1930's and 1940's.

During World War II Flaat had a contract to produce and process potatoes for the War Food Administration on a cost-plus basis. For three years during the 1940's Flaat grew 4,000 acres of potatoes and about 1,400 acres of grain. This might have been the world's largest privately owned potato farm.* The crop land was scattered, and coordination and transportation presented immense problems. During that period a year-round crew of 20 men was needed on the Flaat farm. Normally even more men would have been required if the potatoes had been shipped in sacks, but the Flaats had their own processing plant, and potatoes were delivered in bulk.

During harvest 450 workers were required to pick the potatoes. Because of the labor scarcity, Jamaicans, Mexican-Americans, and German prisoners of war made up the greater portion of the crew. It took 60 men and ten trucks just to haul the crop to storage and 24 to run the diggers. Flaat's regular crew was needed just to supervise and handle the most technical part of the harvest. Six people were required in the office to write checks and keep track of the daily production. The Grand Forks auditorium and two other large buildings in that city were leased for sleeping and eating quarters. Three farm houses were also converted to sleeping quarters. The picking crew averaged about 60 bushels a day,

---

*Henry Peterson held a similar distinction during World War II, for from 1942 through 1944 the Petersons were the nation's largest grower of onion sets. When Americans were encouraged to have victory gardens during the war, they created a big demand for onion sets. The Petersons were an old source of supply (since 1904) and produced most of the seed for onions in the western United States. Peterson said, "We happened to guess right on the demand—that's the breaks."

which Flaat said was a "disappointment because the well motivated American farm kids I had been used to hiring could average
100 bushels a day." Average daily production per worker was 48
bushels during a 26-day season.

Those 4,000 acres produced an annual average of 700,000
bushels for the three-year period. Ole Flaat built a 700,000-bushel
warehouse, which was the largest in the Red River Valley up to
that date. As potatoes were processed, more were purchased from
area producers, so that the warehouse remained full as long as
fresh potatoes were available.

Ole Flaat was asked to comment on what it was like to handle a
harvest crew of 560 men needed to raise 700,000 bushels. He said,
"I felt at the time even though it was a big show that 4,000 acres
were too much to be efficiently operated with the technology of
the 1940's. The burden on top management was too great for the
rewards—there just wasn't enough qualified secondary management around to solve some of the problems. Today [1973] it could
be done because I know 2,400 acres of potatoes can be managed
with a snap."[2]

Flaat's Machine Arrives

The large potato growers were very frustrated during the 1940's
and the early 1950's because of the lack of a mechanical harvester.
Henry Peterson had made a four-row planter in 1936, but in 1941
he purchased a new commercially produced machine. Flaat put
two double-row planters together in 1938 but was not satisfied
with it. He returned to two-row planters until the factory-made
four-row machines became available. Tractor cultivators and
airplane spraying removed many of the potato-growing problems
between the 1920's and 1944; but harvesting, except for the digging, had changed little since ancient times. The two-row tractor-
powered digger requiring two men was more than 50 times faster
than Charles Peterson working with a fork. Besides, it was much
easier.

In the late 1930's Flaats and other growers tried using a lifter
that drove down the field shortly after the potatoes had been dug
and ran them over a chain to knock off more of the dirt. This was
done so that cleaner potatoes could be brought to storage. Housewives were complaining of dirt on Red River potatoes. So in an
effort to produce a cleaner market potato Flaat secured the services of J. W. Lambie, a professional engineer, and with financial
help from the Great Northern and Northern Pacific Railroads,

built a "squirrel cage" wash plant. Flaat was "not aware" of any other washing plants in the Valley in 1940 when this plant was constructed. It was his development of this potato plant and his reputation as an innovative grower that were partially responsible for his securing the large production contract from the War Food Administration.

There were many attempts to eliminate the need of hand picking in potato harvest. During World War II school children, housewives, and underemployed small-scale farmers around the Valley could be motivated to harvest potatoes because of patriotism. After the war the labor shortage was greater than ever because of the migration of people from the rural areas. Schools no longer saw fit to close during potato harvest, and other workers were not attracted to part-time jobs. They detested picking potatoes in the cold fall weather, biting the potato dirt, bending their backs to pick potatoes and carrying 25 empty sacks on their backs for 15¢ a bushel. They could get better paying steady jobs in the urban areas. A harvesting machine had to be developed if the industry was to prosper.

Between 1940 and 1945 there were many individual efforts to develop a mechanical harvester, but only a few farmers expressed the need for another "expensive machine." Without an obvious market no manufacturer would risk the cost.

Historically the farmers in the past had developed machines they needed, and the potato harvester was no exception. Ole Flaat and Henry Peterson, like many of the leading growers, traveled to other American, and even foreign, potato-growing areas in an attempt to find "the machine." In his third year [1946] of raising 4,000 acres of potatoes, Flaat attempted mechanical harvesting. A two-row digger was used as in the past to lift the potatoes and put them into a single windrow. Then a tractor straddled the windrow, pulling an elevator which was little more than an endless chain propelled by a mounted motor. A truck was hooked backward onto the elevator so as the tractor drove down the row of potatoes the elevator loaded the truck from the rear. Pulling a truck in reverse down the field behind a loading elevator required considerable tractor power for that period, besides being quite cumbersome. It was not satisfactory, but Flaat and others at least gained some ideas from the experience.*

---

*Henry Peterson used a commercially produced machine very similar to Flaat's for onions and sugar beets. It was clumsy but released so much hand labor that it remained in use until a better machine was produced.

Flaats went back to hand picking and had the pickers put the sacks in rows as usual, but now the truck was driven down the field with the elevator behind it. Two men on each side of the truck dumped the sacked potatoes into the elevator hopper, while a fifth man controlled a cloth spout in the truck box so the potatoes would not be bruised as they fell from the elevator into the truck. There were some obvious advantages. (1) The men did not have to set full sacks of potatoes onto a high truck box. (2) As the potatoes rolled up the chain elevator more dirt shook off so that the potatoes went into storage much cleaner. (3) When the truckload of bulk potatoes arrived at storage, the truck hoist was raised and the potatoes rolled onto a piling elevator. This saved labor in the unloading process. Each truck hauling from the field in this manner required a driver, but it took only four men in the field to keep two trucks busy because they were no longer needed for unloading. Fewer sacks were required, and because each sack cost 10¢ to 25¢ each, a big saving was realized.

In the meantime the "tinkerers" were busy trying to come up with a machine that could dig potatoes and load them at the same time. J. W. Lambie, the professional engineer employed by Flaat, had been trying many experimental models. In 1948 Lambie and Flaat developed a mechanical digger-loader potato harvester in the Flaat workshop. It was called the Flaat harvester. For the next two years Flaat and Lambie kept making many modifications before they let others study it or were willing to make any for the market. Ole Flaat described the harvester as follows:

This machine had a two-row digger which lifted the potatoes out of the ground, much like the digger we were all using, and then dropped them into a hopper which elevated them by chains into a truck that traveled alongside. We had to put five people on the machine to pick out the vines and dirt lumps because there was nothing the machine could do about the dirt if the lump didn't break up and fall through the chains. On some land, especially on lighter soils, the machine worked well but the dirt lumps were nearly impossible on the Fargo Clay. It became obvious that cultural practices and the organic content of the soil were going to be more important than ever with mechanical harvesters.

Henry Peterson commented about the problems of mechanical harvesting:

I had a farm with very heavy soil up by Kragnes. It was O.K. in the days when all you did was lift the potatoes out of the ground and leave them lay for the hand pickers. To get around the problem of hiring a large harvest crew we placed ads in the Fargo-Moorhead and area papers stating potatoes were for sale, so much a pound "you pick 'em." We sold as much as $6,000 worth of potatoes in a single day. Every car that entered the field pulled over the scale. When it came out it was weighed again and they paid so much a pound for the increase in weight. It was hard for them to cheat with that system. This land was too far away from our big manure supply used for mellowing the soil so we sold it.

After the harvests of 1948 and 1949 the "bugs" were removed from the Flaat machine. Then Ole Flaat and Lambie were willing to produce their invention commercially. Because they knew they could sell several machines it was decided to manufacture them in the Simms Machine Shop in Grand Forks. L. E. and Budd Tibert of Voss purchased the first two machines produced in 1950. They cost about $2,600 each. Then two of Flaat's neighbors, Adolph and Herman Skyberg of Fisher, each purchased a machine. Henry Peterson, long-time friend of Flaat, got the fifth machine.

Selling the harvester was no problem. During 1950 one was sold to a grower in Maine, two in Alabama, two in Canada, and six in Wisconsin. In 1951, led by the Hvidsten's, six machines were put into use by potato growers in Marshall County, Minnesota. During 1951 other potato machinery manufacturers came to look over the Flaat harvester. Ole Flaat reflected:

I was not interested in machinery other than how it could better help me farm so when the men from Dahlman and Lockwood came around we let them take pictures and measure whatever they wanted. The next year they started commercial production and sales. There were lots of interested growers; but as usual, the skeptics were around too. Many of the first harvester owners did custom jobs at $35 an acre because the skeptics could no longer get hand pickers. It seems that all of a sudden there were no hand pickers around.

Budd Tibert recalled the contrast in potato harvest on their farm before and after the harvester:

Dad and I formed a partnership in 1948 and we had 400 acres of potatoes on our 1,920 acres. We needed 66 people to harvest those 400 acres. There were 2 tractor-powered diggers which needed to work only half time to keep ahead of 50 pickers; then there were 6 haulers and 6 in the warehouse. A big day would be 5,000 bushels but 4,200 was probably more normal.

When we got our Flaat harvesters in 1950 I knew of no others in the Valley except Ole's [Flaat]. There were some problems but when Dahlmans came out with a much improved machine in 1952 we could see it was time to expand and we did—very rapidly. It had no de-viners so we needed seven people on each harvester, but the production was there and we needed five trucks to handle what two machines could turn out when we were harvesting close to storage. It took more for long hauls. Potatoes came into the house faster so we had to go seven men in storage to handle the load.

Today [1972] with much larger trucks and machines we operate four harvesters and nine trucks and thirty-two men can harvest and store 14,000 bushels a day with much less physical effort. With big pilers only three people are required in the warehouse.*

Joe Thompson, who had maintained about 350 acres of potatoes during the 1940's, had an annual crew of 45 to 60 during harvest. After he got his first harvester in 1952 his crew was reduced to 13. He recalled:

We were able to reduce our harvest season considerably; plus, the job was done with ease. It saved tremendously on the labor, was a real expense reducer, and caused a lot less bruising of the potatoes than with the hand labor. I was fifty-eight and had no son, but if I had been fifteen years younger I would have really expanded because of the great potential the harvester gave us.

Walter Kroeker of Winkler, Manitoba, who bought his first harvester in 1952 said, "It was out of self defense because of the tremendous competition for labor. We had 160 acres of potatoes in 1951, 300 acres in 1952, and have expanded every year since. Mechanization enabled the farmer to compete with industry for good labor. Potatoes were profitable so you had to figure out how to get volume."

Henry Peterson, who had the first harvester in the southern end of the Red River Valley, never had more than 160 acres prior to 1950. In 1952 he expanded to 450 acres. This was too much for

---

*Tibert's average daily harvest capacity per worker was 63 bushels, slightly higher than that of Hagen, Flaat, or Halls. Part of that may have been caused by the fact that higher-grade labor from the neighborhood was used instead of so many migrants. In 1972 Tibert's 32-man harvest crew averaged 14,000 bushels a day, or 437 bushels per man. That is seven times more per man than in the last year of hand harvesting. The workday was the same.

Budd Tibert had a second interesting contrast describing what mechanization did for his farm. In 1927, his last year of binder and thresh machine grain harvest, he needed three binders and seven shockers to cut and shock his grain. At threshing there were eight bundle haulers, two spike pitchers, two men at the machine, one in the straw pile, three grain haulers, and two cooks. In 1928 he purchased an I.H.C. combine. One man swathed the grain, then a tractor driver and combine operator threshed it, and two men with two trucks were used to haul the grain to storage. The four-man crew did the entire job in the same amount of time in 1928 as it took 10 people to cut and shock the grain alone in 1927. The threshing time for 17 people was extra.

one harvester, so he continued to use hand pickers for another year. Peterson said, "I was concerned that the harvester might not pan out entirely so I hired the crew just in case. We sure finished in a hurry that year but the pickers weren't sad about losing their jobs." He added, "It seems like a nice sized potato operation today [1974] is 800 acres because, except for planting and harvest, two men could handle it with ease."

Vern Hagen was so hard up in 1944 that when a neighbor offered to rent him 20 acres of potato land he was unable to do it on his own. Hagen had to get Joe H. Stroer as a finance partner on a 50-50 basis to buy the seed and pay for the fuel as well as the labor not furnished by Hagen. From those 20 acres Hagen netted $350 for his share, and that was his real start. In 1952 he was farming with his brother Cliff, and they were able to buy their first harvester. Vern Hagen said, "It was a good year and we still had our hand crew just to be safe. But we paid for the harvester and two trucks and potato boxes with that crop. We had a big enough production so we could sell to Old Dutch for processing and that's how I got in with them."

Bill Hall was not too excited about their first year's experience even though they were the Red River Valley's largest growers in 1952. The Halls had 1,000 acres of potatoes in the days of the hand crew, and the picking bill alone varied from $18,000 to $30,000, depending on the yield. They decided to buy five harvesters at once, which amounted to a cash outlay of nearly $25,000. Then they needed 12 potato boxes for their trucks at a cost of $10,800 besides the cost of the trucks.

Bill Hall remembered:

We decided that we had to convert our total harvest line or none at all, but what really forced us was the fact that it was becoming nearly impossible to get pickers, even at $.15 a bushel. I'm not sure that we had an immediate dollar saving the first year because we had lots of problems. We had two mechanics going over those harvesters every night that year to help avoid down time during the day.

The purchase of five harvesters and 12 trucks in the first year of large-scale commercial harvester sales brought national publicity to the Hall brothers' farm. They had one 976-acre field that was repeatedly pictured in newspapers and trade journals, showing the five harvesters and from 5 to 10 trucks lined up alongside the harvesters. The field was reputed to be the largest field of potatoes in the United States, according to one of the trade journals.* The Halls needed only 64 workers to harvest their 1,000 acres in 16 days instead of the traditional 145 to 150 workers in 23 days, as required with hand harvest.

The greatest single obstacle to growth was the length of the harvest season. The five young Hall brothers decided to increase their acreage each year and still stay within the 23-day harvest season. In 1953 they had over 1,250 acres in potatoes and even with increased yields had plenty of time to harvest the crop. About the time the harvesters became available, insecticides, pesticides, and fertilizer also came into heavier use. That, along with better tillage plus improved seed varieties, caused yields to increase rapidly.

Budd Tibert, the Halls, Hagen, Flaat, and the other growers all agreed that while 100-bushel yields looked good in the 1930's they expected 300 bushels per acre in the 1960's. In good years they were surpassing the 400-bushel mark. All of this has caused the total volume of production to go up even more rapidly than the total acres. A big grower in the 1920's had 100 acres; in the 1930's 300 acres; by the 1950's 500 acres; by the 1970's 1,000 acres; and many even exceeded 2,000 acres. It is believed that the Halls were the only growers to exceed 1,000 acres in 1961.

In the 1950's, on 60 acres of land that had been well manured, the Halls achieved yields of up to 200 bushels, of which 50 bushels were sorted out because of bruises or other defects. In the early 1970's they had an average yield of 380 bushels on 2,600 acres, with a much lower percentage of potatoes that had to be culled. Each year the larger growers expanded their acreage as the capacity of the machines increased to handle both heavier yields and more acres. The most significant handicap in 1973 was the limited life of the steel rods in the chain elevator and other moveable parts.

Ronald D. Offutt, Jr., along with his partners Ronald D. Offutt,

---

*In the Red River Valley 160-acre fields are commonplace and considered minimum-sized fields. Fields consisting of a full section (640 acres) are not uncommon, but most farmers keep them somewhat smaller, thus preventing wind erosion by having grain fields interspersed with potatoes. To avoid excessive wind erosion the Halls planted two rows of corn every 15 rods to slow down the wind, separate the potato varieties, and catch snow. A cover crop of oats was seeded directly after potato harvest to help hold the soil in position over winter. A man with a pickup truck loaded with a complete line of tools was stationed in the center of the field so that he could observe all five harvesters. If one broke down he rushed to get it moving again. A portable generator was carried to provide power for welding and lights.

Sr., and John W. Swenson, operating as Perham Potato Farm, was the largest grower by 1973. He indicated that in the sandier soils a harvester needed a complete overhaul after about 350 to 400 acres. Because of that, Offutt needed nine machines to harvest his crop, even after providing four hours per day for maintenance men to work on them during the harvest season. By 1973 the trucks being used by Offutt and most of the other large growers had all tandem axles, with a capacity of 28,000 pounds instead of 12,000 pounds, as in the 1960's. Many growers were even using semitrailers at harvest, with a capacity of 40,000 pounds. The harvesters were so much improved that if there was any distance to storage, at least three of those tandem trucks (at a cost of $14,000 each) were required to keep one harvester busy. Instead of seven people on the harvesters only two were necessary because the machine had become much more automatic. However, dirt lumps in the heavier soil were still the major harvesting problem.*

Vern Hagen was of the opinion that the minimum ideal economic potato unit with the technology of 1974 was 1,200 acres. "This," Hagen said, "would require a six-row planter and three potato harvesters. Of course, a farmer should have at least 2,400 acres in grain to rotate with his potatoes. This way you are big enough to afford the best equipment and big power plus a group of good year-round reliable men."

Henry Peterson, who had been farming since the 1920's, felt that by constantly getting better equipment and making work easier, securing a good supply of labor had become less of a problem each year. Ole Flaat's records indicated that except for those big 4,000-acre crops during World War II he required about 120 men to harvest his crop in the 1940's. In 1972, with more than triple the total production, he was doing the same work in a shorter season with 30 men or less. Flaat added, "and not near the muscle effort." From 1950 to 1972 the Hall brothers had nearly tripled their potato acres, increased their yields by more than 50 percent, cut their harvest from 23 to 20 days, and reduced their harvest force from 150 to 100—an increase in output of more than seven-fold per harvest laborer.[3]

---

*In 1974 the Offutts and Swenson had on order several $38,000 four-row potato harvesters that had sufficient capacity to keep five tandem trucks of 28,000-pound capacity busy hauling potatoes from the field. In a 300-bushel-per-acre crop this machine can harvest 2,250 pounds of potatoes per minute. Bin pilers capable of receiving potatoes from five trucks at a time were being secured to keep up with the harvesters.

After Harvest

Once the harvester was developed other obstacles to large-scale mechanized potato production became obvious. With increased harvest capacity it became popular to plant as many acres as possible within the limits of the harvest season. Greater acres meant bigger planters. The four-row version became commercially available in the early 1940's. Ole Flaat bought his first four-row planter in 1942.

Potatoes have to be hand cut when seeded and should be quartered just prior to planting. This means that a cutting crew has to work fast enough to keep a planter moving. When the first four-row planters appeared, 50-H.P. tractors had difficulties in pulling them. In the 1970's 100-H.P. tractors were pulling planters with ease at even higher speeds.* Speed was limited only by the performance of the planter. In the 1940's six hand cutters were needed to keep up with a four-row planter. As the planters improved, up to eight cutters were necessary. At that point there was renewed interest in a mechanical potato slicer, which had first been used in the 1920's. About 1954 a new improved slicer was brought on the market that was capable of keeping up with the bigger and faster planters.

Later the slicers were modified, and treating equipment was added, combining the treating with the slicing. This greatly reduced the disease problems that had intensified with larger potato production and faster rotation of planting on the land.

Mechanical harvesting also required mechanical potato boxes for the trucks. Sacks and platform truck boxes were discarded for a potato box equipped with a bottom which facilitated the unloading by means of an endless chain. The heavy work of lifting sacks of potatoes onto the truck was replaced by driving along the side of the harvester while the machine filled the box. Unloading required the starting of an electric motor and removal of the false floor by means of a crank, so that the potatoes could fall onto the endless chain. In the old way, four men could haul 500 bushels with horses and wagon or 1,000 bushels a day if they had a short drive with a truck. In 1973 it was possible for one man with a tandem truck to haul 3,500 to 4,000 bushels, if storage was located within a few miles from the field.

With bigger harvesters and larger trucks the problem of han-

---

*Some farmers, like the Kroeker brothers, used their four-wheel-drive, 150-H.P. tractor to pull their planter.

dling potatoes at the warehouse became a serious one. Burying potatoes under a straw pack in a trench had become obsolete by the early 1900's. In the 1890's Henry Schroeder built a large underground storage cellar with a sorting room in which sacked and graded potatoes could be elevated and loaded in railroad cars or wagons. By World War I these warehouses were in use throughout the Valley. Straw was used freely over the roof of these buildings to keep the potatoes from freezing, and with the use of small auxiliary stoves the temperature was maintained at about 34 degrees. In such storage, moisture released by the potatoes was the biggest problem.

It was not until the advent of electrical power, both for lights and for running the grading equipment, that these underground potato warehouses became really mechanized. Lack of fresh air made stoves, lanterns, and gasoline engines unsafe for use in underground storage. Hand operation of the grading machine continued until the 1940's, when electricity changed the situation. But even with the use of electric motors to operate the grading equipment and elevators, potato grading and sacking in underground storage remained primarily hand labor. It took a shoveler, a sorter, and two sackers 12 hours to grade, sack, and load 400 hundredweight of potatoes. If potatoes were of good quality, 50 sacks an hour could be graded and sacked, and about one-half that much time was needed to load them into a rail car. The expense of storage, grading, sacking, and loading could equal 50 percent of the total cost of raising the crop. Frequently an entire crop could be damaged or totally lost by improper storage or from unpreventable spoilage through disease.

A 300-bushel crop of potatoes was equal to 9 tons of a perishable commodity per acre that had to be stored and handled carefully. Winter crews to grade and ship potatoes might be as large in number as the crew that was needed to raise them. With four- and six-row planters, aerial sprayers, and mechanical harvesters in vogue, it became a necessity to mechanize the storage, grading, and shipping facilities for potatoes. Budd Tibert said, "Aboveground controlled ventilation storage, mechanical handling, and bulk shipment have been the recent [late 1960's and early 1970's] big boons to the potato business."

Machines capable of handling 3,300 pounds of potatoes a minute were put into use in 1973. As many as five trucks delivering potatoes from the field could now be unloaded at one time. These potatoes could be graded and loaded into a semitrailer truck fast enough to fill the truck in 12 minutes. Such equipment was necessary when as many as 10 harvesters were working in the fields at one time, requiring 28 to 40 trucks to haul the potatoes to storage.*

Ole Flaat was forced to build storage for 700,000 bushels of potatoes in the early 1940's to handle his 4,000-acre crop. Processing facilities were nearby, and his potatoes were delivered in bulk, greatly reducing shipping expenses. Unfortunately, potato warehouses are not always next to processing plants, and potatoes had to be moved a great distance. Farmers soon came to realize that above-ground potato storage with controlled ventilation was the best way for bulk handling, as well as for providing the kind of potato the processors needed.

Vern Hagen, the Halls, the Ryans, and the Kroeker brothers were among the leaders in large-scale, above-ground storage units using controlled ventilation so necessary to remove the moisture given off by the potatoes. Once again great savings were realized, after an initial capital investment in a million-bushel capacity warehouse. In 1953 the Halls adopted a water flume system to move potatoes from storage bins to the grading equipment. One man controlling the potatoes into the water flume could empty a bin of potatoes in two and one-half hours, while formerly it had required two men 20 hours each to shovel the potatoes with potato forks. Because the Halls specialized in certified seed potatoes and table stock potatoes bagged according to customer specification, they were interested in prolonging the marketing season. Refrigerated storage for nearly a million bushels was the next step in their progressive program.

A strong demand for burlap in the late 1960's forced the price of potato sacks to 30¢. This caused the farmers to reason that if potatoes could be shipped in bulk form the expense of the sacks and the heavy work of handling 100-pound sacks of potatoes could both be eliminated. Soon bulk shipping became the vogue. In 1964 the Ryan brothers built large shipping facilities, where three railroad cars could be loaded indoors at one time, even in the

---

*In 1973 the Offutt and Swenson crew shipped out 67 semitrailer loads of potatoes in a single 24-hour harvest period. These potatoes were loaded from the field trucks over the graders into the semis without going into storage. Very likely this represents a record for shipment for one farm in a single day.

coldest weather.* By 1968 they were shipping as many as 2,500 carloads of potatoes annually. The Ryans also acted as brokers and were doing considerable direct marketing with their own equipment at that time. Since the late 1950's Minnesota and North Dakota have generally produced 10 to 12 percent of the nation's total potato crop. Some of the region's largest growers produce enough potatoes to make a significant impact on the nation's annual supply of potatoes.

In 1967 the Hall Brothers shipped 700 carloads of table stock potatoes from their warehouses in Edinburg, Grafton, and Hoople, North Dakota. The warehouses were nearest to the potato fields because it was much cheaper to move men and machinery than potatoes. The Halls marketed only their own production; and in 1967, using average consumption figures, they provided a year's needs for 350,000 people from their farm. As early as 1960 the *American Vegetable Grower* journal credited the five Hall brothers as having "one of the largest growing and processing operations in the country." In addition to the production of table stock potatoes the Halls also plant 300 acres of certified seed potatoes each year to provide seed for their commercial crop the following year.

Until 1958 the Halls' total production was marketed in 100-pound sacks, but because of demand coming from chain stores they also sell potatoes in 5-, 10-, 25-, and 50-pound mesh or Foto-pak bags. The production line is fully mechanized, using a crew of 15 people to process and market the potatoes during the October through July season. The Halls have their own brand names— Hall's Super Pack and Five Star—but they also market under brands of their buyers. Labor requirements are obviously much greater when preparing sales for the retail market in specialized containers than they are when potatoes are sold in bulk to processors. All potatoes are mechanically washed, but still many laborers are needed to sort out the deformed or bruised potatoes as they pass over the grading table.

Arden Burbidge produces enough potatoes on his farm each

*Gerald and Thomas Ryan had taken over their father's 3,200-acre farm in 1950 and from the start were progressive in the production and marketing lines. The Ryan brothers' father, Gerald Ryan, Sr., had "grown up" in the potato business as a broker and went into the production end in the 1920's. He had been quite active in building a sizeable farm during the 1930's. His sons, Tom and Gerald, continued the pace set by their father, and in 1955 they bought out the estate of their father with a $175,000 loan, the largest from the St. Paul District office of the Federal Land Bank at that time.

year to meet the average annual consumption of 200,000 people. Each member of his permanent labor force produces the potato needs of 15,400 people. He also raises about 1,800 acres of certified seed grains as well as operating a commercial and seed-treating business. Burbidge is also a certified potato-seed producer, and the production from one year's seed grown on his farm would be about three-fourths of 1 percent of the nation's annual crop. By 1968 Burbidge had sold seed in 26 states. He plants thickly, which reduces yields but results in smaller potatoes that do not have to be cut for seed purposes. This aids in preventing disease. Burbidge believes that seed grown north of the forty-seventh parallel is more likely to be virus-free than potatoes grown farther south.

The Vern Hagen Farm at East Grand Forks, Minnesota, set a new company record for loading potatoes in the spring of 1974. A six-man loading crew using a four-wheel-drive mechanical loader, commonly called a Bob Cat, along with high-speed grading machines with belt conveyors loaded 1,056,000 pounds of potatoes in 11 hours. These six men loaded 18 semitrailer trucks carrying 42,000 pounds each and three railroad cars of 100,000 pounds each.* This grading and loading of potatoes at the rate of 16,000 pounds per man per hour is in sharp contrast to the loading of 833 pounds per man-hour in the days when potatoes were graded into 100-pound sacks and hand-loaded. The Hagen farm also owns three mechanical potato loaders powered by "bottled gas" so they can be used within the storage houses. They are capable of operating in very small spaces and can carry about 800 pounds of potatoes in their scoop bucket.**

Large-scale equipment and mechanical power enabled the well managed potato farmer to expand rapidly. Besides being able to handle more acres he also tripled or quadrupled the yield from the 1920's to the 1970's. Ironically the number of farmers producing potatoes has decreased sharply during the same period. In the

---

*The railroads have accommodated bulk handling of potatoes by building insulated hopper cars that can be loaded by conveyor from the top and unloaded by chutes at the bottom. Almost no human effort is required in either loading or unloading.

**These four-wheel-drive small-scale loaders have completely revolutionized the bulk movement of potatoes because of the speed with which they work. Damage done by the machinery is much less than was formerly inflicted by a man using the standard potato fork. The small loader tractor was the dream of Eddie Velo, a large-scale turkey farmer of Rothsay, Minnesota. The first model was made at the Keller Brothers blacksmith shop at Rothsay.

days of large families and an abundance of local labor many farmers raised a few acres; but with each step in mechanization the less efficient potato growers dropped out, accelerating the trend to larger, more economical units.

In the 1920's anyone with 100 acres was a big grower. By 1961, when the full impact of field mechanization was in force, there were 2,200 growers with an average of 90 acres of potatoes per farm. The largest grower had 1,005 acres that year. By 1969 there were 1,619 farms producing potatoes in the Valley, and the average acreage had climbed to 130. By 1969, 500 acres of potatoes for a single farm was common, and many farmers were growing at least 1,000 acres. The largest growers were surpassing 2,500 acres. In 1973 there were only 1,400 growers, but the average acreage had grown to 160. In 1974 the top grower had nearly 6,000 acres, with over 3,000 under irrigation.

All of the potato growers interviewed were asked what they felt was the smallest economic unit to be efficient and successful in a potato operation. Their answers give a candid insight into an industry that uses high-cost machinery. The cost of a standard four-row potato planter alone amounts to $5,500 and up. The cost for a potato harvester varies from $14,000 for the two-row units to $38,000 for the four-row units. The standard tandem axle truck with a self-unloading multi-purpose potato and grain box capable of carrying 28,000 pounds of potatoes from the field retails for $16,000 or more.

Since the simple trench storage in the ground has given way to controlled air-insulated storage, the cost of storing potatoes has become an increasingly important and expensive part of potato production. A typical storage unit in 1974 cost about $2 per hundredweight of storage. This does not include the equipment for grading and loading. A satisfactory warehouse for a 500-acre potato unit averaging 180 hundredweight of potatoes per acre cost from $125,000 to $150,000 complete. The total investment for a 500-acre potato unit exclusive of land cost would vary from $390,000 to $440,000, or about $830 per acre.

But the growers knew that with each expenditure in capital for mechanization they were rapidly decreasing their labor cost per unit of production. They also knew that the more mechanized they were, the more sure they were of getting the job done properly and on time.

Production costs per acre have increased sharply over the last years, but rising yields and greater volume per man have helped to keep the cost of potatoes relatively stable. The cost of seed is generally the greatest variant in the cost of production from year to year. Potato seed in a five-year period from 1970 to 1974 varied from $1.50 to $15.00 a hundredweight. In 1961 the cost of production was $105.15 an acre; in 1969 it was $158.08, and in 1974 projected costs had risen to $410 an acre for several of the Valley growers, who used cost accounting procedures. Seed cost in 1974 was as high as $240 an acre.*

Of the active Valley growers interviewed, their 1973 potato acreage varied from 600 to 4,800 acres. The total cropped acres operated by those growers at that time ranged from 2,500 to 13,000 acres. Ironically, Vern Hagen and Bill Hall, who were among the largest growers, had the smallest minimum goals under which a farmer could efficiently raise potatoes. Hagen stated that "if a man is satisfied to get by with all used machinery he could make a living growing 160 acres of potatoes in rotation with 480 acres of other crops. He would have to contract his potatoes because he would be too small to run any risk." Hagen added, "I could not live that type of life for there would not be enough challenge and certainly not enough profit for living like I do."

Bill Hall, speaking for J. G. Hall and Sons, said:

I think that 160 acres of certified potatoes and 160 acres of certified wheat would be the smallest sized potato operation for our area. I know some operators who aren't doing much more than this and are living well. I think a young man could get a start on a unit of this size. We think the optimum size potato-grain operation is 2,600 acres of potatoes with 5,200 acres of small grain to rotate with it. However, I do know of a very well managed operation of 3,300 acres of potatoes and 6,000 acres in other crops which is a real money maker. Of course, at that size you don't survive if the management isn't tops.

To which Ray and Keith Driscoll added:

We don't know how small or large a potato operation should be. Maybe a father-son team with some help would have the greatest efficiency but their standard of living could not be what ours is. We have pushed our potatoes to 1,200 acres and have not seen any reduction in profit per acre so we don't know how big we could get. We have leveled off because we have several other enterprises including 800 acres of beets and we want to live too. We believe the limit depends on how big a risk you want to

---

*The break-even cost based on 145 hundredweight production per acre in 1961 was 65¢ a hundred and in 1969 was $1.46 a hundredweight. By 1974 the break-even cost on a projected 180 hundredweight of saleable product per acre was $2.27 per hundredweight. If the farmers had not increased their average yields between 1934 and 1974, the cost of production in 1974 based on 1934 yields (2,700 pounds) would have been $15.18 per hundredweight instead of $2.27.

take. We have a four year rotation and look on our grain and beets as stabilizer plus it helps offset our labor cost because it balances our work load.

Earl Hvidsten had the following answer to the question:

I have been at 600 acres of potatoes for over ten years [since 1960]. I have grown more but I am where I want to be. We could grow more and be efficient, but we grow sugar beets and they are competitive with potatoes for labor. We can plant sixty acres a day with one planter and our two harvesters have a snap getting those 600 acres out in about 230 hours of operating time. In case of bad weather we have a lot of insurance with two harvesters because one harvester can do over 500 acres in a good year. I think 2,400 acres would be just as efficient as our unit. Our potato cutter is a good example. It could cut enough to keep four planters like mine going. With the quality of labor I have it would be no problem. We would just need more of them. As it is we plant and harvest for other farmers every year.

Walter Kroeker, who up to 1973 had the Red River Valley's largest potato acreage, expressed his feelings about economy in size as follows:

The smallest economical potato unit would have to be 500 acres. Economies improve rapidly with each acre up to that level but after that the benefits come more slowly. Less than 500 acres you are just not competitive. With processing potatoes it is easy to get to 2,000 acres, but after that management costs rise rapidly. Even at our level we have excellent profit chances so I think a grower could go to 6,000 acres of potatoes without a serious drop off in profits.

Kroeker, who farms in Manitoba, is forbidden by law to increase his potato marketings under control of the Manitoba Control Board. Although he intensely dislikes the law, which was passed in order to limit farm size, it is his opinion that the growers who are going to survive will have to expand to the 500-acre size. The Kroekers have developed a cattle-feeding program to enable them to feed surplus potatoes in a year when their production exceeds their marketing allotment.

Gerald Ryan expressed something similar when he stated that there was no basic physical limit to how large a potato unit could get. Ryan said, "I want to market at a profit and I want to be big enough to command the attention of the buyers."

Large-scale industrialized potato growing also brought about definite changes in the marketing pattern. In the pre-1950 era, when there were many small growers, potato brokers traveled from one farm to the next and offered what they felt was a fair price in order to get potatoes. Most brokers worked on commissions, but some actually bought the potatoes for resale. Because potatoes are so perishable any slight increase in production beyond the immediate demand caused a sharp drop in price. Potato growers had experienced the doubling of the price or a similar reduction in one month's time. Far too many had also experienced having to haul their crop back to the fields in a manure spreader in the spring because there was no market and the potatoes were starting to decay.

With larger production of potatoes per farm the more innovative farmers began to spend a greater portion of their time marketing and less time producing because they anticipated the marketing problem. Frank Kiene was a master marketer and because of that did well in potatoes, expanding his potato production to 1,200 acres by the early 1950's. In 1956 Ell Kiene decided to discontinue potato growing because he disliked the constant problem of marketing. Kiene, like most of the more progressive growers, realized that marketing potatoes was far more difficult than raising them. The larger growers, unlike many of those with fewer acres, were unwilling to accept the offer of the first broker that stopped at the farm.*

Henry Schroeder, the Red River Valley's first sizeable potato grower, was also one of its earliest concerned marketers. There are rumors that he had connections with certain people on the railroad, so he was able to learn by signal from the engineer of the passing trains what the prices and potato movement were along the line. This supposedly helped Schroeder to determine whether he should market or not. Joe Thompson was the oldest active buyer among those interviewed. Thompson related that he had started buying in 1918, the first year he farmed. He said, "I wanted to learn more about the potato business and I was not that motivated by the routine problems of farming as long as I could get men to do the job. Besides, how could a person keep his mind occupied on 320 acres?"

Thompson bought potatoes in the Red River Valley and throughout the territory from Moorhead to Minneapolis. He attributes stability in his income to his buying activities; but he felt that marketing, even in the early days, was the potato grower's greatest problem. But being a buyer gave him an advantage with

---

*Jerry Griffiths, a large-scale grower in several western states, expressed the opinion that marketing and delivery were his greatest problems by far. He felt that the basic problems of managing his 12,000 acres of potatoes were quite routine by contrast to the job of the market manager.

his own potato production. He was quick to add: "I think I would have been more of a plunger and a lot bigger farmer if it had not been for my buying activities. But marketing would still have taken most of my time."

Thompson was a good buyer, and in 1922 he was responsible for most of the 922 carloads of potatoes that were shipped out of Nash, North Dakota, a village of 38 residents. Nash was acclaimed to be one of the largest shipping centers in North Dakota of that date. As a buyer for Leonard-Crosset and Riley, Inc., of Cincinnati, Ohio, in 1926, Thompson was paid $500 a month and given a new car every year. Such a salary was considerable for those days, according to Thompson, but he felt his exposure as a buyer to growers from other regions and the ability to market his own potatoes were even more important. The chance to "pick up some good buys in land" was also a significant side benefit of his buying activities.

Roy Douglas is credited with having established the first marketing and packaging plant for potatoes in the Grand Forks area. But Gerald Ryan, Sr., however, who settled in Grand Forks in 1922, started a potato business that became one of the most innovative and largest in the entire Red River Valley. Together with his father, Dennis Ryan, Gerald Ryan, Sr., had been buying potatoes in the Valley almost from the first year of commercial production. His sons, Gerald, Jr., and Thomas, went to direct marketing of certified seed in 1953. By 1962 they had become the largest seed growers in North Dakota and were the largest seed handlers in Minnesota and North Dakota. Selling under six brand names, the Ryans shipped a million pounds each week. Their combined warehouses held the production from about 3,000 acres. A structure built in 1962 held over one-half million bushels and was the largest single warehouse in the Red River Valley at that time. The Ryans, who operated under several firm names, are brokers, shippers, warehousemen, washers, and packers of not only their own potatoes but those of other area growers too. The Ryan brothers freely admit that if it were not for their marketing business their farming operations would be considerably larger, even though for three decades it has been one of the most extensive in the Valley.

Henry Peterson learned the technique of marketing from his father, who had one of the largest retail truck farms in the Valley. Peterson said with emphasis:

Marketing was the greatest single factor in my success. Through good farming practices I was able to produce a superior quality product and I soon started to guarantee my product. In my last fifteen years of farming I sold all of my potatoes to four chains. When I planted in the spring I knew they were sold in the fall because those quality chains wanted my potatoes and always paid better than average market prices to get them. I have always been amazed at the indifference of the average farmer toward marketing.*

After 30 years of farming, Ralph Hvidsten had reached a potato production at 800 to 1,100 acres. He gave the following reason for his success:

At this level we have ease of marketing. We have a big enough volume to be noticed and with quality potatoes plus a reputation for delivery our customers have come to rely on us. Maybe maximum production efficiency comes at 300 to 500 acres with a one man outfit, but the economy of large warehousing and better marketing more than offsets any efficiency losses in the field. Besides, now we are big enough to have a steady labor force to do an economical job of processing. We have contracted most of our potatoes since the late 1950's; it has done a great deal to stabilize my market. The marketing facilities are greatly over expanded because there are still far more buyers in the Valley than necessary and they all want to make a living.

John Bogestad, who started farming in 1934, quickly learned that marketing of his potato crop was every bit as important as production. He acquired some customers in 1937 who have remained with him. Bogestad's production grows with his customers' demands and he consults them before planting, to be sure he can fulfill their needs. Bogestad's experience with certified seed production is similar to that of the Hall brothers, who market only table stock potatoes. Bogestad has found that even though he does not contract he has been able to sell his potatoes in good years and poor years. The Halls found that chain stores would like to buy direct from them; but there is a difference, because some chains want quality, while others are price cutters. The Halls did not care to deal with "shoppers" so they decided to work through three established brokers.

The Halls have used the same brokers since 1949 and feel that they would have to hire three full-time men to do their own marketing and would thereby sacrifice any savings. Because the Halls are large enough economically to have their own processing and

---

*The Petersons were also strong believers in both manure and commercial fertilizer because it not only increased production but it improved the quality of their product. By 1929 the Peterson farm was using at least a full carload of commercial fertilizer per year besides great quantities of manure that were hauled in from the Fargo-Moorhead area.

packaging facilities they can prepare specialty products requested by the brokers. This enables them to keep their labor force employed for 12 months.

Bill Hall is pleased with their system. He said:

In 1972 and 1973 we were able to market our entire 1972 crop, some at a pretty good price when a large percentage of total Valley production was dumped at a great loss to the growers. It's real proof of the importance of good market outlets and the need to be big enough to command attention. Even though we have maintained our same customers over several decades we feel marketing is the potato farmer's biggest problem. Our second problem is keeping the potatoes in proper condition. Producing them is easy by comparison.

Each spring prior to planting, Vern Hagen contracts his production to seven processors. Hagen said, "I project what I think my minimum production will be and cover myself. This gives me stability and then I know what the problems will be from there. Delivery is important so we have our own refrigerated truck fleet [45 semitrailer units in September, 1974]. This gives us protection to deliver as contracted."

As a 16-year-old boy Hagen started marketing his own product by peddling potatoes with an old truck. He said of that experience:

I quickly learned that the average farmer was not a marketer. That's why it was so difficult to get a potato grower's association going. My brother Cliff found that out. By peddling potatoes I learned I could buy them from farmers who didn't concern themselves with marketing for $.50 a hundred and sell them in Duluth for $2.50. The biggest thing I learned was to evaluate people. It wasn't long and I knew what it was like to really be cheated—I was stung good a few times.

Once I couldn't sell a load so I convinced a used car dealer to give away five sacks with each used car. He bought the load. I also learned that in making the 530 mile round trip to Duluth in 24 hours and bringing a load of fruit back I could make as much in one day as I did in my entire first year raising 20 acres of potatoes. That's the second lesson I learned in marketing.

Arden Burbidge, the Kroekers, the Paul Horns, the Offutts, and the Ryans all spend a great deal of time doing their own marketing or hiring their own full-time marketing specialists. Burbidge said, "We have developed a steady market for our potatoes but our marketing manager travels over almost the entire nation keeping those contacts." The Paul Horns have a separate marketing agency, and if another buyer can pay more for potatoes from Paul Horn Farms, Inc., than Horn Sales, Inc., can they are sold to that buyer. Horn Sales, Inc., has a full-time professional marketer.

The potato farm of Ronald Offutt, Sr., Ronald Offutt, Jr., and John W. Swenson at Perham, Minnesota, had the largest acreage in both total acres planted to potatoes and acres under irrigation on the Minnesota side of the Red River Valley in 1974. Production from the combined Offutt enterprises was equal to 6 percent of the total production of the State of Minnesota in 1973. Its 1974 projection was for 11 percent of the state's total production. Much of the strength of the Offutts and Swenson and the other sizeable operations comes from their ability to buy their total production requirements and to market their produce at an advantage over the smaller units. A new Offutt and Swenson warehouse covers an area equal to two football fields, holds 500,000 hundredweight of potatoes, and makes this a source of supply that attracts the largest of the processors. Employment of specialists in seeds, soils, disease, finance, and marketing on either a full-time or a consultant basis enable the partners to have management equal to or superior to even the best of smaller enterprises.*[4]

French Fries Tomorrow

Some of the major potato growers are afraid that the increase in the processing and storing of processed potatoes will influence the normal fluctuations of the price level in favor of the processor to the disadvantage of the growers. The consumption of processed potatoes, such as potato chips, granules, canned potatoes, hashbrown potatoes, French fries, shoestring potatoes, and other products, has risen sharply since the early 1950's, when the average per-capita consumption of all potatoes hit a low point. But soon processing opened up a new market for snack foods and convenience potatoes which appealed to the working housewife. In 1950 the consumption of processed potatoes was 6⅓ pounds per capita; it rose to 36 pounds by 1965 and 66 pounds in 1973. Well over 50 percent of the potato market was in processed form at that time.

Earl Hvidsten was a late comer in contracting his potatoes, but because he anticipated a shift from a market for table-stock potatoes to one for processing potatoes he started contracting with the processors. Hvidsten felt that the processing industry was a

---

*In 1973 the Barrel O' Fun potato processing plant was established at Perham, Minnesota. This plant was started on the basis of securing its raw product from the potato enterprises of the Offutts and Swenson. In 1976 John W. Swenson purchased full ownership of the Perham Potato Farms. The Offutts then purchased one of the largest irrigated farms in the State of Texas.

big investment and to protect itself would have to go to contracting its annual requirements. This he felt would stabilize the industry, but in the long run could tend to make it less financially attractive to the growers because producing for the process market was a relatively simple business by contrast to providing table-stock potatoes.

From the early packaging plant of Roy Douglas in the 1920's the Valley potato industry has changed considerably in both marketing and processing. The Early Ohio potato had long been a mainstay in the Valley potato production. Unfortunately, there were always growers who cared little about quality production, and this hurt the market image of the industry. H. D. Long, who had been with the North Dakota State Seed Department since 1910, joined the Walsh County Agricultural School in 1928. In 1934, about the same time he decided to go farming, Long developed the Red Pontiac potato. The Pontiac, an excellent table potato, was called the Dakota Chief and provided a quality product for the national market. It replaced the Early Ohio, and total commercial sales of Valley potatoes more than quadrupled in the next decade.

In 1948 the Red River Valley Potato Growers Association was formed in an effort to improve the national image of Valley potatoes. Quality of product and identification with the region were necessary, but they were never really successfully achieved. One can only surmise that it was the independent nature of the growers in the Valley that prevented any unity of purpose among them. Even though the growers had an association, in 1973 a bargaining group under the leadership of Art Greenberg of Grand Forks made a concerted effort to regulate marketing and production. This move came about because the processing industry had become concentrated enough in the Valley to attain a powerful influence in regional and national marketing.

About 1950, potato processing came to the Valley with the organization of King of Spuds, Inc., at East Grand Forks. It produced instant mashed potato granules. In 1958 a flaking plant was established in Grand Forks and a starch plant at Grafton. About the same time, a flaking plant was opened at Barnesville, Minnesota, but because of its limited capacity, under-capitalization, and failure to associate with a major marketing agency, it failed within a few years.

The second significant attempt to start the potato processing industry in the Valley came in 1960. Vern Hagen, a dynamic and

progressive potato farmer who had accomplished his original goal
by age 29, decided to free himself from the tensions of the potato
business. He reflected:

Maybe after traveling around the country a while I thought it would be
fun just to do some grain farming and raise cattle. At that point I had
about $500,000 plus 2,500 acres of land which had very little mortgage on
it. But on that extended trip I saw the western growers bulk load potatoes
into every kind of rail car. Then I saw them processing French fries in
Idaho which excited me, but after I saw California Kennebecs being
processed as Idaho Russetts I realized that the Red River Valley had the
biggest potential of any potato area in the nation. I couldn't wait to get
back to the Valley and start raising spuds again.

Hagen recalled that prior to his travels, Jack Croonquist, the
Scotts, the Ryans, and a few others had tried to start a French fry
plant in Grand Forks, but they were unable to raise the funds
through sale of stock. "That bothered me, but I knew we could do
it," said Hagen, in his persuasive way. "Our problem was to raise
money. Finally we landed the old Great Northern roundhouse at
Crookston and we had a plant site."

Whether the farmers were really sold on the future of potato
processing or whether Hagen and Crookston banker Marvin
Campbell were good fund raisers is difficult to determine, but
$250,000 was secured in the short time between Friday and the
following Monday. In retrospect Hagen said, "I know now we
should have raised a lot more before we started because shortage
of capital was really our biggest problem."

In 1960 only about 1 million out of the 24 million pounds of
potatoes raised in the Red River Valley were processed there.
There was reason for optimism, however, because from 1951 to
1960 the national per-capita consumption of frozen French fries
had increased nine-fold. In 1961 there were 12 processing plants
in the Valley, and only two of them produced French fries.

The new industry in Crookston was named Jiffy Fry, and Vern
Hagen, one-third owner, was almost solely responsible for the
management. The local newspaper labeled Jiffy Fry the "brain
child" of Vern Hagen. Initial projections were that the plant
would require 8,000 to 10,000 pounds of raw potatoes per hour
and 4,000 acres of potatoes would have to be grown locally to supply the plant. Another 2,500 acres would be grown in other areas
of the nation so that the plant could operate the year around.*

---

*Hagen used his own crews and equipment to raise a large volume of potatoes in
Arizona for the Crookston plant. That venture did not prove profitable.

Many technical problems were encountered, and it was three years before Jiffy Fry was in full production. Vern Hagen said of his experience:

I personally put my signature on the line for nearly everything the company bought. At times that was well over a million dollars. I worried more about the other stockholders losing their money than anything else. I even went a couple years without salary and still I heard petty bickering. We eventually got our 4,000 pound frier to produce 7,000 pounds a day and had a couple good years before we sold out to Jack Simplot. I discovered farming was fun compared to the bureaucracy of industry.

By 1969 Jiffy Fry was processing 2,500,000 pounds of raw potatoes to fill five carloads of hashbrowns or French fries every 24 hours. By then 25 percent of the Valley crop was being used for potato chips, which presented another challenge because they required expert handling in storage to prepare the raw potato for the best quality of chips. French fries and hashbrowns were less delicate in that respect.

As area potato-processing facilities expanded, more and more large-scale growers sought to improve efficiency through mechanization and steered away from producing table-stock potatoes, which still required a considerable amount of hand labor. Some, like the Flaats, the Kroekers, the Ryans, and Thompson, had sought and obtained foreign sales for their potatoes. At the time of interview the Kroekers were sending ten 40,000-pound containerized units of potatoes to the European market. The big market still appeared to be, however, in processed potatoes as convenience food and in French fries, for which there is a growing demand at the nation's hamburger stands.

One of the great uncertainties in the future development of the potato production industry is the possibility of governmental interference in an attempt to regulate prices. None of the producers interviewed in the American part of the Valley would welcome such government intervention. Most of the growers remember the years of controlled potato production, and their memories are not fond ones. The Hall brothers have monthly average prices of potatoes dating from many years before government restrictions to the present. Bill Hall said:

Our records prove to us that the overall price without government restrictions is considerably higher than during the days of the programs. In a free market we produce for the demand and in the long run it has been less risky on the free market; besides, it has been more profitable. Under the government programs the farmers produced for the government, meaning, in our way of thinking, a perpetual surplus causing the lower overall price. I think the government almost ruined the potato business; besides, we were constantly threatened by more interference. I am worried whenever we hear talk about a market based on the cost of production instead of a free market where the real profit prospects are.

Most of the potato growers interviewed were market oriented people and rather enjoyed the risks of the free market, which over the years had been exciting and rewarding for those with efficient quality production.

According to Donald Kroeker, the government of Manitoba has already taken action to prevent continued overproduction of table stock and seed potatoes. That market is tightly controlled by the Manitoba Potato Board, but the market for processing potatoes was still free in 1973. The Kroekers—and, it seems, most larger farmers in Manitoba—intensely dislike the control board even though Walter Kroeker is chairman of the pricing committee. Kroeker stated that the Manitoba potato industry would have collapsed if

TABLE 1. Size of Operation of Some Red River Valley Potato Growers in 1973[1]

Total Acres in Farm	Acres of Potatoes	Total Tractor Horsepower	Full-time Employees[2]	Part-time Employees[3]	Other Major Enterprises
5,200	1,000	—	—	32	—
7,200	1,200[4]	1,230	17	50	yes
1,400	450[4]	—	3	15	yes
6,000	2,600	1,540	15	145	—
4,000	800	1,318	7	30	yes
9,500	3,300[4]	2,340	40	100	yes
2,500	900	625	6	26	—
3,800	800	1,250	7	25	—
6,000	1,400	1,345	42	70	yes
6,400	1,200	1,780	11	40	yes
3,100	1,050	—	9	49	yes
3,000	600	610	3	23	yes
3,500	1,400	—	5	40	yes
3,500	600	695	3	20	yes
3,000	800[4]	—	7	20	yes
10,200	4,800[4]	2,130	13	126	yes
23,000	3,500[4]	2,560	40	100	yes
Av. 5,959	1,553	1,452	14	54	

[1]Data from personal interviews.
[2]Owners included under full-time employees.
[3]Does not include full-time employees. In some cases the potato house crew, which works nine months a year, is included in full-time labor column. That accounts for the variation in the number of full-time employees. Much of this is because of a difference in the form in which the potatoes are marketed.
[4]Includes irrigated potatoes.

marketing quotas had not been established. However, since 1964 the industry has been static except for the processing market. Walter Kroeker added, "The larger operations with top market management capacity have the decided advantage in what is left of the free market."*

A leading potato industry magazine survey indicated that 46.2 percent of its subscribers sold over $100,000 worth of produce annually. Over 23 percent of them marketed over $250,000 yearly. All of the active growers in this study easily surpassed that dollar volume. As long as those factors hold true the trend toward larger-sized potato farms in the Red River Valley will continue in the future.[5]

---

*The law in Manitoba restricts any farm selling a volume beyond the sales of the largest producer in the province at the time the law was passed. Kroekers were the province's largest producer of onions and potatoes at that date, so are unable to expand their volume for the commercial market.

NOTES

[1]Stanley N. Murray, *The Valley Comes of Age* (Fargo, 1967), pp. 36, 178, 180, 187, 204-205, 209; Hiram M. Drache, *The Challenge of the Prairie* (Fargo, 1970), pp. 123, 177, 317-318; A. L. Monroe and E. V. Swartz, editors, *A 1970 History of Red River Valley Potato Industry*, pp. 26, 29, 49; Interviews of Henry R. Peterson, Moorhead, Minnesota, December 28, 1969, April 4, 1974. Peterson was an active participant in the development of potato, grain, vegetable, and flower seed varieties for several decades. He worked closely with both the Minnesota and the North Dakota experiment stations and grower associations; *Walsh County Record*, July 25, 1940; Interview of Joe Thompson, Grafton, North Dakota, March 13, 1973; *Minnesota Agricultural Statistics, 1959*, State-Federal Crop and Livestock Reporting Service, p. 70; *North Dakota Agricultural Statistics*, R. F. Engelking, C. J. Heltemes, and Fred R. Taylor, North Dakota Agricultural Experiment Station Bulletin No. 408 (Fargo, 1957), p. 30; Interview of A. N. Midgarden, Park River, North Dakota, March 13, 1973. A handwritten manuscript, "Early Recollections of Potato Growing in Walsh County," by A. N. Midgarden written March 1973. Mr. Midgarden was born in Walsh County in 1891 and came from a family that was active in the potato industry from its beginnings.

[2]Interviews of Ole A. Flaat, Grand Forks, North Dakota, February 15, 1972; January 18, 1972; Laurel D. Loftsgard and Melvin G. Mailer, *Potato Production Costs and Practices in the Red River Valley*, Agricultural Experiment Station Bulletin No. 451 (Fargo, 1964), p. 26; *Land of Plenty*, pp. 58-59; Joe Thompson interview; Henry Peterson interviews; Deere and Webber Company, John Deere Implement price list, February 10, 1916, property of Grant Mattson, Casselton, North Dakota; *History of Red River Valley Potato Industry*, pp. 29, 55; A. N. Midgarden interview; Interview of Bill Hall, Hoople, North Dakota, February 14, 1973. Mr. Hall was the spokesman for J. G. Hall and Sons, which represents the five Hall brothers, one of the leading potato-growing firms in the Valley for several decades; Interview of Arden Burbidge, Park River, North Dakota, January 3, 1970; *Minnesota Agricultural Statistics, 1959 and 1972*; *North Dakota Agricultural Statis-*

tics, Bulletins 408, 408 Revised, and No. 26, 1972; John Deere price list for November 1, 1950, property of R. D. Offutt, Inc., Casselton, North Dakota; Interview of John W. Scott, Sr., Gilby, North Dakota, January 3, 1970, and February 1973; *Fargo Forum*, February 2, 1973; Planting Calculations by Arden Burbidge, Park River, North Dakota; Interview of Vernon Hagen, East Grand Forks, Minnesota, March 29 and 30, 1974; Henry Peterson interview; Telephone conversation with Ralph Mathew, Barnesville, Minnesota, April 22, 1974. Mathew is an aerial sprayer providing service to grain, potato, and beet growers. Mathew has sprayed 500 acres in a "good" day.

³Ole Flaat interview; Henry Peterson interview; Interview of J. Budd Tibert, Voss, North Dakota, March 14, 1973; Joe Thompson interview; Vernon Hagen interview; *History of Red River Valley Potatoes*, pp. 26, 55; Telephone conversation with Ole A. Flaat, April 18 and 19, 1974; Interview of Walter and Don Kroeker of A. A. Kroeker and Sons, Ltd., Winkler, Manitoba, Canada, February 15, 1973; Bill Hall interview; "Potatoes Are Big Business in North Dakota," *Market Growers Journal* (January, 1954, p. 29; *Crops and Soils*, VI, No. 2 (November, 1953); "Harvest 10,000 Bushels a Day," *The Dakota Farmer*, LXXIII, No. 21 (November 7, 1953), p. 20; "Hall Family's Potato Dynasty Is Traced," *The Cavalier Chronicle* (October 24, 1963); *Walsh County Record*, November 1953; James Hammill, "Potatoes from the Valley," *Monthly Review*, publication of the Federal Reserve Bank (July, 1963), p. 3; Interview of Ronald D. Offutt, Jr., Glyndon, Minnesota, January 15, 1974.

⁴Kroeker interview; Flaat interview; Telephone conversation with Ralph Mathew, Barnesville, Minnesota, April 22, 1974; Tibert interview; Hagen interview; Hall interview; *Market Growers Journal*, p. 29; Offutt interview; *Walsh County Record*, October 20, 1969; Interview of Gerald C. Ryan, Ryan Potato Company, Inc., East Grand Forks, Minnesota, February 14, 1973; "The Ryan Brothers Build for the Future," *Produce Marketing*, VII, No. 3 (March, 1964), pp. 17-18; "Spud King Could Feed Chicago," *The Grand Forks Herald*, January 25, 1968; *The Dakota Farmer*, LXXIII, No. 23 (December 6, 1958); "Five Brothers Continue Tradition," *Produce Marketing*, VI, No. 12 (December, 1963), pp. 19-20; *American Vegetable Grower*, VIII, No. 8 (August, 1960), pp. 10-11; Interview of Arden Burbidge, Park River, North Dakota, March 14, 1973; *Monthly Review*, pp. 9, 12; *History of Red River Valley Potatoes*, pp. 15, 18, 26, 49; Interview of Keith and Ray Driscoll, East Grand Forks, Minnesota, January 17, 1972; Interview of Earl Hvidsten, Stephen, Minnesota, January 17, 1972; Interview of Charles Lysfjord, Kennedy, Minnesota, March 14, 1973. Lysfjord is the operations manager for the Kiene Farms; Interview of Jerry Griffiths, western potato grower, at Glyndon, Minnesota, February 7, 1973; Thompson interview; "Town Got into Its Stride in 1922 with Joseph Thompson," *Walsh County Record*, July 25, 1940; Henry Peterson interview; Interview of Ralph Hvidsten, Stephen, Minnesota, January 1, 1972; Interview of John H. Bogestad, Karlstad, Minnesota, March 22, 1972; Interview of Paul Horn, Jr., Moorhead, Minnesota, December 23, 1969; "Profile of a Complete Operation," *Valley Potato Grower*, February, 1974, pp. 10-13; *Grand Forks Herald*, February 11, 1962; Dorothy Holden, "Potatoes Any Way You Like Them," *Dakota Farmer*, LXXXIX, No. 7 (April 5, 1969); "Grower Profile: Ryan Brothers, *The Grower-Packer*, February 15, 1969.

⁵*Monthly Review*, p. 9; *History of Red River Valley Potatoes*, pp. 15, 39; Earl Hvidsten interview; Don and Walter Kroeker interview; *North Dakota Agricultural Statistics*, 1957, p. 29; U.S.D.A. Agricultural Statistics; Hagen interview; "Frozen French Fried Potato Company to Employ 225 Here," *Crookston Daily Times*, October 5, 1960; Offutt interview; *Dakota Farmer*, LXXXIX, No. 7, April 5, 1969; *The Packer-Potato Grower Survey*; Letter from Stan Erickson, East Grand Forks, Minnesota, June 6, 1974. Mr. Erickson is an employee of the Red River Valley Potato Growers Association; 1974 Proceedings of the National Potato Council.

CHAPTER X

Sugar Beets—A Bonus Crop in the Red River Valley

Background

THE RED RIVER Valley in Manitoba, Minnesota, and North Dakota, is North America's greatest concentration of sugar beet production. The treeless, stoneless, fertile prairie with its large fields provides an ideal setting for beet growing with one of the lowest production costs per ton. In 1974 seven sugar beet processing plants existed in the Valley, of which two were used for the first time that year.

The first sugar beets were grown commercially in the area in 1919, but it was not until 1926 that a plant was built at East Grand Forks, Minnesota. Initially farmers were not easily convinced that it was profitable to grow a crop that required so much hand labor. Therefore, it was not until mechanization was introduced to beet production that growers felt safe to contract with the sugar processors for the sale of beets. Since that time beets have been a bonus crop in the Valley—especially in the areas of extremely heavy soil, where potato production is affected by dirt lumps.

Sugar beet allotments have become a cherished commodity for many farmers even though beets are still referred to by some as the "ulcer crop." That term arose because of the intense management necessary for high quality and because of the close beet company supervision of production. This was contrary to the ideals of many a farmer who cherished being his own boss.

Iowa, Minnesota, the Dakotas, Manitoba, and Montana have other beet growing regions, but none are quite the same as those in the Red River Valley. Beets have come and gone in other areas because often they have been unable to compete with other crops, especially corn, or because the processing plants were not

358

economically profitable for their owners. The Valley's beet production was chosen as the chief source of this study because of the presence of some of the pioneer growers who have experienced every step in the gradual mechanization of sugar beet production.

Sugar beet production experienced its first mechanization in the late 1920's and early 1930's. Progress was slow, for it was not until the 1940's that the costly labor problems involved in harvesting were reduced by mechanization. Almost yearly since World War II additional inventions have made beet raising a part of industrialized agriculture.

A Crookston, Minnesota, farmer, Carl Wigand, had a beet plot in his garden in 1918 with seed secured from a source in Michigan. At harvest the beets were sent to Chaska, Minnesota, for evaluation. In 1919 the Minnesota Sugar Company supplied Wigand with enough seed to grow five acres. These beets were also shipped to Chaska. Results that year were excellent, and the company became interested in beet production in the Valley. Eleven acres of beets were grown in the Valley in 1919, which yielded 6½ tons per acre. By 1973 there were 3.5 million tons of beets produced on 231,000 acres, and two plants requiring another 100,000 acres were being built. In 1920 Wigand contracted 10 farmers in the Fisher area to produce 100 acres of beets. Elsewhere about two dozen farmers were contracted, and total Valley production increased to 293 acres, yielding 1,880 tons. At this time Ferdinand H. Ross of Fisher was the largest grower, with 30 acres.

Producing beets was a real struggle, and some of the growers quit after the first year. New growers signed up, but only 273 acres were used for beet growing in 1921. Fortunately, the yield increased to 7.74 tons per acre, and sugar content was 17.4 percent, which rewarded the growers well. R. T. Adams and L. D. Wagner were two of the successful new producers that year. Those two men with the help of Ferdinand Ross and Wigand signed more new growers each year, for farmers were desperately looking for better sources of income. Higher yields and good sugar content encouraged the farmers to put up with the hardships because net profits on beets were greater than from any other crop except potatoes. In addition, beets had one advantage over potatoes that appealed to many—an assured market at an established price.

The turning point for the Valley beet industry came in 1923, when more farmers became interested in beet growing and signed contracts for producing them in 1924. At this time the bulk of the Valley beet crop was found in the area around Fisher and East

Grand Forks. Farmers had to be located close to railroad facilities because of the expense involved in transporting 7 tons or more of produce from each acre. Ferdinand Ross, who became a field man for the American Beet Sugar Company (later the American Crystal Sugar Company), was particularly responsible in encouraging 24 growers in the Fisher area to plant 503 acres to beets.

In 1924 Ross convinced other prominent farmers in the region to grow beets. Among them were John W. Scott, Sr., of Gilby, North Dakota, 20 acres; James Driscoll of East Grand Forks, 20 acres; and Charles L. Ryan, 35 acres. The Elk Valley farm, famous from bonanza days and managed by the Larimores of Larimore, North Dakota, contracted 45 acres. Ferdinand Ross and his 15-year-old son, Walter, of Fisher, held combined contracts for 80 acres, making them the largest single beet growing unit in the Valley.*

The encouraging results spurred the Minnesota Sugar Company in 1924 to search for a site to construct a sugar beet processing plant. Fisher was the first choice, but an active Chamber of Commerce at East Grand Forks convinced company officials that they had a better location. Production did not change much in 1925, but with the certainty that a plant was being built and beets would no longer have to be hauled at great cost to Chaska or Mason City, 10,056 acres were contracted in the Valley for production in 1926.

From that year until the 1940's the sugar company had a constant struggle soliciting farmers to grow enough beets to make the plant pay. The company needed at least 200,000 tons of beets to make the operation of the plant profitable, but by 1930 production was only 113,041 tons, just 37,000 tons more than had been processed in 1926. To meet the plant's needs the company was forced to rent land and go into sugar beet production itself in 1931, 1932, and 1933. Although the plant was still not operating at full capacity in 1933, the company decided to discontinue raising beets because of the difficulties in handling the large work crews needed over a scattered area.

In an effort to get the plant working at full capacity an agree-

*In 1924, 15 acres was the average contract, and 80 acres the largest. In 1965 Robert Yaggie and L. D. Loftsgard determined the average contract to be 86 acres. The largest was over 350 acres. Their study proved a grower should have at least 90 acres for a profitable beet unit. By 1974 the Yaggie family held the largest combined beet allotment, totalling 2,500 acres, and the average of all contracts was 150 acres. An indication of the importance of mechanization in beet production can be derived from the preceding facts.

ment was made with the governments of Manitoba and the United States to permit beets grown in Manitoba to be shipped to East Grand Forks duty-free and sugar returned duty-free.* In 1931 26 Canadian growers, including the farm operations of two jails and a monastery, produced 2,540 tons on 369 acres. This exchange was discontinued after the 1933 campaign because of protests from American farmers. By this time American and Canadian farmers came to like raising sugar beets in spite of the hardships. The $4.90 a ton price on a 7-ton crop made them profitable and palatable. From 1934 on, many efforts were made to build a plant in Canada, but it was not until 1940 that the Manitoba Sugar Company finally succeeded in building one near Winnipeg in time for that year's crop. The Manitoba growers suffered severe losses for several seasons during the 1940's, and by the 1960's sugar beet production was a stable, attractive enterprise in Canada too.

In the early days the sugar beet field man in both countries had to work hard to get farmers to take on beet contracts. In 1924 there were 25 loading stations established throughout the Valley in Minnesota and North Dakota. Most of these were found in an area about 60 miles by 60 miles centered around East Grand Forks, Minnesota. That area contained 265 growers, who raised 4,004 acres of beets, or an average of just over 15 acres per farmer. The largest single grower was James T. Sullivan of Sullivan, Minnesota, with 151 acres. Except for the Rosses there was no other family that had more than 50 acres under contract.

By 1928 there were about 70 freight-loading stations and 390 growers holding 10,353 contracted acres for an average of 26 acres per farmer. From that date on, the number of growers and locations of beet fields in the Valley increased gradually each year. But it was not until the 1940's that there was any noticeable increase in the number of acres per grower.[1]

Primitive Methods of Beet Growing

The standard comment among several of the pioneer beet growers was that everything in the first years was done by horse, hoe, and hand. They did not hesitate to add that much time was spent on hands and knees, crawling through what must have appeared to be endless rows of beets.

---

*This is referred to as transit in bond and did in no way affect the American beet-growing industry. But it did help the sugar company to get a better return on its investment.

Standard equipment for a beet grower in the 1920's was a two-horse four-row drill that could seed an average of 15 acres in a nine-hour day. Merle Allen remembered that on one cool day he was able to put in extra hours with the horse-drawn cultivator, enabling him to complete 17 acres. Allen said, "Conditions were ideal and I had good horses, but I never came close to that figure before or after that." There was no hard work involved in seeding, but the seeding was done at a much heavier rate (20 to 25 pounds per acre) than necessary because of the bulkiness of the seed and the inability to determine the rate of germination. At planting, 50 pounds of 0-20-0 fertilizer was spread on each acre. Walter Ross commented that in the 1920's fertilizer came in 125-pound bags, which represented the most strenuous task in seeding.

The next step in production was cultivating with a two-horse four-row cultivator that was capable of doing about $1^7/_{10}$ acres per hour. Several growers pointed out that when the weather was warm, horses were restricted to about seven hours of field work a day. Horses were sometimes changed during the day for more efficient use of the cultivator; but generally this was unnecessary because the cultivator could work the total acreage of most growers in one or two days. Some of the early growers had a one-horse two-row cultivator, but it did not prove to be satisfactory because it was so hard to steer.

After cultivation came blocking and thinning the beets and hoeing for weeds. A worker went through the field with a short-handled hoe and chopped everything out of the row, leaving spaces between blocks of young beets. The purpose here was to give the remaining beets a chance to grow to a larger size. Because of the use of multi-germ seed as many as 20 to 30 beets were left growing along with the weeds in each of the remaining blocks.

After forming the blocks the worker, on hands and knees and equipped with the short-handled hoe or a knife, thinned each of the blocks, leaving only one beet. Workers wrapped sacks around their knees or had pads sewn into their trousers to protect the knees. Aching knees was a common complaint of everyone engaged in thinning or blocking. Ideal thinning started when beets had six leaves and were about 3 inches tall. Once he spotted "the lone beet" in a block, the hoer would cut out all others around it. Because the young beets were so fragile all weeds had to be hoed from around them. In the early days pigeon grass, wild oats, and mustard were the three most damaging weeds affecting the growth of the beet plant. For every beet that was allowed to grow, 50 or more had to be eliminated. This work represented not only a great waste of seed, but also an extreme labor cost. Daily progress could not be measured in acres, for even the most skillful partner team could not do more than ½ acre a day. In the early years blocking, thinning, and weeding kept workers busy from planting time until the beets became large enough to shade the ground and weed growth was minimized.

John P. Friesen of St. Joseph, Manitoba, was one of the pioneer beet farmers of that province. His father had grown beets in 1940, the first year of the Manitoba Sugar Company, and Friesen himself started producing beets in 1943. Friesen said, "We started farming when it was still considered a way of life, but getting into beets ended that concept, for farming then became a real business. We only had eight acres our first year, but everyone in the family was kept busy hoeing beets that season."

Thinning, hoeing, and cultivating were repeated several times. Most of the hand work was contracted on an acre basis rather than at an hourly rate. Hand labor involved a considerable cash outlay long before the crop was produced, and quite frequently the beet companies had to help out in making advance labor payments. In the early years a single laborer could not handle more than 5 acres of beets throughout the season. Since the development of the electronic thinner, of chemical weed control, and of mechanical harvesters, contract workers are able to handle an average of about 30 acres per season.

When beets were ready for harvesting a two- or three-horse one-row beet lifter was used. This simple machine lifted the beets part way out of the ground. The process was very slow, and all growers remembered that a lifter could do only ¼ to ⅓ acre per hour. Two acres was the maximum for a one-row lifter per day. The capacity of the lifter itself was really not a major concern, for it required a large crew of hand labor from that point on to complete the harvest.

A crew of two followed the lifter. They finished pulling the beet out of the ground, knocked off the excess dirt by bumping the beets together, and tossed six rows of beets into a single row. Next came two to six toppers, who used heavy knives to chop the tops off the beets. When topped, the beets were thrown into piles to make it easier for the loaders to fork them into the wagons or trucks. A good worker could top ½ acre of beets per day if the yield was not over 9 tons.

In some areas a "float" was used to facilitate the loading. It consisted of a pair of 2-inch by 12-inch planks with grader blades attached and fastened together in a V shape. The float was pulled through the field between the rows of beets, making it easier to shovel, because all tops and lumps were pushed aside. Men with forks about 2 feet wide, similar to the large forks found in potato storage or for handling silage, were used to toss the beets into wagons. During the 1920's and even as late as the early 1940's in some cases horse-drawn wagons were used to haul the beets from the field. If the haul to the railroad siding was not too far a two-horse team was used and 2 to 2½ tons of beets could be hauled. For longer distances larger wagons were used to haul 4½ tons, and four horses were used. Frequently the lifting and topping crew helped load so that the wagons were not delayed in the field.

Henry Peterson commented that since his farm was one mile from the railroad siding he used two wagons to haul the beets to the rail cars. Each wagon could haul four 2½-ton loads a day. A third man called a spike pitcher helped the two wagon drivers load the beets. In this manner three men with two rigs could haul 20 tons of beets each day. At the railroad siding the wagons were driven alongside the gondola cars, and the haulers had to fork the beets into the car. Because of the high sides the haulers had to throw the beets higher than their heads to get them into the cars. To make it easier for the men unloading the beets, very high-wheeled wagons were used in the early years, even though this meant a higher throw when loading in the field. A high-wheeled wagon pulled more easily in the loose soil, but in wet conditions an extra team had to be used to help pull the wagon out of the field. When an area became established in beet growing, dirt mounds or loading ramps were built to make unloading easier.

The Rosses used two wagons to haul their 30-acre crop 3 miles to the Fisher railroad siding. Each man hauled two loads of 3 tons each per day. Because of the time-consuming 6-mile round trip the Rosses were able to deliver only 6 tons of beets per hauler per day.

The Friesens at St. Joseph, Manitoba, had to haul their beets 7 miles to the siding at Letellier. They used four-horse teams and hauled 4½-ton loads. The Friesens piled their beets, and when an empty wagon came to the field the lifter, driver, three hand lifters, and eight toppers all grabbed forks and loaded the wagon in 20 minutes. Each wagon was able to make two trips per day so that 18 tons of beets could be harvested and hauled by a 14-man crew for an average of 1²⁹/₁₀₀ tons of beets per man per day.* It was not until 1949 that the Friesens changed their method of harvest.

All of the work was heavy, but probably the loaders had the hardest job of all. A loader was expected to be able to fork at least 1 ton an hour, and 10 tons was an average day's work. A few loaders forked up to 25 tons in a single day. Norman Krabbenhoft, who was a field spike pitcher, commented, "We could load twenty-five tons per man in a day, but at the end we were so tired that we fell over the dirt lumps rather than step over them."

Merle Allen harvested his first crop of beets in 1934 when he raised 18 acres that produced 87 tons of beets. Since a railroad siding (Ruthruff) was adjacent to his farm, the longest distance the beet hauler had to travel was ¾ mile. One man drove the lifter and did the loading and hauling while two others pulled, topped, and piled the beets. This crew, which Allen said were excellent workers, required 18 days and averaged 1⁶/₁₀ tons per worker per day to complete the harvest.

Walter Ross gave a good description of a record hand harvest of beets:

> For twenty-four years we had the family of Jesus Marmalejos working for us. They were excellent workers and fine people. In 1927 I won a prize for rapid harvest of beets. Our crew for thirty acres was a hand lifter, two toppers, and two haulers [this does not include the lifter driver] and they delivered 300 tons of beets from thirty acres in thirty working days, an average of two tons per worker per day. The fact that the beets averaged ten tons per acre made the task easier, but the company considered our workers' productivity amazing.

The Rosses harvested with horse-powered equipment and relied almost entirely on hand labor from 1920 through 1927, when they established the record for harvest by hand labor. In 1928 they experienced their first major change in harvesting when they hired a hauler with a Model T Ford truck that could haul 3 tons to a load. They also purchased an International truck with a three-speed shift with a two-speed auxiliary that could haul 4 tons per load. To make loading easier the Rosses piled beets from several rows into large piles that were about 20 feet apart. The driver loaded the left side, and a second man loaded from the right side. Because the trucks traveled so much faster the Rosses were able to deliver four loads for each truck per day, or 28 tons total. The

---

*It should be noted that while a definite number of workers is cited in various examples, the crew size varied from farm to farm and even daily on some farms. Average crew size was used for the purpose of illustration.

biggest problem with trucks came when they got stuck. Since they were heavier than wagons, horses could not pull them out of the mud, and sometimes beets had to be unloaded to free the truck.[2]

Progress Comes to the Beet Field

The use of the tractor was the first significant step in getting away from horse power and costly hand labor in beet production. The Ross family purchased a Farmall in 1928 that was used in ground preparation for beet production, but since the wheels were not adaptable to the beet rows the tractor could not be used in the beet field after planting.

In 1930 a 15-H.P. Model 10 track-type tractor, the smallest made by Caterpillar, with special narrow tracks to fit the beet rows, was purchased.* The Rosses immediately modified the hitch on two four-row beet drills to attach behind that tractor for seeding. This did not save on labor because one man had to drive the tractor while the other rode the modified eight-row drill. After planting in eight rows the next step was to put two four-row cultivators together so that eight rows could be cultivated at once. This required three men—one to steer each cultivator and one to drive the tractor.

Hot weather is probably the most efficient time to kill weeds. Here the tractor proved itself by working more steadily and far more hours than horses could. Also, the eight-row tractor-powered cultivator was much faster and more reliable. Use of tractor power provided for steadier control of the cultivator, and the protective shields were put closer together so that all but 2½ inches could be cultivated as opposed to horse cultivator shields set at 4 inches. This helped to reduce the amount of hand labor necessary for weeding in the rows.

It was only natural to use an eight-row cultivator in cross-cultivating the beets. The cultivator was adjusted in such a way that most of the undesired beets and weeds in the rows could be eliminated in all but 2½-inch blocks out of each 18 inches of row length. This greatly reduced the time needed for blocking and thinning. Beets were cultivated lengthwise twice and crosswise once before the hand workers entered the field. The growers gen-

---

*In the days of horse power farming beets were raised in 24-inch rows because that was the narrowest that horses could walk in. Some used mules because they were more careful than horses and did not step on the plants. Once tractors came into general use the beet rows were reduced to 18 inches. In the late 1950's beet rows were standardized at 22 inches to accommodate the mechanical harvester.

erally agreed that the process of cross-cultivation increased the output per worker from the standard 5 acres to at least 12 acres. Hand labor was reduced to thinning out the beets deliberately left in blocks by cross-cultivation.

After tractor-powered eight-row drills and eight-row cultivators plus cross cultivation were introduced the beet grower was limited only by the capacity of the harvesting operation. In 1933 a two-row tractor-powered beet lifter was purchased by the Rosses. The one-row horse lifter did 2 acres a day at the best, while the new two-row tractor machine could do 10 to 12 acres a day. It took 30 toppers to keep up with the two-row lifter.

For a two-row lifter and 30 toppers more trucks were required by the Ross family, who had one truck and had to contract the services of two other haulers. An unloading dump was erected at Fisher so that with good luck a single truck could haul ten 5-ton loads a day. Six men besides the drivers were needed to keep the trucks going. If each truck hauled its 10 loads it meant that the nine-man crew averaged a little over 16 tons per day of hand forking beets. When toppers were caught up with topping they were expected to help load the trucks.

The lack of a mechanical harvester was obviously the biggest obstacle to the improvement of industrialized beet production. But as long as there was an abundant supply of low-cost labor there seemed to be little concern about such a machine. Beets were one of the most profitable crops, with a contract price varying from $4 to $5 a ton. Yields increased so that the top growers were able to produce up to 10 tons per acre. By being able to contract labor at $13 to $18 an acre from the first blocking to the end of harvest, there was still enough margin left to provide the grower with a good profit. The per-acre net profit on beets sometimes equalled the gross income on small grain.

In 1934 the sugar beet growers received an added "windfall," as Walter Ross called it, "We were getting $4.75 a ton for our beets and the government payment under the Sugar Act was $1.90 a ton. That came to be our profit as production costs rose, and it was very attractive in those first years."

The United States has been a sugar-deficit country, importing more than half of its needs annually and with its imports making up about 20 percent of the total sugar production in world trade. The major purpose of the Sugar Act was to guarantee a steady flow of sugar for American consumers at a relatively stable price. This was accomplished by giving various countries allocations and contracting the price a year in advance. To maintain its allotment a country had to fill its quota each year, whether the United States price was above or below the world price.* American producers since the 1970's could easily have supplied our domestic needs, but the government wanted to keep our sugar market open to foreign nations for political reasons. As Walter Ross explained it:

We not only profited from the $1.90 a ton extra payment but we also had a good crop in 1934 so we were able to buy back all the land that had been lost by the estate. We got it back from the mortgage company for the amount of the mortgage. Despite the lack of harvesting machinery we decided to increase our beet acreage and by 1937 we had 500 acres. I'm quite sure that made us one of the largest producers if not the largest in the Valley.**

The Ross family was able to harvest 150 tons of beets a day during the 1930's, using 37 men for an average of 4 tons per worker per day. Walter Ross added, "I shoveled my share of beets as did every beet grower because once they were topped and piled it was important to get them off the field as rapidly as possible." There were some cases where growers were penalized if they allowed their harvested beets to freeze in the field prior to delivery. Sometimes beet tops were used to cover the piled beets to keep them from freezing.

Charles Peterson and sons were the first known irrigators of beets in the Red River Valley. Henry Peterson recalled this experience:

We started irrigating in 1928 on the old Bergquist farm which is just north of the Clay County Court House today [1974]. We used a Skinner system that had portable steel pipes that stood about four feet above ground and rolled so it covered a swath of about fifty feet. We had four units each about 350 feet long and each covered about one-half acre to a setting. It took two men one-half hour to reset each unit. The water was pumped out of the Red River. With the electric pumps those four units cost a total of $4,000, but we sold $3,800 in radishes the first year alone by irrigating daily for six weeks.

---

*Ironically it was just 40 years later, in June, 1974, that the Sugar Act was discontinued because of pressure from industrial sugar users who felt they would rather take their chances buying on the international market than be guaranteed a stable price for their sugar. The government payment varied from $.80 to $2.40 per ton dependent upon the import duties on sugar collected during the year. When the government payment exceeded $5,000 for a grower in any given year, the rate per ton was reduced.

**By contrast to the 500-acre allotment to the Ross family in 1937 the largest in the Valley in 1974 appears to be the 2,500 acres of beets by the Leo Yaggie family of Breckenridge. The Yaggies, a father and four sons, operate about 25,000 acres of land. Their diversified farming includes a major hog enterprise in addition to sugar beets.

The irrigation was so successful that the Petersons got a larger unit in 1934 made by the Fargo Foundry. This unit sprayed 50-foot-wide sections for a quarter mile. It was flexible so that two men could reset it in a half hour. In 1939 the Petersons used irrigation on sugar beets with amazing results. Their 123 acres of non-irrigated beets averaged 10 tons an acre, while the irrigated crop went to 30 tons. That fall 22 toppers were required to harvest 132 acres in addition to the 12 men needed for loading and hauling.

The Petersons did two other things in their beet production that were unique in the Valley. Because they needed a large labor force for their truck gardening it was beneficial to keep this force employed with other jobs during periods of low labor-requirement periods. After beet and potato harvest these men windrowed beet tops on the farms of other growers and hauled them to the Peterson farm feedlot. All this work was done by pitchfork. Peterson paid $3 an acre for use of the tops, whether they were grazed or hauled off. Usually enough tops were gathered to provide a roughage supply for three or four months. Henry Peterson grinned as he remarked:

Beet tops proved to be one of the biggest breaks we ever had in farming. In 1936 I bought fifty head of 500 pound cattle at $1.50 a hundred and fed them very little other than beet tops, straw, and some oats. We sold them at 900 pounds for $9.50 a hundred. That $3,500 in 1936 was really a boost because we bought several pieces of new big equipment with it. Sheep were more reliable income than cattle, but that was our biggest single haul.*

The other practice used by the Petersons was hauling manure from the West Fargo Union Stockyards to their land. Petersons had long made it a practice to haul manure from the livery stables and residential barns of Fargo-Moorhead, but with the opening of the Union Stockyards in 1935 the supply of manure became even larger. Each fall, for as long as seven weeks, 8 to 10 trucks were used to haul manure to the Peterson farms. Five men were needed to unload each truck that had been used in beet hauling and were equipped with tip-down sides. Two men with forks walked on each side of the truck with one man in the rear, spreading the

---

*In the 1930's and 1940's most of the beet growers fed the beet tops, either by stacking them on the yard or by letting animals graze them off. Sheep were chiefly used to salvage the beet tops, and they consistently gave an additional profit to the beet grower. Merle Allen grazed about 14 sheep per acre of beets for 90 to 100 days each fall. Each year a $1.00 to $1.50 per head net profit was realized, giving an added return of $14 to $21 per acre of beets grown.

manure as the truck drove down the field. As much as 25 tons of manure were put on each acre.

Henry Peterson recalled that loading at the yards was the only mechanical part of the operation. He developed an overshot manure loader that took a scoop of manure from the pile and tossed it over the top of the D-4 Caterpillar into a truck parked to the rear of the tractor. The wartime labor shortage following 1942 caused them to discontinue hauling manure from the stockyards. Peterson added, "The heavy applications of manure on our land had a lot to do with our high yields during the 1930's. Our land not only worked better but it also held its moisture and gained in fertility. For twenty years we could see to the line and tell it on the soil exactly where we left off applying manure. It certainly helped to ease the squeeze during those lean years."

The Ross family decided to increase their acreage after the passage of the 1934 Sugar Act had made beets so lucrative. However, they had to make several changes in their production methods. In preparation for the 1935 crop they purchased an I.H.C. F-12 rubber-tired tractor for planting and cultivating. This was the second tractor that Walter Ross used in their production of sugar beets. For field preparation drainage work and for small grain production they purchased an I.H.C. TD-35 Diesel track-type tractor. This was a 35-H.P. machine capable of pulling a seven-bottom plow.

The third purchase in 1935 for the Ross farm was a six-row John Deere check-row-beet drill. This worked similarly to the corn planter that dropped several seeds in a single hill. This drill was set to plant in 18-inch rows, and the check wire was also set for 18 inches. The notched wire was strung from one end of the field to the other and attached to the drill so as to trip it every 18 inches, dropping 5 to 20 seeds per hill. The drill was built to operate at 130 checks (or trips) per minute so that top speed was limited to about 2 mph.

The work was tedious, and the seeding of 25 acres was a big day, but the benefits were great. Merle Allen went on a 24-hour-a-day planting schedule and was able to do 40 acres in that period of time if they did not stop and the operator ate while driving. By limiting the seeds planted per acre to 10 pounds instead of 20 the savings in seed alone paid for the planter. But there was more to be gained by hill-dropping. The former practice of blocking was reduced to eliminating the excess seeds per hill which greatly lowered the labor cost. The third savings came with the ability to cross-cultivate more accurately, leaving only the weeds within the block around each beet. The final and most unexpected benefit of check-row planting was an increase in yields because each beet was given an area of 9 inches in every direction in which to develop.

The reduction of blocking and thinning cost was extremely beneficial to the beet grower for it represented the largest or second largest cost in producing a ton of beets. The only costs that ever exceeded blocking and thinning were the combined topping, loading, and hauling expenses. Henry Peterson's 1937 thinning contract called for $10.50 an acre, while the cost of his topping was $7.50 and of the loading $2.50 an acre. The total bill on contract labor was $20.50 an acre on a 10-ton crop that grossed $66.50 per acre. In 1949 John Friesen's thinning contract cost him $20 an acre, while his total harrowing and cultivating bill was only $5.39 per acre, including chemical weed and worm control. Mechanical and chemical weed and insect control obviously was a major factor in savings for the beet grower, for while labor needs were reduced wage rates rose rapidly, and the total bill was not lowered much.

Manitoba sugar beet growers suffered severe financial reverses from 1940 until the harvest of the 1946 crop. A combination of bad weather, insects, labor shortage, and a sugar price war in Canada hurt the Manitoba growers. Instead of the anticipated expenses of $30 an acre, most growers were experiencing costs of $50, which amounted to more than the income derived during several of the previous years. However, a world shortage of sugar in 1946 gave the Manitobans the price break they needed to survive. The Canadian growers generally had smaller acreages than the upper Red River Valley growers and therefore were slower to adopt mechanization. This handicapped their profit potential immensely. Friesen, one of the first Manitoba growers, still did not have a mechanical harvester in 1949 and in that year his topping and loading bill was 28 percent of his total cost of production. Those two costs were $2.61 a ton in contrast to his total net profit of $3.49 per ton. Obviously a mechanical harvester would have given him one of his greatest opportunities to increase profits.

The mechanical loader, which came during the World War II era, eliminated hand loading and was the invention that some growers felt probably saved the sugar beet industry in the Red River Valley. Timing once again was important, for although most growers were happy with the profits from sugar beets, many were beginning to feel that the wartime labor shortage would force an end to their production. Only the strongest and best men could load up to 25 tons of beets a day, but unfortunately many of them were soon drafted. Once again it was the Ross family who made history. They had started with one-row lifters and by the mid-1930's were using three-row lifters. But in spite of the help by the lifter, beets still had to be hand topped and loaded. In 1939 they purchased a Julius Sisch loader that was capable of loading 1 ton of beets per minute. This particular machine picked up the beets from a heavy windrow made by the toppers, who tossed 12 rows of beets together as they topped them. Basically all the work the loader did was to elevate the beets onto a truck driving alongside; but at least loading labor time was reduced from 30 minutes to 2 minutes a ton.*

Even though the difficult job of loading beets was now mechanized, the harvest was not speeded up because topping was as time consuming as ever—one-half acre per man per day. Vern Hagen of East Grand Forks had 75 acres of sugar beets in 1945 and was using the Espe loader. Hagen's total crew for 23 days, which were required to harvest a 750-ton crop of beets, numbered 15. Three men did the lifting, loading, and hauling. A two-man crew was needed to do the lifting, and then three were required to do the loading. If there was no waiting at the plant one truck could easily keep ahead of 12 toppers. For the harvest that year Hagen's crew averaged $2^{1}/_{10}$ tons per man per day. The Friesen crew at St. Joseph, Manitoba, numbered 14 men in 1949, and they were harvesting and delivering only $1\frac{1}{3}$ tons of beets per worker per day because of a longer haul to the railroad.

In 1941 a 12-row beet drill was introduced that cut the planting time in half, but in 1942 there were even greater innovations for planting. That year segmented seed was introduced, greatly reducing the amount of labor necessary to thin the beets.** This eliminated the check system of planting so that planters could travel up to 3½ mph.

By now Walter Ross had increased his cultivating capacity by pulling three 6-row cultivators behind an I.H.C. F-20 tractor. Four

---

*Other types of loaders required the truck to be pulled backward down the field behind the loader. This elevated the beets directly to the truck by use of two chains. The most common make in the Valley was the Espe loader, manufactured at Crookston, Minnesota.

**Segmented seed is seed in which the multi-germ seed balls are separated and each germ is isolated. Generally only one plant grows from each seed, but frequently two or three plants grow and still require some thinning.

men were still required to do 18 rows of beets, but horses were eliminated in favor of the more efficient tractor power. With the adoption of a 12-row planter the handmade three-unit cultivator attachment had to be discarded and a 6-row tractor-mounted cultivator was purchased. Although capable of taking only one-third as many rows, the mounted cultivator was much more flexible than the triple unit pull type. Two six-row tractor-mounted cultivators performed much better than the old 18-row pull-type setup. In 1955 Armin Ross made three-row extensions for each side of his six-row mounted cultivator, making it into a 12-row mounted rig. It proved very workable and became widely adopted throughout the Valley.*

In 1940 Ross started increasing commercial fertilizer applications to his beet crop. Not much experimentation had been done before beyond the use of the standard phosphate application. Ironically, unlimited use of commercial fertilizer did not prove to be practical because it caused the beet to become very large while adversely reducing the sugar content. This was not desirable to the processor.

Since 1932 there had been some experimentation with mechanical loaders, but it was not until World War II that a combined topper, lifter, and loader was developed for complete mechanical harvest of beets. It is believed that the Marbeet double-row harvester was the first commercially produced machine of this type. It appeared on the market in 1943; but it was not accepted by the growers, who were reluctant to experiment with costly machines. In 1944-1945 the Blackwelder Company developed a single-row Marbeet machine that satisfied the American Crystal Sugar Company. The company purchased 40 of these machines and rented them to its growers in California.

In 1944 International Harvester Company placed a one-row experimental model on the Ross farm. This machine had the usual bugs found in any experimental model, but it had only one serious drawback—its inability to separate the dirt from the beets. Otherwise the Rosses were impressed and ordered a harvester for the

---

*In 1961 Ross experimented with a 16-row mounted cultivator for International Harvester but it did not prove so easy to handle and so efficient as the 12-row unit. In about 1928 R. T. Adams of Fisher converted a Model T Ford 1-ton truck into a mechanical-powered four-row cultivator. The unit was made for field work by adapting a rear-axle unit made by Espe Machine Company of Crookston to the Ford chassis.

1945 crop. Jay Wilder bought the other unit shipped into the Valley that year. This one-row harvester had a mounted rotary blade that topped the beets and then lifted them to a chain elevator, which took the beets to a picking table. At the picking table two men caught the beets, knocked the dirt from them, then threw the beets into a tank, from which they were loaded into trucks. Every beet still had to be individually handled, but a tractor driver and two men on the harvester could do 3 acres per day. Under the hand system it would have required 12 to 15 men to do the same job and with far more strain.

The Ross family was able to compare costs in the transition because they phased out of the hand-harvest method over a three-year period, adding a second machine in 1946 and a third the following year. These one-row harvesters were capable of 100 acres of harvest each season, so the Rosses reduced to 300 acres for a few years because they did not care to operate more units. In 1948, with three harvesters, three trucks, and a total crew of 12, they delivered 10 tons of beets per day per worker, just five times as much as under the former hand-harvest system.*

With harvesting equipment a reality it was only natural that the cycle should start over again. In 1948 Ross purchased a 12-row beet planter that was especially designed for segmented seed. This planter (or drill, as it is called in the industry) could seed at the rate of 3 mph and was capable of 70 acres per day with much less effort and greater accuracy than their first planter, that did 25 acres and was extremely wasteful of weed. The ability of that newer planter to control the seeding rate cut thinning labor costs by 25 percent.

By 1950 beet growers were getting anxious to obtain larger equipment to keep up with constantly expanding acres, increasing yields, and bigger tractors. In 1952 Walter Ross visited Chicago in an effort to convince engineers for International to build a three-row harvester. But they would not listen, and for one of the few times in his farming career he purchased a machine made by another company.** In 1956 Ross secured a three-row Opal har-

---

*The Ross production described above was on a 12 ton per acre field with a three mile haul. John P. Friesen in Manitoba with the same equipment but a 7.4 ton yield and a seven mile haul was able to harvest only 5.14 tons per person per day.

**Ross had done considerable farm testing for International Harvester, and they generally followed his suggestions.

vester and increased his acreage again. One single-row harvester was kept to open fields.*

In 1957 a representative for Opal visited the Ross farm and after consultation with Walter convinced his company to build a four-row harvester. In 1958 Ross used the four-row Opal machine that had been especially made for him. Each change of harvester had adaptations that made it better than the previous models. The blade lifter was replaced by a wheel-type lifter that reduced the dirt lump problem. Later a grab roll was added that nearly eliminated the dirt problem.

Men were no longer needed to work on the harvester, but a new problem was created when the multi-row harvesters became capable of lifting beets at unprecedented rates. The beet grower faced a shortage of hauling equipment, a pleasant obstacle when contrasted to all the past problems, but an expensive one to overcome. To keep a four-row harvester going, three tandem trucks capable of 12- to 14-ton loads were required. In a single 12-hour shift one man on the harvester, one man half time on a roto-beater, and three truck drivers could harvest and deliver 294 tons of beets. This was 65 tons per man per day in contrast to 2 tons per man per day in 1927, the year Ross won the company prize for harvesting the greatest tonnage per day for each worker during the entire campaign.**

Two major beet harvest innovations took place as a direct result of the mechanical harvester. Both were necessary inventions to keep up with the virtual flood of beets produced by multi-row harvesters and larger trucks. In the early years most trucks had tip-down sides, and the beets were rolled off the trucks by a large hoist that lifted one side of the box. When the truck returned to the field, the tare (beet refuge and dirt) had to be shoveled off by hand. Both of these methods were time consuming. About 1946

---

*Once harvesters were perfected, farmers converted to them rapidly. In 1946 Merle Allen purchased a harvester that pulled so hard that his I.H.C. TD-14 track-type tractor had difficulty with it. In 1951 Allen decided to try again. This time he purchased six single-row harvesters for $27,000. Ten trucks were needed to make the 10-mile haul to the Moorhead plant. The big capital investment paid off, for the cost of harvest was reduced by 50 percent. In 1956 Allen purchased a four-row self-propelled harvester mounted on a road-maintainer chassis at a cost of $22,000. This machine could easily lift a ton of beets a minute, but it was extremely difficult to steer. Hydraulic row finders have since eliminated that problem.

**In 1962 John P. Friesen had harvested and delivered a 9$^7/_{10}$-ton crop seven miles with a three-man crew using a three-row harvester. This crew averaged 38$^8/_{10}$ tons per man per day, nearly eight times more per man than what Friesens had harvested as recently as 1948.

hydraulic hoists made end dumps feasible for faster unloading of beets at the piling stations, and the tare was dumped on-the-go when the truck returned to the field. Merle Allen had four hoists in service in the 1947 campaign and felt the biggest relief came because of the elimination of hand shoveling the average of 1 ton or more of tare per load of beets: "That dirt was hard, sticky stuff to shovel."

Twenty-four-hour beet receiving at the piling stations doubled their capacity. To avoid a jam of trucks at the pilers all beet growers were placed on one of two shifts—noon to midnight or midnight to noon.* Growers were allocated a specific number of trucks based on their total acres and miles of haul. This was done in an effort to keep a steady flow of trucks at the pilers and also to keep the harvesters going in the fields. About 120 truckloads of beets arrived as scheduled each hour at each plant, and they had to be unloaded with as little waiting as possible, for trucking was rapidly becoming the major cost in beet production. The early fall in the Red River Valley unfortunately restricted the number of good harvest days so that the 24-hour work shift was a welcomed arrangement.

Not satisfied with factory-built harvesters, the Brekke Brothers in Polk County, Minnesota, made two three-row factory-built harvesters into a single six-row machine. The Brekkes commented that the problems with the larger machine were no more serious than with the smaller units, and they were able to fill their trucks in half the time. This meant that less than 10 minutes were needed to load a 14-ton truck in a field yielding 14 tons of beets. This was ¾ ton of beets per minute per worker, an amount equal to one-third of what a worker could do in 12 hours just 40 years earlier.**

In the 1950's the evolution of the monogerm seed by Drs. V. F. and Helen Savisky, two Russian immigrants, gave the beet grower another step to better production with less labor. The monogerm seed, first released in 1954, produced only one beet per seed, and beet drills were refined to control the planting rate to about 1¼

---

*At first farmers volunteered to go to the night shift, and it was discovered that harvesting was so much more efficient because of reducing waiting time at the unloading station that more farmers quickly went to the night shift. By the second year a formal rotation of shifts had to be established by the beet company.

**For many years farmers' wives have helped in beet harvest. In the beet growing areas it is a common sight to see a tandem truck loaded with 14 tons of beets being driven by wives and daughters of the growers. Power steering, power shifts, air brakes, and air-conditioned cabs make it a pleasant job by contrast to horsepower days. This is true even for those on the 12-hour night shift.

pounds per acre. Hand-labor requirements were drastically reduced at this point, but beet growers were still not satisfied.*

In 1964 126 growers in the Valley agreed to grow beets on a no-labor basis. These growers agreed to experiment with an average of 18 acres each, on which all cultivation and thinning would be mechanical or chemical. The results of the first year's experiment showed a yield reduction of $1^7/_{10}$ tons per acre as compared to hand-worked beets. In 1965 the experiment was continued, and one hand hoeing for weeds only was permitted late in the season. Those beets outproduced the hand-worked fields so that beets grown without hand labor were almost a reality.

Mechanical thinners were perfected about 1965, and they effectively thinned the beets so that stands of 110 beets per 100 feet of row were accomplished.** Excessive beets (over one per foot) were reduced by cross harrowing. Chemicals were introduced to spray on weeds and to protect against insects. In some cases the cost of chemicals exceeded the remaining labor bill. In 1969 an electronic beet thinner with an electric eye was developed to eliminate unwanted beets and weeds in the row. This precision tool seemed to be the answer for the beet grower, who was often criticized for not giving proper housing and pay to his migrant workers. Social legislation was forcing the growers to make changes, and now an electrically operated knife was an alternative for them to replace the man with the hoe. The $6,000 to $18,000 electronic thinner was both less expensive and more reliable than "the man with the hoe could ever be."

Each year beet production has increased, but each year it has been done with fewer workers. In the span of one generation the sugar beet industry has come from ancient times to the industrial age. Those who paid labor bills of up to $50 an acre for blocking, thinning, weeding, and topping saw that the machine and the chemicals were the way to progress.***

John P. Friesen of Manitoba and Walter Ross of Minnesota, one a small grower and the other a large grower, could prove that in their careers as beet growers they had increased man's productive

---

*F. V. Owen, a Utah geneticist, worked with the Saviskys.

**The mechanical thinner was a modification of the cotton chopper used to thin cotton stands to obtain better growth.

***Robert Yaggie researched and wrote about the scale of economy in sugar beet production while working on a master's degree in agricultural economics. After school he joined his father, Leo Yaggie, and three brothers in the largest single sugar beet operation in the Red River Valley. Mr. Yaggie feels they could go to 5,000 acres of sugar beets and still be profiting from economy of scale.

TABLE 2. Changes in Productive Capacity of Labor in Sugar Beet Raising Based on the Experiences of Five Growers in 1943 and 1973.[1]

	Man-Hours per 100 Acres	
	1943	1973
	(7 tons per acre)	(13 tons per acre)
Field Preparation		
Fall—moldboard plowing	144	none
Fall—chisel plowing	none	10
Spring—field cultivating	40	5
Harrowing	20	2
Fertilizer application	none	6
Planting		
(Tractors are limited to 3 mph for precision planting.)	80	20
Four cultivations	235	30
Hoeing, including blocking and thinning	2,400	800[2]
Electronic thinner and chemical weeding	none	20
Horse equipment		
Lifting	300	none
Hand pulling and windrowing	900	none
Hand topping and piling	1,000	none
Hand loading, hauling with horses	700	none
One-row harvester tractor-powered and trucks (1944), illustration only, not included in totals	[700]	—
Four-row harvester and three tandem trucks	none	228
Total hours (two-hoeings method)	5,819 for 700 tons, .12 tons per hour of labor	1,121 for 1,300 tons, 1.16 tons per hour of labor
Total hours (one-hoeing method)	5,819 .12 tons per hour of labor	721 1.80 tons per hour of labor

[1]Merle Allen, Robert A. Yaggie, John P. Friesen, Norman W. Krabbenhoft, and Walter Ross production figures. See Yaggie and Loftsgard Bulletin 466, October, 1966, on production figures for various size growers. Their figures are nearly identical to those used in this table.

[2]Many growers have gone to one hoeing, reducing total hoeing labor to 400 hours, as illustrated in bottom line of table.

capacity in the beet field from plowing to harvest more than fifteen-fold. Walter Ross was active in beets from the 1920's to the 1970's. John Friesen had started in 1943; but because he operated on a smaller scale he was slower to mechanize than Ross, for it was not until 1950 that he started using trucks for the seven-mile haul to the railroad siding. Ross actually worked with the

TABLE 3. Size of Units of Sugar Beet Growers[1,2]

Total Acres Operated	Total Beet Acres	Total Horsepower in Tractors	Full-time Laborers (including owners)
1,100	130[1]	300	2
7,000	200[1]	567	4
1,600	300[1]	555	3
1,560	400	515	2
4,000	350[1]	1,318	7
2,400	400[1]	595	2
2,400	800	725	6
2,000	500	800	3
3,500	70[1]	520	4
6,400	800[1]	1,780	11
3,000	160[1]	610	3
3,000	500[1]	900	10
6,000	400[1]	1,540	18
3,560	755	1,015	3
1,920	600[1]	625	3
1,500	500[1]	585	4
8,000	1,000	1,390	8
3,500	600	875	3
6,000	150[1]	1,345	8
3,900	200[1]	905	9
2,500	650	645	5
5,200	600[1]	1,297	7
1,400	280[1]	600	3
3,060	620	1,370	5
25,000	2,500	3,330	14
Av. 4,380	539	986	5.88

[1]Farmers for whom beets are not the major enterprise.
[2]In 1973 the 1,397 growers in Minnesota and North Dakota Valley counties averaged 145 acres.

machines as they were developed, while Friesen sought them only after they were well proven.

The 10-acre beet allotment has given way to the 300-acre allotment just as the 160-acre homestead has given way to its larger successor. In the Red River Valley, with seven existing processing plants, beets will become an increasingly important part of the region's economy.[3]

NOTES

[1]Introduction. Production statements of the American Crystal Sugar Company, 1919-1973; Letter from Aldrich Bloomquist, Fargo, North Dakota, April 26, 1974. Mr. Bloomquist is an official of the American Crystal Sugar Company which is owned by the farmers who produce beets; Interview of Walter J. Ross, Fisher, Minnesota, March 17, 1973. Mr. Ross is a son of Ferdinand H. Ross, one of the

pioneer beet growers in the Valley. Walter Ross is widely acknowledged as one of the long-time producers (since 1930) and one of the innovators in the mechanization of sugar beet production; Robert A. Yaggie and Laurel D. Loftsgard, "Sugar Beet Production Cost and Practices in the Red River Valley," North Dakota State University Agricultural Experiment Station Bulletin No. 455 (October, 1966), pp. 8, 31; Heather Robertson, *Sugar Farmers of Manitoba*, Manitoba Beet Growers Association, 1968, pp. 52, 55; Interview of Robert A. Yaggie, Breckenridge, Minnesota, August 12, 1974. Mr. Yaggie, whose graduate degree is in agricultural economics, with specialization in economy of scale, is a partner with his father and brothers in production of 2,500 acres of beets on their 25,000 acres of crop land.

²Walter Ross interview; Conversations with Hugh Trowbridge, Barnesville, Minnesota, and Norman W. Krabbenhoft, Moorhead, Minnesota, both of whom have had experience in large-scale production; Interview of Henry R. Peterson, Moorhead, Minnesota, December 12, 1969. Mr. Peterson's father started raising beets in 1924; Interview of John P. Friesen, St. Joseph, Manitoba, May 24-25, 1974. Besides being a pioneer Manitoba beet grower, Friesen also had excellent records; "The Walter J. Ross Story," *Crystal-ized Facts*, XXI, No. 1 (Spring, 1967), p. 6; Interview of Merle S. Allen, Moorhead, Minnesota, July 9, 1974. Mr. Allen started raising beets in 1934 and by 1955 in combination with his sons had 700 acres of beets on his 3,000 acres.

³*Crystal-ized Facts*, XX, No. 2, pp. 19-20; XX No. 3, pp. 14-5; XXI, pp. 6-8; Ross interview; Peterson interview; Frederick D. Gray, "That Coffee from Brazil and Other Food Imports," *Food for Us All, The Yearbook of Agriculture, 1969*, Jack Hayes, ed. (Washington, 1970), p. 18; *Fargo Forum*, October 22, 1939; Friesen interview; Vern Hagen interview; Robertson, *Sugar Farmers of Manitoba*, pp. 5, 6, 84, 85; "Walter Ross of the Red River Valley," *Tractor Farming*, XXXI, No. 2 (Spring, 1948); "Brekke Brothers," *The Farmer*, LXXXVI, No. 22 (November 16, 1968), p. 37; "The Red River Valley Untouchables," *Crystal-ized Facts*, XX, No. 3 (Winter, 1966-67), pp. 22, 24, 25; Allen interview.

Several farmers were interviewed for this chapter on sugar beet production. Although major emphasis was given to John Friesen, Henry Peterson, and Walter J. Ross, the supportive information provided by these other growers was essential for sound documentation. Other beet farmers interviewed were:

Herman H. Lee
John H. Reque
Ellery Bresnahan
George Sinner
William Sinner
Gerald and Thomas Ryan
Earl Davison
Arthur Skolness
Tilney Farms
John S. Dalrymple, III
C. H., Leslie and Nolan Underlee
J. G. Hall and Sons

Paul Horn, Jr.
Earl Hvidsten
Keith and Ray Driscoll
Earl Glidden
Frank Farms, Inc.
Virgil Mellies
John W. Scott, Jr.
John W. Scott, Sr.
Harold T. Peterson
Arlo J. Gordon
Ervin Bourgois

Not all of these farmers were from the Red River Valley, but their contributions supported the information given by Valley growers. Some were temporarily out of beet production because of the shutdown in processing plants in their area, but were contracted to resume with beets as soon as new plants were opened. Norman W. Krabbenhoft and Merle Allen served as technical consultants for production figures used in this chapter. See Table 3 for details on the scope of operation of the individual farms involved.

For additional data on economy of scale in sugar beet production see Laurel D. Loftsgard and Robert A. Yaggie, "Sugar Beet Production Cost and Practices in the Red River Valley," Agricultural Experiment Station, North Dakota State University

Bulletin No. 455 (October, 1966); L. C. Rixie and H. R. Jensen, "Cost Advantages in Size of Farm in Red River Valley Farming," Agricultural Experiment Station, University of Minnesota, Bulletin No. 469 (June, 1963); Dale O. Anderson and Donald M. Hofstrand, "Sugarbeet Production Costs and Practices in the Red River Valley," *North Dakota Farm Research*, XXVII, No. 6 (July-August, 1970), pp. 3-5; Allen interview.

CHAPTER XI

Turkeys and Chickens: The Old Flock Is Gone

Early Days

AT ONE TIME nearly every farm in the nation had a small flock of chickens, ducks, geese, or turkeys. These flocks were used to provide subsistence for the farm family in the form of eggs or meat and to provide the farmer's wife with income for household expenses. "Egg money" was commonly used to buy groceries or for special projects in the home.

If strict cost accounting methods had been used, it is doubtful that these small flocks of poultry would have ever showed much of a profit. The farmer usually let his wife have the grain, family labor was never figured as a cost, and the flock was often housed in a shed that was too small for any other purpose. Sometimes the birds ran loose in the barn and in the cooler climates warmed themselves on the backs of cattle. If the flock had its own housing, farm children had the unhappy task of cleaning the roosts on school-free Saturdays. In the summer the collecting of eggs from nests among the trees, in the haymow, under the corn crib floor, or wherever a hen could comfortably lay the eggs was all part of a necessary increase in subsistence, even if it was not an economically efficient business.

As farmers prospered and became more specialized they tended to drop the subsistence enterprises that they considered to be less profitable. Poultry frequently was one of the first segments of diversified agriculture to be dropped. It was easier and more economical to buy eggs from the neighbor's wife, who also made a business out of butchering chickens, ducks, geese, or turkeys for the holiday season.

Production statistics from the various states indicate a steady decline in the number of units involved in poultry enterprises and except for Iowa and Minnesota a decline in the numbers of birds raised. For years Iowa and Minnesota ranked among the top 10 states in either chicken or turkey production. In recent years Minnesota has attained first or second place nationally in turkey production. But in none of the states of Iowa, Minnesota, Montana, North Dakota, South Dakota or the province of Manitoba does poultry produce more than 7 percent of the total cash farm income.

In spite of this general decline, the production of chickens, eggs, and turkeys, however, is still big business for a limited number of producers. In 1959 there were approximately 100,000 farms raising an average of 800 turkeys. By 1964 the number of turkey farms had declined to 42,000, but the average number in each flock increased to 2,500 birds. Five years later 5,400 farms produced an average of 19,000 birds each. By 1972 there were still fewer turkey operations, but the average-sized unit had increased to 30,000 birds. A similar trend took place in the broiler business because poultry production adapted itself well to the greater efficiency that came with an economy of scale with larger units.

Vertical integration and concentration became so conspicuous in the poultry business that much controversy was aroused over the issue. Ralston Purina Company, one of the nation's largest feed manufacturers, became involuntarily involved in the broiler business in the late 1950's. The move on the part of the company was done at the request of its dealers and its farmer customers, who wanted someone to share the risk of a fluctuating market with them. Even though there were many advantages to vertical integration there were even greater disadvantages, most of which were created by a combination of human and business complexities. This caused Purina to withdraw from the industry in the early 1970's.*

As farmers nationwide abandoned the poultry business, the industry became concentrated in a limited number of rather sizeable units of production. The traditional, very uneconomic, small farm poultry enterprises gave way to some of the most efficient and scientifically managed poultry farms in the nation. The larger operations profited from the natural economies of scale and since

---

*The question can be asked, has vertical integration in agriculture failed to become established? Many feared such a move, while others looked to it as part of a natural economic trend. The question, at this point in time, has still not been answered.

1965 have reduced feed requirements per pound of turkey by 25 percent, reduced labor time by half, and sharply decreased the bird mortality rate. Contrary to most aspects of agriculture, over half of the turkeys and nearly all of the broilers are produced under contract. A limited market makes contracting necessary, to reduce risk.[1]

Profiles of Some Modern
Poultry Farmers

Because of the tendency towards concentration in the poultry business, a study of only a limited number of producers is necessary to describe its present state and to speculate on its future. The enterprises studied for this chapter are all very large when contrasted to what is still considered an average-sized farm of the 1970's, but they are family-owned businesses. In each case the poultry business evolved either as a segment of a diversified farm or as a boyhood dream, or was started by adversity.

Three farmers of this study—Pete Hermanson and John Heline, both from Iowa, and Lyall Larson of Minnesota—have developed sizeable turkey-growing enterprises along with other enterprises as part of their general diversified farming. These farms are all well known in their area and have had considerable publicity recently by virtue of their size and their owner's industriousness. When interviewed, these three farmers had a combined turkey population of 430,000. Each farm annually raised over 1,000 acres of corn. In addition one had 275 dairy cows, a second had several hundred head each of beef cows, feeder cattle, and feeder pigs. The third produced over 2.5 million pounds of pork. All the owners were also involved in other commercial activities besides poultry raising as part of their daily life.

Corn growing was considered the basic farm activity by each of the three families. All were equipped with large-scale machinery, and two of them frequently did custom work to help out their neighbors and to reduce the cost of owning this big equipment. Their corn provided the basis for the quality feed needed in the turkey operation. Less than top quality corn in every case was used in other livestock enterprises where quality feed was not so critical to good production. Purchasing feed supplies from neighboring farmers was an established practice by these farmers because their turkey and livestock enterprises consumed more than was produced on the farm. One farm normally purchased two and one-half times as much as it produced.

These farmers expressed the opinion that buying grain was profitable because it was more desirable to invest their money in the fast turnover of turkey and livestock enterprises than it was to become too involved in buying land for greater crop production. This, however, did not preclude them from buying additional land when it became available.

The Hermansons and the Larsons are family farms. A father and two sons are associated in farming and provide the backbone of management. In addition, skilled and proven managers are employed. Heline is the sole owner and overall manager of his farm. The three farms combined have 44 year-round employees, and in peak seasons the combined total of their labor force swells to 188. Their labor bill, including family labor, averages 11 percent of their gross income. All three mentioned that having diversified farms enabled them to employ more profitably a year-round labor force. Good wages and working conditions contributed to the long average tenure of their full-time labor force.

Marketing is a major problem of all three farmers. But Lyall Larson felt that "the big factor to success is internal management—I don't mean to belittle markets, but they are not the reason why farmers fold. A mistake in management can be very costly. If we didn't have a market setback once in a while we would never test our efficiency. In the good year one is inclined to get careless and expenses increase faster than income."

The Hermansons and Heline place more stress on the need to have a contract market. The Hermansons have contract marketed for 20 years. Part of their production is contracted directly, and part is pooled under contract with other producers. The Hermansons are in the process of contracting their sales as much as three years in advance. Pete Hermanson said, "This looks very attractive because it enables us to get real economy of scale without taking tremendous risk." Heline has contracted his turkeys every year since 1940. He contracts 75 percent of his production and sells over a six-month period. Heline said:

A turkey producer absolutely must have a marketing agreement. His first consideration is that his turkeys are accepted and his second consideration is a basic floor price with escalator provisions. Before I would spend more money on expansion I would want a long term sales and production contract with a processor. The investment in turkey production units is great and contrary to the historical trend of land these units do not increase in value over the years.

Although Larson maintained he was a risk taker and did not ap-

pear quite so concerned about contracting the sale of his production, he stated that his birds were committed for slaughter at the
date of purchase. Larson said:

> Our turkeys are marketed with one processor . . . who knows exactly
> when to expect them months in advance. Sometimes we commit our tur-
> key kill when the eggs are hatched. We sell over a period from July
> through January. Because we are able to finance our own production we
> do not contract price, but only have a firm agreement to sell. Our proces-
> sor has treated us well, but in recent years a strike at the company cost us
> a pretty penny.

When these three farmers were asked what the minimum size of
a turkey operation had to be for efficiency they agreed that 30,000
was the very smallest. This, ironically, was the national average of
all turkey enterprises in 1972. Larson commented, "30,000 is a
good family sized unit and has some built in economies because it
is a family operation. A family could average about $30,000 net
throughout the cycles and would not have to overwork to do it."

From a personal standpoint these three farmers were of the
opinion that 250,000 turkeys would have the greatest economies,
and two of them preferred that number because they saw full
economy of scale working efficiently at that level.[2]

The Velos

When asked about the beginnings of his farming experience,
Eddie Velo responded:

> My grandfather came from Norway in 1868 and settled in the Rothsay
> [Minnesota] area in 1870. My dad, Martin, was two and one-half years old
> then. When Dad started farming in 1898 he bought eighty acres of the
> homestead which we still own. Dad gradually acquired 340 acres and had
> ten horses to do the work. The land became so infested with quack and
> thistles by 1926 that the family decided that they had to purchase a tractor
> to control the weeds or quit.

Eddie Velo was 17 when his father, Martin, purchased an I.H.C.
10-20 tractor for $800 in 1926. Eddie became more interested in
farming than ever because he thought the tractor "was marvel-
ous." The Velo boys put old car headlights and a generator on
their new tractor so that they could work day and night. Soon the
Velos were farming 840 acres and were surpassing their less progressive neighbors. A John Deere tractor was purchased a year
later. More land was added and a turkey operation started. Now
the Velos were able to cut down their labor force to one hired man

because the tractors were so much more efficient and time saving than horses.

Martin Velo, like most area farmers, had a small flock of chickens from the time he started farming. In 1917 he increased his flock to about 400 birds and became a commercial poultry farmer with a good record. Because of his success with chickens a local hatchery man asked him in 1928 to carry a flock of 400 turkeys for hatching eggs. Turkeys provided a chance for extra income without increasing the need for extra labor and proved to be "a good deal." All the nests were under shelter and near the farm grove. Lights were strung out between the trees. Wind chargers with their storage batteries provided the current for the lights.

Eddie Velo and his brothers continued to farm as a single family unit until 1940. By then the brothers had married and desired to be "on their own." That year Eddie Velo had 280 acres of his own land, 800 turkeys, 6,000 chickens, 14 milk cows, 4 sows, and a new I.H.C. Model H on rubber tires. Mrs. Velo kept very good records of their farming operation, and it soon became obvious that chickens and turkeys were the most profitable enterprises. So in 1942 hogs and milk cows were dropped. Velo recalled that during all of their years of farming they had operated on a cash basis and that from 1928 to 1954 they did not have a single year with an operating loss. Throughout the 1920's and 1930's the family used its own funds to make real estate purchases—a remarkable achievement in light of the unfavorable situation of agriculture in the nation for those decades.

The Velos continued their steady progress. After dropping other farming enterprises the chicken and turkey flock was increased annually. Their records proved that 1955 and 1956 proportionately were their years of greatest margins. By 1954 they were raising 6,000 laying hens and had four groups of 30,000 broilers plus 4,000 turkeys.

In 1954, 1955, and 1969 either windstorms or fires destroyed their turkey barns filled with turkeys. Velo recalled his banker's comment: "Anyone can build a barn, but it takes a good man to rebuild from disaster. Anyone can build, but only good management can pay for the buildings." Each time the Velos rebuilt their buildings were bigger than ever. In 1961 a market slump and in 1964 a serious disease epidemic among the birds brought heavy losses to Velo's turkey business. A long history of profits, however, plus excellent records enabled the family to overcome disaster, and they continued to grow.

Mrs. Velo had been the record keeper for the farm business in addition to raising a family of eight children. But by the 1960's the oldest daughter had taken over the records, and the boys were helping with the turkeys. By that time the farm had grown to 1,200 acres and 200,000 turkeys and was supporting the 10 Velos besides employing two single men and five family men. Five different crops of turkeys were being produced each year. This included fryers, heavies, and breeders.

One of the biggest problems the Velos had to face was the removing of the manure from the buildings. In the early 1950's they built a 40- by 300-foot building with holes in the floor so that the manure could be shoved through the openings to fall into a manure spreader parked below. Cleaning the barns was such a time-consuming and costly process that Velo said he was kept awake night after night trying to "dream up an idea." Finally he came upon the idea of building a small three-wheeled vehicle with the single hind wheel on a pivot so that it could get around in the turkey houses.

Velo discussed the idea with Cyril and Louis Keller, skilled blacksmiths, at Rothsay. The Keller brothers went to work on the idea and by using spare parts, including old jail bars for loader tines, developed a 1,300-pound belt-driven mechanical scoop shovel. The machine was powered by a 6.6-H.P. Kohler engine and could lift 500 pounds. Total cost of parts and labor for the machine was $1,400. Velo commented that he got the idea when he saw a rear-mounted loader on a WD-45 Allis Chalmers.

The original belt-drive was replaced with clutches, and the next model made by the Kellers was a four-wheel drive for better traction. Later the "Bob Cat" with larger engines became the universal scoop shovel for farm and industry. The Kellers made seven machines before they were employed by Melroe Manufacturing Company at Gwinner, North Dakota. Through this transition the Melroe Company became the first major producer of one of the significant mechanical contributions to twentieth-century agriculture.

As Eddie Velo described it:

That three-wheeled powered scoop shovel was the key to our expansion. Manure was the greatest mechanical problem we had. Our labor costs were reduced by 50 percent. Conservatively it replaced four men with scoop shovels. Barn cleaning was not popular work, but everyone wanted to drive the Bob Cat. It took some skill, but everyone liked the job. I suppose the only real problem I have left is disease. One has to be prepared to act quickly and always expect the worst.

Marketing, an ever present problem for the turkey producer, had bothered Eddie and Leola Velo, too. But in 1956 they minimized this problem when they became associated with about 50 turkey growers and formed the West Central Turkey Cooperative at Pelican Rapids, Minnesota. Velo said, "This gave us a dependable market and eliminated our major bottleneck."

Velo became the first president of West Central Turkeys, which soon experienced a rapid growth in members and production. His turkeys are contracted with the processor on the day he buys his young turkeys, but he also keeps an option open on the date of sale price. Because he sells year around he comes out very well in price. In his words: "If anything ever happened to the co-op I would contract to market my production without a contract price because this would give me maximum freedom to produce without the restriction of an established price."

In the 1940's the Velos were producing only about one-fourth of the total corn needed for their turkey operation. The farm continued to expand, and so did the turkey operation. This caused Velo to become interested in the elevator, storage, and feed-preparation business. His records indicate that it was more profitable to expand turkey production and feed processing than to buy land. Even though it took considerable capital to build an elevator, it gave the Velos the advantage of buying grain during the price lows and helped them to have quality control over their feed supplies. It quickly became apparent to the Velos that raising much of their own crop, plus having an elevator and feed mixing plant in addition to producing turkeys, gave them great internal stability.

Since most of the eight Velo children were interested in farming and because of the size of the combined enterprises, the Velos decided to do the next obvious thing in maintaining a closely knit family farm—incorporate. Three corporations were formed. Evelo, Inc., was the land-holding company and turkey-raising firm. Damart Turkeys, Inc., was the turkey-raising company for the children, who were active in the whole farming operation. Velomix, Inc., the third corporation of the Velo family, was established to keep the feed-buying and -mixing business separate from the other family enterprises. Eddie Velo commented about the corporation, "If family farms are going to survive, families must have some way of passing on what has been built." Mrs. Velo added, "I think being incorporated created more interest and gave more incentive to the children because they could clearly see their part in the operation."

The Velos are a strong family unit, with a history of four generations on the farm that was homesteaded in 1870. In many respects they typify the past, while at the same time portraying what is to come. The Velos believe that "the strength of the family unit can't be beat." Eddie Velo said, "We have been at a plateau and now we think we must expand. Personally, I think there is no upper limit to how large a farm operation such as ours can become as long as you expand gradually."[3]

I Really Enjoy Chickens

When Dean Myhro lost his mother at the age of three he became the ward of his grandparents, who had a seven-section farm near Hamburg, North Dakota. Dean remained with the grandparents through the eighth grade. Then he had to live with aunts and uncles at Deer River, Minnesota, because Hamburg had no high school.

At the age of 13 Dean was fortunate enough to get a job at Peck Hatchery and Feed Store at Deer River to earn some money for his expenses. In his freshman year he became involved in a chick-raising project through a Future Farmers of America unit and his agricultural classes in high school. By his senior year he was sent to a chick-desexing school. Dean had special permission to attend high school three days a week, leaving him three days for desexing chicks at the rate of 1¢ per bird. He could do 1,000 chicks in an hour but admitted that most flocks were so small that he could not get many hours' work.

All through high school Dean maintained a flock of 300 to 500 chickens in a discarded chicken coop near his aunt's house. His hatchery earnings financed the chicks, feed, medicine, and equipment. He also held the job as usher and janitor at the Deer River Theater, which paid $5 a week for seven nights and Sunday afternoon. Dean Myhro gave all of his earnings to his aunt and uncle for room and board. After graduating from high school he left for four years of military service.

Upon completing his military service late in 1945 Dean married, took a job first as a Dakota harvest hand and then returned to Peck Hatchery and to desexing at the same 1¢ rate per bird. In 1948 Dean and Lillian Myhro decided to start in the hatchery business on 100-percent borrowed funds. Mr. Peck, his former employer, loaned him a brooding battery. Next he was able to buy 10 used incubators for $3,000 on the installment plant, and a used pickup truck for $300.

"Hatching was only a three month business in those days," Myhro said. So to earn a year-round income Dean and Lillian went to dressing chickens for customers. They purchased a new pickup and traveled to farms to get the chickens. They charged 10¢ per bird for picking the feathers and 25¢ for picking and drawing. Working together Dean and Lillian could do 300 birds a day. Many of these chickens were returned dressed to the farmers, who kept them for personal use. Others were bought by Myhro and were sold to retailers.

When not occupied with hatching chickens or dressing them Dean attended Bemidji State College for two years. "That $105 a month from the G. I. Bill sure helped on our living expenses," grinned Dean, "but all I could think about when sitting in English class was how does this help me in the hatchery business? I just couldn't get my mind off chickens."

The Myhros were living very inexpensively in a rented farm place about two miles out of Bemidji, with a small building for their hatchery. One day Nick Welle, a Bemidji banker, visited Dean and told him that the old flour mill was for sale and that he would make a loan for the full $10,000 needed to buy it. With this purchase the Myhros found themselves selling processed feeds and operating a feed mill and a hatchery, all housed in the stately old mill.

Because he had no resources to buy his feed and supplies for resale, Dean obtained a warehousing consignment arrangement with the North Dakota Mill and Elevator at Grand Forks, North Dakota. Each Saturday he had to pay for the feed sold during that week. Dean said, "That was the only way I could generate cash flow, but I sure learned the hard way about accounts receivable. I virtually captured all the business of the small dairy and chicken flocks—herds up to fifteen cows and flocks from fifty to 150 birds—and was it hard to collect from those people."

Soon Dean was asked to become the feed manager for his biggest competitor and was given the opportunity to take his hatchery with him. Shortly after, he became involved in a sideline turkey-raising enterprise. After one year his partner, George Guyan, decided to withdraw from the partnership. Even though Guyan had another business he was unable to stand the tensions of the fluctuating turkey business. By luck the business ended with a profit in its first year, and Dean was able to get financial help to carry on alone. From 1951 to 1959 the volume increased each year, so by then he had 2,000 turkey breeder hens for hatching eggs and a 13,000-bird commercial growing flock. In those nine years Dean learned that he could make the poultry business pay when he took care of it personally. His confidence was strengthened by the experience.

In 1959 Myhro became the manager for Red River Hatchery in Moorhead and in 1960, with the help of partner Fillmore Trites, purchased the firm from Ed Moe. Three years later, aided by his banker, Al Severson, Myhro started a commercial pullet operation. Myhro commented on his financing at that time: "There was not another banker in the entire area who would have worked with me like Severson did. Why aren't there more good agricultural bankers like that?"

From 1964 through 1972 Myhro built a new building each year, which meant nine buildings with a capacity of 129,000 chickens. A separate organization, called the Myhro Pullet Farms, had been formed as a business independent from the Red River Hatchery. As his business grew Myhro began to realize potential savings by processing his own feeds. If he could just save a few pennies per bird it would amount to a handsome profit each year. After several years of planning he built a feed mill with a daily manufacturing capacity of about 75 tons. Myhro was happy when he said, "That feed mill saved me. I would be broke today if I had not built it. Remember eggs dropped to twenty-one cents a dozen and then the Russian deal pushed the price of grain way up. There would have been no way of avoiding bankruptcy."

However, Myhro was not interested in enlarging his business by raising his own grain. He said, "We use nearly two semis of corn a day and that's more land than I would want to farm. Besides, I like chickens better." The mill is used to process approximately 900 tons of feed a month for Myhro's 200,000 laying hens, located on the farms of 12 contract producers, and 130,000 pullets in his units. The mill is also used to supply feed to a very limited number of producers and hatcheries that are associated with Myhro, but he is not interested in seeking outside customers. Only two kinds of rations are made, both in bulk; and only one man, Jim Flores, is needed in the completely automated mill to run it at top capacity.

Marketing is the biggest single problem of the Myhro enterprise. Dean Myhro's personal preferences, however, are financing and chickens, and not marketing. "I am husbandry minded; I really like working with chickens. I have the feel for them and understand what has to be done. I prefer to leave marketing to others." Some Myhro eggs are marketed in his own Big Valley Brand, in Minnesota Certified Double A cartons. The remainder are sold directly to egg processors. Myhro does not like selling to large warehouses because his eggs lose their identity and he does not get the benefit of his quality production. All of Myhro's laying hens are sold at the end of a 20-dozen laying cycle, that takes place in just over 12 months. These hens go to a southern Minnesota processor to be made into chicken pies and T.V. dinners.

Thirty-five people are employed in the poultry business operated by Dean Myhro. They use 14 trucks to haul eggs, feed, and chickens. The labor of about 1½ persons is required to feed and care for 60,000 birds in the automated cage systems used on the Myhro farms. It takes four or five employees to clean and maintain those same facilities in the older buildings. In his newer buildings, which use complete automation for feeding, gathering, and packing eggs, only three people are needed for 250,000 layers.

In his building program Myhro always seeks maximum labor efficiency because he finds that there are very few people who really enjoy working with chickens. He has traveled extensively to find a solution to his waste-handling problems. On a European trip looking for answers to his problems, he discovered that manure fermentation pits were working well. Myhro installed them and has not had to clean them in the first five years of use. There is no indication that the pits will need any attention for five to seven more years. This eliminates Myhro's greatest single labor problem. Great care must be taken that no cleaning solutions are dropped into the pits, or the enzyme action is stopped.

When Myhro was asked in 1970 what he thought about the future of the chicken business, he answered: "Like anything else in agriculture you can no longer expect to make a living on a small unit. In the mid-1960's within fifty miles of us there were 32,000 commercial layers in flocks of 4,000 hens or more. Today those people have 350,000 hens. I don't believe I can stop growing."

Myhro thinks a husband and wife team could make a living on 50,000 laying hens, but they would have to work close to a supplier and to a market outlet. Such a small unit would be a risky living, in his opinion, if it were the only enterprise. Myhro knows that the cost of production per dozen eggs rose from 21¢ in 1970 to

42¢ in 1975, and profits became heavy losses for those who were not integrated.

In 1975, five years after the initial interview, Dean Myhro's chicken farm is larger and more integrated than ever, and still growing. He was asked why he was still growing. After pausing for some time with his hands clasped behind his head and leaning back in his desk chair, he said, "Lillian has asked me that many times. My business is fun and not work. I don't know where I could have done as well, especially since Lillian and I started with nothing. I really don't know what has motivated me, but I know you are broke if you stand still. I guess every guy needs a hill to climb."

Dean Myhro, orphaned early in his youth, found challenge in raising chickens in an area where beets, potatoes, and wheat are king. With Myhro, chickens are king. He does have a cattle ranch of some size, but that's just something to keep his mind busy during his free time. Or is it another hill?[4]

A No. 1 Turkey Man

Earl B. Olson, Willmar, Minnesota, the nation's largest independent producer of processed turkeys, is responsible for 6 percent of the nation's turkey production. This makes him a number one turkey producer in the nation's leading turkey state.

Earl Olson's father, Olaf Olson, came from Sweden in 1893 at the age of 19 and quickly established himself in farming near Murdock, Minnesota. Olaf Olson purchased his first farm from profits made in horse trading. During World War I he profited greatly from this business.

Throughout the 1920's the Olsons made a living from farming with horses. They used an eight-horse hitch on a three-bottom plow and milked 25 cows daily. The Olsons also had chickens to tide them through hard times.

While Earl Olson was attending country school he milked six cows daily, before and after school. At the age of 14, eighth-grade graduate Earl went to the Minnesota Agricultural High School in Morris because he wanted to learn more about farming. But Morris was 40 miles from the family farm, and Earl had to live away from home. While attending high school he had a steady job in Morris shoveling coal or unloading lumber from the railroad cars. These jobs earned him 12¢ an hour.

Olson graduated from Morris Agricultural High School in the spring of 1934 and took a job as butter maker's helper at the Murdock Cooperative Creamery at $35 a month. He admitted that he "had pull" to get the job because his father was on the creamery board. Washing cream cans was his major duty when he first started, but the manager had a weakness for weekend parties, and before long the 19-year-old Olson was testing cream, churning butter, fueling the boiler, and keeping records. By the time two years had passed, Olson had worked at every job in the creamery, except the final record keeping. The board saw that the manager was failing in the performance of his duties and increased Olson's salary to $37.50 a month in order to encourage him to stay.

In the summer of 1936 the nearby Swift Falls Cooperative Creamery decided to either hire a new manager or close its doors. Olson was interviewed for the job as manager, and in September, 1936, was offered the position at a salary of $80 a month plus a 1-percent commission on gross sales. After moving to Swift Falls Olson borrowed $500 for six months from Russell Hanson, the banker. He was able to repay that loan in five months, and shortly after he was sure that Hanson would loan him $1,000 if he presented a good proposition.

With that knowledge Olson proposed to the Swift Falls Creamery board that the creamery should put trucks on the road to collect cream and eggs. The board declined because they did not have the funds and were reluctant to consider borrowing. Then Olson asked the board if they would permit him to buy a truck to collect eggs and cream. He was willing to take the risk, for he knew he could borrow the money. He also realized that his commissions would increase as the creamery sales would grow. The board agreed and Olson started a new venture.

The Olsons traded in their family car for a new 1936 ¾-ton pickup that cost $1,200 complete with a box for eggs and cream. Russell Hanson loaned $1,000 on the deal, and Olson recalled paying the money back two months before the note expired. During the winter of 1936-37 Olson got up at 4 a.m. to fire the boiler and go on the egg and cream route before it was time to churn. Once the route was established Olson bought a second truck and hired a man to collect the cream and eggs. Olson proceeded to establish another route of new creamery patrons. Before long three routes were in existence, and the second truck was going full time, making three trips a day.

By then the volume at the creamery was large enough to keep Olson busy all day churning butter and candling eggs. He remembers:

All the time my salary grew because of the 1 percent commission on sales. Next thing I hired another man and purchased a third truck. Acquiring a line of feeds and seed corn was the next step for the drivers could deliver those commodities while making their routes. Finally we started buying poultry and turkeys and the truck route grew almost weekly. I didn't want to buy any more trucks so I turned to contracting with men with trucks hauling cream, eggs, poultry, turkeys, feed, farm supplies, and eventually livestock.

Cold-storage locker plants were being introduced into the rural areas in the late 1930's. The Swift Falls Cooperative Creamery board soon recognized the benefits of such an enterprise and added one to their growing business. By 1939 the creamery was making money in every department and was growing rapidly. By 1942 this once nearly bankrupt creamery was one of the largest in the region and was criticized by its "cooperative competition" as being out of its mind for paying such high prices for eggs, cream, and poultry.

Earl Olson said, "Our competition did not realize that by getting the farmer's egg, cream, and poultry business we also got his locker, feed, seed, fertilizer, and farm supply accounts, all of which were profitable. We just had to break even on the main lines and by increasing the prices on what the farmer had to sell we could get new customers daily."

Success did not come without adversity for Olson, but never did he slow down because of financial or personal reverses. The real test of the man came in 1938 when Olson's butter-making helper accidentally dumped a thousand gallons of scalding water on him. As Olson lay in the hospital bed recovering from burns, he was unable to do anything but think. He realized that his future lay entirely in the hands of the board of directors of the creamery and that they would be satisfied only with steady progress.

Olson's inborn caution developed from observing how his father's farming enterprise paid off in the 1920's and 1930's. Quite early in his work in the creamery he convinced his board to carry insurance for him, and he also carried a personal policy. The creamery board also permitted him to draw salary and commission while disabled in the hospital. During those eight months of dreaming and thinking in the hospital he concluded that turkeys were the surest way to success if he could use his surplus insurance payments to buy turkeys and equipment. It took some time to recover and to formulate exactly how he would approach the creamery board with his proposition.

Once the creamery was making profits in every line, he informed his board of directors that he felt he wanted to go into the turkey business as a sideline. They approved of it because they hoped for an increased feed and turkey business for the creamery. In 1941 Olson raised 300 turkeys and made a profit of a dollar a bird. In 1942 he raised 600 turkeys, and in 1943 5,000 at the same rate of profit per bird. Each year his sideline in turkeys grew until 1948 when it reached 12,000 and his profits zoomed to $4 per bird. That was one of the best years he ever had in the turkey business.

Success went to his head. Olson envisioned that the more birds you grew the more money you made. So why not get bigger? In 1949 he ordered 35,000 day-old poults, but he was able to get only 30,000 because all the other turkey growers wanted to expand too. That year Olson experienced a loss of a dollar a bird. This lesson gave him a new perspective of the business. For even though his base salary as creamery manager was only $1,200 a year his commissions of 1 percent on gross sales amounted to $25,000. He was the creamery's biggest customer and always paid full price for his purchases.

Olson was not defeated by the $30,000 loss in his sideline turkey venture. Instead he was determined to recapture his losses because he believed that great commercial opportunities still existed in the turkey business on a large scale—so he forged ahead. By this date he had purchased the creamery at Marietta with Gilbert Ahlstrand because they both knew the creamery business and expected to profit from its operation.

In 1947 he had convinced the board of directors of the creamery to establish the only processing plant exclusively for turkeys in a five-state area. This completely automated plant was very efficient. By 1949 more turkeys were available than the plant could process, and the profits from processing were good. Olson attributes much of the early success of the Swift Falls turkey plant to the availability of good labor recruited from nearby farms. In 1949 the turkey plant had 100 employees in Swift Falls, a village of only 75 people. Farmers anxious for a second job made excellent employees and seized the opportunity for alternative income as farms mechanized and required less manpower.

In 1949 All State Supply of Willmar wanted to sell its turkey processing plant. Olson encouraged the Swift Falls Cooperative board to buy the plant, but the board felt that it would be expanding beyond the limits of its constitution if it did so. Instead they encouraged Olson to buy the plant and lease it to the creamery. Olson called on his banker friend, Russell Hanson, at the Swift

County Bank at Benson, to borrow $75,000. The bank's loan limit was $15,000, and Hanson declined. Olson was driving away from the bank when Hanson called him back and suggested that he and Olson should try to get a loan from the Midland National Bank in Minneapolis. The next day Earl Olson contacted that bank and with the help of Russell Hanson quickly obtained a $75,000 loan.

The loan was made for 10 years and was based on the strength of a lease with Swift Falls Creamery, which was to pay 1ᶜ a pound of turkey processed plus all building expenses. In the first full year of operation Earl Olson's $75,000 investment returned $40,000 paid on 4 million pounds of turkeys dressed. Olson's creamery salary at this time was still $1,200 plus $26,000 in commissions, besides income from a fleet of trucks used in hauling to and from the plants. Encouraged by his success he continued to expand his turkey production.

The catastrophe in the turkey business of 1949, in which Olson lost considerably, proved to be another turning point in his career. Because turkey growers were financially desperate that year they demanded cash for their turkeys rather than selling them on consignment. The creamery board would not consider paying cash, and Olson was forced to seek his own financing to pay for live turkeys on delivery if he wanted to maintain his processing business. This disagreement with the board caused him to leave his position as manager of Swift Falls Cooperative Creamery. Two years later the creamery was bankrupt. Olson had secured all of their business.

By 1952 there were 400,000 turkeys on Earl B. Olson Farms, and his processing plants were growing rapidly because they were competitive bidders for as many turkeys as they could handle. In addition, his plants were more efficient than those of many of his competitors because his personal turkey production was structured to be marketed during the months when farmers traditionally did not have turkeys to sell. In the 1950's most live turkey production was marketed and processed during the last four months of the year. Olson realized that if he could market his own turkeys at other periods he could avoid the glut on the market and reduce the fixed unit costs in his plants. His theory proved correct, but he had to develop a year-round market for processed turkey.

As Olson's processing facilities grew he cut deeper into the market of his less efficient competitors. In 1957 he assumed operations in the plant of a former competitor at Litchfield. The defunct Litchfield plant was for sale, and Armour and Company were the only known opposing bidder. The Armour people were so sure that they were going to buy the plant that they had started to move supplies to it. Olson and his consultants calculated exactly the price Armours were bidding on the Litchfield facility and they jumped the bid by $800 to $200,800 and were the winners. By 1959 the Litchfield facility was in full production, enabling Olson to process two million turkeys, 500,000 of which were produced on his farms.

At this time Al Huisinga of Willmar Poultry and Egg Company cooperated with Olson by specializing in hatching turkeys and selling them to him. This freed Olson to concentrate on raising and processing turkeys. By 1959, 60 million pounds of turkeys were processed in Willmar, which was considered the world's leading turkey center. Willmar Grain Terminal, also a part of the Olson enterprises, had a million bushel capacity and manufactured 20,000 tons of feed for turkey producers in the area.

Overproduction and low turkey prices in 1951 again dealt a severe blow to turkey producers. Earl Olson was not immune to it and experienced one of the greatest financial reverses of his career. However, Olson found new cost savers and recovered from a near-fatal financial blow.

In 1960, when Olson was just getting back on a sound financial basis, another competitor, the Melrose Produce Company, was having financial trouble and wanted to sell. Donald Sonstegard of Paynesville, deeply involved in the Melrose firm, wanted very much to sell Olson that business. Within less than a week Olson and Sonstegard were able to complete a purchase agreement. Olson purchased the Melrose Produce Company for "a penny down and ten years to pay" and increased his processing capacity by another one million turkeys. In 1963 his farms produced 700,000 birds, and his processing capacity had increased to 3 million.

With the enlargement of his business greater total integration was deemed necessary to have better control of the turkey production as well as reducing risks. Therefore, in 1963 Earl Olson took the inevitable step by getting into feed processing on a large scale. His 20,000-ton manufacturing capacity at the Willmar Grain Terminal could no longer supply his needs. In 1963 the feed mill at Swift Falls was purchased to be used exclusively for turkeys on the Olson farms. In 1965 the Farm Service Elevator at Willmar was purchased, and in 1966 a feed mill and buildings of a large chicken-growing facility at Brainerd were added. The Brainerd purchase enabled Olson to increase his turkey flock to one million.

Expansion, however, came at an inopportune time, for 1967 was another difficult year for the turkey industry. Olson lost nearly a dollar per bird on his million turkeys and also experienced financial setbacks in his processing business. To overcome his reversals Olson made three major decisions in 1967, all of which proved to be very profitable. To improve the margin of profit in his processing plants Olson decided to bypass the traditional marketing channels of brokers and commission firms and trade directly with the chain stores. The Earl Olsons had one daughter, and after considerable discussion the family decided to use her name for the Olson brand turkeys—Jennie-O-Foods. Probably more processed turkeys have been marketed under the Jennie-O name than under any other brand.

At first Olson had difficulty marketing because only the very largest grocery chains could handle a semitrailer or carload of turkeys. To be an efficient marketer Olson had to sell full truckloads at a single stop. This forced the adoption of another innovation. Departing from the traditional practice of selling whole frozen turkeys wrapped in plastic, he merchandised processed turkeys in the form of turkey rolls and turkey parts. Marketing processed turkey products instead of just the traditional "holiday bird" required selling the stores on the idea of year-round turkey sales. Consumers had to become accustomed to the idea of eating turkey 12 months a year instead of just during the holiday season. By 1970 processed turkey caught on, and Olson was selling full semiloads of Jennie-O products to chains that had as few as 10 stores. Now Jennie-O brand turkey products were being sold in one-half of the chain stores in the United States.

In 1972 Jennie-O brand made up 70 percent of the 105 million pounds of processed turkey products sold by the Olson-owned firms, and the other 30 percent was custom packaged for large retail outlets. Consumers had taken to the idea of year-round turkey eating by 1972, for only 45 percent of the annual sales were made in the four traditional turkey months.* Over 6 million of the 19 million turkeys grown in Minnesota were processed in Jennie-O plants, and Olson's farm production provided nearly one-half his total processing output.

Anticipating the changing consumption pattern, Earl Olson

---

*The traditional turkey months are October, November, December, and January.

shifted his turkey-production cycle to take advantage of the new market. From the day he had owned his first processing plant he had geared his farm's production to the off season to improve efficiency of that plant and other plants that he had acquired. Olson's records proved that every day his plants were idle they cost him $15,000, so the secret was to keep them going. By 1968 he geared those farms to year-round production by going to turkey confinement. In this manner production facilities were used full time, and fixed cost per bird was reduced by 40 percent. Olson reasoned that if southern growers could afford confinement to keep their birds cool, northern growers could afford to build structures to keep them warm.

Other turkey growers have changed their production patterns; but because nearly all production is contracted for when the young turkeys are purchased, it is possible to anticipate the slack period for the processing facility. Earl B. Olson farms are flexible to the point that they gear their production to the time when birds are most needed by the plants.

The adverse year of 1967 offered Olson another opportunity to test his judgment. He reasoned that many turkey producers were so demoralized by their 1967 losses that they would quit or cut back production in 1968. He decided to go ahead and expand his turkey numbers as much as possible for the 1968 season. He increased from one million birds in 1967 to 1.5 million in 1968 and reaped the profits which he had anticipated because of defeatism in the industry. Some people call this luck, others call it good psychology, and some say it was sheer guts. Production on the Earl B. Olson farms increased to three million by 1973 in addition to one million birds in 13 other production units in which Earl Olson has a part of the ownership.

Processing turkeys for sale in many different packages 12 months a year increased the labor cost by 300 percent per bird. In spite of the most automated equipment in use in Olson's plants, hand labor requirements are still high. Only more modern processing plants' facilities could offset that increased cost. Olson explained:

Something has to pay the $4 million payroll of the 1,100 employees in the processing plants. I think a family farm operation using three people to produce 100,000 birds is a nice little unit of production with real money making possibilities. We need four people to produce 100,000 birds. Dr. Peter Poss aided by Dr. Jim Johnson are in charge of our growing, research, nutrition, feed formulation, medication, and experimentation with equipment. We need 30 people to process 70,000 tons of feed and we do all of our own veterinary work so it takes more people. It seems like we are not as efficient as the family unit, but we do a much greater part of the total production because we are integrated. This gives us flexibility and some immunity to the uncontrollable variations in the industry. Of course, the total investment is much greater.

At the time of this comment Earl B. Olson's family enterprises were grossing $30 million a year. They were adding $3.75 million expansion of facilities that would employ 200 additional employees but would increase gross sales to $60 million annually. Olson said, "I think we are successful because of our size, rather than in spite of it. We can make decisions rapidly and don't have to spend time in committee meetings as larger firms do."

The combined Olson operation has over one hundred trucks to haul feed, equipment, and live and processed turkeys. Forty-five tractor-trailer units are used to haul turkeys from five states for processing at the central Minnesota plants. The additional volume of turkeys hauled in boosts the total production to about 7 percent of all turkeys processed in the United States. Olson's farms alone produce nearly 3 percent of the nation's annual output of turkeys, and he thinks there is still room for expansion.

His current goal in processing is to develop a handy small package so that consumers can purchase a limited amount of turkey just as they do other meat products. Developing a direct market with the chain stores and institutions is part of the overall problem facing Olson.

Olson's turkeys are produced on 25 farms totalling 4,000 acres, and not a foot of sod is turned for crops. Under such conditions disease is the major concern of the management. Therefore, not more than 500,000 birds are concentrated in an area of 5 square miles. Under present land ownership it would be possible to expand to 5 million turkeys before the saturation point would be reached. Olson admits that neither he nor his management team have any hesitation about growing to that volume. "We find as we keep on growing that there are new efficiencies that we had not anticipated coming into our operation. Our volume keeps pushing the price of our inputs down and also increases our ability to market."

Basically the Olson enterprises have only two missing links in a totally integrated turkey enterprise. They do not retail, nor do they raise their own crops. There is no interest in selling turkeys directly to the consumer because a large investment is required to sell their product and the margins are not large enough to justify it. Production of corn and soybeans to enable them to provide their turkeys with home grown feeds is also not considered practical. According to Olson:

At one time we could have purchased elevators for country buying of corn and soybeans, for except for the government storage programs they were not faring too well. Those were in the days when we had little money for such ventures. Today because of the business we have given them country elevators in our area have done well and are not for sale.

Turkeys require quality corn and beans, and no farmer can expect to produce that top-quality feed each year unless he has constant good weather and excellent management. Olson reflected:

Considering the investment and skilled management needed for crop production we think that is a whole new ball game. Except for 1973 and 1974 I doubt that we could have produced our own feed at a profit—so why bother? Area farmers who once resented us because we grew and processed our own turkeys now love to see us coming because we buy over 3 million bushels [the production of more than 30,000 acres] from either them or their elevators. Our area has changed from a corn surplus to an importing area. The market has reversed from 10 cents under Minneapolis to 5 cents over. We have not only given their families jobs, but have increased their profits from their traditional cash crops.

When one of the nation's largest turkey growers and its leading independent turkey processor was asked how a poor immigrant's son had achieved such a high position in American agriculture, Olson retorted:

Having eight months' time to think about my future from the hospital bed was the turning point. Then there were always those negativists who said the little guy could never compete with big industry. That alone was enough to challenge me. I learned exactly what records could do when I was still a buttermaker's helper and business-wise I have always been able to determine what I was doing; this has kept me on good terms with the bankers.

Russel Hanson, my lifelong friend and banker, deserves lots of credit for my ability to borrow money, which to this day is probably my biggest job. Anybody can take money to the bank, but getting it from the bank in the first place is another ball game. We have the financial doors open, but they require constant attention.

Olson admitted, "Probably the real reason for this growth is the challenge—just seeing if that Swedish kid from the farm could do it." Mrs. Dorothy Olson grinned and agreed that to be challenged was the fun of it all. The Reverend James McBride, one-time pastor to Earl and Dorothy Olson, said, "Earl was always positive, whether things were good or bad. He was very analytical and when he moved you could bet it would be the correct move. And you knew Dorothy was backing him all the way."[5]

NOTES

[1]"Turkey Growers' Wish," *Agricultural Situation*, LVII, No. 10, November, 1973, pp. 2-3; Ann Burckhardt, "Americans Gobble Up Willmar's Production," *The Minneapolis Star*, April 19, 1972; Norman M. Coats, "Why Purina Got Out of Broilers," *Farm Journal*, XCVII, No. 4, April, 1973, p. A-8.

[2]Interview of John Heline, Pierson, Iowa, February 23, 1972. Mr. Heline started farming in 1940; Interview of Pete A. Hermanson, Woodland Farms, Story City, Iowa, February 22, 1972. Hermanson is general manager of a family corporation farm in which his parents and the Hermanson brothers make up all the corporate officers and of which they are the sole owners; Interview of Lyall P. Larson, Sargeant, Minnesota, March 26, 1972. Mr. Larson and his two sons, Vance and Larry, operate the several family enterprises involved in the Larson farms.

[3]Interview of Eddie A. Velo, Rothsay, Minnesota, February 26, 1973; *Dr. Salsbury's Turkey Views*, IX, No. 1, March, 1958, pp. 12-13; *Norbest News*, XXV, No. 2, Fall, 1960; "A Family Owned Corporation," *The Farmer*, LXXXVII, No. 15, August 2, 1969, pp. 9, 19.

[4]Interviews of Dean Myhro, Moorhead, Minnesota, January 8, 1970, and June 11, 1975; Milton R. Dunk, "I Am Holding Status Quo," *Poultry Tribune*, 80, No. 11 (November, 1974), pp. 12-16.

[5]*West Central Daily Tribune* (Willmar, Minnesota) Centennial Edition February 25, 1971; Interview of Earl B. Olson, Willmar, Minnesota, April 24, 1972; "Earl Olson," *Frozen Food Age*, [n.d.] "Earl Olson—Farmers Produce Company of Willmar, Minnesota," *Processing Equipment News*, XIV, Nos. 5-6, May-June, 1959, pp. 5-10; Greg Lauser, "Minnesota Turkey Producer-Processor Turns Family Firm into $30 Million Business," *Feedstuffs*, (January 3, 1972), pp. 28-29; Virgil Smith, "Willmar Looks to the Future," *St. Paul Pioneer Press*, April 16, 1972; Interview of Rev. James McBride, Barnesville, Minnesota, January 26, 1975; Letter from Earl B. Olson, June 2, 1975.

CHAPTER XII

The Four-Wheel Drive

Its Background

THE 1970 Yearbook of Agriculture, *Contours of Change*, which had as its theme the mechanical, technological, and managerial changes that had taken place in American agriculture, contains the following quote:

> Up to about 1920 the nature of farming had not really changed very much in 100 years. Most farmers were still largely self-sufficient with respect to what they needed for production. Horses and mules were the chief source of power except for threshing. . . . Some folks imagine farming is still like that.

The advent of the tractor in the second decade of the twentieth century has probably been the most striking single change in farm machinery of the first half of that century. Horse and mule numbers reached their peak at about 26 million head in the period from 1915 to 1920 and then declined steadily to about 3 million head in 1959. In 1915 there were only 25,000 tractors. By 1920 there were 246,000, and by 1925 there were 549,000. Nationwide each time a tractor was purchased by a farmer 4⁴/₁₀ horses were displaced and 850 man-hours were saved annually.

Farm tractor numbers increased yearly until about 1965 when 4.8 million machines were in use on American farms. After that date the horsepower per tractor climbed rapidly, and the tractor numbers declined accordingly. The farmers, in an effort to gain greater efficiency and reduce their total labor requirements, were demanding ever-increasing horsepower per tractor. In the meantime agriculture's share of the labor force decreased from 31 percent in 1910 to 4 percent in 1973. At the same time, the farm worker's capacity to provide for others increased from seven persons in 1910 to 56 in 1973.

Man-hours required in agriculture dropped from 23 billion in 1910 to less than 6 billion in 1973, while the population of the nation was increasing from 75 million to 210 million. The tractor, the truck, and electricity had been the greatest direct time savers for the farmers. As early as 1913 farm authors were stating that mechanical horsepower would force the same reorganization on the farms that it did in the factory. "This is a day of bigness and the big way is the cheapest." A 25-H.P. tractor was claimed to have been as enduring as 100 horses and as expensive as 10. With the full development of horse-drawn equipment the horse's thermal efficiency was only 2 percent, and there was no way to increase it.

To those with vision it was clear that the horse era was doomed, but only a few had that vision. As late as the first quarter of the twentieth century the agricultural colleges were still writing bulletins on the efficiency of the horse. A six-horse hitch was the solution to making a quarter section of land a one-man farm, according to Circular 324, published by the agricultural extension of the University of Illinois in 1914. Thomas Cooper of the University of Minnesota, however, warned farmers in 1911 that the cost of maintaining a farm work horse ranged from $75.07 to $90.40 per year and was too high. From 1902 to 1910 Cooper's records proved that the cost of horse labor per hour ranged from 7¢ to 12¢. He warned that most farmers failed to recognize the cost involved. Horsepower was 45 percent of the total cost of growing an acre of corn. Economists pointed out that horsepower cost varied from $450 to $750 per 160 acres of land. Many farmers were guilty of overstocking their farms with horses and also guilty of overfeeding them. Both factors needlessly increased the cost of production. On some farms horses worked only 500 hours per year, while on others they worked 1,300 hours.

The *1926 Yearbook of Agriculture* carried several articles that clearly indicated a change in farm-power use. H. R. Talley noted that farm efficiency rose rapidly after 1910 because of extreme growth in available horsepower on the farms. At the same time, a warning was issued that horse production was declining rapidly in the nation because of the near-total replacement of horses by automobiles and trucks on city streets and on some farms. "With improvements being made in tractors, it is difficult to foresee the extent to which tractors will eventually replace horses on American farms, but it is not likely that the horse will ever be entirely displaced. At least one team will be necessary on most farms," said C. F. Sarle of the United States Department of Agriculture in 1926.

By 1930 the trend toward tractors was so evident that professional agriculturalists were writing about the impact of the new power revolution. One such authority in Minnesota, Professor Cavert, wrote: "These changes in farm power indicate that the family farm of the future will be larger and will require more capital. The farmer will be more dependent on market prices as the farm will become a less self-sufficing unit. He will be more of a specialist than at present. These changes will require an abler and better trained farmer."

It was estimated that between 28 and 34 million acres of crop land had already been displaced by the decrease in the horse population by 1929. Cavert continued, "Evidently, the decreasing amount of feed required for horses and mules has been a more important factor in the agricultural depression than has generally been realized. Also, it is a factor that will operate for several years in the future, as a further material reduction in the number of horses and mules on farms is in prospect."

The tractor revolution was a proven reality by 1920, but it had not come without suffering. Farmers were forced into bankruptcy or were displaced when their more progressive and successful neighbors sought the ever-increasing efficiencies that mechanical power offered. In 1920 over 87 million acres were still required to feed the horses of the nation, even though the requirements to feed horses from the cities had dropped by 6 million acres. In 1945 only 35 million acres were required; by 1960 only 12 million. At that time most of the horses were kept for luxury purposes. The farmers striving for efficiency had freed 81 million acres from horse-feed production to human-food production. This caused prolonged suffering in agriculture while the consumers benefited from continuous surpluses and a relatively decreasing cost of food.

Every tractor purchased helped to decrease the farm labor force, which dropped from 13.5 million in 1910 to 4.2 million in 1973. At the same time, the average farm size increased from 138 to 385 acres.[1]

The Search for All-Wheel Power

The mechanical power revolution during the first half of the twentieth century brought an irreversible change to American agriculture and to society in general. The facts of attaining cheaper and more food were clear, and the nation was better off for it. A large portion of the American farmers, however, sentimentally resisted the change.

During the period after 1910 while the great majority of farmers were slowly becoming accustomed to mechanical power, a few visionaries were looking ahead to the day when a large engine mounted on an efficient all-wheel-drive tractor would power the nation's farms. Such a tractor was slow to become a reality. The 1960 Yearbook of Agriculture, *Power to Produce*, stressing the great technological revolution in agriculture, did not have a reference to four-wheel-drive tractors. Nor did *Land of Plenty*, a 1959 publication of the Farm Equipment Institute, refer to four-wheel-drive tractors. *Men, Machines, and Land*, a 1974 publication of the same organization, had only one small paragraph on four-wheel-drive tractors, but it did contain several pictures of them in action.

The four-wheel-drive tractor was in the process of development for nearly a half century before it became a reality. In 1912 the Nelson four-wheel-drive tractor was produced in three models. This tractor, which also possessed four-wheeled steering, transferred power through a double-driving sheave via hardened chains to the front- and rear-axle units. The three models were the 15-24, which had an engine speed of 1,000 rpm; the 20-28, operating at 900 rpm; and a 35-50, working at 800 rpm.

A few years later Olmstead produced a four-wheel-drive, 28-H.P. tractor powered by an engine that operated at 400 rpm, using force-feed oiling. The four wheels were operated by a chain drive. To keep the weight more properly balanced a 75-gallon cooling system was placed over the front wheels. A tubular radiator with a fan in back of it was used for cooling.

The Samson Iron Horse made its appearance in 1919. This four-wheel drive was chain driven, with independent control for each side of the tractor. Levers were used to tighten the belts that set the tractor in motion and were also used for steering. Reins were fastened to the levers, enabling the driver to operate the machine from a distance. As in the case of its predecessors, little else is known of the Samson Iron Horse.

The Massey Harris Company produced a four-wheel-drive, general-purpose tractor in 1920 that came in four widths to adapt to different row spacings. Like the Samson Iron Horse, this tractor was steered with lever-operated brakes that controlled both wheels on one side or the other. The company claimed that the tractor could be turned within a 6-foot circle. Like its forerunners, this tractor quietly died, but not the idea, for in 1936 Massey Harris introduced the Challenger. This was a four-wheel-drive, rubber-tired, general-purpose tractor in the two-three plow power range. It received considerable publicity and was more widely accepted than previous four-wheel drives. Some thought this might be the long-sought breakthrough. Oswald Daellenbach, Clay County, Minnesota, Extension Agent, recalled that in the very wet year of 1950 a Norman County, Minnesota, farmer was able to work his land long before anyone else in the area because he had a Challenger. The Challenger soon became a sought-after museum piece as it disappeared from the fields, probably more the victim of a depressed agricultural economy than any mechanical weaknesses.

In 1949 the Detroit Tractor Corporation introduced four-wheel-drive tractors in 16-, 35-, and 45-H.P. sizes. The Detroits used stick steering like the track-type tractors and could travel at speeds up to 17 mph.

The following year the National Implement Company produced the Harris Power Horse, a four-wheel drive with a 96-H.P. engine. This large-size tractor had an independent axle for each wheel, with roller chains uniting the front and rear wheels. That year General Tractor Corporation marketed the Powerbuilt, a 36-H.P. four-wheel drive that included a three-point hitch attachment that could be mounted either front or rear. Little resulted from either of these efforts.

During the mid-1950's widespread interest arose on both the farms and at the agricultural colleges over the adaptation of conventional farm tractors into tandem-tractor hookups. Generally the front wheels were removed from both tractors, which were then hooked in tandem in such a manner that they balanced on the rear axle units. The driver rode the rear tractor and via extension levers controlled the clutch, gear shift, throttle, ignition, and steering. Sometimes the front axle of the front tractor was left intact, and the driver operated the tandem unit from the front tractor.

As is true of most innovations, many individuals were working on the same idea at the same time, so it is difficult to determine who actually invented the original model. Cornell University, Iowa State University, and Michigan State University were all studying the idea of tandem-tractor hitches in 1957, and published articles about them.

James A. Garman wrote a master's thesis on "Tandem Tractor Studies" at Cornell in 1957. Garman, like many others familiar with the Corn Belt, had concluded that tractors in excess of 90 H.P. were too expensive for the Midwest row-crop farming. This was because tractors of 50 to 90 H.P. were all that was needed for most of the work. If farmers could put two of their smaller tractors in tandem they could do plowing, discing, drainage, and land leveling, the heaviest jobs, with a 50-percent reduction in labor, and then separate their tractors for lighter work.

Garman established that tandem hitches were practical with two-plow tractors on farms as small as 118 acres if the labor cost exceeded $1.50 an hour. He also concluded that the cost per horsepower using tandem-hitch, wheel-type tractors was nearly as inexpensive as the larger-horsepower, track-type tractor. In addition, with the tandem-tractor unit the farmer had the economy and the advantage of two smaller tractors for lighter jobs. Other agricultural engineers agreed that two "medium-sized" gasoline tractors were less costly than one large conventional four-wheel drive of equivalent horsepower.

Garman, working an experimental tandem arrangement under the direction of Professor William Millies at Cornell University, proved that two tractors in tandem had about 10 percent more power with about 30 percent savings in fuel. The increased efficiencies came because no tractionless wheels had to be pushed through the soil, both sets of drive wheels worked in the same track, and there was less slippage. The only difficulties encountered were in steering, controlling the engine speeds, and engaging the clutches at the same time. The major disadvantage was that more weight was placed on the drive wheels and they lost much of their function as a slip clutch under heavy loads. This tended to put more strain on the engine.

However, these minor mechanical problems were quickly solved, and tandem tractors worked very well. Steering became so easy and positive that even after combining the tractors and larger equipment the headlands did not have to be increased in size. Farmers liked operating the tandem units. Five to seven pins had to be removed, and the tractors could be separated in about 45 minutes.

In 1958 Garman went into producing the full tandem-tractor units that sold for about $2,000. Half-tandem units, which allowed for the front axle to remain in place on the lead tractor, sold for $200. In the next six years Garman was able to sell all that he could produce in a small factory, but he was unable to get a large company to agree to either produce or market the tandem hitch.

The large companies did not consider the market potential large enough for them to become involved. The greatest problem in marketing was that most farmers wanted to have a hitch that would adapt to the tractors they already owned. This meant that nearly every hitch had to be custom-made.

Garman made no hitches after 1964, but individual farmers continued to make and use tandem-tractor hitches after that date. The Wiese brothers of Comfrey, Minnesota, made a hitch for their two four-plow tractors for $300 plus their labor. One man operating this rig could plow 3 acres or disc 10 acres an hour. As soon as the heavy work was completed the brothers separated the tractors and used them on individual jobs.

The Wiese brothers were still using two tandem-tractor units in 1975 in addition to a large four-wheel-drive tractor. Donald Wiese felt that their tandem units and the four-wheel-drive had strong and weak points, but it was his opinion that the tandem-tractor units were superior in heavy traction operation. The Wieses won several tractor-pulling contests, including the state championship, with one of their tandem units. Despite the availability of proven four-wheel-drive tractors, the Wieses still intend to build a third tandem-tractor unit, using two of the largest conventional two-wheel-drive tractors.[2]

The Four-Wheel Drive Succeeds

From 1912 each attempted four-wheel-drive tractor or tandem-tractor unit quickly disappeared, but farmers and inventors never stopped striving for a successful machine. Each successive model or adaptation of the four-wheel drive came one step closer to a practical model. The widespread publicity and adoption of the tandem-tractor idea proved that farmers wanted bigger tractors to reduce their production costs. Machinery companies kept building bigger tractors, but never so quickly as the innovative farmers wanted them. The progressive farmer was constantly searching for the power unit that would get the job done on time at the lowest possible cost, for he understood the efficiencies involved.

Bill and Gene Schmidt, machinery dealers in Ohio, constructed a four-wheel-drive tractor in 1947. The machine, which they called the "Big Bertha," had a 200-H.P. potential, but it was never weighted down or used with dual tires so it could obtain maximum results. The Schmidts felt there was no need to do so with the machinery they were using at the time.

John Steiger homesteaded in Montana in 1917 and after successive crop failures from drought quit farming to take a job in Minneapolis. In 1928 Steiger started farming on his father-in-law's 160 acres of "sand dunes" north of Red Lake Falls, Minnesota. He used a big Rumely to clear the land and pull six-bottom plows to break new sod. By the late 1930's John Steiger was farming 640 acres and operating a blacksmith shop, where he did "a lot of improvising" on farm machinery, or building new machines to suit his needs. Steiger looked ahead, and when his neighbors quit during the depressed 1930's he would have liked to take over their land, but he was held back by his more cautious wife.

Fortunately when his sons, Maurice and Douglas, reached their teens, Mrs. Steiger did not oppose them when they wanted to buy land. By the 1950's the two youthful farmers were operating 5,200 acres, of which 3,300 acres were owned and 1,900 were rented. Like their father, the Steiger brothers were adventurous, but they too ran into an obstacle—the lack of adequate power. To clear land in 1949 they built a dozer to attach to their Oliver Cletrac. This was their first mechanical invention. They owned large tractors, including a track-type model for field work, but they still felt they were hampered by not getting their crops seeded or harvested on time.

From 1946 through 1957 the Steigers experienced four nearly total crop losses. The wet fall of 1957 proved to be the final blow, and the search for bigger power became a necessity. First they tried a Caterpillar D-W-15. This was a 170-H.P. two-wheel-drive tractor that was used in the road construction business for pulling road scrapers. Even though the D-W-15 was powerful and fast, it was made for construction work, not farming. Next the Steigers sought to obtain a Wagner four-wheel drive, which had been a success in the logging industry. The brothers, working with some of their neighbors, decided that they could sell two Wagners besides buying one for themselves. They bargained for a dealership with Wagner but were rejected.

Over the years Maurice and Douglas Steiger had traveled considerably in an effort to find a solution to their problems and came to the conclusion that they had two alternatives. They could go to small compact tractors like the 8N Fords with a three-point hitch and use very low-grade labor, or they would have to build one very large tractor that would be capable of working two sections of land with one skilled driver. The Steiger brothers, persuaded by their 63-year-old father, decided to build a large tractor of their own.

Their first hurdle was credit. Even though they were going to furnish all the labor and build as much as possible from used equipment, the size of machine they envisioned was costly. The Steigers did not consider themselves to be "really good farmers," but they knew they were more progressive than their neighbors in what the brothers felt was not "the best farming area." Fortunately, their banker at Union State Bank of Thief River Falls, Minnesota, encouraged them to build their tractor. Douglas Steiger said, "It seems that our banker encouraged us most when we were down and our balance sheet less than zero."

After they had arranged financing late in the fall of 1957, the Steigers sold their milk cows to raise more money and to use the dairy barn as a shop. In retrospect the brothers joked: "We feel we really did not know what we were getting into when we started building the tractor. But we were determined to build a big one and get it in operation by the spring of 1958. We did not have tools to build a thing and were either too proud or too embarrassed to ask others to help us. Now we laugh about how little we knew and at the lack of equipment we had."

The Steigers were motivated by one central thought—build a big tractor: "Our purpose was to build one [a tractor] that would pull a real load at low speeds. Speed was not our concern. We wanted to pull twelve 16-inch plows and felt that it would take 28,000 pounds of weight to get the necessary traction. We did not realize that horsepower developed traction."

They were fortunate in one respect—they were familiar with the iron mining activities of northern Minnesota and knew where they could locate used mining machines that would provide component parts for their big tractor. Without any precision tools they were forced to use sledge hammers and their blacksmith forge to heat and pound component parts into fitting. With that crude method they were even able to bend the heavy bell housing adapter to within $1/10$ inch of being true. A hand file was used to get a perfect match.

Pictures from farm magazines or snap shots they had taken of the Wagner four-wheel drive enabled the Steigers to "model" the type of tractor they wanted. They were experienced mechanics and recalled that they could pull their WD-9 apart, put in rear-end bearings, and have it back together in two hours. Maurice said, "We do not consider ourselves good mechanics, just patchers who had learned how to keep machines working. We have made our share of mechanical boo-boos."

After the used parts were assembled, a few new parts were purchased. The new parts were the 23.1- by 26-inch tires, which were considered big at that time, hydraulic components, and some raw steel plates and rods. The engine was a 200-H.P. Detroit diesel that had been used in an HD-14 Allis Chalmers track-type tractor. There was no steering wheel, only a stick that operated double-action hydraulic cylinders, which always had to be centered or the tractor would continue to turn. There were no brakes, even though it had a top road gear of 17 mph.

When the neighbors heard about the big four-wheel-drive tractor that was being built in the Steiger's cow barn, they came over to watch the process. The brothers remembered that there was more criticism than encouragement. The critics were particularly bothered by the hinge in the middle; they suggested a large chain underneath so that if the hinge broke, the driver would not be pinched between the front and rear sections of the tractor. Some even suggested laying rails so the tractor would not sink out of sight. Others said machinery would never follow it in the field. But through all travails their 63-year-old father encouraged them to keep building.*

Only 57 days after starting their machine the 28,000-pound tractor was completed. Its cost amounted to $9,200 for materials, with no charge for labor. Everyone in the Steiger family was eager for spring to come so that they could put their new machine to work. John Steiger, who was 64 in the spring of 1958, had the greatest pleasure plowing with 12 bottoms, and in one 15-hour day the Steiger brothers field-cultivated 300 acres.

They were very pleased with their invention. When they got into wet spots in the fields and became stuck the Steigers discovered that the jointed tractor could walk itself out through a bending action. This made it far superior to the track- type tractor they had been using. "We knew our four-wheel put us way ahead of the crowd."**

---

*Judging from comments made during an interview session, many of the remarks the Steiger brothers heard about their tractor were similar to what critics said when they first saw steam-traction engines in action in the 1880's.

**In the early 1960's John Steiger and two of his grandsons, using three later-model Steiger tractors—a Wildcat, a Bearcat, and a Tiger, totalling 720 H.P.—were able to field-cultivate 160 acres in two hours. Another time John Steiger assembled 85 feet of harrow behind one tractor and harrowed 160 acres in 1½ hours. But because of the lack of a good method of folding the harrow it was not used again. In 1973 Versatile Manufacturing Company commercially sold 85-foot harrows, which became an instant success in the Prairie Provinces and the Great Plains.

Douglas Steiger described that experience:

Our goal with the tractor was to farm 1,200 acres with one man; but our sixty-four-year-old dad did all the plowing and field preparation in 1958 while we got the planting done with the standard tractors. The next thing we knew we were out doing custom work with that big rig because we could work so much more cheaply than people with smaller tractors. We sensed the secret of the four-wheel drive with its constant traction effort and greater fuel economy. We could see within weeks that 1,200 acres was not near enough to keep our tractor busy so we just kept using it until we hit the limit of one big power unit. Our 5,200 acres seemed about right. A farmer just cannot understand what a four-wheel drive will do until he has driven one; it is so much more efficient.

Some flaws showed up as the tractor was put through every possible test. When the driver forgot to pull the steering stick back after a turn the big tractor just kept turning, sometimes nearly climbing onto the machinery it was pulling. Other times for the same reason the tractor went into the ditch. Through those experiences it was learned that the tractor had great stability, for it never overturned. The four-wheel drive has since become recognized as the most stable of all wheeled power units. This has made it extremely popular in hilly farming areas.

After about 20 days of intensive field work the power divider broke down. The next divider lasted about a year. When it wore out, the Steiger brothers went to a major gear producer and attempted to explain exactly what they wanted. Various individuals at the gear company laughed at them, but by 1973 about one-third of that company's production was going to Steiger tractors.

As time passed, other defects appeared and modifications had to be made. Because they were good mechanics, Maurice and Douglas Steiger and their father were never discouraged when something failed. Eventually most "bugs" were eliminated, and the more progressive farmers in the area began to pay serious attention to the performance of the 200-H.P. four-wheel-drive machine. As the Steigers recalled, "It wasn't the local neighbors because very few of them were concerned or prepared for the concept of the four-wheel drive. It was the more progressive operator who drove twenty to fifty miles to see what our tractor would do."

The Steigers had never been concerned about developing a tractor which would be sold to others, for their interest was centered entirely on making their farm operation more efficient. By 1962, after four full years of farming with the 200-H.P. unit, the brothers decided that they wanted a smaller model to pull a 48-foot grain drill. Their big tractor had the power to pull a hundred feet of grain drill, but small fields and narrow bridges made a drill of that size somewhat impractical.

They also knew that many farmers were convinced of the advantages of the four-wheel-drive tractor, but most of them preferred a smaller size than the original Steiger. Two area farmers agreed to buy four-wheel-drive tractors if the Steigers would build them in 100-H.P. range. Three tractors of 100 H.P. each were made for the 1963 crop year.

The Steigers displayed their new smaller tractor at the February 1964 Crookston, Minneosta, Winter Shows. Earl Christianson, a machinery dealer from Elbow Lake, Minnesota, saw it and became excited about the potential of a true four-wheel-drive power unit. Christianson said, "In 1963 I had seen a Wagner four-wheel drive and was sold on the idea at once. It could outpull large horsepower models of two-wheel drives and the tougher the going the more advantage the four-wheel drive had. When I saw the Steiger model, which was made for the farm, I knew it was the real thing." For several years Christianson had observed the excellent performance that road contractors were getting from 200-H.P. rubber-tired power units, and he felt that within a short time farmers would be using machines of the same size. Christianson toured Montana, stopping to visit with farmers who owned big four-wheel drives. Most of them were using 200-H.P. units because they felt they needed the added efficiency to overcome their smaller production per acre as compared to production in the more humid areas.

Christianson came to the conclusion that if Montana farmers had an efficiency advantage with the big power units, the farmer in the humid area could benefit even more. With that faith and renewed confidence he set out to sell four-wheel-drive tractors and was not overly concerned about building smaller models because the progressive farmers he called on were more interested in cutting the cost of production than they were in the initial cost of the tractor. He had encouraged the Steiger brothers to build a 100-H.P. model, however, because it had a greater market potential.

In February, 1975, Christianson recalled how farmers at the "local bar" always wanted to make fun of the first purchasers of a four-wheel drive in any community. "Sometimes I wondered if I was really on the right track because it seemed like there were so many people against the idea of four-wheel drive. But ten years later I find them still using the same argument. The alert, progressive farmer comes into my business all intent on learning how much he can do with one. Now most of them are pushing for ever increasing sizes even in our area where we have a very high ratio of row crops."

The Steigers calculated the price of all component parts needed for mass-producing 100-H.P. four-wheel-drive tractors and discovered it would cost so much more to build them than two-wheel tractors of the same horsepower that they would not be able to effectively compete. After arranging financing with their banker they decided to build four 200-H.P. machines during the winter and spring of 1964. They could no longer turn to the iron mines for component parts. They had to go to the manufacturers, most of whom seemed to think that such a tractor "was ridiculous for farmers." These all-new material tractors sold for $24,000 each. Earl Christianson received a dealer's commission on each of them and encouraged the Steigers to manufacture more in 1965. That year they manufactured nine of the 200-H.P. models and three 170-H.P. units, which they called the 1700. In 1966 the newly incorporated Steiger Manufacturing, intent on hitting a larger market, added the 1250, a 125-H.P. model that sold for $12,000.

The Steigers were inventors and mechanics, not promoters and financiers, and they preferred to let Christianson handle those details. Even though their tractor had an excellent reputation it was more difficult to sell to dealers than to farmers. The chief reason was that dealers were not accustomed to investing such a large amount of money in a single machine for floor planning purposes. It was up to Earl Christianson to keep the Steigers motivated on building and perfecting while he sought to expand the market.

An early, exciting sale took place when a farmer from Spain with over 3,000 acres under production came to see the Steiger perform. A demonstration took place on the Art Skolness farm at Glyndon, Minnesota. Skolness had one of the first three units built by Steiger and used that as a pattern to build an even larger unit in his own shop. Skolness' homemade unit, called the "Valley Bee," became well known in northwestern Minnesota because he entered it in many tractor-pulling contests at local fairs. That tractor stirred much interest in four-wheel drives in an area of very progressive large-scale farmers. After the Spanish farmer had watched the Steiger perform on the Skolness farm he asked for an order blank on the spot. The only paper available was a dinner napkin, and an order for a 2200-model Steiger for $24,000 plus shipping charges was written up and signed.

J. Budd Tibert, Art Greenberg, and John W. Scott, Sr., of the

Red River Valley area all purchased Wagner 14's in 1957 at a list price of $15,000 each. Tibert said:

> We liked the four-wheel drive at once. It was so much more versatile than our track-type tractors, besides being quicker and far more comfortable. The Wagner did much more work than even the salesman said it could do. We had big machinery for our track-types so we just hooked them to the Wagner which traveled faster and was much more efficient. We have kept our track-types around, but hope we never have to use them again. Today with two of the 2200 models we can work the ground and seed 300 acres a day.

Tibert's experience with his Wagner 14 was so successful that he decided to purchase a Steiger 2200 in 1964. This transaction was doubly successful for the new tractor promoters because it exposed the Steiger tractor to the progressive and large farmers of the rich potato-growing region of Grand Forks and Walsh Counties, North Dakota. The Wagner 14 that Tibert traded in was sold to Robert Kruer of Mapleton, Minnesota, in the heart of the corn and soybean belt of that state. This was probably the first four-wheel drive sold in that row-crop region and provided unexpected exposure in an area where it was thought that there would be little sales potential for the four-wheel-drive tractor. Kruer upgraded his four-wheel-drive tractors regularly and by early 1973 had a 320-H.P. model. Each time he traded, his older model was sold in his home vicinity. By 1973 four-wheel drives were being sold extensively in the corn area extending into Iowa, Illinois, and Nebraska.[3]

The Market Creates Interest

Maurice and Douglas Steiger remembered that on either the second or third day of their first four-wheel drive trials in the fields in the spring of 1958, products engineers from John Deere were around to observe and to take pictures. Wayne H. Worthington, who assembled and drove the first American tractor sold to Czarist Russia and had extensive engineering experience with several major American tractor manufacturers, was appointed Director of Research for Deere and Company in 1948. Among other responsibilities Worthington worked on the development of larger tractors. The John Deere "R" at 47 H.P. was the "world's largest two-wheel-drive tractor," and under Worthington the company sought to develop a model that was double the size of the "R."

Early in 1951 about 30 Kansas wheat farmers converted their 44-H.P. Case tractors to 78-H.P. models by installing 371-inch

General Motors diesel engines. Worthington and other engineers at Deere and Company concluded that the experience of those farmers had proven the workability of larger tractors and the eagerness of the farmers to use them. So Deere built a "field laboratory" tractor powered with a 108-H.P. General Motors diesel. This was 14-H.P. larger than twice the size of the Model "R."

This experimental tractor was taken to northwestern Minnesota, to both Dakotas, Montana, Arizona, and California. At each of the farm places where the tractor was used the farmers were asked to hook onto their biggest implements and set the depth and speed at which they would like to operate. John Deere test officials measured the draft and speed and in every test found that farmers wanted at least 25 percent more draft and 50 to 75 percent more speed than they had in 1952. This was a clear indication that farmers wanted power units 225 percent greater than anything in existence. Ironically, when asked before the test about how much power they wanted, none of the farmers had suggested any tractor as large as the test model. Once they were given the opportunity to test such a large tractor they reacted differently.

In the spring of 1953 Dr. Worthington spent several months in Europe studying farm conditions and tractor plants. On that tour he saw the world's first fully articulated four-wheel-drive tractor, designed by Hans Huber in the early 1900's. Huber was the chief engineer for the Lang Machinery Works in Germany. This 12-H.P. semi-diesel tractor was mounted on steel wheels with low lugs commonly used on steam engines of that period. From the Lang Museum Worthington went to Paris, where he observed two other articulated four-wheel-drive tractors—one made by Renault and the other a German machine.

Upon his return Deere and Company decided that the largest of their new general-purpose tractors should develop approximately 80 H.P. This was not so large as the field tests had indicated the farmers wanted, but the decision was not made without long and bitter debate. At the same time the Deere Plow Works was instructed to develop moldboard, disc, and chisel plows capable of working at speeds of up to 8 mph.

Quite by accident another step toward four-wheel-drive tractors came in the spring of 1954. Several hundred small John Deere crawler tractors were being used in the lumber industry in northern Quebec to drag logs over the muskeg to the loading station. The track-type or crawler tractors constantly became stuck in the muskeg. This prompted the men at a local repair shop to remove the tracks and rear sprocket. A chain drive was used from the rear axle to the front axle, and large rubber tires were mounted. Even though steering was done completely by clutching and skidding, their performance, according to Dr. Worthington, "was amazing."

These crawlers converted to rubber-tired four-wheel drive not only could pull out a crawler that had been mired up to the operator's seat but could then travel across the area that had been dug up by the crawler and not become stuck. A group of company directors flew to the location deep in the heart of northeastern Quebec to see those converted machines perform. The lumbering people proved that "four-wheel drive performance was an absolute necessity and initial cost was secondary."

Deere and Company officials realized that if they produced a four-wheel drive they would also be obligated to build a complete line of larger implements. Field tests of all types relating to speed, draft, and compaction had to be made. Tire companies had to be convinced to build a larger tractor tire than the 14-inch tire available in 1954.

The year of decision was 1955. Worthington spent considerable time testing industrial loaders, such as Hough and Clarke, which dominated the market of that time. Logic and experience convinced him that to be successful a four-wheel drive would have to be fully articulated and have the weight equally distributed on the four wheels.* To make the tractor even more efficient it was decided a three-point hitch (integral) should be used on associated tillage tools.

Next Worthington decided to observe performance and interview owners of existing four-wheel-drive Wagner tractors then being used in Montana. These tractors ranged from 90 to 160 H.P. and far outperformed the 150-H.P. John Deere two-wheel-drive experimental tractor. Steering around hills up to 20 percent in grade was one of the superior points of performance of the Wagner. Dr. Worthington rented a Wagner and placed it alongside the Deere field laboratory tractor. He then arranged for five company directors to observe the results which he knew would prove the merit of the four-wheel-drive unit.

Through interviews Worthington and his associate, Robert Tweedy, learned that even though the Montana farmers had en-

---

*Most four-wheel-drive tractors carry about 60 percent of their weight on the front wheels and 40 percent on the rear wheels. This gives almost even balance when the tractor is pulling a load which adds weight to the rear wheels.

countered many failures with their Wagner four-wheel drive they would buy another if they could buy one of top quality. These farmers also volunteered that they would not be willing to pay much more for a four-wheel drive than they would pay for an equal-horsepower conventional-drive tractor.

After thoroughly studying the four-wheel drive history, Worthington concluded that a four-wheel-drive tractor should be in excess of 150 H.P. and had to be fully articulated. There was a market for smaller four-wheel drives in specialized operations, for the four-wheel drive definitely had advantages where soil compaction was a problem; and four-wheel-drive tractors would, in most cases, displace track-type tractors in agriculture. The four-wheel drive probably offered no advantages in row-crop cultivation because of economic reasons, and Deere and Company would basically have to make its own tractor rather than buy component parts from other manufacturers. This final point was essential to control quality and to keep cost down.

Early in 1957, after nearly a year of discussion, the research and engineering staff was authorized to develop a large four-wheel-drive tractor. The only restriction was that such a machine would have to live up to company standards of quality and performance. Another group of researchers was instructed to develop a line of machinery suitable for such power units.

Nine months later a 218-H.P. tractor was ready for field testing. After being disguised it was taken from farm to farm throughout the country and put through every possible test. Performance was so successful that after only six months of intensive testing Deere and Company was ready to build three pre-production models. Serious heating of the transmission oil appeared to be the greatest defect of these units; but after that was corrected, performance was rated as highly successful.

In 1960 Deere and Company formally announced its four-wheel-drive tractor to its worldwide dealer organization. W. H. Worthington, the man most responsible for that tractor, had retired shortly before the announcement. Worthington had no doubt that the four-wheel drive would hasten the industrialization of agriculture.

Not all major machinery manufacturers were convinced of the total four-wheel-drive concept. At least three of the major tractor makers built large conventional models with small front-wheel-drive attachments. Although a step in the right direction, they did not prove as successful as the full four-wheel drive.

Acceptance of the early 210-H.P. Model 8010 John Deere was less than the company had anticipated. Company officials felt it probably was too early for a widespread acceptance of the four-wheel-drive concept. There was, however, a second factor responsible—the early four-wheel drives were expensive because many components came from manufacturers geared to precision low-volume production. This tended to make the tractor too high-priced for commercial success, even though technically it gave excellent results. Most of the 8010 tractors were sold to non-agricultural customers.

Production of the four-wheel drive was discontinued, as Deere and Company preferred to market Wagner tractors as model WD-17 and WD-19. It was not until 1971 that the company decided to re-enter the manufacturing of four-wheel drives with the Model 7020. This time sales caught on, and Deere and Company quickly became one of the largest manufacturers of four-wheel-drive tractors.

Robert Tweedy, now an Allis Chalmers official and vice president of the American Society of Agricultural Engineers, commented:

Farmers on the average have been increasing horsepower at the rate of about 4 percent per year compounded and the big farmers have been doing so at an even faster rate. It is very likely that in the not too distant future farmers will insist on 500 horsepower tractors. This will enable them to continue to grow in acres, but still basically enable them to remain as family farm operations. They want super power so they can do several hundred acres a day and not have to hire much labor.[4]

The Mass Market Arrives

Henry Ford's Fordson, developed in 1917, made the gasoline tractor popular among American farmers, who had to be convinced of the merits of a small tractor before they would buy a big one. The eventual mass adoption of the four-wheel-drive concept a half century later was no different. A few innovative farmers willingly sought out large, costly four-wheel-drive tractors, but the big market did not develop until a smaller, mass-produced four-wheel drive became a reality.

Roy Robinson, along with his brother-in-law, Peter Pakosh, and Mike Stasiuk, a long-time friend and an experienced production designer, first recognized the idea of a practical four-wheel-drive tractor in 1957. The three men, who in 1974 were the ranking officers of Versatile Manufacturing, Ltd., had read an article in a farm magazine about the tandem tractor tests at Cornell University. Roy Robinson recalled:

Two two-plow tractors could pull six bottoms because pulling power increased 18 to 48 percent depending on soil conditions and the fuel savings corresponded with work accomplished. Peter, Mike, and myself felt that four-wheel-drive tractors were definitely something we wanted to produce. We knew our first obstacle was to convince some of our management people. They felt such a tractor was too complicated for us to build.

The three of us stuck together; we visited suppliers of components of other four-wheel drive manufacturers and learned about the problems. Most component parts came from big trucks or from road building and construction equipment and was not entirely suited to the farm tractor which works under much steadier pulling needs. The four-wheel drive required special components.

We studied Steigers, MRS, Wagner, Case, and the 8010 John Deere as well as talking to farmers and dealers. It appeared to us that initial cost was still a factor and that a 100 horsepower four-wheel drive would be necessary to test the market potential. We reasoned that financial timing and size of plant to get mass production in order to keep the price down were the keys to success.

When these basic concepts were developed, the men from Versatile approached the Steiger brothers with the intent of buying them out. By then the Steigers had become convinced that a smaller-model tractor would hit a broader market and were working on a 125-H.P. tractor that would sell for $12,000, only half of what their 220-H.P. unit cost.

When the Versatile officers concluded that they would not be able to purchase the Steiger Company, they went to work on their own tractor. Robinson, Pakosh, and Stasiuk decided that they should buy the best-known quality components for their tractor to make it easier to sell their machine. The popular Spicer 5000 truck transmission was slightly altered for Versatile's needs. Ford Motor Company proved very cooperative and assured Robinson that they were developing a diesel that would be adequate for the 100-H.P. tractor.

The greatest single difference of the original Versatile, in contrast to most other four-wheel drives, was that it had a longer wheelbase. Robinson explained that this was done to get less weight transfer from the front to the back wheels under pulling conditions. It gave better rides and it greatly reduced the "jump out" experienced by four-wheel drives when pulling heavy loads on packed surfaces. Robinson said:

We were positive that the innovative farmers understood the advantages of the four-wheel drive and if we produced a competitively priced machine it would sell. We got a scare, however, when a market analyst told us that the market was limited to a hundred four-wheel drives a year which would sell at a cost of $30,000. Fortunately, I was neither a professional engineer or market analyst and was unaware of the hangups. All I had was a hunch we could sell small four-wheel drives.

In February, 1965, Versatile decided to build a prototype tractor. On June 27 it was placed on the Oscar Rennsch farm, 30 miles west of Winnipeg. Stasiuk stayed with the tractor for several weeks. Soon five farmers were at the factory wanting to buy such a tractor. Five prototypes were made by November. As in the case of other manufacturers, the tractor was basically sound except for some defective components. Stasiuk lamented, "We feel those pieces were oversold to us."

In 1966 it was decided to make 100 D-100 models, which were rated at 100 H.P. A total of 106 were manufactured and sold that year. The eager acceptance by the farmers encouraged the company to build 500 in 1967. That year several innovations were made. Versatile started production on the 145 model, which was a 145-H.P. unit equipped with a Cummins diesel.* Because some of the major components were not performing up to expectations, Versatile developed the transmission drop box. This reduced the cost to the farmer about 6 percent. Next they made their own axles, especially designed to stand the continuous strain of field work. Robinson expressed the opinion that the development of their own axle was one of the major reasons for the success of the Versatile four-wheel drive. The final innovation was the adoption of a cab.** Apparently the farmers who were willing to buy four-wheel-drive power units were also willing to buy cabs for the comfort and protection of the operator.

The first 100-H.P. D-100 models sold for $9,600 to $9,800. This was about the same price as conventional two-wheel-drive tractors of the same power. The early 145's, with 145-H.P. ratings, sold for $12,000. Many refinements were made in the next couple of years. By making many of their own component parts in addition to mass production the price of a 1969 gasoline-model 125-H.P. unit with 18.4- by 30-inch tires sold for only $9,210, and the 145-H.P. unit with a V-8 diesel engine and 23.1- by 30-inch tires sold for only $12,890.

In 1967 and 1968 dealers were very eager to get tractors, and the company took this as a sign that farmers were interested. Because

---

*By 1974 Versatile had become one of the largest customers of Cummins Diesel, which formerly had very little production geared to the agricultural market.

**Because of the roll bar requirement by law, the company had decided by 1974 that certain models of its tractors would all be produced with cabs.

of that reaction production increased to 4 tractors a day in 1967 and was stepped up to 12 by 1969.

Success did not come without adversity for the young tractor company for in 1969 and 1970 the Canadian wheat farmers were staggering under stockpiles of wheat that had no market. It became virtually impossible to sell a tractor unless a dealer took wheat in trade. This forced the company to look for possible foreign markets. Outside of the United States, France became the first outlet for Versatile four-wheel drives. The U.S.S.R., Australia, Morocco, and Algeria quickly followed. The Russians were using track-type tractors exclusively for large power units, but after their first experience with a four-wheel drive they decided to adopt that type of power for their farms.

Versatile also became more aggressive in marketing throughout the row-crop areas of the United States and other countries, discovering that there was an unanticipated market for four-wheel-drive tractors. Farmers in those regions wanted live power take-offs to help in harvesting their row crops and hydro-mechanical transmissions that provided infinite speed changes ranging from 1¼ to 7½ mph in the field. Three-point hitches also became necessary if the four-wheel drive was to compete fully in the row-crop area. Foreign acceptance, like that of the American Corn Belt, was far beyond the original expectations; and this was fortunate, for with the Canadian farmer in a desperate financial plight Versatile needed a savior.[5]

The Four-Wheel Drive Creates
a Revolution

The Steiger brothers remembered the Canadian wheat glut of 1969 and 1970. They compared that to the glut in the four-wheel-drive market of the same time caused by the mass production of Versatile, which appeared to have flooded the market. Versatile and Steiger, as well as the other big tractor producers, all felt the market squeeze of those years. Ironically, they were all caught short. However, when the farm outlook changed in 1971 a virtual explosion in demand took place. Even though sales were slow in 1969 and 1970, the farmers were waiting for their next prosperous years to buy.

Gregory Weiland of Euclid, Minnesota, had a 3300-model Steiger that created a sensation in the lower part of the Red River Valley. The 330-H.P. tractor could pull twenty-one 16-inch plows at the rate of 4¾ mph and could plow 160 acres in 12 hours. This moldboard plow turned 56 feet of sod in each round. Fuel consumption was just under 1 gallon per acre. Douglas Steiger said, "It is performance and economy like Weiland's which makes the four-wheel drive inevitable. We knew it had to come about as soon as we started using our first tractor, but because we were not promoters we did not know how to make the public aware of it."

Carl Bergstrom of Jefferson, Iowa, working on heavy Clarion-Webster soil, was able to pull a nine-bottom plow and turn an acre of sod every seven minutes. His large 300-H.P. machine was so economical that his neighbors hired him to custom plow their land. Bergstrom was able to do the job at a lower cost than his neighbors could do with their two-wheel-drive tractors, even if they did not figure a value for their own labor. Other farmers soon discovered that a 145-H.P. four-wheel drive was capable of doing twice as much work in a fixed period of time as was a 110-H.P. two-wheel-drive tractor. Farmers in the Corn Belt, with as little as 900 acres, were buying four-wheel-drive tractors to enable more timely planting and harvesting in case of adverse weather conditions.

It was assumed that the Great Plains grain farmers would be the leaders in wanting four-wheel-drive power. After many of the horse-power farms had succumbed to the hazards of Great Plains farming and were replaced by mechanized farms, the region gradually developed a big-power-machinery reputation. Montana led in the adoption of track-type tractors and four-wheel drives, but North Dakota was not far behind.

By 1972 sales of tractors over 100 H.P. in Dakota amounted to 70 percent of all tractors sold, as compared to 29 percent nationwide. The most rapid increase in sales for all tractor sizes was in the over-140-H.P. models. Texas, Kansas, Montana, and Illinois were the next four states in order of number of tractors purchased over 140 H.P. Even more significant is the fact that about 80 percent of total horsepower sold in North Dakota in 1972 came in tractors of 100 H.P. or larger. Changes came even faster with the sharp rise in prices accompanied by increased foreign sales of farm products.

In 1975 Canadian farmers purchased 1,000 four-wheel drives, Americans 8,000, and the rest of the world about 1,000. There were only about 50 such tractors produced in 1965. The revolution was well on its way, and some machinery dealers in the midwest were predicting that most Corn Belt farmers with 500 acres or more would come to own a four-wheel drive between 1978 and 1980. Boyd C. Bartlett, vice president of marketing for Deere and Company, predicted that by 1980 the most popular-sized tractor would be 200 H.P. His prediction is a safe one, for the progressive farmer has outpaced most predictions about what the future held in store. The farmer with the four-wheel drive had now arrived at the point Thomas D. Campbell had talked about when he started farming with Aultman-Taylors in 1917. In 1974 Campbell's farm had seven four-wheel drives in operation, and many other farms throughout the five states and Manitoba had four to six of them. The progressive farmers, as usual, rushed to take advantage of four-wheel-drive economy and were once again setting the pace toward reduced production costs. By now several of the four-wheel drive owners could prove labor reduction costs of as much as 50 percent in their crop production. That alone made the surge for four-wheel-drive tractors irreversible.[6]

NOTES

[1]Warren R. Bailey and Donald D. Durost, "What's Happened to Farming," *Yearbook of Agriculture, 1970: Contours of Change*, ed. Jack Hayes (Washington: U. S. Government Printing Office, 1970), pp. 2, 3; Martin R. Cooper, Glen T. Barton, and Albert P. Brodell, "Progress of Farm Mechanization," Misc. publication No. 630, U.S.D.A. (Washington, 1974), pp. 2, 24, 78, 79, 85; Casson, et al., *Horse, Truck and Tractor . . .*, pp. 2, 3, 256; E. T. Robbins, *Big Teams on Illinois Farms*, Agricultural Extension Service, University of Illinois, Circular No. 324 (Urbana, 1914), p. 16; Thomas P. Cooper, *The Cost of Horse Labor*, Department of Agriculture, University of Minnesota Extension Bulletin No. 15 (St. Paul, 1911), pp. 5-6; M. R. Cooper and J. O. Williams, *Cost of Using Horses on Corn Belt Farms*, U.S. Department of Agriculture Farmers Bulletin No. 1298 (Washington, 1922), pp. 1, 2, 13; H. R. Tolley, "Efficiency of U.S. Agriculture Is Increasing," *The Yearbook of Agriculture, 1926*, ed. Nelson Antrim (Washington, 1927), pp. 319, 324; C. E. Sarle, "Horse Production Falling Fast in U.S.," *The Yearbook of Agriculture 1926*, ed. Nelson Antrim (Washington, 1927), p. 437; W. L. Cavert, *Sources of Power on Minnesota Farms*, University of Minnesota Agricultural Experiment Station Bulletin 262 (St. Paul, February, 1930), pp. 4, 33; *Historical Statistics of U.S.*, p. 280; *Minnesota Crop Reporter*, Bulletin No. 505A (January, 1973); William J. Promersberger, "More Time to Live," Faculty lectureship, North Dakota State University, February 15, 1972.

[2]Royal Fraedrich, "Politicians Can't Repeal the Impact of Fertilizer, Hybrids, and Horsepower," *Big Farmer*, XLIV, No. 6, October, 1972, editorial page; Gray, *Development of the Agricultural Tractor . . .*, Part 1, pp. 22, 28, Part II, pp. 23, 46, 48, 52; Conversation with Oswald Daellenbach, Clay County Extension Agent, Moorhead, Minnesota, January 27, 1975; James A. Garman, "Tandem Tractor Studies," Unpublished Master's Thesis, Cornell University, September, 1957, pp. 11, 19; James A. Garman, Eureka, Illinois, letter to the author, April 24, 1974; Wesly F. Buchele and E. V. Collins, "Development of the Tandem Tractor," *Agricultural Engineering*, XXXIX, No. 5 (April, 1959), pp. 232, 233, 234, 236; Robert G. Rupp, "Homemade Dual Tractor Has Four-Wheel Drive, *The Farmer*, June 20, 1964, pp. 14, 15; Letter from Donald Wiese, Comfrey, Minnesota, February 8, 1975.

[3]"A Priceless Collection of Old Tractors," *Farm Profits*, XX, No. 2 (January and February, 1975), p. 17; Interview of Maurice and Douglas Steiger, Red Lake Falls, Minnesota, at Fargo, North Dakota, February 28, 1973; Interview of Earl Christianson, Elbow Lake, Minnesota, February 26, 1973, and February 12, 1975; Interview of J. Budd Tibert, Voss, North Dakota, March 14, 1973.

[4]Steiger brothers interview; Correspondence with Dr. Wayne H. Worthington, Cedar Falls, Iowa, January 25, 1974. Dr. Worthington was Director of Research and a member of the Deere and Company policy committee from 1948 until his retirement in 1959. He has 38 American and foreign patents and has 18 published papers in professional engineering journals. Worthington served as vice president of the American Society of Agricultural Engineers in 1930-31 and was its president 1955-56; Interview of Robert H. Tweedy, Manager, Product Planning, Allis Chalmers, Milwaukee, Wisconsin, at Fargo, North Dakota, January 29, 1975. Mr. Tweedy was senior engineer in the advance research department of the Deere and Company tractor division.

[5]Interview of Roy Robinson and Mike Stasiuk of Winnipeg, Manitoba, on January 10, 1974, at Fargo, North Dakota. Mr. Robinson is president of Versatile Manufacturing, Limited, and Mr. Stasiuk is a vice president of the firm; Interview of John L. Eckmire and Peter Pakosh at Winnipeg, Manitoba, January 11, 1974. Mr. Eckmire is Vice President of Finance and Secretary Treasurer of Versatile Manufacturing and Mr. Pakosh is chairman of the board; Steiger brothers interview.

[6]Interview of Henry Weiland, Euclid, Minnesota, March 17, 1973; The 3300 is one of the early model Steigers; Steiger brothers interview; Gordon W. Erlandson, "Trends in Sales of Farm Tractors," *North Dakota Farm Research*, XXXI, No. 1 (Sept-Oct. 1973), North Dakota Agricultural Experiment Station, pp. 21-22; Interview of Carl Bergstrom, Ogdon, Iowa, January 26, 1974; Mike Hood, "They'll Never Go Back to Two-Wheel Drive," *Successful Farming*, LXXII, No. 2 (February, 1974), p. 31; "Off and Running Agriculture's New Era," *Successful Farming*, LXXI, No. 12 (November-December, 1973), pp. 34, 36; David Glinz interview; Curnow interview.

CHAPTER XIII

Farming Is a Business

FREDERICK JACKSON TURNER, the frontier historian, in writing about the American agricultural frontier, described it as the safety valve for laborers and immigrants of the nineteenth century. According to his theories the frontier was the safety valve in the sense that the restless and dissatisfied city laborers had an option to quit their job and take a homestead if they wanted to improve their way of life by farming. Turner's theories were accepted by many of his followers, but more recent historical studies have presented a viewpoint quite contrary to his theory. They point out that a large portion of the homesteaders were sons of farmers from the adjacent eastern areas, and most of the rest were immigrants from foreign countries. For each city person who left his job to become a permanent homesteader and farmer, at least 20 children of farmers left the farm to become city laborers.

G. F. Warren, in his book, *Farm Management*, published in 1913, pointed out that the smaller the farm the more likely the tendency was to leave the farm. Warren concluded that farmers' sons were six times more likely to leave farming if they came from farms under 30 acres than they were if they came from farms over 200 acres. Farmers' children provided the necessary labor force for America's rapidly increasing urban industries. Still it was the dream of many pioneer farmers to become successful enough in farming to start each of their sons on a separate quarter-section farm. For example, Fingal Enger of Hatton, North Dakota, was able to leave each of his 10 children a couple sections of land. But more often the homesteaders' children were lucky if the parent let them share the use of machinery while they were starting their own farm. Instead of a quarter section, a more likely wedding present from the farmer to his son or daughter was a wagon, cow, or team of horses. Therefore, in most large farm families children left the farm as soon as they could break away.

Lack of education prevented many farm children from finding good alternative ways of making a living in an effort to avoid the drudgery of farming. Many farmers felt "farming was a greater purpose . . . than book learning"; and as Theodore Roosevelt's Country Life Commission concluded, they were skeptical about having their children educated beyond the basics. Some of the more progressive farmers, however, felt otherwise. As Mrs. Ole Overby said, "I hope at least one of you boys can find something else to do than to follow the tail of a horse." Eight of the 11 children of R. M. Probstfield, a prominent Clay County Minnesota farmer, left the farm as soon as they were old enough to do so. This was a natural trend with most of the pioneer families.

The constantly diminishing numbers of farms since the 1920's have caused much concern about the farmer in the future. Perhaps there is no need for pessimism because there are many innovative, progressive farmers who look at farming as a good business and an excellent way to make a living even in the future. Their sons and daughters will carry on the tradition, but they will also have competition from outsiders who have discovered that farming, if properly managed, can be just as profitable as any other business. Modern farming, like any other business, is a matter of mechanization, money, and management; and good businessmen from non-farm families can see farming as being as challenging and as rewarding as any other opportunity open to them.

Who entered farming in other than the conventional manner? This chapter deals with two such men. They had careers open to them in non-farming businesses and were successful in them, but they eventually decided that farming was what they most wanted to do. Neither of these men, like the modern farmer, typifies the homesteader of an earlier era except that each liked the freedom of work and the challenge so apparent in farming. To them the phrases plow jockey, sod buster, hick, hayseed, poor farmer, or dumb farmer were terms in popular usage and not an image of reality as so many have probably held.[*1]

David G. Drum—"An Honest Name and a Lot of Hungry"

Dave Drum's mother had a favorite expression about what her offspring inherited from home—"We left our children with an

---

*The writer remembers as a high school student at Owatonna, Minnesota, in the days just prior to World War II, that these phrases were used by city students when referring to country students. He was a "country student."

honest name and a lot of hungry." Dave's father, Jay Gould Drum, homesteaded at Jordan, Montana, in 1914, in the heart of the region that Montanans have since named "The Big Dry"—so named because dry weather forced almost every sod-breaking homesteader to quit after a few years of futility in farming. Jay Drum experienced nine successive crop failures before he gave up raising wheat. For a while he managed to keep his cow herd alive with Russian thistles.

But after a lot of hungry days, and after three years of farm failure at Jordan, Jay Drum decided to hedge his future by getting involved in an International Harvester dealership at Miles City. Shortly thereafter he married Frances Frazee, a Northwestern University student who had come to Montana to visit her brother and to "look for a husband." Later the Drums added a hardware store to their enterprise.

In spite of hardships at the beginning, the Drums were able to build a beef cow herd on the ranch along with building their other businesses in town. Adversity struck hard in 1930 when Jay Drum's banker called a loan even though he had promised he would not do so for another year. Drum was forced to sell his cow herd at a loss and had to liquidate other property to pay the balance of the note. At that point they were almost wiped out. Dave Drum remembered that his father repeated the phrase "never trust a banker" to his dying day. To which Dave added, "He disliked the Democrats nearly as much as bankers."

Jay Drum spent the remainder of his life working for the machinery dealership, and Mrs. Drum taught declamation. By frugal living the family pulled through the depression years. Gradually the Drums paid off all debts, and it was during that period that the phrase "honest name and a lot of hungry" became the favorite expression of Mrs. Drum. There is little doubt to those who know David G. Drum that that expression left a deep imprint on his life.

Dave Drum's "hungry" made him a hustler. While in college at Montana State he was manager for a fraternity house and also had a sandwich route serving all the fraternity and sorority houses on campus. He averaged $315 a month from this work besides his $120 a month veteran's benefit derived from his service in World War II. His first job out of college was as an oil company salesman at $175 a month. As soon as Dave discovered that the top salesman who had been with the company for 15 years was making only $300 a month he quit his job.

A succession of sales jobs gave Dave Drum good experience. In 1947 he earned $12,000 as the top salesman for a wholesale hardware firm. But his success was dimmed by an $11,000 loss from the water follies show that he had promoted as president of the Billings Junior Chamber of Commerce. This was not to be the last of his promotional failures, but with his "hungry," a failure was only a diversion to overcome.

Selling was Dave Drum's life, and next he became involved with Tom Felt in selling and erecting steel buildings that sold for $10,000 and more. This proved to be a good apprenticeship for Drum, who in this job discovered that there was an art in making and "closing a big sale on the first call and leaving the customer as a friend." Drum said:

> Felt and I were good salesmen, but not good detail men, so even by working seven days a week we did not do well. While traveling I had lots of time to think. It occurred to me that one does not deserve success until you first experience failure. I have done many things wrong but I have a good memory. One cannot get down when things are tough because success goes to the problem solvers.

Soon Drum opened the Drum Farm and Garden Center, which the family nicknamed the "Drum Starvation Center." However, banks were not inclined to loan money for the purchase of a volume inventory, and without that lever Dave realized that the independent store had little future against chain store competition in that line. As he recalled:

> I got out of the Garden Store in the nick of time and got into the anhydrous ammonia business in 1953, which was just starting in Montana. During the winter season I sold liquid feed to the feedlots. Working on the farm and ranches kindled the fire to get involved in the cattle feeding business. A salesman has a great opportunity to learn if he just keeps his eyes and ears open. I became aware of the cattle feeding opportunities in Montana. It appeared to me that the feeder calf salesmen and the bankers had managed to keep the ranchers from feeding their calves just to keep themselves in business. We had an abundant supply of some of the cheapest feed in the nation, besides we were exporting about a million and one-half of the best feeder calves each year. Why not finish them in Montana?[2]

The Birth of T-Bone Feeders and KOA

So far Dave Drum had learned from each of his mistakes, for he knew some day he would come up with the right idea and be successful. Drum calls himself a promoter of ideas and credits his brother-in-law, Robert Boorman, an accountant, with the ability to stop many of his wild ideas before they are put into operation. In 1961 Drum came up with two ideas that were acceptable to Boorman. Both are significant enough to give Drum a permanent place in the business history of Montana and the nation.

The people of Billings, Montana, wanted to capitalize on the tourist traffic traveling through the state over U.S. 10 and 12 en route to the Seattle World's Fair of the 1960's. There was great pressure for a public campground along the route, but the city council balked at the idea of competing with established businesses. After considerable research Drum, Boorman, and John Wallace became convinced that there was a need for independent, standardized camping facilities. They formed Kampgrounds of America, Inc., and using only borrowed funds went into business. By 1974 there were 800 KOA camps in operation throughout the nation, surpassing the National Park Service and U.S. Forest Service in number of campsites. Once the idea and the initial promoting were over, Drum left the details of operating KOA to others while he turned to a venture more to his liking—cattle feeding.

Drum's thoughts and dreams about cattle feeding in Montana appealed to fellow members of the Billings Jaycees. In 1961 seven members decided on selling ranchers and other interested persons on the idea of establishing a cattle feedlot, which they called T-Bone Feeders. As Drum remembered this adventure:

We got ranchers to put up the money and build a feedlot at Shepherd, and because of poor management we ended up having a bundle of trouble. There was dishonesty in the setup. I learned some lessons. I stayed clean and the real problem rose to the top. We had too many people with limited voice in the business and no one in a majority position to take command. In a setup like that you are bound to make mistakes. Since then I have only gone into enterprises where we can be innovative and I have a say in the operation.

It seemed everyone including the stockholders had lost faith in T-Bone Feeders. We had lots of trouble, but we had a good bunch who could see the bright side. What we needed was a dominant figure to make sure that the operation was going to go and grow. He would have to take us through the sweat period which any business must have. He has got to be tough, call the shots, and not be worried about being first in a popularity contest.

It wasn't long before all the stockholders realized that they had lost "a bundle of money" in T-Bone Feeders. Drum tried to think of a time when he had been in a worse position, and he recalled his combat days on Okinawa. For three days he had eaten no warm food, had endured a steady rain during the same period, and suffered from an around-the-clock hand grenade serenade by the Japanese. He said, "My business problems during those critical days of T-Bone Feeders were not so bad even if the banker would not help me out."

The low point for T-Bone ʻFeeders came in 1963 when many stockholders wanted to get out. "I had faith that it would go and I found a banker who had faith in me and the feedlot," said Drum, "so I bought the stock from a large stockholder for five cents a share." Over a three-year period he accumulated a majority paying up to $1 per share.

A search for an experienced manager resulted in the hiring of Leon Miller, who had been active in farm and feedlot operations on a large scale for over 30 years. Miller liked the facilities of T-Bone Feeders and agreed to lease the lot on a tonnage basis if he could get financing for cattle, or could get individuals who wanted cattle fed for them. In order to gain control of the corporation Drum had to use all the funds he could borrow. Miller's finances were also not in the best shape. He possessed not more than his "ability, experience, and credit at the First National Bank." He had experienced a serious reversal of his fortune as a result of heavy losses in both turkey and cattle business in 1963, when the markets declined sharply.

The first consignment of 50 head arrived at the feedlot in March, 1964, but it took until the fall of 1965 before the lot was filled to its 2,500-head capacity. While Miller was concentrating on keeping the lot full of cattle, Drum turned his attention to buying most of the remaining stock of the corporation.

The longer Drum was associated with the cattle business, the more disturbed he became over the fact that his feedlot was full during the winter months, but nearly empty in the summer. Ranchers preferred to run their cattle on grass when it was available and used the feedlot only during periods of short grass supply and in winter. Cattle buyers seemed to know how to talk ranchers into selling their calves rather than feeding them to butcher. Drum was concerned for two reasons. To be successful his feedlot had to be full 70 to 80 percent of the time. Secondly, he was convinced that Montanans had many natural advantages in the three stages of the livestock industry from the cow to calf to fattening and were missing a great opportunity.

Following his instincts as a promoter Dave Drum rose to the challenge. He had to find investors and risk takers to prove that his theories were correct. If local ranchers did not want to feed cattle, then he would have to find other progressive cattlemen who wanted to do so. Contracts with local farmers and ranchers would have to be made in such a way that a dependable supply of cattle, silage, hay, and grain would be available. Drum knew that a large, well organized feedlot could serve as the floor for the advancement of agriculture in the area around it because of the demand for feed and cattle. He reasoned that any large successful feedlot would profit the cash grain farmer, the silage producer, the rancher, the local laborers, the users of manure, the local service enterprises, and the local bankers. Drum commented, "You make lots of people happy when they can benefit from your success. Eventually we had more of an impact on Montana grain prices than government programs. In fact our feedlot even had an impact on much of western North Dakota."

Under Leon Miller the management of the T-Bone Feeders Corporation at Shepherd, Montana, improved, the long-range outlook brightened, and the ghost of bankruptcy vanished. In four steps the facilities were improved and enlarged so that by late 1968 as many as 12,500 head could be fed at one time. Drum and Miller visited the best lots in the country to learn every management technique that could be applied to T-Bone Feeders. Miller's long background as a feedlot manager, his excellent records, and his good feeding results proved Drum's theories about the feasibility of successful cattle feeding in Montana.

Since Montana is quite distant from the world's markets the freight rates are relatively high. Therefore, the price of grain in the state is reduced proportionately, providing a lower feed cost for local cattle feeders. Drum and his associates reasoned that by taking advantage of the lower feed cost the best way to reduce feed cost for the consumer is to export only the finished product. Defatted and deboned beef weigh much less than the grain and the roughage used to produce it, so why not export beef?

About 1968, the time of the last enlargement of the T-Bone lot, Dave Drum's brother, William, suggested that outside investors be sought for the purpose of keeping the lot full. Articles for a limited partnership were drawn up that encouraged individuals to invest funds for a seven-year period. This provided a more stable clientele and a more balanced investment through a normal cattle cycle for the investor. The partnership funds are used to buy cattle, grain, and roughage on a monthly basis that provides dollar averaging throughout the year. Cattle are sold on the same basis.

By 1970 Drum and Miller had worked T-Bone Feeders into a

sound financial position and were prepared to expand farther. It appeared that the best way to expand would be to erect an entirely new lot away from T-Bone Feeders. The T-Bone #2 Feedlot was organized and located at nearby Huntley, a site regarded as ecologically and economically sound. A total of one million dollars was borrowed from the Connecticut Mutual Life Insurance Company to finance the construction of a first-class lot.

The next step was to secure financing for the purchase of cattle. Drum went to the Production Credit Association because he felt they had the best potential for the large amount of credit necessary to buy cattle for the new lot. After credit arrangements were made for $18 million, cattle were purchased and trucks started hauling them to the new lot. Just as the first truck loads of cattle arrived a call was received from Production Credit Association stating that the loan had been stopped. PCA regulations apparently did not permit such a loan to a partnership. Drum approached Fred Marble, his banker in Billings, with his predicament. Marble advised him to keep on buying cattle even though he as a banker could not give Dave long-range credit.

In the meantime every stockholder of the corporation tried every bank in the area, and all received the same answer—they did not want to loan money for cattle feeding in Montana. There remained only as a last recourse an approach to the Chase Manhattan Bank in New York. As Drum tells it:

It seemed like we were at a dead end. North West Banco said they were not even interested in talking to me, so I could not even get them to say no. They are willing to take money from Montana agriculture, but reluctant to loan it back. . . . We heard Chase Manhattan was loaning to Monfort of Colorado so Fred Marble, my banker, made a contact with Chase. We met with Dan Klingenberg of their agricultural department and in six weeks they had a complete study of the Montana cattle feeding business. They decided to get in on the ground floor of what appeared to them an industry with a future. We got $400,000 to finish our feedlot, then one and one-half million. Next we were up to four and one-half million, and eventually received a commitment for fourteen million.

Chase Manhattan Bank made the loan and arranged it through a Billings bank that received one-half percent for handling the business. Dan Klingenberg, who had been the agricultural technical service advisor for Chase Manhattan at the time the loan was made, was hired by Montana Beef Industries, the holding company of T-Bone Feedlots. Klingenberg was the type of man Drum needed to oversee the total management of the feedlots.

Shortly after financing for the Montana Beef Industries was arranged Dave Drum purchased the 146-section (93,440 acres) KRM south of Malta, Montana. To meet the packer's demands Dr. Ray Woodward, animal scientist, was engaged to crossbreed animals. Plans were also made for irrigation of the land so that the number of cows could be increased from 3,000 to 10,000 and calves could be kept on the ranch until needed in the feedlot. Drum envisioned an integrated operation from brood cow to packing plant, or possibly direct to the retailer. Drum felt that such an integrated operation of raising cows, calves, and feed on irrigated land plus a superior type of cattle would produce at least $50 more return per brood cow. It was Drum's opinion that his system would cut cost along the line and the consumer would be the final beneficiary. The cattle boom of 1973 caused some alteration in his plans, however, because three neighboring Montana ranchers offered Drum a price for one-half of the well-known KRM Ranch, an amount too tempting to decline.

As soon as the feedlots at Huntley, Shepherd, and Vaughan, Montana, were operating at capacity, Dave Drum concentrated on hiring the best possible management team. As he explained:

I am motivated by the fun of putting things together. I can get something going, but it takes another to work out the details. This makes partnerships fun because you use the potential of both partners. After promoting to get things going you can sit back and watch others do the job. The best executive is the one who does the least work. I hire efficient hardnosed managers to run a tight ship so I can be free to do my job in the area of imagination.[3]

The Problems of Size

Any business the size of Montana Beef Industries and associated companies takes very tight management to keep costs in line with production and to compete with an efficiently operated one-man farm. Dave Drum was not afraid of this challenge. His attitude toward larger agricultural enterprises is best summed up in this manner:

Never in the history of agriculture has there been so much opportunity as there is today [1972] for those who practice good management. We try to incorporate all the new ideas that we can get hold of and use. . . . We may make a few mistakes from our innovations, but generally we profit and in the long run the entire nation will profit from our experience. We think the secret of success lies in integration. We note the banks are promoting larger farm units because they have come to recognize where the profits are.

According to Drum, management in agriculture is no different from that of any other enterprise. He feels that the basic rules of management apply to all industries. The first need is to establish a budget and then set strict guidelines to live within that budget. He feels a budget is the best way to establish a track record so necessary for the purpose of financing. Financing, so critical to any industry, is often dependent upon purely personal considerations. Drum mentioned that at one time he had a loan that had been approved by a local banker only to be rejected by the bank's directors. After inquiry into the rejection he discovered that the banker had concluded that a loan to Drum, if it were not 100 percent solid, could have possible reflections on his chances for promotion to bank president. On that basis the banker recommended to the directors that they decline the loan.

Securing good management personnel along with financing is another key to the success of any enterprise. Montana Beef Industries, M & D Feedlots, T-Bone Feeders, and Feeders Supply are no exception to the rule. Dan Klingenberg, former loan officer of Chase Manhattan, is the executive vice president and general manager of the feeding enterprise that fattens nearly one-third of the 235,000 head of cattle finished in Montana. Klingenberg's knowledge of the money market and financing fits into Drum's first criterion for sound management.

Klingenberg is ably supported by Max Henthorne as general manager of the feedlots. Capable managers under him at each of the lots are D. E. Deitchler, a CPA who is treasurer and controller, and Greg Wallander, who together with his staff is responsible for securing feed and livestock. According to Klingenberg, "It takes top management to run a good feedlot and we feel the key to our success is having good people. We are backed up three deep with knowledgeable people, from the cowboys in the pens to managers in the office. The opportunity lies with those willing to practice good management."

A staff of 41 people is required at T-Bone Feedlot alone when it is filled to capacity. There are over a thousand cattle on feed for each employee. This staff includes office personnel, truck drivers, cowboys, mill operators, and managers. These people are responsible for securing, processing, and feeding 310 tons of grain, 95 tons of silage, and 20 tons of supplement at one feedlot daily. Contrary to the practice of many other agricultural enterprises, Montana Beef Industries attempts to hire as many college graduates as possible. Unlike former times, when a farmer hired the first transient who walked onto his place looking for a job, David Drum wants his prospective employees to be thoroughly screened so that only the most educated and ambitious people with high standards are employed.

These people are expected to do a job with the minimum of supervision. Hiring good people and paying them good wages has earned the feedlot a fine reputation. Thus a steady supply of trained people are in line for employment. "Labor is a problem because it is a people problem," said Leon Miller. "We have to look upon them as people, not tools. The esprit de corps is good because these people know they can do better as a single unit than if they were running forty-one separate feedlots."

With intelligent employees who have the desire to do their best it is profitable to use automation wherever possible. Automation serves to challenge the good employee, and historically this has reduced costs even though top-notch personnel are required to do the job. This enables Dave Drum's feedlots to compete with any industry for top-quality labor.

Purchasing feeds and cattle to maintain the lots at 55,000 head of cattle requires strict budgeting. The very size of operation presents a major logistics problem. This is especially true relative to silage for it is not considered practical to haul silage more than 10 miles. However, some has been hauled as far as 50 miles because of the limited precipitation. Irrigation is essential for good corn production in Montana. About five dozen farmers make contracts annually to produce over 90,000 tons of silage. For an eight-year period, income from silage increased at the rate of about $10 per acre per year. This was attributed to the use of 15 to 20 tons of manure from the feedlot plus improved management practices by the farmers, who became skilled in silage production. The farmers who contract silage have priority on the manure, and their trucks haul it from the feedlot.

In 1972 over 2,400 acres of corn silage, about 7 percent of the state's total production, was contracted with farmers within 50 miles of the feedlots. Their production varied from 14 to 18 tons per acre—not high by Corn Belt standards, but far above average for the shorter growing season in Montana. That year the price paid varied from $1.35 a hundred pounds dry-matter basis for high-moisture silage to as high as $1.75 a hundred pounds for 50-percent-moisture silage. Farmers received $11.00 to $17.50 a ton for their silage plus hauling charges. This paid them over $200 an acre besides giving them the privilege of obtaining manure from the feedlot. This made contracted corn silage the highest gross-income crop for the farmers in the area. Only sugar beets and specialty crops would compete. Dryland wheat in Yellowstone County, where two of the feedlots are located, would have to be priced at more than $7.50 a bushel to produce comparable income.

The company has storage capacity in trench silos for 150,000 tons. In the Vaughn area, where corn is not available, alfalfa silage and chopped straw and chaff are used as roughage. Every available bit is purchased from farmers as distant from the lot as it is economically feasible to haul. Silage, haylage, and chopped straw must be purchased and put in storage during the harvest season. This means that a large amount of capital is required to purchase feeds that are to be fed during the next 12 months. Max Henthorne is of the opinion that the logistics of roughage is a major factor on the limitation of the size of any feedlot.*

Even though the beef animal is a ruminant and converts roughage efficiently, the logistics of roughage handling plus the availability of low-cost grain up to 1972 made it economically attractive to use a high portion of grain to fatten beef.** T-Bone Feedlots started their cattle on an 85-percent roughage ration and reduced them as rapidly as possible to 30-percent roughage. Barley and wheat, both plentiful in Montana, are the major grains used to supplement the roughage.***

Drum's estimation of his feedlots' impact may appear to be high, but each year 3 million bushels of grain were required to finish the cattle in his lots. This was the equivalent of 2,500 semitrailer loads. As far as Montana production was concerned, his feedlots used about 2 percent of all wheat produced, or 5 percent of the state's barley. The total Yellowstone County production of barley and winter wheat would not have been adequate to support the T-Bone Feedlots in 1972.

---

*Dr. Duane Flack, general manager for the 230,000-head feedlot of Monfort of Colorado, verifies Henthorne's opinion on the logistics of roughage. Monforts contract all their roughage within 10 miles of each of their feedlots.

**Probably in the future, ruminants might be fattened on a decreasing amount of grain as roughage production is scientifically advanced and becomes less dependent on labor. It is probable that future grain prices will remain relatively high because of world food demands and this will force the use of more forage in livestock production.

***Montana had one of the lowest price-support bases of any of the states affected by the federal government pricing programs. Remoteness from population centers and foreign markets are two of the factors having the greatest impact on grain prices.

Purchasing all grain locally probably would be the most feasible arrangement, but normal marketing channels do not work that way. Most grain, however, is purchased within a 175-mile radius of the feedlots. To guarantee a steady supply of grain, two large storage units measuring 420 by 100 by 26 feet, each capable of storing one million bushels of grain, are needed. Each of these units can receive or ship over 7,000 bushels of grain per hour. By being able to buy and store grain at harvest time about $2 a ton can be saved on the purchase price of two million bushels. With nearly one-half-year reserve available fewer trucks are needed to bring in grain from farmers and country elevators.

The railroads deliver multi-car units of grain to storage at a greatly reduced cost. In 1971 the railroad was preparing to ship 128 carloads of grain into storage at Huntley from bins that were formerly part of the Holly Sugar Plant at Hardin. In one 6-week period during the previous year, the railroad had delivered 485 carloads of grain to the Huntley site. This grain had been accumulated in northeastern Montana and northwestern North Dakota. With grain stored only 9 miles from two of the feedlots, only three double-bottom semitrailers were required to haul grain to the feed processing center.

Initially, purchasing grain from individual farmers located close to the feedlots appeared to be the practical way to guarantee a steady supply. Although contracts are made annually with about two dozen large grain producers, who can supply from 100,000 to 700,000 bushels per year, direct buying does not work so well as anticipated. Having to make several calls on individual farmers plus the great amount of paper work involved for each purchase forced Montana Beef to revert to larger grain outlets. Contracting ahead with country elevators, even though they must make a profit, helps to reduce the paper work and acquisition costs. Large purchases are made with only a single phone call and the writing of one check. Multi-carload shipments generally reduce the freight costs in contrast to a fleet of semitrailers that were needed for on-the-farm pickups.

Each year Montana ranchers sell from 1.5 to 1.8 million beef calves, of which all but 250,000 are sold to customers outside the state. Only two counties in the state feed more cattle than they produce. One is Yellowstone, where Montana Beef Industries is located. Yellowstone County is the largest beef calf county in the state, and adjacent Big Horn County ranks among the top four counties in beef calf production. Their total production of 200,000 to 250,000 calves annually would be enough to keep MBI's feed-lots filled.

Under ideal conditions buying cattle directly from a rancher located near a feedlot would reduce acquisition costs as well as minimize the death loss on those cattle. Any feedlot operator would prefer to buy grass-fresh cattle direct from the rancher, and Dave Drum definitely expressed a preference for direct buying. Drum likes to inform the rancher about his long-term needs in hopes that he can secure cattle on a contractual basis each year. Drum also believes that once the rancher understands the needs and desires of the feedlot operator, he would produce cattle to fill those needs. Like many of the progressive cattlemen, he hopes that someday cattle may be followed from the ranch through the feedlot to the rail. In this manner traits could be bred into the cattle that would be of benefit to all involved—rancher to consumer. Feedlot manager Henthorne also agrees that the better ranchers have been most receptive to direct marketing as well as responsive to a breeding program based on the feedlot needs.

Unfortunately, the established power structure of the auction markets together with the grip of tradition over many ranchers has made direct buying of feeder cattle a slow process. MBI has five established order buyers throughout Montana who attempt, if possible, to buy nothing but ranch-fresh cattle. When their buyers are forced to go into the established auction markets they are quick to sense the "squeeze play" and are often forced to pay up to a dollar more over the market just to be able to fill their needs. MBI buyers would prefer to pay the ranchers that premium for better cattle fresh from the ranch. They feel there are no soft spots in purchasing, because information is so widespread that every alert rancher knows exactly the value of his cattle. Buying is a matter of psychology, according to the management of MBI. It is being able to buy without letting the seller know how seriously you need his cattle. Being able to buy 2,000 to 4,000 calves at a single ranch without adding either selling or buying commissions represents a savings for both parties involved.

Even though size creates certain problems for the large feedlot operator, owner Dave Drum realizes that it also has many factors in his favor. He is aware that each year an ever-increasing number of cattle are being fed in the large-sized lots because there are so many economies of scale favoring the bigger operations. His general manager, Max Henthorne, says:

Our top management are all college people. We have a good crew of experienced people who know how to get the job done. They are a real idea machine and are not hung up with tradition. Each of them has their responsibility and they are free to make their moves to get the job done. There is no reason for any delay in decision making as far as central management is concerned because the on-the-spot manager is expected to act.

Dave Drum strongly reaffirms Henthorne's statement on responsibility:

A good manager is responsible for the men under him and the people problem is the greatest problem with any organization even if it is only two individuals. But the manager must handle it. I cannot divert his loyalty by interfering. It cannot arise at meetings with my management people who are there for we must talk about problems of the managers and their ideas and cannot be bogged down in details. I am a believer that you seek out the smartest people you can hire, pay them a good wage, and in time cream as well as the bubbles will rise to the top.

Buying is the only place that I really watch. Our policy is that bids must be secured on all items over $2,500. There are too many ways that sellers use to induce management personnel to swing a purchase—a trip to the islands, a Stetson hat, or a banquet. The established policy on bids keeps the temptations under control.

The final step in any feeding program is the ability to sell the cattle. Former partner and general manager Leon Miller, commented on the selling program: "We are not order takers to deliver; we sell our product and we have some control over our market." It may be said with justification that MBI made a market. Prior to the establishment of MBI's large operation there were only two small packing plants in the state of Montana. By early 1972 both of the plants located at Billings were planning on doubling their capacity. By attracting fat cattle packer buyers to Billings other local feeders have also been helped because the buyers will also call on them when they are in the area. Previously the small feeder had to rely strictly on the limited local market.*

Acquisition costs as described in the case of grain, roughage, and cattle for the feedlot are also a problem for the meat packer. It is at this step that the large lot has a definite advantage, especially if there are a thousand head or more that can be sold at each deal. Packers avoid traveling around the country looking for a few cattle at each farm if they can find them in large lots. Drum said, "I

---

*Dave Drum doubts that Montanans will overcome traditional practices of selling their calves to the Midwest feeders; and because of that, the packing industry will not be able to increase much beyond its present size.

think this could be the death sentence of the small feeder because the packer simply cannot afford to give him the necessary attention. On the other hand we have often sold by phone because the packer knows that our cattle will perform on the rail like we tell him."

MBI has every available connection with current market conditions of livestock and any factors likely to affect the livestock outlook. LFM (Livestock, Feed, and Meat) and Cattle Fax teletype systems give them the last-minute markets and conditions throughout the working day. In this manner the large feedlot operator knows everything that the packer knows. On one occasion MBI had 9,000 cattle ready for sale, and seven packer buyers from several states flew to Billings to place bids. One packer purchased 7,100 head the first day, the next day another packer bought a thousand, and on the third day a foreign packer took the remaining 900 head.

Dave Drum, looking to the future, hopes to break Montana of its isolated position by becoming an exporter of beef to other nations. By operating on a sufficiently large scale he will be able to export at least five carloads of containerized beef in a single shipment, enabling him to get freight rates reduced enough to make Montana products competitive on the world market. Volume is definitely in his favor, and the day is not far distant when his dreams will become a reality.[4]

Les Melroe Dreams of Beef

Lester W. Melroe, grandson of Norwegian immigrant homesteader, Olaf Melroe, is far from being typical of a third-generation midwestern farm boy. Les Melroe's father, Edward G. Melroe, had learned to cope with the hazards of farming on the North Dakota prairie and had succeeded. Although not afraid of hard work, E. G. Melroe constantly sought to make farming less drudgery and a more profitable business, using his inventive genius for perfecting better farm machinery. Les Melroe, the eldest of the E. G. Melroe children, knew hard work and poverty too and was aware of his father's near bankruptcy on three occasions during the 1920's and 1930's. But he also realized that while others gave up under similar conditions his parents just sacrificed a bit more and worked harder than ever to make things go. E. G. Melroe was interested in machinery and was quick to employ his knowledge in order to enlarge his farm whenever it was possible to do so. By the time the farm prosperity of World War II arrived,

E. G. Melroe, still in the prime years of his life, was operating 3,520 acres of land and was in a position to reap the rewards of big volume and of high prices.

The income from the farm came just at the right time for E. G. Melroe, who was also an inventor. He used farm profits to provide capital for the manufacture of some of his inventions. His manufacturing firm prospered and expanded, and soon its needs for more capital far exceeded the farm's ability to generate new funds. When his son, Les, returned from the military in 1946 he took over the farm while E. G. Melroe was concentrating on manufacturing. In 1952 the Melroes went to Fargo to borrow $250,000 to build a factory for manufacturing farm machinery at Gwinner, North Dakota. One of the Fargo banks declined the loan request, for in their opinion Gwinner was a poor location for a manufacturing plant because it did not even have a hard-surfaced road leading into town.

The business was growing rapidly and the Melroes knew that somehow they had to raise money for a new factory. In desperation they went to a Chicago bank for money, and instead of getting the $250,000 they had asked for in Fargo the Melroes received $300,000. After thoroughly investigating the agricultural market, the Chicago bank knew the potential for the Melroe product, so they were willing to make the loan.

In 1952 E. G. Melroe's firm in addition to other products was producing the coil spring harrow which was useful for weed control in corn and soybeans. To reach a better market Les Melroe was sent to Minneapolis to be closer to an area of corn and soybean production. For the next 14 years he lived and worked in the Twin Cities. The family's manufacturing firm met with one success after another. The earliest success was with the Melroe swath pickup, then with the coil spring harrow, followed by the Bobcat skid loader that was developed by Eddie Velo and the Keller brothers. This became popular in the construction industry, and then among farmers; it eventually was sold in foreign countries. Even though the firm grew beyond all expectations Les Melroe still wanted to get back to the land and to farming.

In 1967, upon the suggestion of Earl Christianson, Les Melroe tried to encourage the Melroe Company to become involved with the Steiger brothers in manufacturing and distributing their four-wheel-drive tractor. The board of the Melroe firm declined to become involved in this deal, and Les had to wait for another opportunity. In 1969 the family sold its manufacturing firm to Clarke

Equipment Company. Now Les Melroe, as one of the sellers, had enough money to live comfortably the rest of his life. It was not comfort that he sought, but an opportunity to put some of his dreams to work.

Having grown up on a rather large mechanized farm that specialized in small cash grain production, Les Melroe, early in life, realized the tremendous task involved in large-scale farming. Thousands of acres of small grain were produced annually in North Dakota, and only the grain part of the crop was used—all of the chaff and straw were wasted. Melroe knew that many corn farmers were plagued by the same problem. Only about 60 percent of the farm's total corn production was being utilized. He knew of Garst's experiments with corn cobs and cellulose in Iowa, so he thought why not do the same for small grain.

In his opinion farmers needed big power, better roughage-handling equipment, and a different breed of cattle to fully realize the potential of world agriculture. The first thing Melroe did was invest heavily in land. He purchased more than 4,000 acres and leased another 8,000 acres. Many in his home community of Gwinner thought of him as the home town boy with lots of money, who could afford to lavish it on a model farm. The home town critics did not know that Les Melroe had invested in several other industries and had come to realize that the profit potential in agriculture was as good as any other business.

Melroe was looking ahead in 1969 and felt that from his knowledge of the world food situation the beef industry had to play a greater role in it. That potential could perhaps be based on the ability of the ruminant animal to convert roughage to meat and the ability of a farmer to make use of a large amount of roughage with as little labor as possible. In 1969 there were still many basic obstacles to Melroe's idea. Although most farmers realized the potential of using chaff, straw, and other roughages in producing beef, few of them made any concerted effort to do so. Basically many of them did not feel a strong economic need for going into this extra work because of the absence of well tested roughage-salvaging machinery. The large grain purchases commencing in 1972 plus starvation in underdeveloped countries might change the whole outlook on beef feeding.

In 1969 Melroe also realized another of his dreams. He invested in the production of the Steiger tractor and purchased control of the struggling tractor firm because he strongly believed that the greatest profit potential in agriculture rested in the hands of those who used the largest possible machinery. Large machinery called for large power units, and by 1969 many progressive farmers of the nation had recognized the potential of four-wheel power as found in the Steiger and Versatile 145 tractors. The Versatile 145 had done for four-wheeled tractors what Fordson had done for two-wheeled tractors 50 years ago. Les Melroe wanted to push his newly purchased firm into building tractors of 200 H.P. and over because he felt that that was where the market potential was best.

He advised fellow board members to "invest in the farmer of tomorrow with big four-wheel-drive tractors and let the full line of machinery implement manufacturers produce the little hundred horsepower two wheeled power units." But he had to reconsider and let Jack Johnson, president of Steiger, decide what he could best market at that time. The company was growing rapidly, and it needed above all else to generate cash flow to stay alive. Melroe did not object; besides, he wanted to spend his time in farming.

There were several reasons why Les Melroe wanted to return to farming.

First of all, farming is what I have always loved since my boyhood. Those fourteen years in the Twin Cities really reconfirmed my desire to farm. I get a greater feeling of accomplishment from farming than anything else I have done.

Then, I have always had a mission about this thing of finding more practical ways of growing beef. I would like to relate feed production to big machinery. If I accomplish anything besides making more money, which we know we are doing, it is to coordinate crops, machinery, and cattle. Making money in this business of farming is always our prime concern, but to do so we have to keep striving to produce more food for less cost. In that way everybody profits, both the farmer and the consumer, plus it helps our export potentials.

Going back to using more roughage which we can so economically produce in this area [eastern North Dakota] we can reduce the cost of production. Our roughage, most of which is now wasted, can be the byproducts of our small grain farming. Or it can be legumes which we can economically produce; besides, they help rebuild our soils and enable us to reduce the amount of petro-chemical fertilizers which we need. Recycling with manure helps us in the same manner.

Les Melroe is almost a fanatic on his visionary concept of the integration of machinery, roughage, and beef. He has a 1,100- cow herd that has been fed a great deal of the straw saved from 3,000 acres of small grain. A single stacker unit works continuously from June through fall on alfalfa, prairie hay, and straw. Two forage choppers are used during the summer and fall seasons. The straw provides about 50 percent of the total ration of the cow herd. In the winter it is blended with chopped alfalfa and corn silage. The calves are fed the same mix of straw, alfalfa, and silage except that they are given a small amount of grain to speed their growth.

The cows on Melrosa Ranch are mixed breeds, but they are bred to a special strain of bulls that Melroe and his associate, Truman Kingsley, are developing. Kingsley, who is well known in the animal breeding world, was formerly in charge of the herds of Cyrus Eaton, the Canadian industrialist. The two partners have developed bulls which weigh 2,600 pounds at two years. The bulls are used for artificial insemination of about 650 of the best cows on the Melrosa Ranch. Each year the cows that produce the poorest calves are heavily culled so that Melroe and Kingsley can achieve their goal of having a herd that will produce 1,000-pound calves in 365 days. In the sixth year of their program they are satisfied that they are within reach of their goal.

Better beef, using salvage from his grain farming, is not the total scope of Melroe's operations. With his managing partner and farm neighbor, Donald Hartness, he has established a very large "farrow to finish" hog program. The purpose of this program is to fully use the grain produced on the farm. A farrowing house with 36 stalls provides the backbone of a system where 400 to 425 litters are produced each 12 months.

Because pigs are born 12 months a year there are finished hogs available to market every month, thereby allowing the farmer to avoid being forced to sell one or two times annually. A totally automated confined environment protects the hogs from the extremes of North Dakota climate and saves all the manure for recycling. Besides being very economical in regard to feed and manure, it is possible for two men to handle all the hogs from breeding to farrow to finish.

Les Melroe is totally convinced that the American farmer wants a 500-H.P. four-wheel-drive tractor. He is also convinced that the consumer will continue to insist on the lowest cost for food. The farmer then has no choice but to adopt the largest possible machinery and to salvage former waste products to produce whatever food he can derive from that material. As a prime mover of the Steiger tractor Melroe will see that the farmer of the future gets the tractor he wants.* However, health problems, complicated by the impairment of his vision and complex labor problems, have forced Les Melroe to quit his farming operation.

---

*Currently the Steiger company is field testing a 750-H.P. tractor.

Drum and Melroe serve as examples of the type of men who are being drawn into agriculture from the industrial world. These men all had other opportunities; they were financially secure, but they had a desire for higher profits and saw a challenge in farming. To them agriculture was not the industry for those who could not make it elsewhere. To them, being a farmer did not imply hick, hayseed, dumb farmer, or plow jockey. To them farming meant a challenging business with a future as great as that of any other industry. College-educated, alert young men might look to agriculture as one of the growth industries of the future. Gordon Anderson, chief appraiser of Federal Land Bank in northern Minnesota and North Dakota, put it so well: "For the man who is determined enough to dig for it there will be the capital necessary to farm. That phrase, lack of capital to farm, is an excuse more than an economic fact."[5]

NOTES

[1]Warren, *Farm Management*, p. 268; Drache, *The Challenge of the Prairie*, pp. 218, 295. Other citations on education and migration off the farm are found in that book as well as in *The Day of the Bonanza*; Fite, *The Farmers Frontier 1865-1900*.

[2]Interview of David G. Drum, Billings, Montana, May 16, 1972, March 19, 1973, and April 28, 1975; Drum is the major stockholder in Montana Beef Industries, M & D Feedlot, T-Bone Feeders, and Feeders Supply Corporation, all of which make up the largest cattle feeding complex in Montana. He also is involved in several other small businesses not directly related to agriculture; "Spotlight on Montana," *Agricultural Situation*, LVI, No. 5 (June, 1972), pp. 5-6.

[3]David Drum interviews; "KOA—A Home on the Open Road," *DKQ Review*, IV, No. 3 (Fall, 1972), p. 3; Interview of A. Leon Miller, Shepherd, Montana, May 16, 1972. Miller interview; Interview of Max Henthorne, Billings, Montana, March 21, 1973. Henthorne was general manager for the three Montana Beef Industries feedlots at Huntley, Shepherd, and Vaughan; Jerry Sinise, "Big Sky Country Is Flexing Its Feeding Muscle," *Beef*, IX, No. 4 (December, 1972), pp. 13-15.

[4]Drum interviews; Miller interview; Henthorne interview; *Beef*, IX, No. 4, pp. 14, 15; Interview of Dr. Duane Flack, Greeley, Colorado, March 12, 1975; *Montana Agricultural Statistics*, XIII (December, 1970), pp. 42-43, 73, 78, 79; *1972 Annual Report*, Yellowstone Valley Electric Cooperative, Inc., "Feeders Supply Corporation"; *Montana Wheat and Barley 1952-1967*, Supplement to Montana Agricultural Statistics, XII (March, 1969), pp. 31, 59, 60, 79, 98; "The Successful Cattle System," Montana Beef Industries, Inc., brochure.

[5]Interview of Lester W. Melroe, Gwinner, North Dakota, February 11, 1972, and March 21, 1975; Robert W. Karolevitz, *E. G. Inventor by Necessity* (Aberdeen, 1968), 16-24; Conversations with Donald (Tom) Hartness, operations manager and independent land owner who handles crop and hog production on Melrosa Ranch; Gordon Anderson interview.

CHAPTER XIV

The Farmer's Banker

The Historical Setting

LAND, LABOR, and capital are the three basic essentials of any production, including that of agriculture. Historically land generally has been the cheapest and, hence, the most expendable of the three, particularly for the American farmer. In general the cost of land was minimal until the time of World War I. By that time most of the free land of the Western frontier had been taken. In the early 1900's some of the farmers began to realize that they were facing a new age, when land was no longer an expendable commodity, and any future expansion would have to be purchased for hard cash. This was a depressing thought for farmers, who had long lived under the happy illusion that there would always be an abundance of free land available to them.

Free land under the Homestead Act had enticed far more people into agriculture than were absolutely essential to feed and clothe America. But the main goal to settle the wide open spaces of the West had been accomplished. The cost in broken minds and bodies associated with the overexpansion of the agricultural frontier will never be known, but the toll was great. Agricultural historian Gilbert Fite wrote, "The surprising thing is not that many pioneers failed to establish commercial farms . . . but the remarkable thing is that so many of them succeeded."

The overexpanded agricultural base in the nation was a blessing to nearly every one but the farmer. The farmer paid for the rapid settlement on free land through more than a half century of reduced income. The problem of an overexpanded agricultural population has been partially solved because since 1920 successful farmers have enlarged their operations through the process of annexing farms of those who have retired or quit. This evolution

450

was hastened by the replacement of horse-power farming with mechanical power. Economics obviously dictated the change.

The abundance of land challenged the farmer, for it had to be used to be of any value. Labor, the second ingredient of production, could best be provided for farming by his having a large family. Much has been written about the 160-acre farm and the farm family with a half dozen or more children. Eight to 12 children seemed to be the most practical number of children necessary to operate an Iowa or Minnesota quarter section—fewer were needed farther west. The unpaid labor of women and children prolonged the life of many farm enterprises. When times became exceptionally difficult, belt tightening and the continued denial of household conveniences helped to balance the budget. Under such conditions of marginal subsistence the family farm became a symbol of endurance. Commercial farmers, who had to pay their workers a living wage, had to compete with farm families that were willing to see the farm succeed at any personal sacrifice.

Better education through school consolidation in the first decades of the twentieth century provided new alternatives of livelihood to farm children. In addition, two world wars helped to weaken the desire of the farm family to pay the high cost of keeping the homestead alive. The advance of industrialization also provided a constant opportunity for the farm youth to escape the drudgery of farm work. After World War II even the farmer's wife found city jobs increasingly available and rewarding. With the disappearance of "free family labor" or underpaid neighbors' children, many farmers realized their choice was three fold—pay a living wage, mechanize, or quit.

In the day before industrial mechanical power was available to the farmer he was unable to pay industrial wages. But after World War II the tractor and its complementary equipment provided the farmer with a lever enabling him to bargain for labor at industrial rates. Through mechanization it became economically possible to pay higher wages. But there was one drawback to mechanization—it required capital.

Capital had long been the least readily available of the three basic needs of production for the American farmer. Historically most capital in agriculture had been generated within the industry. In periods of prosperity the demand for and the amount of external capital increased. But farming, with a history of poor management, did not easily attract venture capital, except for speculative purposes. It is estimated that in the first 50 years of the twentieth century $79 billion were invested in agriculture. Of that amount $17 billion, or 22 percent, came from financial institutions or other lenders; the rest was generated on the farms.

It is difficult to determine precisely whether or not agriculture suffered from an actual shortage of capital in the late nineteenth and early twentieth centuries. There was, however, the Greenback, the Free Silver, and other Populist movements that served as protests against the rates and terms at which money was available. Few farmers secured as much credit as they needed. The overexpansion of the agricultural base and resulting surplus production was accomplished through liberal use of donated homesteaded land and much personal sacrifice. If capital had been more readily available in that period it probably would have caused an even quicker expansion in an industry already growing too rapidly. However, in no way did the nation suffer because of any lack of capital investment in agriculture.

The boom psychology so prevalent on the frontier was often the basis for liberal extension of credit by those who had the means. Many homesteaders found the local merchants and bankers more than willing to lend credit. Big charge accounts at local stores and money for first and second mortgages from eastern sources probably hurt farmers as much as it helped them. The high rate of failure among the homesteaders was the principal cause of the high bankruptcy rate among local merchants and country banks.

Farmers were generally optimistic about what they could do with new land on the frontier, but they were also reckless. Many had the attitude that if they failed on one location they could always move west and start over again. The banker, the local merchant, and even the farm machinery companies helped to foster this attitude. Surely some would succeed and pay their bills; and if others failed, the mortgage holder was bound to cover his losses through rising land values.

Ole D. Larson of Hendrum, Minnesota, was a typical local merchant who liberally extended credit to his farm customers. From the mid-1880's until 1924 Larson conducted a highly successful hardware store, undertaking business, McCormick Deering agency, and automobile agency. He did a large business for that period and accumulated considerable wealth. The members of the Larson family did whatever they wanted because money was never a problem for them. When the children went to college they were given complete and free use of the checkbook. By 1920 Larson had 5,120 acres of land paid for, was a major stockholder in five area banks, held considerable stock in a large hardware wholesale house, and had several thriving businesses in Hendrum.

Larson made the mistake of extending unlimited credit to farmer customers and making personal loans to farmers in the normally prosperous Red River Valley. When he sensed a decline in economic conditions in the mid-1920's he and his two sons started collecting accounts receivable and outstanding loans. For the next three years the Larsons spent much of their time trying to collect cash; but they had little success, for most of the farmers were worse off than the Larsons. Soon the banks with which he was associated failed, then he lost his land when he tried to pay his hardware and machinery accounts, and by 1924 his businesses closed. Larson was broke when he died in 1932, for even his customers in the rich farming region adjacent to the Red River would not or could not repay their debts to him. At this time many country merchants met a similar fate.

Peder L. Solberg came to Milton, North Dakota, in 1894 and taught school until 1901. He homesteaded at the same time as did other members of his family, and by building a house over quarter lines they were able to prove up three quarters. By 1901 the Solbergs owned the entire section. Solberg and a cousin operated a country store and the post office at Merl, North Dakota. After four years they had to close the store because accounts receivable had grown to $4,000.* In 1905 Solberg went to Fairdale, North Dakota, which was just being founded but already had two banks, three hardware stores, two general stores, two livery stables, two lumber yards, and five elevators. Solberg worked at various stores and banks in Fairdale until 1911, when he was able to purchase one of the elevators for $5,000. He borrowed the full amount from Ole Haug, a local farmer, and at the end of one year was able to repay Haug the entire amount and also to buy new cars for Haug and himself. Later external conditions forced him to sell his elevator.

In 1916 Solberg invested his earnings in a chain of 15 locally owned banks in North Dakota. Solberg said:

We were real flush at all of our banks and besides, I had quite a bit of my own money to invest so I made personal loans to local farmers because it looked better than any other investment. The next thing I knew I was unable to collect either interest or principal on my personal loans and

---

*In 1970 Solberg recalled: "And all but $500 of that $4,000 is still outstanding. We had to borrow on our farm to keep the store going. We stocked everything including machinery. I'm still holding the notes."

was being assessed on my bank stock to save the banks. I stayed with the Buxton bank until 1927 and then went to Mekinock; but by 1930 all of us, as directors, were broke so we decided to close the banks.

At age fifty-six I went back to farm work at a dollar a day. In 1933 I was hired by the Federal Land Bank at $7.00 a day and expenses to appraise farms for refinancing. In 1934 I worked for the government making feed loans in Richland County. Farmers would borrow from the government to get feed for their cattle and later I would find out they had plenty of feed, but they were using the government for operating. Of course, I also bought a lot of fully grown cattle for about $17.00 a head in Cavalier County which were really of little or no value to anyone.

Such were the experiences of Peder Solberg, who, with liberal credit, had helped to open the frontier for farming. Initially he made a quick fortune only to lose it again through default when farmers were unable to repay their debts.

Levor B. Garnaas left Norway for America in 1886 and worked for farmers and small town merchants before he settled at Sheyenne, North Dakota, in 1895. With a 100-percent loan he started a small store there and soon after became a dealer for McCormick Deering machinery. He had to extend credit, but was careful to collect his loans every year at harvest. By 1906 Garnaas sensed the need for a bank in Sheyenne. At that time the 36-year-old merchant had $5,000 in assets. As he remembered at the age of 100:

> Nels Faar, a bachelor farmer, put money in my bank in 1909 when he could see that I was going to succeed. I really got the idea of starting a bank because so many people asked me to keep their money in my safe in the store. Even the Indians from Fort Totten had money in my safe. I enjoyed doing business with them. . . . From its start in 1906 to 1970 our bank never closed out a farmer. Somehow we avoided trouble in the 1920's. I do admit that the R.F.C. helped a few in the late 1920's and later the Farm Security Administration saved many. Our farmers seemed to be quite cautious. Once in the early days I had to spend $5.00 for attorney fees to collect a late note.*

Gilman Klefstad started a cream and egg station and grocery store in 1918 at Hillhead, South Dakota, which had a population of 30. For a start he used $145 of his own money, borrowed $500 from the Kidder, South Dakota, bank and $360 from area farmers, and charged $500 in groceries at Leach and Gamble in Wahpeton. He operated on a 25-percent margin. He lost money in the early

---

*Many of the farmers in the Sheyenne area have grown too large for the Sheyenne Bank. However, Alf L. Garnaas has financed some of the largest cattle feeding operations in North Dakota. In the 1970's the insurance agency has become a major source of income for the bank and helps to make the bank profitable.

1920's when two of his buyers, a cooperative creamery and a Fairmont, North Dakota, poultry-buying firm, went broke. Both owed him "a couple hundred dollars," and Klefstad said, "that really hurt in those days." Individual charge accounts often got as high as $500 in a year's time at the Klefstad store, and the annual loss on those open accounts was about 10 percent. The biggest account he ever lost was $500 because he had a policy of not letting anyone charge more than that amount.

For a period of years the Hillhead store remained a profitable venture for Klefstad even with an annual 10-percent loss on charge accounts. In 1919 the Liberty State Bank was established at Hillhead. According to Klefstad: "Two men from Veblen started it, but there was no justification ever for a bank at Hillhead. It lasted for seven years and when it went bankrupt in 1926 the town started to die. I was lucky for I had put about $800 in the store property and I sold in early 1929 for $4,000. If I had not sold when I did I probably would never have gotten rid of it."

According to Klefstad, farmers were too optimistic in the Hillhead area in 1918 and 1919. "Things were just too good and land was bid up to $125 an acre by early 1921. When the drop in prices came land values fell steadily hitting a low of about $5.00 an acre early in 1937. . . . Then it rained and when the farmers saw water in the sloughs once again they became optimistic. By 1940 the sloughs were full and the war had brought the prices up again."

On January 1, 1930, Gilman Klefstad invested in the Havana State Bank, Havana, North Dakota. The bank had been in operation since 1900, but it had been hurt by the declining farm economy. Klefstad and his fellow investors agreed to put $15,000 into the bank in return for the depositors' agreement to accept 50¢ on the dollar for their deposits. The bank remained solvent and by 1936 had repaid its depositors 87¢ on the dollar. Realizing that Havana with only 200 people did not have much of a future, Klefstad nevertheless worked his bank into a very liquid position. Later he moved the bank to Forman, a larger and more prosperous community.

Klefstad recalled:

In seven years at Havana the bank had lost only $15.00 on loans. We never closed out a farmer, but some walked out on us. One left a note at the bank during the night saying all of the personal property was at the farm and a herd of Holstein cows was standing in the barn. We had no choice but to haul them to West Fargo where they didn't bring much. We really hoped to get out of the Havana bank and thought the Bank of Sargent County at Forman might buy us, but to our surprise they went broke. We seized the opportunity and moved our bank to Forman where we had a future. It was a good move.

When asked to sum up the banking history of his area from the 1920's to the 1970's, Klefstad said:

I don't think the country bankers in our area were really to blame. They had started in the late '90's or early 1900's and things constantly prospered until the summer of 1921. They did not expect such a long period of depressed agriculture caused by low prices and drought. Our area was not hit too badly until the 1930's because our people were quite diversified and operated close to the belt. Who would have guessed in 1928 that lower prices and such a prolonged drought would still be ahead? I often think how lucky I was to sell my store in Hillhead in early 1929—it was sheer luck.

If my memory is correct North West Banco shares were $99 in 1929 and I believe were only about $3.00 in 1937. They were no better off than the independent country banker.

I probably would not have started in banking in 1930 if I had known there was going to be such a long dry spell. But I was lucky I came into banking fairly near the bottom and I looked at all those poor notes the previous banker was hung with. Many of those notes were no good the day they were written and both the farmer and the banker knew it. Sometimes bankers extended notes only on the hopes that things might change.

Most bankers stuck with the farmer if the farmer was game. About 45 percent of our people just like the pioneers hung on and many of those really did well. I saw enough of this that I must say I don't think the average country banker would have closed out a farmer then or would do so today if there was any chance that he could make it. But then as today there are people who have no hope and just quit. Honestly, the quicker some of those quit the better off they were and so was their banker.

Ole A. Olson, who was a farm manager for 160 farms financed and repossessed by the National Life Insurance Company of Vermont, agreed with Klefstad's conclusion. Olson has this comment about his experiences:

To keep farmers from abandoning farms we would provide them with seed and livestock feed. This was practical because it gave the farmer a chance to recover if he was a good manager and it prevented loss through vandalism which inevitably came when places were vacant. Most of the people who kept faith and stayed on their farm eventually repurchased them from our company.

My company wrote off $192,000 on farms in one year because they were sold for that much less than we had loaned on them. One thing that really hurt was that the water table fell and many shallow wells went dry. This forced many farmers to sell their cattle or drill a new well right at a time that the money pinch was worst.

Quack also became a major problem. My co-worker had forty-three quarters under his control that had quack so bad no one wanted them. We discovered that unless a farmer had a tractor there was no point to put him on that land. Horses could not work effectively enough on hot days to kill quack.

North Dakota, which is so heavily wheat oriented, experiences some of the greatest fluctuations of income of any state in the Union. In 1904, when the state was still in the process of being settled and the farm economy was strong, local papers boasted of having the highest per-capita income of any state in the Union. By 1932 Dakota ranked forty-first, with a per-capita income of $176. With the prosperity of World War II Dakota ranked only behind New York and Nevada. In 1973, as a result of the great export market in wheat, North Dakota farmers had realized a net income of $21,238, a net surpassed only by the farmers of Nevada and California.

North Dakota and Montana both had a very large increase in farm numbers during World War I. Drought in Montana in 1919 and the sharp price drop of 1921 caused a very high rate of bankruptcy among those farmers who had not enough time to get established. In 1920, 71 percent of North Dakota's farms were mortgaged, the highest rate in the nation. In 1930 the rate had declined to 67 percent, by 1940 it was still 64.8 percent, and by 1950 34.6 percent of the state's farms were mortgaged. In each period this was the highest rate in the nation. Between 1930 and 1944 one-third of the state's farmers lost all of their equity.*

A study on farm mortgages by Paul A. Larson indicated that the three leading causes for their increase during the 1920 to 1950 period were: expansion, purchase of tractors, and purchase of automobiles. Larson found that many farmers were unable to borrow directly for a car, so they mortgaged their farms to buy one.

Declining land values and short-term loans at high interest rates were the two major factors for farm foreclosures. Average land values that were $41.10 state-side in 1920 and as high as $150 an acre in the Red River Valley had declined to $12.92 by 1940. Even with the sharp rise in income during World War II land prices were only $28.86 by 1950.

Henry C. Wallace wrote in 1924 that in 15 Midwest states, 8.5 percent of the farmers lost their farms from January, 1920, to March, 1923, another 14.5 percent of farm owners were insolvent

---

*In 1935, 1940, and 1945 North Dakota farmers were 55, 73, and 11 percent delinquent on their loans with the Federal Land Bank. This compares with 33, 22, and 6 percent for the nation at the same period. At this time the life insurance companies held only 13 percent of all farm mortgages in the state.

but not forced out of business, 14.5 percent of the tenants were
insolvent and forced off the farms, while another 20 percent of the
tenants were insolvent but allowed to remain on the farms they
were operating.

The banks suffered with the farmers, and in the four years ending June 30, 1924, these five states experienced some of the
greatest numbers of bank failures. North Dakota led with 250
banks failing; Montana and South Dakota had 150 failures each;
and Iowa and Minnesota experienced 100 failures per state. Each
year after 1924, 32 or more banks failed in North Dakota. Out of
898 banks 573 failed between 1920 and 1933. In 1947 only 150
banks were left.

The banks in part had caused their own downfall because there
were far too many of them in a state of only 600,000 people. They
had become extremely competitive in their loan procedures.
Farmers could get loans so easily at the banks that they never
bothered to ask for long-term loans from other institutions. When
the economy slackened and the depositors wanted to withdraw
some of their savings, the banks could not renew their farm loans
and had to make forced sales. All of the banks that failed had made
loans in excess of 100 percent of their deposits. The state banks
had averaged loans of 120 percent of their deposits, while the national banks varied from 47 to 285 percent of their deposits. The
permissive optimism of the banks not only caused their own bankruptcy but often caused many of the farmers to go bankrupt also.

The role of frontier banking in North Dakota was similar to the
role of banking in other frontier areas of the nation. Fortunately,
tighter banking laws have reduced such irresponsible activity, but
banking is still subject to human errors. Some prospects will continue to be overrated, and others will be underrated; but there is
no room for extreme frontier optimism in financing industrialized
agriculture. In the day of highly capitalized agriculture the banker's judgment of the management factor is more important than
ever.[1]

The Changing Financial Needs

O. B. Jesness, long-time professor of agricultural economics,
wrote in an essay in 1969:

Commercial agriculture is dynamic, not static. Changes are continuous.
Among the striking changes important place may be assigned to the consolidation into fewer and larger family units and to the decline in numbers of the farm population. . . . The overall effect of consolidation of

farms . . . and off-farm migration is to raise the net farm income. . . . These changes tend to improve utilization of production resources, especially manpower [and] so strengthen the economy.

Former President Lyndon B. Johnson stated in 1965 that the number one farm problem was how to liquidate 2.4 million farmers. It was still the number one problem even though 2.6 million farmers had left the land in the previous 20 years. Great resentment was caused by those statements, which indicated that top leaders in government clearly felt that the American society could best be served by less than a million farmers, while there were well over 3 million in operation. Agriculture was proclaimed an industry with overexpanded resources in labor and capital.*

Many people resented these statements, and some criticized the popular American idea of trying to equate bigness with goodness. E. W. Mueller of the National Lutheran Council said, "There is nothing sacred about bigness. Nor is there anything sacred about smallness. . . . The determining factor is how adequately the needs of the people are met." According to Mueller rural America had been going through a purposeless transition since about 1900, and "aimless change gets people nowhere." But change was inevitable, and it was affecting the total concept of society.

Agriculture, which in the century up to World War II had effectively used surplus land and labor to expand, now turned to capital as the most easily available factor for continued growth. Until World War II, agriculture had relied heavily on internal financing for most of its growth, but by the mid-1940's greater reliance was placed on external sources. There was surplus capital in agriculture, but it was widely disbursed among millions of small units of production and was not being effectively used. Fortunately, enough new deposits in the country banks were being generated to provide new capital for the expanding farmer. The greatest source of those deposits, according to Leslie Peterson, an agricultural banker, "came from the liquidation of farm assets by those retiring or being forced out of farming." Peterson added, "A source which is diminishing each year and new channels will have to be provided if the country bank is going to be in a position to help the farmer of tomorrow with his ever increasing capital requirements. . . . Many farmers don't get all the financing they

---

*Secretary of Agriculture Orville Freeman stated before the Senate Agricultural Committee on June 16, 1965, that only about 400,000 of the nation's more than 3 million farms had anything near parity income. Those were the farms that were large enough to provide full employment for their operators.

want or need, but many get more than they are capable of managing. Collectively farmers have received more credit than they need because they are always able to come up with a surplus and they manage to get the job done. Most farmers, however, receive funds from more than one source which indicates that credit may not be dispensed in the most efficient manner."

Governor Andrew Brimmer of the Federal Reserve Board wrote in 1967 that country banks were having two problems with farm financing: first, from 1957 through 1967 rural bank deposits had increased 70 percent, but farm loans had increased 120 percent and the trend was continuing. Brimmer's second point was that credit use per farm was increasing so greatly that the line of credit many farm units required was beyond the limits of the country bank.

Leslie W. Peterson, addressing the Midwest Banking Institute, said at a later date on the same problems: "With credit needs of agriculture increasing almost twice as fast as rural bank deposits growth, will there be enough loanable funds to satisfy the demand? With the average credit per farm increased by three times in the last ten years, will we in unit banks be able to triple our capital accounts in the next ten years to provide the necessary loan limits so funds will be available?"

Under Secretary of Agriculture Phil Campbell's remarks to the American Society of Farm Managers and Rural Appraisers in January, 1970, were as follows: "What we need is new ways to finance farming. It now takes an investment of anywhere from $100,000 to $300,000. But by 1985 it will require a cool $1 million for a young man to launch an efficient farming operation."

Robert Rupp, in an editorial in *The Farmer* entitled "More Credit Needed," wrote that farm credit needs will expand 9 percent per year because there is constant pressure on the land from farm and non-farm users. The basic reasons for the pressure came first because the average farm was still too small to be efficient, therefore many farmers would have to expand to lower their production costs. Most farmers will continue to substitute capital for labor, and many of the more progressive farmers have a changing attitude toward indebtedness as a permanent feature of their operation.

Nationwide, the farmer was allocating about 6 percent of his expenditures to payment of money, but he was learning that money spent for machines, chemicals, and fuel proved to be a reliable necessary investment. It appears that a greater portion of the operating budget will go for capital in the future because it is both more available and more reliable than land or labor.

In 1969, when 177 farmers in 35 Iowa counties were surveyed, more than 90 percent expressed the opinion that neither land nor labor was available at current rates, but credit was readily accessible. The average indebtedness of 162 of these farmers was $21,427, and they indicated a willingness to increase their liabilities to $50,240 if the opportunity to expand their operation was available. Another 15 farmers of those surveyed indicated they had no personal upper limit on the amount of credit they would use if the opportunity for expansion presented itself.

Credit was the smallest problem of the three tools of production to those Iowa farmers who in 1969 had an average of 269 acres under their management, just slightly above the 234-acre state average. When asked how large they would like to become, 167 farmers indicated as a group that they would like to increase to 544 acres; another 10 indicated that there was no upper limit to the number of acres they would like to operate. Only 4 percent of those 177 farmers indicated that they had any real apprehension about getting adequate credit. The conclusion of A. Gordon Ball, Professor of Economics at Iowa State University, was that this was a clear indication why we were moving toward a capital-intensive agriculture. "Farmers find credit easier to obtain than labor [or land] and much more flexible." Through capitalization and mechanization farms will grow without increasing the amount of labor required.

Ironically, the United States Department of Agriculture concluded a study in 1973 indicating that the one-man farm was technically the most efficient farm unit. The U.S.D.A. determined that such a one-man operation with optimum efficiency would be 800 acres in row crops and require $480,000 in land and $130,000 in other capital. Obviously the Iowa farmer with a state-wide average of 249 acres in 1970 would not fare well in total efficiency by comparison. For a number of reasons farmers were not using their total resources available. With a total agricultural debt load nationwide in 1973 representing only 18 percent of gross assets, it was clear that farmers were not maximizing their capital opportunities, and the Iowa farmers were no exception. For the nation's farmers total indebtedness was about three times their annual net income. Financially agriculture was in an enviable position.

The Agricultural Outlook Conference for 1973 projected that by 1985 there will be 830,000 farms with gross sales of $20,000 or over. These farms will account for 90 percent of the nation's commercial farm production. By that year average total assets per farm will be $246,000, and the average debt will be $71,200. These commercial farms will have an average gross income of $87,108.* Thomas R. Smith, an Iowa banker, stated in 1972: "The report card of management is profit. We think $100,000 in annual sales is a reasonable goal for a family commercial farm and $50,000 is a minimum for good profit opportunity." Smith, a very aggressive banker, was obviously looking far beyond the farmer of 1985 with only $20,000 in gross sales.

The ever-increasing need for purchased contributions to production has changed the financial requirements of most farmers quite drastically. In early America the farmer and his family supplied nearly the total production needs of the farm; only salt, gunpowder, plow lays, and a few tools had to be purchased. By the late 1800's the government recommended that a farmer should have $800 to $1,000 to be able to start a frontier farm. R. M. Probstfield, who farmed 400 acres in Minnesota in 1874, listed $1,328.63 in total expenses, of which $690.13 were for family living. This was in Probstfield's sixth year of farming. Ironically, $62.90 of that amount was for tobacco and saloon expenses and $62.70 for purchased shoes and clothing.

The average machinery investment per farm in Illinois in 1890 was $143, in Iowa $181, and in six counties of the Red River Valley of Minnesota and North Dakota $287.** Many quarter-section homesteaders probably did not gross more than $300 per year on the average during their first six years. Their capital structure was based on the family labor used to erect buildings and to break sod. Cash income earned by members of the family from off-the-farm employment was expected to be used toward the support of the farm.

Family labor remained the chief contribution toward farm capitalization. As late as 1940 family labor was still 59 percent of total production requirements. Mechanization, forced on the farmer by the World War II labor shortage and accompanied by good prices, hastened a rapid decline in the importance of labor to agricultural production. By 1950 labor's contribution was 42 per-

---

*In 1973 total average farm assets were $135,673, and the total liabilities $26,032, leaving net assets of $109,641 per farm. This indicates a net asset of 80.8 percent of total assets. Few industries could match that financial strength.

**The Red River Valley area contained the bonanza farms which relied heavily on machinery to enable them to compete with the homesteader.

cent and capital's was 39 percent. Only a decade later labor's share had fallen to 30 percent, and capital made up 61 percent. In 1950 a farmer had to have $700 in assets to create $1 in income. By 1963 he needed $14 to do the same job. By the 1970's a $20 investment was necessary on many farms to generate $1 in income. Farming had changed from being a nearly self-sufficient industry to one that in the late 1900's was purchasing 75 percent or more of its production needs from off the farm.

Capital needs became intensified with this trend and actually were compounded, because generally the larger the farm the greater the portion of capital input used. Larger farmers appeared to be much more willing to risk a higher ratio of indebtedness than smaller farmers. Farming to the large farmer was no longer a way of life; it was a healthy competitive business that if professionally managed could be very profitable.[2]

Changing Farmer Attitudes

Richard Retz, a rural Iowa banker, remarked: "Farmers are bringing bankers into the twentieth century." Retz, who had farmed before he became a banker, felt strongly that the farmer with an outlook to the future had to educate and inform his banker to get the kind of financing he needed. The reason, Retz felt, was that all too often the banker did not thoroughly understand the needs of the modern farmer. Generally the banker was willing to supply standard credit to the average manager as long as he had security, but he was often unwilling to accommodate the farmer who really wanted to advance no matter how sound his proposition was.

In 1940 American agriculture required the expenditure of 13.6 billion man hours; by 1970 that figure had declined to 6.5 billion hours, with further decline anticipated by 1980 to 3.6 billion hours. At the same time, real estate costs were projected to remain relatively stable. Obviously capital, the third factor of production, would be responsible for the increase in total farm outlay. From 1940 to 1980 the projected increase in fertilizer and chemicals would be seven-fold, power and machinery three-fold, and purchased feed, seed, and livestock four-fold. Invested capital per person engaged in farming increased about 10 percent in each of the early decades of the 1900's. From 1940 to 1950 the rate of increase was 31 percent, while in the decade of the 1950's the rate jumped to about 38 percent. The capital investment per worker increased even faster in the 1960's as capital was used to substitute for the steadily declining number of workers.*

Wylie D. Goodsell, writing in the *1960 Yearbook of Agriculture*, stated: "I may point out that farmers are now generally more efficient than they were previously and that in this respect they are in a better position to withstand a price-cost squeeze and to meet certain costs." Goodsell was referring to the greater efficiency of the capitalized mechanized farm of the 1960's in relation to the labor oriented farm of the 1920's and earlier. Farm management studies of the early 1900's recommended that no tenant should start farming with less than $2,500 capital and no owner with less than $5,000. Farms with capital investment of $2,001 to $3,000 had an annual labor income of only $145. The farm with $4,001 to $5,000 capital returned a labor income of $570, or just over 11 percent yield on investment, the highest income for labor of the average-sized farms. Even in the early 1900's the very large farm, a combined owned and rented unit with over $20,000 capital investment, resulted in the highest labor returns of $2,269.

During periods of high net farm income, such as experienced in 1972 and 1973, many farmers expressed a strong desire not to expand their operation. Some even objected to the Department of Agriculture's study indicating that a one-man, 800-acre Midwest corn-soybean farm and a one-man 1,920-acre Great Plains small-grain farm were the most efficient units. "One man cannot farm 800 acres. He might run over it, but farm it—never. . . . Anyone who can't make a good living on 300 acres won't make it on more than that either. . . ." Another farmer made this criticism: "I don't think one man could handle 1,920 acres of wheat and sorghum properly." But many farmers had already proven that they were capable of handling even larger acreages.

Progressive farmers long ago had learned how to use financial leverage for the betterment of their farming. In doing so some were reckless, hoping that inflation would rescue them if they could not make it through farm profits. Others felt if they failed they could always start over as a renter or move west to new land. Even such prominent farmers as Oliver Dalrymple of bonanza fame was open to criticism for his daring financial policies. James

---

*The total agricultural labor force declined from 10,450,000 in 1929 to 4,500,000 in 1969. This accounts in part for the sharp increase in capital invested per worker. The hired labor force declined from 3.4 million to 1.2 million during those years. From 1940 to 1960, making no allowances for inflation, the amount of capital invested in the average farm increased seven-fold.

B. Power wrote: "Mr. D[alrymple] purchased [land] from Gen. Cass which may lead to expansiveness but which I fear will end in financial bust, for without backing the loan will be too heavy, interest will eat large holes in future profits."

Dalrymple did have difficulty financing through regular loan agencies, but he managed some way or another to get the capital he needed. In one case, he owed McCormick Harvester Company $21,185 for wire and twine purchased in 1879. In December of 1885 the company warned him that if he did not pay his debt they would take legal action. Dalrymple evaded a lawsuit and in 1888 still had not paid that bill.* Dalrymple's method of financing disturbed many, but by the early 1900's this once almost penniless lawyer owned far in excess of 25,000 acres.

In Dalrymple's bonanza days of the 1890's there were 389 farms in North Dakota over 1,000 acres. By 1910, shortly after Dalrymple's death, there were 2,416 such farms, and in 1930 the state had 5,433 farms over 1,000 acres. Mechanization had made its impact on Dakota farmers, and by 1970 there were 14,384 farms of 1,000 or more acres.

It is frequently said in rural circles that the only way to start farming is through marriage or inheritance. From surveys and studies relating to the opportunities of starting in farming from the 1870's through the 1970's we can come to the conclusion that it is probably no more difficult to get into farming in the 1970's than it was a century earlier. The free homestead was really not that "free." Ken Thomas, extension economist at the University of Minnesota, answers in a frank way the cynics who believe that it is too difficult to get into farming in the mid 1900's:

There is plenty of opportunity in farming today [1967]. The fact that entry into farming is difficult suggests that resultant earnings will likely be higher. There is much more variation in earnings among farmers than between farming and other careers. Therefore, the young man starting out who gains control of plenty of resources and uses them effectively can expect to make a respectable living. Our records show this.

Bankers in general are aware that young men have to start in farming, and progressive bankers frequently allocate a certain portion of their available funds to such young farmers. John Chrystal,

---

*Farmers have often felt that they were permanently tied to their banker. This is best expressed by Thomas R. Smith, an Iowa banker: "Farmer Eber Johannsen left written instructions for his community banker to be his pallbearer at his funeral. . . . He has carried me most of my life; he'd just as well finish the job."

banker at Coon Rapids, Iowa, said of a young farmer getting started:

I object to the statement that there is not sufficient capital available for existing or beginning farmers. If a man has difficulty getting money maybe he had better look at himself. Iowa agriculture is not handicapped by a shortage of funds either short or long term. If a man has management ability he can attract money because farming is not as risky as most other businesses—for example the filling station, Ma and Pa corner store, a local hardware store, a small town restaurant, or many phases of the recreation industry. Farming is built on two of the most stable aspects of our economy, appetites and land.

O. M. Jorgenson and Eugene Coombs, two Montana bankers, think that the family desire to keep the farm in operation should not be overstressed when looking at the urge and capacity of a young man who wants to farm. Both bankers said:

We are always looking for young farmer prospects. One of the first indicators we search for is how willing is he to learn from others. Next is how well has he done with what he has worked with up to the present. We do not look at the father to determine if the son is going to be a good farmer. We are interested in the soundness of the young man's goals. Does he have the desire and does he like the challenge? The clincher is determining if he will work well with us if we take a risk on him.

To enforce the points just made, the opinions of some agricultural financiers are expressed here. Gordon Anderson, chief appraiser for the Federal Land Bank in Northern Minnesota and North Dakota, has this to say:

A man might say he can't get started because of the lack of capital and another might say he cannot expand for the same reason, but I maintain this is not true. If a person has the desire he will overcome the road blocks of capital shortage. All too often the phrase "lack of capital" is the excuse because an individual does not have the desire and that comes because he fears his own lack of management ability. In my opinion lack of desire, fear, and management limitations all preclude the lack of capital. I have seen many make a real go of it and they started on a shoestring.

Frank Larson, a North Dakota banker, has this word of caution:

I am not very anxious to finance a young man who grew up on a farm and expresses that it is his God given right to be a farmer. Many farmers seem to expect this, but I don't know of other vocations where they assume this. I am sentimental about the family farm, but the farmer has to learn that you make it or you don't just like any other business.

We recently started with a young farmer who was not getting cooperation from his former banker and had $12,000 more in liabilities than in assets. He had 1,200 acres of land, but he needed livestock to get enough volume. We went to work with him and in three years he had a $100,000 positive net balance. It was just a case of recognizing the man's management ability even though he was young and way undercapitalized.

One of the other bankers who was interviewed, and who was previously cited in this chapter but will remain anonymous in this instance, had this dramatic story to tell:

Twelve years ago in March I had a young man who had just been discharged from the service and decided he wanted to farm. He leased a very poor farm near town and used his service savings to buy used machinery. When he needed more money he came to me and wanted to borrow $1,500 for operating capital.

My immediate reaction was no, this was surely money down the drain for the bank. While he was talking about what he was planning to do I remembered here was a boy who was the son of the town ne'er-do-well, who had worked at odd jobs to support a family and not too well at that. Here was a chance to take a long shot against the bank's reserve for bad debts and I told him he could have the $1,500.

The following July I drove around the country looking over our farm accounts and was amazed when I saw the crop coming on this poor, run down farm. When I drove on the place the young man blushed and said he had been thinking about coming to the bank for more money to get hogs and get ready for harvest expenses. I encouraged him to call on my assistant, and explain what he wanted. When I got back to town that night I called my assistant and informed him to let this young man have whatever money he needed. I had to be out of town the next day.

Well, to make a long story sweet, that young man was just in the bank settling with us today. After eleven crops and nearly twelve years of farming he has a net worth of $280,000. And there are some people who say you can't start from scratch in farming.

Obviously many of the young people who are determined to go into farming pay too much attention to these cynics who have said for decades it is too difficult for a young man to start farming. Achievements of young men in the programs of the Future Farmers of America and the Outstanding Young Farmer are positive proof that an intelligent, determined young manager can start farming and make a real success of it.

In 1967 the *Farm Journal* analyzed the records of the 41 state winners for the OYF program of that year. The results show: "The average of them started with $12,500 net worth, then built this to $211,000 in eight years of farming. Seventeen families had less than $5,000 net worth when they started full-time farming. We found almost no relationship between starting net worth and annual rate of growth." Milton Hertz of Mott, North Dakota, started farming while going to college so he could pay his own college expenses. In his first year he operated only 90 acres. After he graduated from college he operated 160 acres, along with teaching school.

In 1959 Hertz was in his fourth year of teaching and then decided that there were greater challenges and opportunities in farming than in teaching. The 24-year-old youth told his father that he wanted to become a partner in the family farm operation. His father, Gotthold, refused to consider a partnership, but gave his son one-half section of land that was worth less than $10,000 and an eight-year-old tractor worth about $800. Hertz then informed his school board that he wanted to go into farming, but would like to continue teaching if his wife, Carol, could substitute for him in September and May. This was agreeable to the board, and he signed a $3,150 contract to teach during 1960 and 1961.

At once Hertz mortgaged their half section for all they could get to buy equipment, seed, fuel, and fertilizer and to make partial rental payments on 480 acres of leased land. While Carol substituted for him in school Hertz put in the crops and harvested them. They continued this double activity until the spring of 1963.

Carol and Milton Hertz did not have it easy even though they could rely on the teaching income for household expenses. In 1961 drought and hail caused an operating loss; in 1963 they lost 600 acres of wheat to hail; in 1964 hail destroyed another 1,000 acres of wheat, and they lost 80 cows in a blizzard. But they accepted the challenges, and at the end of 1968, after nine full years of farming, Carol and Milton Hertz owned 240 beef brood cows, had 300 beef cattle on feed, rented 5,960 acres, and owned 5,360 acres. At the same time, both had been very active in community and civic functions.

Hertz readily admitted that being able to buy a $200,000 ranch and grain farm early in his career with only 5 percent down was a big break. He was also surprised at the lack of aggressiveness of other farmers in his area, and he felt that this was part of the reason why it was easier for him to get ahead. Having cattle and grain proved to be a strong feature of the Hertz operation because one or the other seemed to pay out each year. Probably the strongest single asset to Hertz in his early years was that he kept excellent records and had much less difficulty convincing his bankers to extend credit to him than might otherwise have been the case. Having a degree in business administration plus a semester of graduate work in business management also proved its value to this young farmer.

In 1968-69 Carol and Milton Hertz were recognized as one of

the nation's four outstanding young farm couples. Milt Hertz was quick to state that good records caused him to discontinue operations on his farm that he liked but that were not profitable.* His records caused him to cull many of his best-looking cows, but by doing so he increased weaning weights 100 pounds in three years. His farm continued to grow and prosper so well that he added additional livestock units each year and generated more income. With meager beginnings Hertz, at the age of 37, had developed one of the largest farm operations in North Dakota.

There were 42 state winners at the 1971-72 National Outstanding Young Farmer Congress. These men, all of them less than 35 years old, had been farming for an average of 10 years. Their gross assets averaged $290,700, and their net worth averaged $168,280. On the average these young farmers had gained $14,800 net worth per year from sales of just over $200,000 per farm. That year only about 65,000 farms out of 2.8 million had grossed over $100,000 in sales. These young men each started with about 110 acres and after 10 years owned 486 acres and rented another 862 acres per man.

While the farmers nationally had borrowed only 18½¢ on each dollar of assets, the 42 OYF's had an average debt of 42 percent. In addition, they were leasing far more land and equipment proportionately than the average of the nation's farmers. Financial leverage was one of the chief reasons they grew so rapidly. Banker Thomas R. Smith said of those younger successful farmers, "They are opportunists; they are intelligent with a tremendous urge to succeed. They are not afraid to borrow money and know how to accept risk, which is quite contrary to the thinking of the average farmer."

In 1973 the OYF's had their annual congress in St. Paul. At this time their average age was 32, and after farming about 10 years they had accumulated a net worth of over $250,000 each. Even though the initial worth of those 40 young men was $18,000, several had made their start with a 4-H calf or an acre of land in an FFA project, and six of them began their farming career with a negative financial balance. These 40 participants belonged to an average of 12 civic and agricultural organizations. Robert Rupp, a farm magazine editor who helped judge that contest, noted that the challenge and independence of farming were the two most

---

*Joe Bohlen, sociologist at Iowa State University, stated: "Some of the romance is going out of farming, as farmers look at each enterprise to see if it's paying its way."

significant reasons why those young men chose to make farming a career.

It is significant that the four top winners for national Outstanding Young Farmer in 1972-73 had a distinctive five-point approach to money management: "1. They borrow money to make money. 2. They're good cash flow managers. 3. They watch debt interest closely to cut out unnecessary expense. 4. They put limits on how much money it is safe for them to borrow. 5. They use money to 'buy' profitable skilled help."

Generally these men were big renters of land, preferring to use borrowed money for expansion of operation, not ownership. One of them, Donald Keil of Montana, said, "You have to find bankers with imagination about the potentials in agriculture, but your past performance is your best asset in finding one." All of these top four winners also paid good salaries to get excellent employees. Donald Keil employed two specialists with Ph.D's and a third who was a Ph.D. candidate.

When the Farm Credit System invited 100 top young farmers to a conference in early 1974 to get their ideas on modern agriculture of the 1970's, it discovered what innovative managers could really do. These 100 farmers, whose average age was 33, had been farming an average of 11 years. They farmed 840 acres of land, had two years of college, and controlled $600,000 in resources, which included $278,000 of leased property. On the average they had increased a $14,000 net worth from 1965 to $190,000 by early 1974. In addition all of them were very active in agricultural and civic organizations besides doing experimental work for many agri-oriented companies.

These young farmers were heavily in debt and "were not particularly worried about it." They preferred operating on a larger scale and living better to having a smaller farm and less debts. One of the suggestions they had was that the Federal Credit System should make it easier for beginning farmers to get capital. "Lenders want to share in the agricultural wealth, but they don't want to share in the agricultural risks."

Steve Proebstel, a board member of his local PCA, expressed the following opinion:

We have 1,000 borrowers in our PCA, each with an average of about $24,000 borrowed. We have a $40,000 bad debt reserve and have used it only once. We have lost only $500 since the beginning of the operation. To me that is pathetic. If PCA had placed that $40,000 into a bad loan every single year and lost it all, that would still mean only $40 from each of us. I think I could stand that for some guy coming into farming. Doesn't it make more sense to finance one beginning farmer properly than four or five marginally?

Of interest also is the bankers' point of view about making loans to young farmers. Lindley Finch of Continental Bank of Chicago made the following comment:

When I was in the country bank we had a policy of advancing credit far beyond accepted standards each year to a selected group of young farmers. Examiners rightfully criticized such loans, but the few mistakes we made were outweighed by the benefits. I know a loan officer who had no loans go sour for some ten years. He was asked to resign. He had to be turning down some good credit risks to make such a record. He was too cautious.

Frank Larson, with three banks in North Dakota, said:

Many of our rural bankers are critical of the Twin City banks and that is because so many of our bankers are not doing anything either for the farmer or the state. I think many of the locally owned banks are really negative. The chain banks are doing a better job of banking but treat Dakota as a captive. It seems like Dakotans prefer being taken by outsiders. If locally owned banks had done the job there would not have been such a great opportunity for the chains to come in.

Frank Bauder, chairman of the Central National Bank of Chicago, stated in a report that money is available for the farmer who has a detailed balance sheet, records of five years of operating results, cash flow statements, and future projections. To this Bauder added:

These are the kind of financial reports other industries have taken for granted for years. Yet, all too many requests for credit from agricultural producers are not supported with adequate money management information. The "young tiger" of today who becomes a financial success tomorrow will know how to use these tools as well as he knows how to use his machinery, fertilizer, and other production tools. . . . The key is to maintain control of the business while utilizing the capital to maximize efficiency and return.

Each time farmers have reduced labor and increased use of capital, total production has increased. The young farmers cited above were short on capital but long on the willingness to risk borrowing and managing money. That was a major reason for their success. They understood what Edward Higbee wrote: "A farm long on labor but short on capital is virtually obsolete as an efficient unit of production. . . . It is more profitable to employ capital than people for machines improve and can be depreciated and both represent gain while men have improved little and it is costly to pension them."

Farmers have long frowned on keeping good records. Many of them have preferred to do hip pocket or check book record keeping. Often farmers seeking new lines of credit are asked by the prospective banker to present a household budget and projections for their farming operation for the coming year. These farmers were unable and unwilling to do either of the tasks requested by the bankers. Poor bookkeeping habits were probably practiced by farmers deliberately in the past to avoid having to make an accurate accounting either to their banker or to the Internal Revenue Service. Those days are past.[3]

Who Is Financing the Farmer?

When farmers settled much of the Midwest, including the Dakotas, it was often a practice to heavily mortgage their farms and if things went bad to walk out. The question could be asked why were financiers and bankers so willing to make loans to homesteaders? Their reason might have been that except for the areas within the railroad land grants it was very difficult for non-homesteaders to secure more than a limited acreage of undeveloped public domain. By loaning money to homesteaders, financial institutions—anticipating some foreclosures—hoped to acquire public domain. After 1891 "commuting" for the homestead rather than "proving up" became the popular way of acquiring public domain.* A great number of homesteaders who acquired their homestead by commuting sold out within three months after the commuting date. For many, commuting was a way of generating cash through use of public domain, and it also made it easier for speculative non-homesteaders to gain possession of land.

Through the process of uncontrolled expansion, agriculture became overdeveloped and suffered from what North Dakota historian Elwyn B. Robinson called the "too much mistake." Since the 1920's, however, continual readjustment in the agricultural sector has taken place in an effort to get investment in line with the resources. It was in the Dakotas and Montana where much of the final expansion of agriculture took place as late as World War I. These states were some of the areas of greatest suffering because the readjustment came almost immediately after the last settle-

---

*Commuting was the practice of paying cash to the government for the land rather than spending time to "prove up" the homestead.

ment had taken place. A mass exodus of farmers occurred here within a few years after the greatest surge in homesteading. The 1920's through the 1970's experienced a steady withdrawal of the overexpanded resources in farming in an effort to get a better balance between resources and needs. President Lyndon Johnson's National Advisory Commission on Food and Fiber declared in 1967 that American agriculture still suffered from surplus investment and labor. The commission advised that maximum production could be achieved with 31 percent less manpower than was still available that year. It also pointed out that if farmers were allowed to operate at full efficiency the 1980 prices could be $1.27 for wheat (1967 dollar value), 75¢ for corn, and $1.11 for soybeans. To be able to sell at those prices farmers would have to operate at considerably different levels of efficiency than they did in 1967.

With the national picture as fluid as described before, but with many individual farmers demanding more capital to expand their operation, the country banker was caught in the middle of this dilemma. The issue was not new, for G. F. Warren had written about it in 1913: "Bankers are studying the farmer's increased need for capital. The necessity of a more readily available capital for carrying on the farmer's business is granted by everyone who has given the least attention to the matter. But how is this to be done is the point at which the interests of bankers and farmers are likely to conflict." Warren suggested that farmers should become represented in management and control of financial institutions. His advice was eventually heeded by many commercial banks, which placed farmers on their boards of directors, and by the Farm Credit System's institutions, whose boards are made up almost entirely of farmers.

By 1971 the average capital investment per farm worker had surpassed $56,000, and the average investment per farm exceeded $200,000. With potential individual farm-loan requests exceeding the individual loan limits of the bank, the country banker faced another serious problem. Leslie W. Peterson said of the problem:

> Rural bankers will have to make and adjust to changes in order to survive and prosper in years ahead. They will have to recognize and meet the increasing credit needs of agriculture. Otherwise, they will lose customers to other lenders by default. . . .
> With the size of farm loans increasing to forty, fifty, and even one hundred thousand dollars, is the man behind the desk qualified and capable of dealing with the progressive farmer on the other side . . . ? Will the unit banking system . . . in the Midwest be able to do the job of financing agriculture . . . ? In Minnesota with about 400 small banks in

towns of less than 2,000 in population, I'm afraid there are many banks that cannot answer all the preceding questions in the affirmative.

There are some banks that can't answer any in the affirmative, and some banks whose management doesn't give a D--- whether they can or not.

Peterson, who was strongly devoted to the cause of a rural single-unit country bank, had spotted the real problem. His bank, which was and is 90 percent loaned to agricultural customers, was doing its job. He used correspondent banking whenever his bank could not assume the entire loan. He operated differently than many country bankers, who have been in the habit of telling customers that they cannot handle large loans. So does Lindley Finch, Vice President and Agricultural Consultant of the Continental Bank in Chicago, who is of the opinion that country banks can continue to handle farm loans regardless of size: "Any farmer who thinks his local bank can't handle the size loan he needs is listening to an old wives' tale. If you qualify for credit in excess of your local bank's capacity and your bank won't get it for you, you have the wrong bank or your banker is suffering from managerial inertia, which is a polite way of saying he's lazy."

In 1972 Ed Clausen of Farmers State Bank, Schleswig, Iowa, had $6.3 million of his $10 million outstanding loans on purchase-price feeder-cattle loans. In defense of such a high ratio of loans in one type of enterprise he said: "You have to believe in farm programs built around livestock. Our area isn't suited to straight grain farming. With roughage cattle feeding comes naturally. . . . Beef is a basic industry in this area. . . . We've never done anything but lend money for the full feeding period with no interest due on the note until the sale, even if that's 14 months, past an annual accounting."

Not all country bankers think like Smith, Larson, Peterson, or Clausen; however, most banks make some agricultural loans. Nine out of 10 commercial banks, representing over 13,500 institutions, made loans to farmers in the late 1960's. Their total loans to farmers had increased about 260 percent in the decade of the 1960's. Production Credit Associations had increased their loan volume to farmers nearly 700 percent during the same period. The farmers were getting the money they needed.

Credit institutions are abundant in America, but they also have problems. Stewart D. Gager, Vice President of Chase Manhattan Bank, expressed his view of the problem as follows:

Agriculture's biggest problem is fragmentation and this makes it difficult to finance properly. This causes sloppy financing. Besides, too many country bankers are too limited in their vision. We do not object to a negative cash flow in the early years of an operation if things are going according to schedule. We find country bankers are too asset-liquidation minded, partly because they do not properly appraise the cash flow basis. They are interested in a collateral based operation rather than potential income. A farm is too big to be financed on a typical short term basis. But country bankers shy away from cash flow because they want to be cleaned up each year.

John Chrystal, an Iowa banker, warned fellow bankers that they had to create better correspondent relationships or larger farmers would move to larger banks. Country bankers had a job to do in educating many farmers about proper security arrangements so that they could work with correspondent banks. At the same time, Chrystal cautioned the banker not to become too involved in "saving the poor manager in this changing agriculture." He advised that bankers often do a disservice by sticking with borrowers who are not adjusting to the new agriculture and who are letting their equity position deteriorate:

I think more and more that the failure of many farmers who are moving out of agriculture with no equity and no money available for a new start is more the fault of the banker than the borrower. The banker is supposed to be educated and sees a larger number of people in such financial situations than the borrower does. The banker is unwilling to make the harsh move to advise the man he has neither the management ability nor the equity to stay in a business that requires a high capital investment.

I think the lender can do a greater disservice to the farmer's family than the farmer can do himself. . . .

I see some troubles ahead for very small country banks. There are several in Iowa now that are in existence only as adjuncts to insurance agencies, and this is wrong.

Merle W. Marshall of the Farmers and Merchant Bank in Huron, South Dakota, contended, "The need for community leadership . . . has been overlooked too long by too many bankers. But if you want the banking industry to continue the way we know it today [1969] you must meet this challenge." Marshall, long active in the South Dakota Bankers Association, urged his fellow bankers to take active roles in their communities.

Leslie Peterson, addressing the same group of Midwest bankers, warned that there were country banks who gave no thought to management succession or hiring trained personnel. He added, "The first step is to convince the management in some of our banks that there is a problem—that changes have and are taking place in agriculture and its financing needs. [We must be] educating and training ourselves and our staff to the level of competence needed to serve our customers and to make the bank as profitable as possible."

It is the general consensus of many bankers that if a bank does not have a well trained agricultural consultant it does not deserve the farmers' business.* Frank Larson, a Dakota banker, made this point:

When you finance farmers everyone has a special problem, but you must have personnel with an overall knowledge of agriculture to help them. Our purpose is to help the farmer get the job done. Too many of our bankers, especially those in the chains, are too job conscious. I think this makes them poor loan agents.

A good example of this is the advent of irrigation. There are eleven irrigation systems being financed by our banks. This forces both change for the farmer and the banker. We have pushed irrigation because it is virtually a lead pipe cinch if you have a manager who isn't afraid to learn. We financed every proven manager who wanted to go into irrigation and we will finance any qualified farmer for the same venture. But I notice not many bankers are looking at the opportunity to finance irrigators because they think it is just another headache.

We have two agricultural men in each of our banks and I would like to hire another man just to hold seminars, to stimulate our farmers, and keep them up on the latest in agriculture. If we as country bankers can't do a better job of serving our farmers we don't have much reason to exist. Our individual banks have loan limits of less than $100,000, but we are carrying accounts as high as $500,000. This proves that when you want to you can finance anyone.

The 1959 Farm Credit Act repealed the $200,000 loan limit by the Federal Land Bank to a single borrower. That act also permitted the Land Bank to make unamortized or partially unamortized loans. The last feature recognized the need for a farmer to use his income to maintain or increase his capital assets rather than to reduce debts. Unamortized loans introduced the concept of perpetual indebtedness to American agriculture. This concept, widely accepted in European agriculture and in the industrial sector, is contrary to the thinking of a great proportion of the American farmers who historically had a goal of a paid-up farm.

Howard Elson, one of the nation's four Outstanding Young Farmers in 1972, said, "I don't plan ever to be out of debt. . . . But indebtedness isn't the real issue. I feel the only real measure of

---

*The use of agricultural loan officers is a relatively recent development in the commercial banking field. It was the influence of Production Credit Associations as much as any factor that made commercial banks respond to changing times. In 1959 only about 1,200 of the 14,000 banks loaning to farmers had agricultural consultants.

progress in business is whether or not your net worth statement is growing.... We use plenty of OPM (other people's money)...but we also rely heavily on OPB—other people's brains."

Shortly after World War II the Farm Credit System, through the Production Credit Associations, started making loans on an intermediate-term basis. This provided better financing for the hog farmer who raised and fed all of his hogs. The same was true with a cattle feeder who needed two years from the start of the planting of his crop until it was sold through the cattle. A dairy-man needed such financing for his milking facilities and for the development of his cow herd. The more innovative commercial banks have adopted these policies, but many similar institutions, as of 1974, still prefer one-year notes, with an oral understanding of renewal. This gave all the options to the banker and few to the borrower. All too often the banker's attitude toward renewal was based on whether or not his bank was hard pressed for cash at the renewal date.

Because of the commercial banker's unwillingness to extend intermediate credit, many farmers who wanted to expand were forced to go the route of contract farming. This greatly reduced the risk of the individual farmer, but in many cases it relegated him to a position not too different from that of a laborer, because the suppliers made all but the minor management decisions. Machinery and farm facility companies have sought to fulfill the same need in farm financing. Traditionally merchants, suppliers, and individuals have always provided a large portion of agricultural credit. In 1962 they provided 39 percent of the farmers' total financial needs. In 1972 manufacturer-suppliers alone provided $34.4 billion, or 37 percent of the total short- and intermediate-credit needs. It is estimated that by 1980 that figure will be $64 billion. Indirectly this has put agriculture in a position where it has tapped the major money markets for the first time.*

Robert E. Hamilton, chairman of the Central National Bank of Chicago, contends that it will be difficult for farmers to tap the major money supplies. These are his reasons: "Managers of finan-

---

*The A. O. Smith Harvestore Division has been one of the real innovators in offering manufacturer's financial help to buyers of their products. The company guarantees the loans to the lender, which, in the case of many midwestern buyers of Harvestore products, is the Bank for Cooperatives. This has been one of the most progressive movements by any company to help finance intermediate-term loans for facility improvement.

cial institutions, not just banks in money market centers, do not really understand [agriculture]. . . . They have agricultural specialists, but the specialists do not control the allocation of the institutions' funds—the top management does. . . . A primary source of agricultural credit, the manufacturer-supplier, . . . is not prepared to compound his investment indefinitely."

Leslie Peterson, a Minnesota banker, suggested that farmers are going to have to consider syndicates, partnerships, or other joint ventures to generate capital for expansion purposes and to avoid becoming the captive of the manufacturer-supplier or of other contractual arrangements. If farmers do not take this route it may become necessary, Peterson added, to establish some kind of system of federally insured loans to reduce the risk to the lender. This would be in the form of either a much improved federal crop insurance program or an arrangement similar to what the Federal Housing Administration has for home builders. Peterson admitted that government farm programs had done much to reduce risk to the lenders in the past, but he did not foresee nor desire a continuation of such programs.

Thomas Smith, an Iowa banker, speaking to the Mid-America Livestock Forum, said: "Farm borrowers [are] using a credit volume quite often exceeding their net worth, and risks are high because of the high volume business and lack of diversification. . . . Today, it is most important that a mutual understanding is developed between the lender and the borrower."

Smith also presented a four-point program which he hopes will help reduce risk for both lender and borrower and also will keep the relationship on a consultation, not confrontation, basis. First, a financial statement showing normal asset and liability figures; second, a cash flow statement of projected income and expense by months for the year; third, a budget; and finally, a profit-and-loss statement. As he stated in more detail:

> If you can't get the financing you need . . . something is wrong with (1) your records, (2) your management, or (3) your past dealings with your lender. . . . The borrower needs [these] . . . as a tool in planning. We need it to draft a repayment schedule, recognizing that family living expenses and principal payments on term loans have to be built into cash flow so they won't be overlooked. Our correspondent banks require it on all over-line loans—loans over the $100,000 credit our bank can extend. . . .
>
> The records . . . actually show your community banker and a correspondent banker what kind of manager you are and, therefore, how good a risk you are. That's important, since lenders today may have a larger investment in your business than you do.

Lindley Finch, of Continental Bank of Chicago, stresses the significance of cash flow:

I can't emphasize enough the use of the cash-flow plan. Beware of the bank that doesn't demand them, especially if you are seeking an overline. A cash-flow plan is not only important for charting your year's business. It is common language among credit people for appraising money needs. And don't believe those people who say that farmers can't make accurate cash-flow projections because of the uncertainties in agriculture. Other people have to contend with weather and unpredictable prices too.... A good banker wants you to make financial progress. I think that says it best. Beware of the banker who is interested only in your equity, only in your ability to pay back a loan on time. He should appraise your request equally on how much progress you will make.

Banker Finch candidly acknowledges that the farmer's ability to manage money may be even more important than his agricultural know-how.

Leslie Peterson felt that it was the lender's responsibility to determine the management ability of his farmer borrowers. Peterson said: "There are 160 acre farmers and 1,600 acre farmers," and it is up to the banker to help each set his priorities. Some managers cannot stand the risk of big debt without losing sleep and obviously should not be encouraged.

All bankers warned that there was one problem which, if properly handled, could be anticipated. Lindley Finch said, "Family problems often lead to credit trouble." Leslie Peterson asked, "Does the wife concur in proposed changes? Or are there other family considerations?" L. A. Dickerson, assistant administrator of the Federal Housing Administration, indicated that a commitment from both husband and wife is necessary. He recalled from his experiences that "most of our unsuccessful loans trace back to marital problems or poor judgment in management...." Thomas Smith always advised a farm client who wanted to contract a large loan to make an appointment and bring his wife along. Obviously family considerations have to be built into the total cash-flow projections. Peterson felt that management appraisal is one of the fundamental tasks of the banker and family relations are part of that appraisal.[4]

The Bankers Take Notice

Fortunately or unfortunately country bankers have had a big role in shaping the destiny of rural America. Some are aware of the role and are doing a good job for society. Others finance consumer goods or invest in government securities but neglect the financial opportunities in agriculture.

George Donahue, a rural sociologist, speaking at a Midwest Agricultural Conference in 1965, warned independent bankers:

The concept of freedom and individuality, which permeates the institutions including the banking system of the rural areas, is a concept born of self-sufficiency rather than a concept born of interdependence. This concept was indeed consistent with subsistence farming and the rural community structure associated with it in the past. . . .

The rate at which the rural community changes depends in large part upon the awareness of the leadership structure to the social and economic trends. Yet in times of crises it is often the leadership who feel that by changing they have the most to lose and hence become the most resistant to change. Rather than looking to their position of leadership as a public trust they are inclined to look to their position of leadership as a vested interest.

The majority of our community leadership has a greater fear of change than the average resident; this is particularly true when the prospects of change appear to imply a merger of institutional structures such as churches, banks, and businesses as well as farms.

Donahue understood the social and economic problems of the rural Midwest well. Frequently he admonished farm organization leaders for taking stands which had no other goal than preserving the past. He warned that leadership of that nature only served to keep the rural population divided. He finished his address to the Midwest Bankers in 1965 with the following words:

Reorganization of the rural community into a dynamic functioning system based on social and economic innovations and compromise with the needs of other interest groups will not be accomplished without human and material costs. Whether these costs are high or low appears to depend largely upon leadership with the vision to design community institutions consistent with the "Jet" rather than the "Jenny" age—not upon programs which treat the symptoms of change.

Ten years after Donahue had uttered these words of warning some bankers and some farm organization leaders were still thinking of the Jenny age. Fortunately for rural America, the power of those people was being eroded, for progressive bankers and farmers were working in the Jet age.

Agriculture, as in the past, will be financed, for rural America— and in more recent times urban America—has bankers who see the profit potential of industrialized agriculture. These bankers see agriculture as a growth industry, for as the number of small farms is reduced, fewer but more economic units are being created. Better managers, frequently with an excellent education, who keep records to prove the profitability of their business, are able to cooperate with enlightened bankers to help open a new era. Ironically, it took governmentally created institutions such as are found in the Farm Credit Systems to force the hand of many bankers to change their policy toward farm financing. A popular saying in rural America quite well indicates what happened. The saying roughly is that wherever there is a poor farm banker the Production Credit Associations and Farmers Home Administration have made great inroads. Where the bank has done a good job the PCA's have had a hard time getting much business. The bankers, by default, had created their own competition.*

Duane R. Gjervold, Regional Vice-President for Federal Intermediate Credit Banks, said in 1972:

> There is no difficulty securing available dollars for financing of agriculture. The people responsible for ag-financing are more knowledgeable than they have ever been. We are in a transition period now and we are questioning how much capital for any unit should be borrowed. . . . We look at net operating income history to determine how far we can go. We will loan any amount to a man as long as we can see his net grow. There is no shortage of successful operators. . . .
>
> Presently agriculture is creating sufficient funds for its own capitalization. Some "poor mouthers" won't agree with this and neither will the man who wants to expand too fast, but the money is here for sound planned expansion. But there seems to be a tendency to siphon funds away from agriculture and I believe this hurts the industry.

Reuben W. Jacobson, a professional farm manager, said of the financial problem, "If there is a shortage of capital in a farm enterprise it is frequently the lack of desire to obtain it by the individual. We don't encourage landlords to finance their young tenants because there is financing available for the right man. . . . When we are scouting for new renters we don't get too concerned about debts because some of the fastest growing operators with good profit records also have the biggest debts, but it is not a problem with them."

Gjervold agreed with those conclusions, expressing the thought that P.C.A.'s major concern was long-run positive cash flow and its second concern was the profit-and-loss record, from one year to

---

*The *1962 Yearbook of Agriculture* has an excellent account on the growth of the Farm Credit System in financing rural enterprise.

the next. He added, "We hopefully do not want the liabilities to exceed net worth."

N. R. Lake, vice-president and agricultural representative for the First National Bank of Fergus Falls, Minnesota, implied that in his opinion the chief problem in agricultural financing was the lack of package financing. Lake said, "There are banks who will loan a farmer money to improve his milking facilities, but then they will not loan money for cows to repay the loan." Mr. Lake is an aggressive banker in getting long-term and intermediate credit for his farm accounts. It is his belief that too many farmers are "under the gun" to repay their loans too quickly and this too often causes them to deny their family a proper living. "To get the living standard his family requires a farmer needs volume and profits and these are the first objectives we try to shoot for when he comes to us for credit," said Lake. "We find too many beginning farmers who think too small so we have to show them what size unit is realistic."

Eugene E. Coombs, a Billings, Montana, banker, said that his bank has a higher ratio of record keepers among the younger farmers than the older men. Coombs explained that the older men with proven management ability have a profit record and are less of a security risk: "Young farmers with limited assets have to keep records to help the banker determine his management ability. If after keeping good records and cooperating with us it becomes obvious that a farmer is not going to be a successful farmer we encourage him to either quit or look elsewhere for credit. . . . You have to have this person to person successful relationship between the banker and the customer to have a happy banker and a happy farmer customer."

Kenneth Just, President of the First National Bank of Barnesville, Minnesota, agreed with Coombs, expressing the opinion that the personality and communication relationship of the banker and the borrower sometimes is 90 percent of "any hang up that may exist between the two parties."

Record keeping was considered a vital part of good management practices as specified by the bankers. Farmers often replied that they did not have time to keep records, to which Coombs answered, "The good farm manager should spend time on his records until it will be more profitable to spend it elsewhere." According to Coombs, bankers disliked hearing farmers say they did not have enough time to keep records. It was his finding that on farms where someone kept records the family members became curious as to what the results were, and this caused them to be more enterprise-minded.*

The bankers in general agreed that credit is more essential than ever to farming, and the banker's ability to recognize good management is the most important factor in his decision to finance the farmer. Eugene Coombs summed up what the farsighted bankers are thinking: "Banks as well as other lending institutions will no doubt determine to a great extent who will be our future farmers. Farmers unable to gain control of enough capital to operate a farm large enough to be an economical unit are going to be forced out of farming. Management and credit will determine their destiny. The good farm manager knows he must move fast or perish."[5]

What the Bankers Say

Size of farm operation has become an increasingly important factor in evaluating the potential of a farm loan. In his survey of 150 rural bankers Coombs determined that most bankers placed size as one of the top 7 of 17 criteria listed in his study as an important characteristic of agriculture borrowers. Bankers saw three alternatives open to low-income farmers. The first was to insulate themselves from the changing society and continue to farm on a small scale with a low income. The second alternative was to look for full-time off-the-farm employment where a low-income farmer could benefit from the managerial ability of others. The third choice was to reorganize and expand the farming operation to improve net income.

The ideas of these bankers even in 1963 was not original, for farmers historically had been doing just that. George Donahue, addressing a Midwest bankers' group, pointed out that the most rapid migration from the farm took place in times of greatest prosperity—times when off-the-farm alternatives were best and also when farm income was highest, because small farmers had the best opportunities to sell out to expanding farmers. The pattern has been consistent even though it appears countrary to popular opinion and obviously contrary to many who argue that higher farm prices will reduce the trend toward farm consolidation.

When bankers were asked what they thought the proper size of

---

*Record keeping has not been popular with farmers. Gjervold revealed that out of 46,265 P.C.A. borrowers in the St. Paul region only 4,753 had the P.C.A. record system—AGRIFAX statements on file. The record keepers had average loans of nearly triple the non-record keepers.

farms should be they all had ready answers, because bankers are basically economically oriented.

N. R. Lake, a Minnesota banker, suggested that when a farmer wants to get financing from his bank, first his desired standard of living is to be established and then the actual cost of the family needs can be determined. After this has been determined Lake and his client work from a projected net profit to establish the necessary farm size. Lake used a specific example of a family desiring a $12,000 a year net income. In order to attain the desired net income a livestock-general-support farm of 320 acres of good crop land was necessary. In this case the minimum established farm unit was substantially above the average-sized farm found in the locality.

Thomas R. Smith, an Iowa banker, commented that the farmers who have come to him for financial help and have brought their wives along in order to determine the family budget are the ones who have been the most successful. Smith added that generally these people had life goals considerably above the average.

Most farm families greatly underestimate the actual cash needs for family living. One banker indicated that generally, without working out a budget, farm families estimate their living cost only 60 percent as high as they should. A study conducted at Iowa State University of 244 Iowa farmers covering 1969 to 1972 indicated family cash living expenditures of $6,590 in 1969 to $6,935 in 1971. Because farm income for those 244 families varied from a low of $8,459 in 1970 to a high of $15,841 in 1969 the percent of farm income used for living varied from 43 to 82 percent.

AGRIFAX record returns for 4,753 above-average farmers borrowing from P.C.A.'s in the St. Paul region indicated that family cash living costs varied from $5,224 for Minnesota hog farmers to $10,751 for North Dakota cash-crop farmers. The average family cash costs were about $8,500. The percent of gross farm income that is used for family living costs varied from 3.6 percent for Minnesota livestock feeders to 18 percent for Minnesota dairymen and 21 percent for North Dakota crop-livestock farms. Cash living expenses were about 12 percent of gross farm income for all AGRIFAX members. If these are the above-average farmers for the regional PCA's, it becomes understandable why the smaller units are forced to contribute far too much of their total income supporting the family and hence hindering sound amortization schedules.

Banker Coombs reported that in a survey of 225 farm families in his Montana area about 65 percent of the net income of the farms was needed to support the families. The great danger here lay in the fact that any sharp change in farm income could vitally affect the family living pattern. At the same time, poor management of the family budget was a direct cause for the downfall of many small farm operations. Coombs was of the opinion that in such operations it was imperative that more than one member of the family have direct knowledge of record keeping for both the family and the farm.

Dale W. Anderson, a Sioux Falls banker and chief agriculturalist for Northwest Bank Corporation, indicated that in some cases clients refused to cooperate in any way to keep the family budget in line with the farm income potential. Anderson added:

These people leave the bank very angry, but frequently they are on a one way road to financial disaster. When farmers ask me for a line of credit I like to work from production cost per acre and per animal unit. Next we project income and obviously the difference goes to amortization and living expenses. There is a minimum scale in any type of operation. In our area a cash corn operation should be 850 to 1,000 acres. A crop-feedlot combination can get by on 500 head of cattle, but this is really not an efficient operation. I would rather see a crop farm with an annual feed-lot turnover of 2,000 cattle. There are no upper limits to any of these operations with the proper management. About the only hang up is the logistics of hauling roughage for a big feedlot. . . . I have no upper limit as far as size of loan is concerned, but many small banks just cannot do the job required by large-scale farming.

Tom Smith of Iowa said:

A farmer must gross at least $50,000 for each person working on the farm, but really should have a $100,000 volume. Iowa State indicated it takes 525 acres to equal one man year, but I think opportunity only comes if you are bigger than that. . . . To make a good living a farmer should figure on 10 percent of his gross sales as net income. How big does he have to be to earn the living his family wants? I think a farmer-owner can easily earn 10 percent.

It appears to me that there are two types of farmers who have economic opportunity—the part-time farmer who has a full-time job in town and the commercial farmer who has at least one and one-half man years per farm. There has to be more work than one man can do or there is underemployment.

Stewart Gager of Chase Manhattan Bank was of the opinion that $150,000 gross volume might produce a big enough margin for a farm family to live on. But in his experience of financing large-scale farming enterprises he was fully aware that the larger producers "can pull those production costs way down and will force the small operator out of business."

Gager added this bit of economic philosophy:

In any enterprise economy of scale relates directly to return on investment. In a capital-short world money seeks the smaller operations for the
market forces the trend toward large-scale farming. This is true because
the market mechanism will force the economies of scale. In agri-business
things are no different than any other industry for the market favors the
least-cost operator.

When these bankers were asked if they would rather work with
the larger farmers than with the small ones they agreed on one
point. They stated that the larger operator had a decided advantage over the smaller farmer in equipment and power. Gjervold of
P.C.A. said, "We feel there are no more risks in farming than any
other business and your exposure grows as you expand, but so
does the safety factor of being spread out." O. M. Jorgenson of
Billings, speaking from 50 years of experience said, "Both large
and small suffer from bad weather, but when the weather im-
proves the larger farmer recoups much faster than the small
operator because he has excess margin beyond his family living
requirements. With 65 percent of net operating income going to
family expenses it takes nearly two years to get living expenses
paid while the larger farmer can do it easily in his first year."
Iowan Smith said, "The big farmer is generally not only better
equipped, but more alert so he does a better job quicker than a
neighbor who operates on a smaller scale just across the road.
Over the years we have observed that more small farmers get delayed by weather than the larger ones."

Frank Larson, who had had considerable farm experience before he became a banker, gave his opinion on farm size:

I don't know how much fun it would be to make a living on only a
section of land because unless there is a tremendously vertical situation
what would be the challenge? How many would be satisfied with the
income produced from such a unit? We don't seem to have many leaders
in two of our banking communities because our farmers are too comfortable. In those communities I can name only one real innovator. In our
third community we have considerable excitement because we have
larger farmers who are on the go.

In an interview bankers were asked if farmers had received a
fair return for their time and investment over the years. Once
again it became obvious that management was the key factor.
Gilman Klefstad, a Dakota banker since 1930 and a large-scale
farmer, said:

We have had farmers doing business with us since the 1930's and they have never had a year where they lost money. On top of the profit picture they have also experienced excellent appreciation of property values.

I had one sizeable farmer who was always trying out new ideas and he was virtually bankrupt three times, but I had faith in him and we just kept it to ourselves. The last time not even his wife knew how pressed he was, but he came through to be one of the wealthiest people in the state because when the breaks came he was big enough to capitalize on them.

Too many of our farmers think too small, but now loan companies just will not back a man unless he has some size. In our area 960 acres and a cattle enterprise make a good one-man unit and you need this much to make a living.

Gordon Anderson of the Federal Land Bank said this about fair returns:

If we speak in terms of average farmers they are probably getting a fair return on either their labor or their investment, but not both. However, if he has been good enough to hold out for twenty years, appreciation has made up enough so that his accumulation of net worth plus family living has given him a reasonable return on both labor and investment. Much of this comes from forced economy of living and forced reinvestment of income plus the appreciation.

This still applies to the top farmers, but a good farmer applies himself and his family with greater desire. For them farming has been as rewarding as any industry and certainly this type of man is far better off than if he had decided to seek employment working for someone else.

I must add that I have noticed the tremendous advantages of family closeness among the successful farmers I have done business with over the years. This is an outstanding plus benefit.

Dale Anderson replied, "There is no doubt that the large-scale farmers I work with have had more than a generous return on their time and investment. These are the successful ones, but there are so many farmers who just cannot manage and obviously they have not fared well."

O. M. Jorgenson of Montana observed from his experiences: "I started in banking during the 1920's and found even then those who were good managers and were determined to stick to their guns made a fair return. There is a real intangible feature in agriculture that makes it pay well in the long run. Basically, in spite of all the talk to the contrary, I doubt if there is a difference in profits between the farmers we finance and any of the other businesses."

Tom Smith of Iowa answered:

Yes, a fair return for our top farmers. These men are making 10 percent on their investment. Many of my top farmers could run big industrial complexes. They are doing very well. But less than 20 percent of our farm customers are really making money and our big farmers make up most of these. The other 80 percent look at it as a way of life. Maybe the husband or wife has an off-the-farm job or they inherited the farm and they are just riding it out.

A study by Everett Stoneberg and Herb Howell at Iowa State University found that production cost on a 160-acre Iowa farm in 1972 was $133.42 per acre as contrasted to $76.55 per acre on a 640-acre farm. The return to management on the 160-acre farm was $3,042.10, while management earned $30,052.45 on the 640-acre unit. The greatest savings for the large farm were in machinery, taxes, insurance, and depreciation. In addition labor was more than twice as productive on the 640-acre farm as the labor on the 160-acre farm. A four-fold increase in capital investment brought ten-fold increase in management returns. Major efficiencies came through the use of more and larger equipment.

Frank Larson thought that many of the farmers miss the boat by not taking advantage of the opportunities available to them. "We have too many operators who are resting on their assets; once they were aggressive, now they are just sitting back waiting for things to move. We think that overall our farm accounts are receiving more than ample return for the effort put forth."

Gager of Chase Manhattan said:

There are lots of ways to measure return on investment. . . . My opinion is that return to agriculture is much higher than farmers indicate. . . . A factory built twenty years ago is worth less than when it was built, but land is one capital investment that goes the other way. If return were not better than what farmers in general say more of them would be selling out because they could invest to advantage elsewhere. A good way to trap some of these complainers is to ask them what they would pay for their neighbor's farm. They will give a price above market value if they could buy it. I know there are some realistic 25 to 30 percent returns on investment in agriculture.

In another question the same bankers were asked what special trait or characteristic they noticed about the farmers who were larger and more successful than average among their clients. N. R. Lake said:

Half of my customers are so satisfied with what they have that they have no desire to take advantage of the opportunities open to them. Some of these people come in for a loan without really giving me a sound proposal and if I say no they walk away perfectly happy.

The people who I notice that do the best are those who are quality minded. They like to do things the right way and with the best equipment. They are good detail men. These men set high goals for living and farming and they achieve them.

Frank Larson had a reaction similar to Lake's. Larson noted that many of his smaller-scale operators will come in to discuss various alternatives of enterprises, and at the end they will make conclusions almost entirely based on emotions. Dale Anderson concurred with that observation and added that the successful larger operators are very good with the figures and will let the facts of economics determine their decision. Anderson felt that failure to work in a logical manner probably caused small farmers to overinvest in machinery, which eventually got them into trouble.*

Gordon Anderson, who is involved in farming as well as being an appraiser with the Federal Land Bank, felt that the greatest single difference between the growing successful farmer and the average operator was the constant need for a challenge. He noted that most of the successful men had suffered setbacks in farming. But these setbacks only strengthened their desire to succeed. They had the ability to strive without letup until they achieved their goal.

It was the general opinion of these bankers that the bulk of the successful farmers were natural competitors—they liked outdoing others, they liked the process of risk taking, and they liked setting difficult goals. Gjervold noted than even though they are risk takers, some of the larger farmers really fear failure and this sometimes causes them to "overdo it" when they do get into problems. Tom Smith said, "They are intense and are poor losers, but you can't stop them."

Eugene Coombs noted one trait in particular which was backed by the other bankers, and that was that the more successful farmers were much superior marketers:

Small farmers make me furious. They work 364 days to produce a crop but sell it in one day with almost no thought given to the process. This has got to be one of the biggest differences.

The successful farmer is on the market constantly. Some of them have a greater volume to sell than is handled by many of our country elevators.

Reuben Jacobson, professional farm manager, said that when his firm looks for a renter for its clients he visits a prospective renter and observes how he does in the fields and in the cattle yards. If

---

*Eddie Dunn, an extension economist with North Dakota State University, is involved in a great amount of program planning with farmers at all levels. Dunn feels that often when all the economic facets of a program are discussed many farmers will let emotions overrule the facts and make their decisions accordingly.

Jacobson is satisfied with the appearance of the fields and yards he looks at the equipment.* These are some of his suggestions:

Today we are looking at the man who is already operating a fairly big acreage and using four-wheel-drive power because we have learned that if he is not afraid to step out it is a real plus in our favor. A good operator is profit minded from farming and isn't all wrapped up in the benefits of what inflation will do to land values. Our renters on the average are operating bigger every year, especially the younger ones.

Many times a man's letter of application to rent is enough to disqualify him. Often the wife will write the letter and this is fine because with 150 renters under our management we have learned that woman's influence is great. We never make a final deal until we get into the house and have a chance to observe how it is kept up and are able to determine the attitude of the spouse.

The final factor which sometimes is the first we can determine is the man's open mindedness and willingness to accept new ideas. We like the organized man who is always looking for better ways to do things. This helps us increase profits from the farm which makes a farm management firm look good and landlord and renter easier to work with.

Eugene Coomb's survey of 130 farmers established as an interesting fact that there is a direct relationship between the bank's classification of a farmer as a manager and the time the farmer spends reading informative agricultural publications. A good manager spends 5.2 hours per week reading specific agricultural information, an average manager 3.8 hours, and a poor manager 2.7 hours.

Coombs also found that 54 percent of his top managers consulted at least once a month with experts in agricultural extension, with bankers, with ASCS, and with other agri-business men. Only 42 percent of average managers and 21 percent of the poor managers sought similar consultations. His conclusion was:

The below average managers seem to seek little outside advice nor do they spend nearly as much time reading as do the above average and average managers. This can be expected and is surely a factor in why the poor manager has his problems in decision making. Apparently very little thought—neither his own nor anyone else—and [or] planning goes into his decision-making in his business. . . . The better managers seek more information and appear to be more capable of interpreting the information given them than do the average and below average managers.

---

*Jacobson's comments are included with the financiers' because he works very closely with bankers both as a consultant and in managing property for banks. His remarks summarized what the bankers in general said about potential loan prospects, for that is the way they looked at new accounts.

Jacobson stressed that according to his observations once poor managers noticed that they could not compete with others they seemed to give up and quit in contrast to the successful man who kept growing. Gager noted that growth became easier when the farms got larger and generally it took less energy proportionally to handle each addition, so that the successful operator just kept growing.* At the same time, the poor manager was blaming. everyone but himself for his problems.

The bankers were asked what they felt might be the greatest danger in allowing farms to grow larger and what effect a greater concentration in agriculture might have on borrowing. They expressed two fears, but neither of them seemed to make much of an impact on lending and therefore would not reverse any general trend. The first general reservation of the financiers was that their own ability to judge thoroughly the farmer's managerial capacity would suffer. All bankers, however, felt that so much progress had been made through the employment of agricultural specialists by the banks that this problem would be reduced in the future.

The second reservation was a fear of the lack of back-up management on the expanding farms. If the primary manager receiving financial help should die, the risk for the bank increases significantly. Tom Smith even expressed the feeling that banks should go together and hire a trouble-shooting manager who could be placed into any position when a farm manager was removed from the scene. In case of death Smith suggests that the bank's temporary manager could keep the farm going until proper liquidation could take place without impairing the estate. Or the manager could stay until new management was found or developed by the remaining family members. Smith cited actual cases where the farmer's widow took full control and did an excellent job of running the estate.

Gager emphasized that once any operation got large enough to have several managers, then the danger was greatly lessened. He, too, cited actual accounts where strong enough management

---

*Jacobson had a significant observation about some of their poor renters. "We have had many renters who we have had to kick off the farm. Later these people have thanked us because they got jobs in town where someone else managed their time and they were living better than ever. On the other hand we have sometimes stuck with a poor renter rather than kick him off when we could see he could not do the job and ended up with more misery than ever. But there were family influences or other emotional ties." Jacobson added, "Quite often their farm organization tended to fortify their poor renters' opinions and this doesn't do a thing to solve the problem."

existed on given farms so that even the removal of two managers would not cause a major risk factor for the bank. Gager noted that depth of back-up management was a significant determining factor when the bank made loans to larger farming enterprises.

Size of units will obviously not limit financial assistance from the banks. In fact, financiers may even encourage farm mergers as a means of economizing on the total dollars needed in agriculture. Generally the financiers felt there was no limit to the size of enterprise an individual could manage. Tom Smith suggests that for the sake of efficiency in operation and in order to secure back-up management at the same time, it might be valid for as many as 8 to 10 competent farmers to create a joint venture out of all their individual farming operations. This practice has been used by many groups in agriculture with considerable success. Joint ventures provide economy of scale, reduce the total capital required, give better management, and lessen the burden of the individual partners.

Grant Mattson, a pioneer in farm machinery leasing, expressed the opinion that many small farmers could have continued to farm if they had learned years ago to cooperate with their neighbors and had gone to leasing large machinery jointly rather than each attempting to own a full line of smaller-scale equipment. Mattson felt that one or two men could have operated the land, leaving three or four others free to develop other enterprises or take a job in town.

Duane Gjervold commented that quite frequently it was apparent that farmer directors on local boards of financial institutions in the Farm Credit Systems balked at loaning more to certain large farmers in their area. Gjervold said, "The common defense used by these directors was that the farmer involved was already too big." Economics and sentiment are at obvious opposite poles in such cases, and time will prove which will be the winner.[6]

NOTES

[1]Fite, *The Farmer's Frontier*, 1865-1900, pp. 45, 119; Fred L. Garlock, "Financing Capital Requirements," *Yearbook of Agriculture, 1960: Power to Produce*, ed. Alfred Stefferud (Washington, 1960), p. 375; Philip S. Brown, "Money," *Yearbook of Agriculture, 1962: After a Hundred Years*, ed. Alfred Stefferud (Washington, 1962), p. 563; Interview of Mrs. Oscar P. Jordheim, Fargo, North Dakota, March 28, 1970. Mrs. Jordheim was the daughter of Ole D. Larson; Interview of Peder L. Solberg, Gilby, North Dakota, April 18, 1970. Mr. Solberg was born April 20, 1874, and was still active in business at the time of interview; Interview of Levor B. Garnaas, Sheyenne, North Dakota, December 11, 1970. Mr. Garnaas was born in Norway, July 1, 1870, and came to America in 1896; Interview of Gilman Klefstad,

Forman, North Dakota, October 7, 1971. Mr. Klefstad is president and major stockholder of the Sargent County State Bank and two other area banks. He was born near Kidder, South Dakota, April 26, 1897; Ole A. Olson interview; *Grand Forks Herald*, June 26, 1904; Farm Bureau News, LIII, No. 19, May 12, 1974, p. 87; Larson, "History of Farm Mortgage ..." pp. 1, 14, 19, 40-42, 66, 113, 116, 121; North Dakota Agricultural Statistics Bulletin 408 Revised (September, 1972); Henry C. Wallace, *Our Debt and Duty to the Farmer*, Century Co., New York, 1925, pp. 36, 42; Elwyn B. Robinson, *History of North Dakota*, Lincoln, 1966, pp. 375-378; Probably the best book on agricultural financing is Alvin S. Tostlebe, *Capital in Agriculture, Its Formation and Financing Since 1870*, Princeton, 1957.

²O. B. Jesness, "Poverty Among American Farmers," *Rural Poverty and Regional Progress in an Urban Society*, ed. Task Force on Economic Growth and Opportunity, Fourth Report (Chamber of Commerce of the U.S., 1969), pp. 235-236; Harry Rash, "The Challenge of Our Rural Communities," *The Independent Banker* (August, 1965), a special report, p. 5, 20; E. W. Mueller, "Every Man Is Worthy of His Hire," *The Independent Banker* (August, 1965), p. S-3;; Leslie W. Peterson, "Address to Federal Crop Insurance Agents, January, 1972. Mr. Peterson of Trimont, Minnesota, has been a very active rural banker and has been an outspoken supporter of better financing for the commercial farmer; Richard Krumme, "Money Management," *Successful Farming* (January, 1970), p. 16; Robert Rupp, "More Credit Needed," *The Farmer*, 90, No. 11, June 3, 1972, p. 5; Leslie W. Peterson address to the G.O.P. Task Force on Agriculture, January, 1970; "The Farmer's Production Dollar: Where It Goes," *Agricultural Situation*, LVII, No. 7 (August, 1973), p. 5; A. Gordon Ball, "How Much Do Farmers Want to Expand?" *Iowa Farm Science*, XXV, No. 3 (November-December, 1970), pp. 3-687, 4-688; A. Gordon Ball, "Land, Labor, and Capital Availability for Agriculture," *Iowa Farm Science*, XXV, No. 2 (November-December, 1970), pp. 16-700, 17-701; *Iowa Annual Farm Census 1971*, Bulletin 92-AG, p. 9; "Efficiency Winner," *Agricultural Situation*, LVIII, No. 2 (March, 1974), pp. 2-4; "Briefings," p. 12, "Statistical Barometer," p. 15; Mike Walsten, "Growth Guidelines to Keep You in Farming," *Farm Journal*, 97, No. 4 (April, 1973), p. 32D; "Profits Records Required to Compete for Financing," *Feedlot Management*, XIV, No. 12 (December, 1972) p. 52; Don Kendall, "Average 1985 Farmer May Sell Out, Pay His Bills and Pocket $175,400," *Fargo Forum*, March 5, 1973; Victor Chou, "Basic Trends in the Farm Economy," *The Independent Banker* (August, 1965), p. S-27; "Farm Assets," *Farmers Union Herald*, 48, No. 4 (February 18, 1974), Drache, *The Challenge of the Prairie*, p. 321; Murray, *The Valley Comes of Age*, p. 149; H. B. Howell, "The Modern Farm Business," *The Independent Banker* (August, 1965), p. S-15; Leslie W. Peterson, "Address to Minnesota Farm Bureau State Convention," November 20, 1972.

³Interview of Richard Retz, Agricultural consultant and Trust officer, Home State Bank, Jefferson, Iowa, January 26, 1974; Earl O. Heady, "The Future Structure of U.S. Agriculture," *American Agriculture: the Changing Structure*, ed. Wyn F. Owen, p. 125; Wylie D. Goodsell, "Technology and Capital," *Yearbook of Agriculture, 1960: Power to Produce*, ed. Alfred Stefferud (Washington, 1962), p. 370; Vernon W. Ruttan, "Agricultural Policy in an Affluent Society," *Journal of Farm Economics* (December, 1966); Philip S. Brown, "Money," *1962 Yearbook, A Report*, p. 566; Warren, *Farm Management*, p. 311; Gene Logsdon, "Expand My Farm? Only If I Have To," *Top Op*, VI, No. 4 (April, 1974), p. 36; J. B. Power to Israel Lombard, December 16, 1878, NDIRS 309; Fite, *Farmer's Frontier* ... p. 87; Thomas R. Smith, "Lead or Follow: Stick Together," *Commercial West* (May 27, 1972), p. 27; Paul Larson, *Farm Mortgage* ... p. 14; North Dakota Agricultural Statistics 408, Revised, September, 1972, p. 3; "Your Sons ... How Will They Farm?" *The Farmer* (June 17, 1967), p. 9; Interview of John Chrystal, Coon Rapids, Iowa, June 14, 1972. Mr. Chrystal is finance manager for the Garst Farms of Iowa, president of the Iowa Savings Bank at Coon Rapids, and past Iowa Superintendent of Banking; Interview of Eugene E. Coombs and O. M. Jorgenson, Billings, Mon-

tana, May 16, 1972. Jorgenson is the retired chief agriculture officer for Security Trust and Savings Bank, the seventy-first largest agricultural bank in the nation and the largest bank in Montana, North Dakota, and Wyoming. Coombs is current chief agricultural officer of the Security Trust; Interview of Gordon A. Anderson, Lisbon, North Dakota, March 7, 1973. Mr. Anderson has been appraiser for the Federal Land Bank since 1956. He also operates a beef brood cow herd and 3,400-acre farm in partnership with his brother; Interview of Frank Larson, Valley City, North Dakota, March 14, 1975. Mr. Larson owns three typical country banks at Bowbells, Ellendale, and Oakes, North Dakota; Interview of Thomas R. Smith, Perry, Iowa, February 22, 1972; Mr. Smith was president of the First National Bank of Perry, Iowa, and is currently president of the bank in Marshalltown, Iowa, which is a member of the Brenton bank group. At the time of interview Mr. Smith was president of the American Bankers Association Agricultural and Rural Affairs Division; *Farm Journal*, April, 1967, p. 11; Interview of Milton Hertz, Mott, North Dakota, April 7, 1972; "Milt Hertz Wants to Do Better Than Dad," *The Bismarck Tribune*, April 19, 1969; "Agriculture's New Era Off and Running," *Successful Farming*, LXXI, No. 12 (November-December, 1973), p. 29; Neil McFadgen, "North Dakota's Outstanding Young Farmer," *Dakota Farmer*, LXXIX, No. 7 (April 5, 1969), p. 24; Royal Fraedrich, "Who Says a Young Man Can't Succeed in Farming," *Big Farmer*, XLIV, No. 4 (April, 1972); Robert Rupp, "As Things Look to Me," *The Farmer* (March 17, 1973), p. 10; Mike Walsten, "Finance Your Way into Farming," *Top Op*, V, No. VI (June-July, 1973), pp. 20-25; Richard Krumme, "Agriculture's New Elite," *Successful Farming*, LXXII, No. 6 (April, 1974), p. 10; Bill Gnatzig, "Farm Credit," *The Farmer*, XCI, No. 7 (April 7, 1973). p. 4; Edward Higbee, *Farms and Farmers in an Urban Age*, N.Y. 1963, pp. 4-5; Gene Logsdon, "Young Farmers Tell Off Their Bankers," *Farm Journal*, XCVIII, No. 6 (June-July, 1974), p. T04; Letter of J. B. Power to Benjamin Cheney, October 23, 1877, NDIRS File 309; Gene Logsdon, "Do You Have the Right Banker?," *Top Op*, IV, No. 7 (August, 1972), p. 20.

⁴Fred A. Shannon, "The Homestead Act and the Labor Surplus," *American Historical Review*, XLI, No. 4 (July, 1936), pp. 639-641; "U.S. Said to Have Too Much Invested in Agriculture," *Fargo Forum*, August 20, 1967; Warren, *Farm Management*, pp. 300-302; Harold K. Street, "Farm Finance—Vantage Point," *The American Banker*, July 31, 1969; *Top Op* (August, 1972), pp. 20-21; Frank Larson interview; Dick Seim, "Something Every Feeder Needs, . . ." *Farm Journal*, XCVIII, No. 3 (March, 1974), B-14; "Growing with Agriculture," *The Farm Picture* (November, 1967), p. 3; Leslie Peterson, FCIC Speech; Interview of Steward D. Gager, Vice President, Chase Manhattan Bank, Billings, Montana, March 14, 1973. Mr. Gager is a specialist in corporate lending for agriculture; John Chrystal, "View of Agriculture Solvency," *The Independent Banker* (August, 1965), pp. 5-8; Garlock, *Power to Produce*, pp. 378-379; Gnatzig, *The Farmer*, April 7, 1973; p. 10; *Feedlot Management*, December, 1972, p. 52; Leslie Peterson, "Task Force Speech"; "I Don't Ever Plan to Be out of Debt," *Successful Farming*, LXXI, No. 11 (October, 1973), p. G-6; Laura Lane, "Will Your Banker Say Yes This Year?" *Top Op* (February, 1972), pp. 23, 25; Leslie Peterson, M. F. Bureau talk, November 1972; Logsdon, "Young Farmers, . . ." *Farm Journal*, p. To5; Leslie W. Peterson, "Comments" National Agricultural and Rural Affairs Conference, Minneapolis, November 12, 1973.

⁵Interview of Oscar L. Olson, Kent, Minnesota, January 29, 1972. Mr. Olson and his brother, L. L. Olson, owned the bank at Kent and for 63 years operated it; Financial statement of Kent State Bank, Kent, Minnesota, for years 1906 to 1969; Interview of Lewis L. Olson, Barnesville, Minnesota, March 6, 1971, and January 6, 1972. Mr. Olson was active in the banking business from 1906 to 1972. Prior to his entry into banking he had worked in a restaurant, grocery store, and hardware store; Interview of Walter R. Olson, Fergus Falls, Minnesota, March, 1971. Mr. Olson owned several country banks in Minnesota and Dakota; Leslie Peterson,

FCIC speech, January, 1972; Interview of Franklin Page, Hamilton, North Dakota, February 12, 1971. Mr. Page was active in banking from 1924 until the early 1970's. He had seen the full growth, bloom, and slow death of his community because of changing times. He was elected secretary of the Pembina County Fair in 1918, and it was with great pride that he volunteered that he was still its secretary in 1971; George A. Donahue, "Population Changes and Government Policy," *The Independent Banker* (August, 1965), p. S-24; Conversations with George A. Donahue at Twin Valley, Minnesota, February, 1968, and Moorhead, Minnesota, March, 1969; Interview of Duane R. Gjervold, Regional Vice President for Federal Intermediate Credit Bank of St. Paul, St. Paul, Minnesota, September 13, 1972; Interview of Reuben W. Jacobson, partner Northwestern Farm Management Company, Fergus Falls, Minnesota, May 1, 1974. This company manages about 60,000 acres in the Midwest; Interview of N. R. Lake, Vice President and Agricultural Representative, First National Bank, Fergus Falls, Minnesota, February 25, 1974; "The Difference Between Those Who Prosper—and Those Who Don't," *Successful Farming*, LXXI, No. 7 (May, 1973), pp. 10-11; Letter from Eugene E. Coombs, Vice President, Security Trust and Savings Bank, Billings, Montana, January 23, 1973; Conversation with Kenneth Just, President, First National Bank, Barnesville, Minnesota, June 2, 1974; Eugene E. Commbs, "Evaluating Management Factors in Agricultural Credit," unpublished thesis, Pacific Coast Banking School, University of Washington, March, 1963, pp. 6, 8, 9, 26, 29, 35, 37, 39, 66, 68, 72, 74, 75, 77-84; Letter from B. G. Crewdson, Agricultural Development Director, Upper Midwest Research and Development Council, to Eugene E. Coombs, Billings, Montana, February 25, 1955; Jerome E. Bambenek (compiler), "Financial Summary of 1971 Agrifax Members" of the St. Paul region, pp. 1, 3, 4, 9. Outstanding State Young Farmer documents of Mr. and Mrs. Cyril Donkers, Faribault, Minnesota, submitted to the author June 19, 1974.

[6]Coombs, "Evaluating Management . . ." survey sheet, pp. 11-13, 37, 42, 56, 57; Donahue, "Population Changes . . . ," *The Independent Banker*, August, 1965; p. S-22; "Statistical Barometer," *Agricultural Situation*, LVII, No. 3 (April, 1973), p. 15; LVIII, No. 5 (June, 1974), p. 15; "Farmers Earn More from Farm Sources," *The Forum*, August 3, 1973; N. R. Lake interview; Thomas R. Smith interview; AGRIFAX, 1971, pp. 4, 8, 9, 10, 13, 15, 16; Interview of Dale W. Anderson, Sioux Falls, South Dakota, June 12, 1972. Mr. Anderson is President of Northwest Agricultural Credit Company, a subsidiary of Northwest Bank Corporation. In his capacity he is chief agriculturalist for this banking chain, which covers the area of study for this book; Stewart Gager interview; Frank Larson interview; Gilman Klefstad interview; Gordon Anderson interview; O. M. Jorgenson interview; "Management Returns: Large Farms' Ace," *Successful Farming*, LXXII, No. 3 (February, 1974), p. 29; Duane Gjervold interview; Conversations with Edward Dunn, Fargo, North Dakota, February 6, 1974; Reuben Jacobson interview; Eugene Coombs interview; Interview of Grant Mattson, Casselton, North Dakota, January 2, 1973. Mr. Mattson was one of the nation's earliest and largest leasors of farm machinery and pioneered the way through the maze of Internal Revenue obstacles for a sound leasing program; Lindley Finch, Vice President, Continental Illinois National Bank, Chicago, Illinois. Mr. Finch is one of the nation's leading advocates of rural banks' using large city correspondence banks to help supply farmers with the necessary credit. Mr. Finch understands the dominant position of capital in contemporary American agriculture. He predicts that the next major change for agriculture will be the structural change, which will reshape agriculture more profoundly than the technological phase did. The structural change will see greater injections of outside influence than ever before in history.

Bibliography, Glossary, and Index

Bibliography

Articles and Books

Bailey, L. H. *The Harvest of the Year to the Tiller of the Soil.* New York, 1927.

Baker, O. E., et al. *Agriculture in Modern Life.* New York, 1936.

Bogart, Ernest Ludlow. *Economic History of American Agriculture.* New York, 1928.

Brown, Philip S. "Money," *The Yearbook of Agriculture, 1962: After a Hundred Years,* ed. Alfred Stefferud (Washington, D.C., 1962), 562-566.

Campbell, Thomas D. *Russia: Market or Menace?* New York, 1932.

Casson, Herbert N., R. W. Hutchinson, Jr., and L. W. Lewis. *Horse, Truck, and Tractor: The Coming of Cheaper Power for City and Farm.* New York, 1913.

Compendium of History and Biography of North Dakota. Chicago, 1900.

Currie, Barton W. *The Tractor—And Its Influence upon the Agricultural Implement Industry.* Philadelphia, 1916.

Dalrymple, John Stewart. *No. 1 Hard: Oliver Dalrymple, The Story of a Bonanza Farmer.* Minneapolis, 1960.

Danhof, Clarence A. *Change in Agriculture: The Northern United States 1820-1870.* Cambridge, 1969.

Davidson, J. Brownlee, and Leon Wilson Chase. *Farm Machinery and Farm Motors.* New York, 1908.

Dieffenbach, E. M., and R. B. Gray. "The Development of the Tractor," *Yearbook of Agriculture, 1960: Power to Produce,* ed. Alfred Stefferud (Washington, D.C., 1960), 25-45.

Drache, Hiram M. *The Challenge of the Prairie.* Fargo, 1971.

———. *The Day of the Bonanza.* Fargo, 1964.

Durost, Donald D., and Warren R. Bailey. "What's Happened to Farming?," *Yearbook of Agriculture, 1970: Contours of Change,* ed. Jack Hayes (Washington, D.C., 1970), 2-9.

Eastman, E. R. *These Changing Times: A Story of Farm Progress During the First Quarter of the Twentieth Century.* New York, 1927.

Fite, Gilbert C. *The Farmers' Frontier: 1865-1900.* New York, 1966.

Freeman, Orville L. "Malthus, Marx, and the North American Bread Basket," *Agricultural Policy in an Affluent Society*, eds. Vernon W. Ruttan, Arley D. Waldo, and James P. Houck (New York, 1969), 282-298.

Garlock, Fred L. "Financing Capital Requirements," *The Yearbook of Agriculture, 1960: Power to Produce*, ed. Alfred Stefferud (Washington, D.C., 1960), 375-379.

Garst, Jonathan. *No Need for Hunger*. New York, 1963.

Gittins, Bert S., ed. *Land of Plenty*. Chicago, 1959.

Goodsell, Wylie D. "Technology and Capital," *The Yearbook of Agriculture 1960: Power to Produce*, ed. Alfred Stefferud (Washington, D.C., 1960), 370-375.

Gray, Frederick D. "That Coffee from Brazil and Other Food Imports," *Yearbook of Agriculture 1969: Food for Us All*, ed. Jack Hayes (Washington, D.C., 1970), 18-19.

Gray, R. B., compiler. *Development of the Agricultural Tractor in the United States, up to 1919. Part I*. American Association of Agricultural Engineers, 1954.

Heady, Earl O. "The Future Structure of U.S. Agriculture," *American Agriculture, the Changing Structure*, ed. Wyn F. Owen (Lexington, 1969), 113-128.

Hibbard, Benjamin H. *Effects of the Great War upon Agriculture in the United States and Great Britain*. Preliminary Economic Studies of the War, No. 11. New York, 1919.

Higbee, Edward. *Farms and Farmers in an Urban Age*. New York, 1963.

Jesness, O. B. "Poverty Among American Farmers," *Rural Poverty and Regional Progress in an Urban Society*, ed. Task Force on Economic Growth and Opportunity, Fourth Report (Chamber of Commerce of the United States, 1969), 229-247.

Karolevitz, Robert W. *E. G. Inventor by Necessity*. Aberdeen, 1968.

Larimore, North Dakota 1881 Diamond Jubilee 1956. Larimore Diamond Jubilee Book Committee, Larimore, 1956.

Malin, James C. *The Grassland of North America: Prolegomena to Its History with Addenda*. Privately printed. Lawrence, Kansas, 1961.

Modern-Farming, the Passing of the Hoe. Minneapolis, 1908.

Murray, Stanley N. *The Valley Comes of Age: A History of Agriculture in the Valley of the Red River of the North, 1812-1920*. Fargo, 1967.

Pinches, Harold E. "Revolution in Agriculture," *Yearbook of Agriculture, 1960: Power to Produce*, ed. Alfred Stefferud (Washington, D.C., 1960), 1-10.

Robertson, Heather. *Sugar Farmers of Manitoba*. Manitoba Beet Growers Association. Altona, Manitoba, 1968.

Robinson, Elwyn B. *History of North Dakota*. Lincoln, 1966.

Rogers, Everett M. *Diffusion of Innovation*. New York, 1962.

Sarle, C. E. "Horse Production Falling Fast in U.S.," *The Yearbook of Agriculture, 1926*, ed. Nelson Antrim (Washington, D.C., 1927), 437-439.

Tolley, H. R. "Efficiency of U.S. Agriculture Is Increasing," *The Yearbook of Agriculture, 1926*, ed. Nelson Antrim (Washington, D.C., 1927), 318-324.

Wallace, Henry C. *Our Debt and Duty to the Farmer*. New York, 1925.

Warren, G. F. *Farm Management*. New York, 1913.

Public Documents

Agricultural Situation: The Crop Reporters Magazine LVII. Washington, D.C., August, 1973.

"Average Prices Received by Iowa Farmers—Beef Cattle, 1908-1969," Iowa Crop and Livestock Reporting Service.

"Average Prices Received by Iowa Farmers—Corn, 1908-1969," Iowa Crop and Livestock Reporting Service.

Cooper, Martin R., Glen T. Barton, and Albert P. Brodell. "Progress of Farm Mechanization," Misc. Publication No. 630, U.S. Department of Agriculture. Washington, D.C., 1947, 1-101.

"Corn: Acreage, Yield, and Production by States, 1866 to 1943," U.S. Department of Agriculture, 1954, 1-67.

"Efficiency Winner," *Agricultural Situation: The Crop Reporters Mazagine* LVIII, No. 2. Washington, D.C., March, 1974, 2-4.

Engelking, R. F. , C. J. Heltemes, and Fred R. Taylor. *North Dakota Agricultural Experiment Station Bulletin No. 408*. Fargo, 1957.

Handbook of Agricultural Charts, 1969. U.S. Department of Agriculture, Agricultural Handbook No. 373, 7.

"Hay, Alfalfa, and Alfalfa Mixtures—Acreage, Yield and Production, by Counties," *Minnesota Agricultural Statistics*, 1950, 1954, and 1973.

Historical Statistics of the United States from Colonial Times to 1957. The U.S. Department of Commerce. Washington, D.C.

Historical Statistics of the United States, 1971. Bureau of Commerce. Washington, D.C., 1971.

Hopkins, John A. "Changing Technology and Employment in Agriculture." Bureau of Agricultural Economics, U.S. Department of Agriculture (May, 1941), 1-182.

Hopkins, John A., and Eldon E. Shaw. *Trends in Employment in Agriculture, 1909-1936*. W.P.A. Research Project No. 8-A (1938).

"Iowa—All Corn—Acreage, Average Yield per Acre, and Production," Iowa Crop and Livestock Reporting Service.

Iowa Annual Farm Census, 1971. Bulletin 92-AG, Iowa Department of Agriculture and U.S. Department of Agriculture, 1-22.

Loftsgard, Laurel D., and Melvin G. Maier. *Potato Production Costs and Practices in the Red River Valley*. Agriculture Experiment Station Bulletin No. 451. Fargo, 1964.

Minnesota Agricultural Statistics, 1959. Federal State-Federal Crop and Livestock Reporting Service.

Minnesota Crop Reporter Bulletin No. 503A (November, 1972, January, 1973) Minnesota and U.S. Department of Agriculture cooperating.

Montana Agricultural Statistics, XIII (December, 1970).

Montana Wheat and Barley 1952-1967. Supplement to *Montana Agricultural Statistics* XII (March, 1969).

Nikolitch, Rodoje. "Our 31,000 Largest Farms," Agricultural Economic Report No. 175, U.S. Department of Agriculture. Washington, D.C., March, 1970, 40.

"Spotlight on Montana," *Agricultural Situation: The Crop Reporters Magazine* LVI, No. 5. Washington, D.C., June, 1972, 5-6.

"Statistical Barometer," *Agricultural Situation: The Crop Reporters Magazine*. LVII, No. 3. Washington, D.C., April, 1973, 15; LVIII, No. 5 (June, 1974), 15.

Taylor, Fred R. *Statistics of North Dakota Agriculture*. Fargo, 1972.

"The Farmer's Production Dollar: Where It Goes," *Agricultural Situation: The Crop Reporters Magazine* LVII, No. 7. Washington, D.C., August, 1973, 5-6.

The Resources and Opportunities of Montana, Montana Department of Agriculture and Publicity, 1914.

"Turkey Growers' Wish," *Agricultural Situation: The Crop Reporters Magazine* LVII, No. 10 (Washington, D.C., November, 1973), 2-3.

U.S. Department of Commerce. *Historical Statistics of the United States from Colonial Times to 1957*. Washington, D.C., 1959.

U.S. Department of Commerce. *Statistical Abstract of the United States: 1962*. Washington, D.C., 602.

Unpublished Material

American Crystal Sugar Company Production Statements, 1919-1973.

Anderson, Dr. Oscar, President of Augsburg College. Letter to author. Minneapolis, August 2, 1967.

Bloomquist, Aldrich. Letter to author, April 26, 1974.

Burbidge, Arden. Calculations on potato planting and related field work.

Campbell, Thomas D. Letter to General of the Army, George C. Marshall, May 22, 1951.

_____. Letter to Leonard Sackett, File 876, NDIRS, April 11, 1956.

Coffman, Albert. "The Story of the Famous Adams Ranch." Unpublished Master's thesis, Michigan State University, 1968.

Contracts between Frank Kiene and Wilson H. Hubbard, dated December 30, 1920, plus Kiene Farms, Inc., records.

Coombs, Eugene E. "Evaluating Management Factors in Agricultural Credit." Unpublished thesis, Pacific Coast Banking School, University of Washington, 1963.

_____. Letter to author, January 23, 1973.

Crewdson, B. G. Letter to Eugene E. Coombs, February 25, 1955.

Dalrymple, John S., III. "Oliver Dalrymple and His Bonanza: An Essay on a Western Entrepreneur and the Operation of a Bonanza Farm." Yale University, 1970.

Donkers, Mr. & Mrs. Cyril. "Outstanding State Young Farmer." Documents submitted to author, June 19, 1974.

Erickson, Stan. Letter to author, June 6, 1974.

Finch, Lindley. Letter to author, Chicago, August 15, 1974.

"Fingal Enger Family History." Privately printed, Fargo, 1961. Copy in possession of Mrs. Delmer Nystedt.

Fuller, Orville M. "Farm Organization as Applied to the Baldwin Farms, Dickey County, North Dakota." Unpublished Master's thesis, North Dakota Agricultural College, 1924. NDIRS File 1116C Baldwin Farms North Dakota Lands 1918-1930.

Garman, James A. Letter to author, April 24, 1974.

_____. "Tandem Tractor Studies." Unpublished Master's thesis, Cornell University, 1957.

Garst, David. "Address to West Coast Farm Institute." February 11, 1963.

_____. "Agricultural Progress and Opportunities in the 1970's." A speech delivered N.A.A.M.A., October 1, 1969.

_____. "Agriculture's Opportunity Knocks." Typed manuscript dated January, 1966.

_____. "Cows and Calves a Companion Crop for Feed Grains." Manuscript for speech, January, 1969.

_____. Letter to author, September 30, 1971, January 5, 1972.

Garst, Roswell. "Cows and Calves Can Add Profits." Manuscript in Garst farm files.

_____. "How to Have Cow Feed That Costs Less Than Nothing." Typed manuscript dated March 28, 1972.

_____. Letter to Lane Palmer, editor of *Farm Journal*, May 27, 1972.

_____. "The Growing Demand for Beef Means We Need More Beef Cows." Typed manuscript dated January, 1972.

Guy, Governor William L. Letter to author, August 14, 1967.

Hampton, Robert J. "What Management Expects of the Engineer." An Address to the American Society of Agricultural Engineers (December, 1967).

Humphrey, Hubert H., Senator from Minnesota. Letter to author, August 4, 1973.

Isaacson, Asbjorn B. "Farm Mechanization in the Red River Valley, 1870-1915." Unpublished Master's thesis, University of North Dakota, 1949.

Kent State Bank, Kent, Minnesota. Financial statements, 1906 to 1969.

Kiene, Ell. Letter dated, September 12, 1969, found in the files of the Kennedy Trading Company, a subsidiary of Kiene Farms, Kennedy, Minnesota.

Larson, Paul A. "A History of Farm Mortgage Indebtedness and Direct Farm Mortgage Relief in North Dakota from 1920 to 1950." Unpublished Master's thesis, University of North Dakota, 1963.

Midgarden, A. N. "Early Recollections of Potato Growing in Walsh County." Handwritten manuscript dated March, 1973.

Miller, A. C. Kennebec, South Dakota. Letter to Ray Jarrett, March 7, 1951. NDIRS 876.

Olson, Earl B. Letter to author, June 2, 1975.

Olson, Ole A. "Some Hazards of Farming During the Dirty Thirties." A typed manuscript based on Olson's experiences as an agricultural representative.

Parker, William N. "The Productivity of the American Farmer in the 19th Century." Unpublished manuscript, Yale University, 1958.

Peterson, Leslie W. "Address to Federal Crop Insurance Agents." January, 1972.

_____. "Address to G.O.P. Task Force on Agriculture." January, 1970.

_____. "Address to Minnesota Farm Bureau Federation State Convention." November 20, 1972.

_____. "Comments." National Agricultural and Rural Affairs Conference, Minneapolis, November 12, 1973.

Power, J. B. Letter to Benjamin Cheney, October 23, 1877. NDIRS File 309.

_____. Letter to Israel Lombard, December 16, 1878. NDIRS File 309.

"Proceedings (1974) of the National Potato Council."

Records of Augsburg College, Minneapolis.

Records of Kennedy Trading Company, Frank Kiene, Inc., Kiene Farms, Inc., and other business accounts of Kiene enterprises.

Reed, Walter R. Financial statements 1900 to 1925, in possession of Max Dahl.

Reese, Herbert R. *Seventy Years down the Road: The Life Story of Herbert R. Reese, Sr.* Privately printed, 1971.

Robinson, A. J. Photo Book and Record Book of the Schermerhorn Farms.

Sabe, Oscar. Letter to author, April 14, 1973.

Scott, Mrs. Ralph. A postcard of Minneapolis Grain Exchange Prices, October 29, 1932.

Stoa, A. E. Letter to Ray Jarrett, August 20, 1957. Minneapolis.

"The Successful Cattle System." Montana Beef Industries, Inc., Brochure.

Urevig, Edgar M. "The Decade Ahead." Typed manuscript based on experience as a farm manager.

Vyzralik, Frank. Letter to author, October 25, 1967.

Wangberg, Louis M. "The Historical Geography of Selected Farms in the Larimore, North Dakota area." Unpublished Master's thesis, University of North Dakota, 1964.

Wiese, Donald. Letter to author, February 8, 1975.

Willson, George. Personal papers in possession of Percy Willson.

Worthington, Dr. Wayne H., Director of Research, Deere and Company. Letter to author, January 25, 1974.

Young, Eugene M. "Life of Eugene M. Young." Typed manuscript in possession of author.

Personal Interviews

Allen, Merle S. Moorhead, Minnesota, July 9, 1974. Served as consultant.

Anderson, Dale W. Sioux Falls, South Dakota, June 12, 1972.

Anderson, Dennis L. Rapid City, South Dakota, April 9, 1972.

Anderson, Gordon A. Lisbon, North Dakota, March 7, 1973.

Askegaard, Arthur D. Comstock, Minnesota, May 15, 1967.

Askew, Clarence. Fargo, North Dakota, June 20, 1972.

Ballard, Mrs. C. D. (Florence). Odebolt, Iowa, June 15, 1972.

Banks, Clarence. Vernon Center, Minnesota, March 1, 1973.

Bell, Viv. Glidden, Iowa, June 15, 1972.

Bergstrom, Carl. Ogden, Iowa, January 26, 1974.

Bogestad, John. Karlstad, Minnesota, February 26, 1972, March 22, 1972.

Bourgois, Ervin. Bismarck, North Dakota, March 10, 1972.

Bouvette, Cliff B. Hallock, Minnesota, January 29, 1972.

Bresnahan, Ellery. Casselton, North Dakota, December 28, 1969, June 21, 1973.

Burbidge, Arden. Park River, North Dakota, January 3, 1970, March 14, 1973.

Card, Charles. Britton, South Dakota, December 16, 1972.

Carley, Mr. & Mrs. Earl T. Casselton, North Dakota, January 6, 1967.

Chaffee, Mrs. Gertrude A. Bismarck, North Dakota, July 27, 1966.

Christianson, Earl. Elbow Lake, Minnesota, February 26, 1973, February 12, 1975.

Chrystal, John. Coon Rapids, Iowa, June 14, 1972.

Coombs, Eugene E. Billings, Montana, May 16, 1972.

Curnow, Les. Hardin, Montana, August 8 through 11, 1971, and several letters on file since that date.

Daellenbach, Oswald, Clay County Extension Agent. Moorhead, Minnesota, January 27, 1975.

Dahl, Jack. Gackle, North Dakota, February 10, 1973.

Dahl, Mrs. Max. Chaffee, North Dakota, February 24, 1967.

Dalrymple, John S., I. Casselton, North Dakota. With Leonard Sackett, August 28, 1953, August 9-18, 1955, NDIRS File 549.

Dalrymple, John S., II, and Mary. Casselton, North Dakota, March 27, 1967.

Dalrymple, John S., III. Casselton, North Dakota, March 7, 1973.

Davison, Earl. Tintah, Minnesota, December 8, 1970.

Dixon, Homer E. Detroit Lakes, Minnesota, February 2, 1967.

Donahue, George A. St. Paul, Minnesota, February 11, 1968, March 9, 1969.

Driscoll, Keith. East Grand Forks, Minnesota, January 17, 1972.

Driscoll, Ray. East Grand Forks, Minnesota, January 17, 1972.

Drum, David G. Billings, Montana, May 16, 1972, March 19, 1973, April 28, 1975.

Dunn, Edward. Fargo, North Dakota, February 6, 1974.

Eckmire, John L., Vice President of Finance, Versatile Manufacturing, Ltd. Winnipeg, Manitoba, January 11, 1974.

Engen, Mr. & Mrs. Otto. Minot, North Dakota, April 3, 1972.

Flaat, Ole A. Grand Forks, North Dakota, January 18, February 15, 1972, April 18 and 19, 1974.

Flack, Dr. Duane, General Manager of Feedlots of Monfort of Colorado. Greeley, Colorado, March 12, 1975.

Frank, William. Hector, Minnesota, March 16, 1972.

Friesen, John P. St. Joseph, Manitoba, May 24 and 25, 1974.

Gager, Steward D. Billings, Montana, March 14, 1973.

Garnaas, Levor B. Sheyenne, North Dakota, December 11, 1970.

Garst, David. Coon Rapids, Iowa, June 12, 1972.

Garst Farm Management Staff—Roswell Garst, Stephen Garst, David Garst, John Chrystal, Bob Henah. Coon Rapids, Iowa, June 13, 1972.

Garst, Roswell. Coon Rapids, Iowa, June 14, 1972.

Garst, Stephen. Coon Rapids, Iowa, June 15, 1972.

Gjervold, Duane R. St. Paul, Minnesota, September 13, 1972.

Glidden, Earl. Hallock, Minnesota, January 17, 1972.

Glinz, David. Pingree, North Dakota, June 24, 1971.

Gordon, Arlo. Murdock, Minnesota, March 15, 1972.

Grant, Mr. & Mrs. Charles. Plentywood, Montana, May 8, 1972.

Griffiths, Jerry. Glyndon, Minnesota, February 7, 1973.

Gummer, F. A. Havre, Montana, May 11, 1972.

Gustafson, Larry. Britton, South Dakota, March 2, 1972.

Hagen, Vernon G. East Grand Forks, Minnesota, March 29 and 30, 1974.

Hall, Bill. Hoople, North Dakota, February 14, 1973.

Hanson, Bert A. and Irene. Vernon Center, Minnesota, January 11, 1971, February 25 to March 2, 1973.

Harkins, Mrs. Myrtle. Vernon Center, Minnesota, March 1, 1973.

Harris, Amelia. Hardin, Montana, August 8 through 11, 1971.

Hartness, Donald (Tom). Gwinner, North Dakota, February 11, 1972.

Heline, Joan. Pierson, Iowa, February 23, 1972.

Henah, Bob. Coon Rapids, Iowa, June 14, 1972.

Henthorne, Max. Billings, Montana, March 21, 1973.

Hermanson, Pete A. Woodland Farms, Story City, Iowa, February 22, 1972.

Hertz, Milton. Mott, North Dakota, April 7, 1972.

Horn, Paul, Jr. Moorhead, Minnesota, December 23, 1969.

Hvidsten, Earl. Stephen, Minnesota, January 17, 1972.

Hvidsten, Ralph. Stephen, Minnesota, January 1, 1972.

Jacobson, Reuben W. Minneapolis, Minnesota, May 1, 1974.

Jarrett, Donald. Britton, South Dakota, March 2, 1972.

Jarrett, Ray S. Britton, South Dakota, March 2, 1972, December 14 through 16, 1972.

Johnson, Walter E. Courtenay, North Dakota, March 8, 1972.

Jordheim, Mrs. Oscar P. Fargo, North Dakota, March 28, 1970.

Jorgenson, O. M. Billings, Montana, May 16, 1972.

Just, Kenneth. Barnesville, Minnesota, June 2, 1974.

Kiene, Mrs. Ell. Kennedy, Minnesota, January 17, 1972.

Klefstad, Gilman. Forman, North Dakota, October 7, 1971.

Knapp, Elizabeth Ann Campbell. Hardin, Montana, August 8 through 12, 1971.

Koebbe, Art R. Hardin, Montana, November 7, 1973.

Krabbenhoft, Norman W. Moorhead, Minnesota, August 10, 1974. Served as consultant.

Kroeker, Walter and Donald, A. A. Kroeker & Sons, Ltd. Winkler, Manitoba, February 15, 1973.

Lake, N. R. Fergus Falls, Minnesota, February 25, 1974.

Larimore, Jameson, II. Larimore, North Dakota, December 1, 1970.

Larimore, Jameson, III. Larimore, North Dakota, August 20, 1971.

Larson, Frank. Valley City, North Dakota, March 14, 1974.

Larson, Lyall P. Sargeant, Minnesota, March 26, 1972.

Larson, Winslow. Hallock, Minnesota, March 23, 1972.

Lee, Herman H. Borup, Minnesota, February 17, 1971.

Lee, Mrs. Roy. Hatton, North Dakota, August 12, 1967.

Lund, Oscar and Melvin. Twin Valley, Minnesota, August 20, 1972.

Lysfjord, Charles. Kennedy, Minnesota, January, 1972, March 14, 1973.

Mathew, Ralph, Jr. Barnesville, Minnesota, April 22, 1974.

_____. Casselton, North Dakota, January 4, 1973.

McBride, Rev. James. Barnesville, Minnesota, January 26, 1975.

Meek, Henry. Stephen, Minnesota, January 28, 1972.

Mellies, Virgil. Hector, Minnesota, March 16, 1972.

Melroe, Lester W. Gwinner, North Dakota, February 11, 1972, March 21, 1975.

Metz, Joe. Lewisville, Minnesota, March 30, 1972.

Midgarden, A. N. Park River, North Dakota, March 13, 1973.

Miller, A. Leon. Shepherd, Montana, May 16, 1972.

Myhro, Dean. Moorhead, Minnesota, January 8, 1970, June 11, 1975.

Noy, Bill A., Jr., and Shirley. Vernon Center, Minnesota, February 28, 1973.

Offutt, Ronald D., Jr. Glyndon, Minnesota, January 15, 1974.

Old Coyote, Henry. Hardin, Montana, August 11, 1971.

Olson, Earl B. Willmar, Minnesota, April 24, 1972.

Olson, Lewis L. Barnesville, Minnesota, March 6, 1971, January 6, 1973.

Olson, Ole A. Fargo, North Dakota, March 4, 1970, January 12, 1972.

Olson, Oscar L. Kent, Minnesota, January 29, 1972.

Olson, R. D. Fargo, North Dakota, April 28, 1972.

Olson, Walter R. Fergus Falls, Minnesota, March 5, 1971.

Owens, Johnny, and Rex Wemple. Hardin, Montana, August 10 and 11, 1971.

Page, Franklin. Hamilton, North Dakota, February 12, 1971.

Page, William R. Grand Forks, North Dakota, February 15, 1972.

Pakosh, Peter, Chairman of the Board, Versatile Manufacturing, Ltd. Winnipeg, Manitoba, January 11, 1974.

Parsons, Mrs. Robert. Owatonna, Minnesota, July 15, 1970.

Pazandak, Ferd A. Oakes, North Dakota, December 16, 1972.

Peppel, Max. Barnesville, Minnesota, February 20, 1973.

Peters, Robert. Jamestown, North Dakota, July 11, 1973.

Peterson, Harold T. Murdock, Minnesota, March 30, 1972.

Peterson, Henry R. Moorhead, Minnesota, December 12 and 29, 1969, April 4, 1974.

Reitan, Oscar S. Comstock, Minnesota, February 20, 1973.

Reque, John H. Redwood Falls, Minnesota, March 15, 1972.

Retz, Richard. Jefferson, Iowa, January 26, 1974.

Roach, Pat. Hardin, Montana, August 12, 1971.

Robinson, A. J. Mahnomen, Minnesota, June 6 and 22, 1971.

Robinson, Roy. President, Versatile Manufacturing, Ltd. Winnipeg, Manitoba, January 10, 1974.

Romain, Clarence. Chester, Montana, May 12, 1972.

Romain, John. Havre, Montana, May 11, 1972.

Ross, Walter J. Fisher, Minnesota, March 17, 1973.

Ryan, Gerald C. East Grand Forks, Minnesota, February 14, 1973.

Ryan, Thomas. East Grand Forks, Minnesota, February 14, 1973.

Sabe, Oscar N. Gascoyne, North Dakota, April 8, 1972.

Schnad, Erwin. Hardin, Montana, August 11, 1971.

Schutt, George. Harwood, North Dakota, February 10, 1967.

Schwartz, Earl. Kenmare, North Dakota, April 4, 1972.

Scott, John W., Jr. Gilby, North Dakota, November 15, 1972.

Scott, John W., Sr. Gilby, North Dakota, January 3, 1970, June 16, 1971, February 10, 1973.

Sebens, William P. Bismarck, North Dakota. With Leonard Sackett, May 11, 1955. NDIRS File 520.

Selley, Roy. Odebolt, Iowa, February 23, 1972.

Serum, Mrs. Orlando. Halstad, Minnesota. With Leonard Sackett, August 25, 1955. NDIRS File 645.

Sinner, George. Casselton, North Dakota, December 28, 1969, June 21, 1973.

Sinner, William. Casselton, North Dakota, December 28, 1969, June 21, 1973.

Skolness, Art. Glyndon, Minnesota, February 17, 1971.

Smith, Thomas R. Perry, Iowa, February 22, 1972.

Solberg, Peder L. Gilby, North Dakota, April 18, 1970.

Stasiuk, Mike, Vice President, Versatile Manufacturing, Ltd. Winnipeg, Manitoba, January 10, 1974.

Steiger, Douglas. Red Lake Falls, Minnesota, February 28, 1973.

Steiger, Maurice. Red Lake Falls, Minnesota, February 28, 1973.

Sulerud, Allen C. St. Paul, Minnesota, March 20, 1970.

Sundberg, Roy. Hallock, Minnesota, January 30, 1972.

Sutton, John, Sr. Onida, South Dakota, April 11, 1972.

Taylor, J. R. Memphis, Tennessee, September 11, 1971.

Thomas, Charles and Bertha. Coon Rapids, Iowa, June 15, 1973.

Thompson, Joe. Grafton, North Dakota, March 13, 1973.

Tibert, J. Budd. Voss, North Dakota, March 14, 1973.

Trowbridge, Hugh. Barnesville, Minnesota, August 9, 1974.

Tweedy, Robert H., Product Planning, Allis Chalmers. Milwaukee, Wisconsin, January 29, 1975.

Underlee, C. H., Nolan, and Leslie. Hendrum, Minnesota, January 19, 1972.

Urevig, Edgar M. Lewisville, Minnesota, March 29, 1972.

Velo, Eddie A. Rothsay, Minnesota, February 26, 1973.

Viker, D. E. Halstad, Minnesota, June 12, 1967.

Warren, Floyd Darroll. Hardin, Montana, February 21, 1972.

Weiland, Henry. Euclid, Minnesota, March 17, 1973.

Whitman, Ward. Robinson, North Dakota, March 7, 1972, February 2, 1974.

Wije, Reuel. Moorhead, Minnesota. With Donald Berg, July 6, 1966.

Willson, Percy. Wimbledon, North Dakota, June 24, 1971, March 2, 1972.

Wright, Charles R. Larimore, North Dakota. With Leonard Sackett, March 22, 1954. NDIRS File 242.

Yaggie, Robert A. Breckenridge, Minnesota, August 12, 1974. Served as consultant.

Young, Eugene M. McLaughlin and Rapid City, South Dakota, April 9, 1972, August 9-10, 1972, August 26 and 29, 1973.

Young, George. Fargo, North Dakota, August 29, 1973.

Bulletins, Newspapers, and Periodicals

Address to English Agricultural Leaders, *Survey Mirror*. Red Hill, England, May 9, 1941.

"A Family Owned Corporation," *The Farmer*, LXXXVII, No. 15, August 2, 1969, 9, 19.

Agricultural Markets, January 14, 1948.

Albuquerque Journal, July 19, 1941, July 15, 1946, March 22, 1949.

America Vegetable Grower, VIII, August, 1960, 10-11.

Anderson, Dale O., and Donald M. Hofstrand. "Sugarbeet Production Costs and Practices in the Red River Valley," *North Dakota Farm Research*, XXVII, July-August, 1970, 3-5.

Anderson, Ray. "The Farm—and Farmer—Mr. K. Chose to Visit," *Farm Journal*, LXXXIII, October, 1959, 39, 86.

Angly, Edward. "Thomas Campbell: Master Farmer," *The Forum*, LXXXVI, July, 1931, 18-22.

"Apostle of Bread," *Rocky Mountain Empire Magazine Denver Post*, October 10, 1948.

"A Priceless Collection of Old Tractors," *Farm Profits*, XX, No. 2, January-February, 1975, 16-17.

Ball, A. Gordon. "How Much Do Farmers Want to Expand?" *Iowa Farm Science*, XXV, No. 3, November-December, 1970, 3-687-3-688.

————. "Land, Labor, and Capital Availability for Agriculture," *Iowa Farm Science*, XXV, No. 3, November-December, 1970, 16-700-17-701.

Ball, Carleton R. "The History of the American Wheat Movement," *Agricultural History*, IV, April, 1930, 48-71.

Bambenek, Jerome E., compiler. "Financial Summary of 1971 Agrifax Members," F.I.C. St. Paul office.

Beal, George M., and Joe M. Bohlen. "The Diffusion Process," Special Report No. 18, Cooperative Extension Service, Iowa State University of Science and Technology (Ames, 1962).

Beef, July, 1973.

Beef Producer. Broken Bow, Nebraska, June 22, 1965.

Billings Gazette, April 20, 1946; February 12, December 18, 1948; April 20, 1949; April 20, 1956.

Billings Gazette, July 19, 1949.

Black, J.D., R. H. Allen, and O. H. Negaard. "The Scale of Agricultural Production in the United States," *Quarterly Journal of Economics*, LIII, May, 1939, 329-370.

"Bonus to Spur Wheat Sales," *Minneapolis Star Journal*, April 30, 1946.

(Bradford, England) *Yorkshire Observer*, April 30, 1941.

"Brekke Brothers," *The Farmer*, LXXXVI, November 16, 1968, 37.

Buchele, Wesly F., and E. W. Collins. "Development of the Tandem Tractor," *Agricultural Engineering*, XXXIX, No. 5, April, 1959, 232-236.

Burckhardt, Ann. "Americans Gobble up Willmar's Production," *Minneapolis Star*, April 19, 1972.

Campbell, Thomas D. "The American Farm Problem," *Mechanical Engineering*, L, October, 1928, 745-753.

————. "The Industrial Opportunity in Agriculture," *Magazine of Business*, LIV, December, 1928, 656-57, 728.

————. "What the Farmer Really Needs," *Magazine of Business*, LIII, June and December, 1928, 724-727, 752.

Cavert, William H. "The Technological Revolution in Agriculture, 1910-1955," Agricultural History, Vol. 30, No. 1, January, 1956, 18-27.

Cavert, W. L. *Sources of Power on Minnesota Farms*. University of Minnesota Agricultural Experiment Station Bulletin 262.

Chou, Victor. "Basic Trends in the Farm Economy," *The Independent Banker*, August, 1965, S-25-S-28.

Chrystal, John. "View of Agricultural Solvency," *The Independent Banker*, August, 1965, S-6-S-8.

Cincinnati Post, February 7, 1948.

Coats, Norman M. "Why Purina Got out of Broilers," *Farm Journal*, XCVII, No. 4, April, 1973, A-8.

"Commodities—Freedom at Work," *Time*, L, No. 20, November 17, 1947, 91.

Coon Rapids [Iowa] *Enterprise*, 1930-1959.

Copper, M. R., and J. O. Williams. *Cost of Using Horses on Corn Belt Farms*. U. S. Department of Agriculture Farmers Bulletin No. 1298 (Washington, 1922), 1-14.

Copper, Thomas P. *The Cost of Horse Labor*. Department of Agriculture, University of Minnesota Extension Bulletin No. 15 (St. Paul, 1911), 1-17.

Copy of Campbell letter to Senator Milton Young, *The Fargo Forum*, February 28, 1954.

"Cows Can't Be Changed from Forage Eaters," *Implement and Tractor*, December 1, 1962.

"Crisis in Wheat—Can It Be Ended?," an Interview with the Nation's Largest Grower—Thomas D. Campbell," *U.S. News & World Report*, XLVI, June 1, 1929, 66-71.

Crookston Daily Times, October 5, 1960.

Crops and Soils, VI, November, 1953.

Crystal-ized Facts, XX, 14-15, XX, 19-20.

Cutting, Malcolm C. "A Manufacture of Wheat," *Country Gentleman*, XCI, August, 1926, 18-19.

Dakota Farmer, LXXIII, December 6, 1958.

Denver Post, October 10, 1948.

Des Moines Register, January 28, 1963, April 12, 1964, August 27, 1967, January 18, 1968, August 11, 1968.

Donahue, George A. "Population Changes and Government Policy," *The Independent Banker*, August, 1965, S-21-S-24.

Dos Passos, John. "Revolution on the Farm," *Life*, XXV, August 23, 1948, 95-98, 101-104.

Dr. Salsbury's Turkey Views, IX, No. 1, March, 1958, 12-13.

Dunk, Milton R. "I Am Holding Status Quo," *Poultry Tribune*, LXXX, No. 4, November, 1974, 12-16.

"Earl Olson—Farmers Produce Company of Willmar, Minnesota," *Processing Equipment News*, XIV, No. 5-6, May-June, 1959, 5-10.

"Earl Olson," *Frozen Food Age* (n.d.).

Erickson, Erling A. "A North Dakota Farm Auction in the Great Depression," *North Dakota Quarterly*, XXIX, Winter, 1971, 37-45.

Erlandson, Gordon W. "Trends in Sales of Farm Tractors," *North Dakota Farm Research*, North Dakota Agricultural Experiment Station, XXXI, No. 1, September-October, 1973, 20-22.

Fargo Forum, June 7, 1925, September 8, 1963.

"Farm Assets," *Farmers Union Herald*, XLVIII, No. 4, February 18, 1974.

Farm Bureau News, LIII, No. 2, May 12, 1974.

Farmer-Globe, Wahpeton, North Dakota, January 25, 1965.

"Farmers Earn More from Farm Sources," *The Forum*, August 3, 1973.

"Farmers—Relief Rebus," *Time*, XI, January 9, 1928.

"Farmers: Something for Nothing," *Time*, XXVII, April 20, 1926, 18, 19.

"Farmfest Site—1400 Acres Owned by 14 Farmers," *The Farmer*, XC, September 2, 1972, 67.

"Farmfest, U.S.A.," *Hormel Farmer*, XXXVI, June 15, 1972.

"Farm Innovator Received Award," *Minneapolis Star*, June 12, 1968.

"Farm Project in New Mexico," *Business Week*, No. 1033, June 18, 1949, 52, 56-57.

"Feeders Supply Corporation," *1972 Annual Report*, Yellowstone Valley (Montana) Electric Cooperative, Inc.

"Five Brothers Continue Tradition," *Produce Marketing*, VI, December, 1963.

Fraedrich, Royal. "Politicians Can't Repeal the Impact of Fertilizer, Hybrids, and Horsepower," *Big Farmer*, XLIV, No. 6, October, 1972.

————. "Who Says a Young Man Can't Succeed in Farming?," *Big Farmer*, XLIV, No. 4, April, 1972.

"From the University of Experience," *Better Crops with Plant Food*, LV, Winter, 1971-72, 6-7.

Garst, Roswell. "Be Patient, Beef Consumers," *The New York Times*, May 9, 1973.

————. "Bob Garst's Latest Crusade," *Successful Farming*, July, 1962, 33, 70-71.

————. "Forty Exciting Years 1930-1970, A History of Garst and Thomas Hybrid Corn Company," Privately printed, 1970.

————. "Growing Demand for Beef Means We Need More Beef Cows," *Big Farmer*, XLIV, August, 1972.

————. "The Agricultural Prospect Before Us," *Farm Quarterly*, Spring, 1951.

Gnatzig, Bill. "Farm Credit," *The Farmer*, XCI, No. 7, April 7, 1973, 10-12.

Grand Forks Herald, June 26, 1904.

Grand Forks Herald, February 11, 1962.

Grand Forks Herald, Green Section, June 8, 1967.

Great Falls Tribune, December 26, 1949.

"Grower Profile: Ryan Brothers," *The Grower Packer*, February 15, 1969.

"Growing with Agriculture," *The Farm Picture*, November, 1967.

Hagen, Dick. "The Setup That Works Year Round . . . And the Price Is Right," *Farm Journal*, April, 1964, 55-56.

"Hall Family's Potato Dynasty Is Traced," *Cavalier Chronicle*, October 24, 1963.

Hammill, James. "Potatoes from the Valley," *Monthly Review*, July, 1963.

Hanson, Dick. "Editorial," *Successful Farming*, LXXI, January, 1973, 4.

Hardin Herald Tribune, May 1, 1946.

Hargreaves, Mary W. M. "Dry Farming Alias Scientific Farming," *Agricultural History*, January, 1948, 39-55.

"Harvest 10,000 Bushels a Day," *Dakota Farmer*, LXXIII, November 7, 1953.

Hayami, Yiyiro, and V. W. Ruttan. "Factor Prices and Technical Change in Agricultural Development: The United States and Japan, 1880-1960," *The Journal of Political Economy*, Vol. 78, No. 5, September-October, 1970, 1120-1128.

"Haylage Takes the Ache out of Alfalfa," *Pioneer Corn Profits*, Summer, 1967. Pioneer Hi Bred Corn Company.

"He's Always First," *The American Magazine*, CLV, January, 1953.

Holden, Dorothy. "Potatoes Any Way You Like Them," *Dakota Farmer*, LXXXIX, April 5, 1969.

Hood, Mike. "They'll Never Go Back to Two-Wheel Drive," *Successful Farming*, LXXII, No. 2, February, 1974, 30-31.

Howard, Joseph Kinsey. "Tom Campbell: Farmer of Two Continents," *Harper's*, Vol. CXCVIII, March, 1949, 55-63.

Howell, H. B. "The Modern Farm Business," *The Independent Banker*, August, 1965, S-15-S-16.

"How Tall The Corn?" *Fortune*, LX, November, 1959, 118, 120.

Hughes, Edward. "Rural Robots," *The Wall Street Journal*, January 19, 1929.

Hugoton [Kansas] *Herald*, November 14, 1947.

Hunter, William G. "The Baldwin Farms," *North Dakota History*, XXXIII, Fall, 1966, 399-419.

"I Don't Ever Plan to Be out of Debt," *Successful Farming*, LXXI, No. 11, October, 1973, G-6.

"John Deere Implement Price List February 10, 1916," Deere and Webber Company.

"John Deere Price List November 1, 1950," Deere and Webber Company.

Johnson, A. N. "The Impact of Farm Machinery on the Farm Economy," *Agricultural History*, Vol. XXIV, No. 1, January, 1950, 58-62.

Kansas City Daily Drovers Telegram, July 25, 1947.

Kendall, Don. "Average 1985 Farmer May Sell Out, Pay His Bills and Pocket $175,000," *Fargo Forum*, March 5, 1973.

Kittson County Enterprise. September 11, 1935, found in file 554, North Dakota for regional studies.

"KOA—A Home on the Open Road," *DKQ Review*, IV, No. 3, Fall, 1972, 3-6.

Krumme, Richard. "Agriculture's New Elite," *Successful Farming*, LXXII, No. 6, April, 1974, 10.

_____. "Money Management," *Successful Farming*, January, 1970, 16.

Lane, Laura. "Will Your Banker Say Yes This Year?," *Top Op*, February, 1972, 23-25.

Lauser, Greg. "Minnesota Turkey Producer-Processor Turns Family Firm into $30 Million Business," *Feedstuffs*, January 3, 1972, 28-29.

Letters to the Editor, *Time*, L, No. 23, December 8, 1947, 11, 12.

Lier, John. "Farm Mechanization in Saskatchewan," *Tijdschrift Voor Economische en Sociale Geografiex*, May-June, 1971, 183-189.

"Like Schoolboys on Picnic," *Salt Lake Tribune*, August 10, 1958.

Lingren, Wilfred L. "Superfarmer," *The Northwestern Miller*, CCXXVI, June 25, 1946, 24-25, 28, 29.

Lockhaven Express, October 6, 1941.

(Lodi, New Jersey) *Messenger*, January 15, 1948.

Loftsgard, Laurel D., and Robert A. Yaggie. *Sugar Beet Production Cost and Practices in the Red River Valley*, North Dakota State University Agricultural Experiment Station Bulletin No. 455, October, 1966.

Logsdon, Gene. "Do You Have the Right Banker?," *Top Op*, IV, No. 7, August , 1972, 19-21.

_____. "Expand My Farm? Only if I Have To," *Top Op*, VI, No. 4, April, 1974.

_____. "Young Farmers Tell Off Their Bankers," *Farm Journal*, XCVIII, No. 6, June-July, 1974, TO-3-TO-5.

Mackenzie, Stuart. "The Greatest Wheat Farmer in the World," *The American Magazine*, XCVI, October, 1923, 36-39, 166.

MacLeish, Archibald. "Grasslands," *Fortune*, XII, November, 1935, 60, 65, 186.

"Management Returns: Large Farms' Ace," *Successful Farming*, LXXII, No. 3, February, 1974, 29.

McFadgen, Neil. "North Dakota's Outstanding Young Farmer," *Dakota Farmer*, LXXIX, No. 7, April 5, 1969, 24.

"Milt Hertz Wants to Do Better Than Dad," *The Bismarck Tribune*, April 19, 1969.

Milwaukee Road Magazine, April, 1951.

Minneapolis Journal, Editorial Section, November 22, 1931.

Minneapolis Star, March 18, 1968.

Minneapolis Sunday Tribune, August 25, 1947.

Minneapolis Tribune, January 14, 1949, March 4, 1951, August 18, 1957.

Mitchell, Jonathan. "Trade with Russia Becomes Respectable," *Outlook and Independent*, CLII, No. 11, July 10, 1929, 407.

Monroe, A. L., and E. V. Schwartz, eds. *A 1970 History of the Red River Valley Potato Industry*. Grand Forks, 1970.

"Montana Claims the Greatest Wheat Farmer in the World," *Current Opinion*, LXXV, November, 1923, 542-544.

"Montana Urges Farmers to Sell Wheat and Save World," *Ipswich Chronicle*, May 9, 1946.

Moulton, Robert H. "Is This the Biggest Farm in the World?," *The Scientific American*, CXXI, August 23, 1919, 183.

_____. "200,000 Acres and Not a Single Horse," *Everybody's Magazine*, Vol. XLI, July, 1919, 47.

Mueller, E. W. "Every Man Is Worthy of His Hire," *The Independent Banker*, August, 1965, S2-S5.

"My Day" by Eleanor Roosevelt, *Lexington* [Kentucky] *Herald*, May 14, 1944.

"New Hope for the Man with the Plow," *World's Work*, XX, July, 1910, 13, 116.

New York World Telegram, January 18, 1949.

Norbest News, XXV, No. 2, Fall, 1960.

North Dakota Institute for Regional Studies, Fargo. File 562.

Odebolt [Iowa] *Chronicle*, March 8, 1951, September 9, 1954, December 2, 1954, October 27, 1955, March 5, 1964, October 5, 1967, March 21, 1968, February 5, 1970.

"Off and Running Agriculture's New Era," *Successful Farming*, LXXI, No. 2, November-December, 1973, 27-37.

Oppedal, Al. "Legume Hays Are Wasted Energy," *Feedlot Management*, XIV, September, 1972, 60-62, 74.

Owatonna [Minnesota] *People's Press*, July 18, 1972.

Patrick, Tom. "Encounter with Garst: Challenger of Tradition," *Big Farmer*, XLIV, April, 1972.

Peterson, Arthur G. "Governmental Policy Relating to Farm Machinery in World War I," *Agricultural History*, XVII, January, 1943, 31-40.

Peterson, William J. "Iowa Weather Statistics—1873-1968, of Time and Weather," *The Palimpsest*, L, January, 1969, 61-65.

Porter, W. R. *Cost of Producing Farm Crops*. North Dakota Agricultural Experiment Station Bulletin No. 104, Fargo, 1913.

"Potatoes Are Big Business in North Dakota," *Market Growers Journal*, January, 1974.

Price, Guy. "Crossbreeding by Computer," *Big Farmer*, XLIV, February, 1974.

"Profile of a Complete Operation," *Valley Potato Grower*, February, 1974, 10-13.

"Profits-Records Required to Compete for Financing," *Feedlot Management*, XIV, No. 12, December, 1972.

"Program on Crop Support Blasted by Wheat King," *Albuquerque Tribune*, 1960.

Promersberger, William J. "More Time to Live," Faculty Lectureship, North Dakota State University, February 15, 1972.

Rash, Harry. "The Challenge of Our Rural Communities," *The Independent Banker*, August, 1965, 5-20.

"Red River Valley Untouchables," *Crystal-ized Facts*, XX, Winter 1966-67, 22-25.

Rixie, L. C., and H. R. Jensen. *Cost Advantages in Size of Farm in Red River Valley Farming*, Agricultural Experiment Station, University of Minnesota, Bulletin No. 469, June, 1963.

Robbins, E. T. *Big Teams on Illinois Farms*, Agricultural Extension Service, University of Illinois, Circular No. 324, Urbana, 1914, 1-16.

Ross, Earl D. "Retardation in Farm Technology Before the Power Age," *Agricultural History*, Vol. 30, No. 1, January, 1956, 12-26.

Rumeley, Edward A. "The Passing of the Man with the Hoe," *World's Work*, Vol. 20, No. 4, August, 1910, 13246-13258.

Rupp, Robert. "As Things Look to Me," *The Farmer*, March 17, 1973, 10.

_____. "Homemade Dual Tractor Has Four-Wheel Drive," *The Farmer*, June 20, 1964, 14-15.

_____. "More Credit Needed," *The Farmer*, XC, No. 11, June 3, 1972, 5.

Ruttan, Vernon W. "Agricultural Policy in an Affluent Society," *Journal of Farm Economics*, Vol. XLVIII, No. 5, December, 1966, 1100-1120.

Ryan, Bryce, and Neal Gross. "Acceptance and Diffusion of Hybrid Corn Seed in Two Iowa Communities," Agricultural Experiment Station, Iowa State College Research Bulletin 372 (Ames, 1950), 661-708.

Saloutos, Theodore. "The Agricultural Problem and Nineteenth Century Industrialism," *Agricultural History*, XXII, July, 1948.

Schmidt, Louis B. "The Agricultural Revolution in the Prairies and the Great Plains of the United States," *Agricultural History*, VIII, October, 1934, 169-195.

Seim, Dick. "Something Every Feeder Needs . . . ," *Farm Journal*, XCVIII, No. 3, March, 1974, B-14 and B-23.

Shannon, Fred A. "The Homestead Act and the Labor Surplus," *American Historical Review*, XLI, No. 4, July, 1936, 637-651.

Sinise, Jerry. "Big Sky Country Is Flexing Its Feeding Muscle," *Beef*, IX, No. 4, December, 1972, 13-15.

Smith, Thomas R. "Lead or Follow: Stick Together," *Commercial West*, May 27, 1972.

Smith, Virgil. "Willmar Looks to the Future," *St. Paul Pioneer Press*, April 16, 1972.

Soth, L. "The Iowa 'K' Will See," *New York Times Magazine*, September 6, 1959, 6-7.

"Soviet Agriculturalist Studies Montana Farming," *Great Falls Tribune Montana Pride Magazine*, August 17, 1958.

Spence, Clark G. "Experiments in American Steam Cultivation," *Agricultural History*, XXXIII, July, 1959, 107-116.

"Spud King Could Feed Chicago," *The Grand Forks Herald*, January 25, 1968.

Stedman, Alfred D. "Montana 'Know How' Sets Wheat Record," *St. Paul Pioneer Press*, August 25, 1947, 1, 8.

St. Paul Pioneer Press, February 25, 1951.

St. Peter Herald, June 24, 1954.

Street, Harold K. "Farm Finance—Vantage Point," *The American Banker*, July 31, 1969.

Strohm, John. "A Farmer with a Message—And a Method," *Reader's Digest*, May, 1960, 148-158.

"Taming of a Bull," *Omaha World Herald*, February 14, 1948.

Taylor, M. C., et al. *Prices Received by Montana Farmers and Ranchers 1910-1952*. Montana State College Agricultural Experiment Station, No. 503 (Bozeman, 1954).

The Albuquerque Tribune, October 12, 1949.

The [Billings] *Herald*, August 26, 1948.

The Butte Miner, July 5, 1926.

The Chicago Sunday Tribune, October 9, 1949.

"The Difference Between Those Who Prosper—And Those Who Don't," *Successful Farming*, LXXI, No. 7, May, 1973, 10-11.

The Fargo Forum, April 12, 1920, October 22, 1939, September 29, 1957, February 2, 1973.

The Globe, Wahpeton, North Dakota, August 8, 1900.

The Grand Forks Herald, December 29, 1943.

"The Hanson Story," *Journal of National Polled Cattle Club*, April 1, 1953.

"The Real Story of Khrushchev's Farm Failure," *U.S. News & World Report*, LII, April 12, 1962, 78-80.

"The Ryan Brothers Build for the Future," *Produce Marketing*, VII, March, 1964, 17-18.

"The Use of Corn Cobs, Corn Stalks, and Grain Sorghum Stubble for Cattle Feed," Garst and Thomas Bulletin No. 5.

"The Walter J. Ross Story," *Crystal-ized Facts*, XXI, Spring, 1967, 5-10.

"The World's Greatest Wheat Farmer," *The Buffalo Courier Express Pictorial*, October 12, 1947.

"Thomas D. Campbell, Business—Farmer," *World's Work*, Vol. XLIX, January, 1929, 256-259.

Thompson, Louis M. "Iowa Agriculture, World Food Needs and Educational Response." Center for Agricultural and Economic Development, Iowa State University (Ames, 1965).

————. "World Food Situation in 1973." Iowa State University (Ames, 1973).

Thorton, W. B. "The Revolution by Farm Machinery," *World's Work*, VI, August, 1903, 3766.

Tolley, Clarence H. "Fingal Enger, King of the Goose River," *North Dakota History*, XXVI, Summer, 1959, 107-122.

"U.S. Said to Have Too Much Invested in Agriculture," *Fargo Forum*, August 20, 1967.

Walsh County [North Dakota] *Record*, July 25, 1940, November, 1953, October 20, 1969.

Walsten, Mike. "Finance Your Way into Farming," *Top Op*, V, No. 6, June-July, 1973, 20-25.

————. "Growth Guidelines to Keep You in Farming." *Farm Journal*, XCVII, No. 4, April, 1973.

"Walter Ross of the Red River Valley," *Tractor Farming*, XXXI, Spring, 1948.

Washington Post, April 6, 1946.

Washington Post, October, 1972.

West Central Daily Tribune (Willmar, Minnesota) Centennial Edition, February 25, 1971.

Western Livestock Reporter, January 17, 1955.

"What Russia's No. 1 Man Will Learn on an Iowa Farm," *U.S. News & World Report*, XLVII, September 14, 1959, 52-55.

"Wheat Harvest Ends at Campbell's Farms," *The* [Billings] *Herald*, August 26, 1948.

White, Owen P. "Such a Relief: An Interview with Thomas D. Campbell," *Collier's* LXXXV, March 22, 1930, 10-11, 72.

————. "Wheat on a Grand Scale: An Interview with Thomas D. Campbell," *Colliers*, LXXXVI, December 20, 1930, 63.

Wik, Reynold M. "Henry Ford's Tractors and American Agriculture," *Agricultural History*, XXXVIII, April, 1964, 79-86.

Wilson, M. L. *Big Teams in Montana*. Montana State College Extension Service Bulletin, No. 70 (Bozeman, 1925).

————. *Dry Farming in the North Central Montana "Triangle,"* Montana Extension Service, No. 66 (Bozeman, 1923).

Wilson, M. L., and H. E. Murdock. *Reducing the Cost of Montana's Dry Land Wheat Harvest*. Montana Extension Service in Agriculture, No. 71 (Bozeman, 1924).

Winters, Glenn R. "Wheat and Justice," *Journal of the American Judicature Society*, XXXXI, No. 2, August, 1948.

Wool Sack, August, 1972.

"World Ploughing Contest," *Minnesota Valley Breeders Association Bulletin VII*, August, 1972.

"World's Greatest Wheat Farm," *Minneapolis Sunday Tribune*, August 15, 1948.

"Your Sons: How Will They Farm?," *The Farmer*, June 17, 1967, 9.

Glossary

ALL-PURPOSE TRACTOR—A tractor with a tricycle-type wheel placement or with adjustable wheels to enable adoption to row-crop work.

ARTICULATED (steering)—Refers to a four-wheel-drive tractor that is jointed in the middle for the purpose of steering and following the contour of the land.

BIN PILERS—An elevator that uses large rubber belts to take potatoes from trucks and place in storage.

BLACKLEG—A potato disease and/or a livestock disease.

BLACK STEM RUST—A disease in grain caused by a fungus. Oats is most susceptible.

BLOCK OF BEETS—The beets left standing in a cluster after initial hoeing or cross-cultivating. The block must then be thinned to a single beet.

BLOCK FARMING—Tillage of square blocks of land as contrasted to strip farming, where narrow strips are alternated.

BOB CAT LOADER or **SKID STEER LOADER**—A small-scale power unit that has four-wheel traction and a mounted hydraulic loader.

BONANZA FARM—A large-scale farm of the late 1800's that used entirely hired labor and horse-powered machinery, and practiced one-crop farming.

BOTTOM—The main plowing mechanism of a plow.

BREEDERS—Turkeys raised for breeding purposes.

BROILERS—Turkeys or chickens used for meat production.

BROODING BATTERY—Artificial brooders used to hatch eggs for chickens or other fowl.

CATERPILLAR ("CAT")—The trade name for a popular track-type tractor.

CERTIFIED SEED—Inspected and tagged seed signifying approval for sale for seeding purposes.

CHISEL PLOW—Plow with long shanks capable of penetrating the soil without turning it upside down as in the case of moldboard plows.

CHOPPER (silage)—A harvesting machine that cuts a forage crop into short pieces for livestock feed.

519

COMBINE—The machine that threshes the grain crop in the field, separating the grain from the straw. It replaces the former binder and thresher.

CONTRACT PRODUCTION—Raising crops or animals on a contract basis for a processor.

COVER CROP—Planted in the fall to produce a growth on the soil to prevent winter wind erosion.

CRAWLER—(See track-type tractor.)

CULTIVATOR—Corn—Cultivates weeds out of the soil from within the corn rows. **Field**—A shallow shank implement for tilling the soil and removing weeds prior to seeding or on fallow soil.

CUSTOM WORK—Done for hire for farmers. Can be any specialized task on the farm up to the total production of crops or animals.

DEEP-FURROW DRILL SEEDING—Seeding done by a hoe drill which mounds soil to hold water.

DETASSELING—Removing the tassels from corn used for seed purposes. This prevents crossing from the male plant.

DIGGING POTATOES—The actual harvesting of potatoes. The term digging refers back to the days when all the machine did was lift the potatoes out of the ground and lay them in a row.

DOUBLE-TREE—Generally a wooden evener fastened to an implement to transfer power from the horse or mule. A single-tree from each horse was fastened to the double-tree.

DRESSING CHICKENS—The process of killing, cleaning, and gutting chickens before processing for food or retailing.

DUMP RAKE—A two-wheeled implement that pulled mowed hay into bunches for the purpose of hand loading. This was operated by a foot pedal to control the tripping of the rake.

ECONOMY OF SCALE—Operating to the level of efficiency as provided by the available equipment, technology, and management.

EVENER—(See double-tree.)

EXTENSION AGENT—Often called the county agent. A government employee who advises on agricultural problems, systems, production, and outlook.

FALLOW (summer fallow)—Idling land for the purpose of controlling weeds or conserving moisture.

FARGO CLAY SOIL—A very heavy, difficult to work soil, found along the Red River. This is very fertile soil, high in organic matter.

FIELD CHOPPER—An implement that chops forage crops into small pieces, generally called silage, for animal feed.

FLIGHT ELEVATOR—Implement with an endless chain containing evenly spaced paddles (flights) to elevate grains or bales into storage.

FOUNDATION SEED—The first seed released by experiment stations to approved farmers for the purpose of increasing the seed supply. The following year it is certified seed.

FOUR-WHEEL DRIVE—A tractor in which all four wheels provide traction. It can be either an articulated or a rigid frame.

FRYER—A fowl raised for human food.

FUTURES—Method of marketing, in which the purchase or sale price is determined in advance.

GANG PLOW—A plow with more than one bottom.

GOVERNMENT PAYMENTS—Payments made for farmers having to idle land or otherwise reduce their production.

GRAHAM HOHME—A trade name for a field cultivator with long, stiff shanks, for the purpose of soil mulching.

GRAIN AUGER—An endless screw grain elevator.

GRAIN DRILL—A seeder which placed the grain into the ground, dropping it between disc soil openers.

GREEN MANURE—A legume crop plowed into the soil for the purpose of nitrogen or organic matter.

GUIDE WHEEL (on tractors)—A wheel fastened to the front of early-model tractors, that followed the furrow and eased the job of hand steering.

HAY BUSTER—A trade name for what is commonly called a tub grinder (see tub grinder).

HEADER—A cutting device which removed the heads from grain straw to be threshed.

HEADLANDS—The end of any field where the implements turn around, usually quite compacted soil and frequently not so productive as the remainder of the field.

HEAVIES (turkeys)—Turkeys grown for meat production.

HYBRID SEED—Specially grown seed of single, double, or triple cross, that produces a superior crop. Requires considerable care in production.

JUMP OUT (of tractors)—Sometimes a trait of a four-wheel-drive tractor when it gets under an extremely heavy load. It will start to jump rather than spin the wheels.

LIFTING BEETS—The term applied to the actual harvest of sugar beets. Formerly all one machine did was lift the beets out of the ground.

LOADER TINES—The pointed steel projections on the end of tractor loaders to aid in breaking loose the manure pack or other material.

MOUNTED PLOW—A plow mounted directly onto the power unit and lifted completely off the ground when not plowing.

MULCHING—Digging the ground to let moisture penetrate, but at the same time leaving the residue above ground.

MULTI-GERM SEED—A seed ball containing more than one germ (embryo).

NEBRASKA TEST—The standard tractor test used by most manufacturers to compare all models of power units.

ONE-WAY DISC—A tilling implement used to cut the soil and mix it with crop residue to reduce wind and water erosion.

OPEN-POLLINATED CORN—Natural pollination by crossing two or more unrelated strains.

PADDLE ELEVATOR—Old-style tubular elevator that had paddles on an endless chain to elevate grain.

PARITY RATIO—The ratio between the value of farm products and the value of what farmers must buy.

PILING STATION—A hard-surfaced area where farmers pile their sugar beets at harvest to avoid a costly, time-consuming haul to the beet factory. The beets are hauled to the plant as needed during winter months.

PLOW BOTTOM—The cutting edge and moldboard that turns the soil completely. A plow may have one or several bottoms.

PLOW DOWN—Refers to a crop plowed down for green manure or weed-control purposes.

POT HOLE—A low spot in a field that can only be worked in very dry years unless it is artificially drained.

POWER TAKEOFF (PTO)—A power shaft extended from a power unit (tractor) for the purpose of operating an implement.

POULT—A young turkey.

PRESS DRILL—A grain drill with press wheels to make a firm seed bed to aid production and reduce erosion.

QUARTER SECTION (quarter)—One-fourth of a section, containing 160 acres.

REAPER—The machine which cut the grain but did not swath or bind it. When the binding mechanism was perfected it became the binder.

RECYCLING—In this case, it is the reuse of animal and poultry manures as animal feed.

RIGID-FRAME STEERING—A four-wheel-drive tractor with a solid frame, which is steered by turning each of the wheels. Sometimes called crab steering.

ROTO-BEATER—An implement with rotating blades which are used to destroy potato vines or sugar beet tops prior to harvest.

ROUGH-LOCK CHAIN—A chain blocking device used on the wagon trains to prevent the wagons from rolling.

ROW-CROP TRACTOR—(See all-purpose tractor.)

RUMINANT—An animal having four stomachs, which makes it an efficient forage converter.

SCAB (potato)—Potato, grown in very high organic or diseased soils, which develops a rough scab skin.

SCREENINGS—Weed seeds, broken kernels, and other impurities that are removed from grain prior to sale or processing.

SELF-PROPELLED—Any machine, particularly a combine, swather, or forage chopper, that provides its own propulsion as well as performs its function.

SHARE RENT—Rent a farm or production facility for a definite percentage of the gross profit.

SHOCK LOADER—A device used to load grain shocks into bundle wagons for delivery to the threshing machine.

SHOCKS—An arrangement of 7 to 12 bundles of grain placed together in an upright position for drying and curing prior to threshing.

SKID STEER LOADERS—(See Bob Cat loader.)

STUBBLE GROUND—Land from which the grain crop has been harvested, with the stubble left standing to slow down wind and water erosion.

SWATHER—Cuts standing grain or hay and places the product in a swath for curing to be combined or chopped.

THREE-POINT HITCH—Tractor attachment which enables machine to be made a direct projection of the power unit for ease of control. The resulting machine is then referred to as mounted or semi-mounted.

TRACK-TYPE TRACTOR—A power unit which has metal tracks instead of rubber-tired wheels for traction. Popularly called crawlers or "cats."

TUB GRINDER—A huge tub-like unit with an internal grinding mechanism. Large bales or other bulky materials are fed into this tub, and finely chopped feed is produced.

UREA—A synthetically produced protein used in livestock feeds. It cannot be used for hogs or fowl.

VERTICAL INTEGRATION—Developing all steps of production and processing of a commodity within one firm, such as land to crops to livestock to processing to retailing.

VOLUNTEER GROWTH—A regrowth of a previous year's crop, caused by the fact that part of the grain is lost on the ground during harvest and seeds itself.

WINDROW (SWATH)—A long row of grain or forage that has been cut and then placed on the ground for curing.

WORM GEAR—An endless gear such as is found on the end of a steering mechanism to turn other gears in either direction.

Index

Abandoned farm, 219
Accidents, with grain wagon train, 126
Accounts receivable, 391
Acra plant, 276
Acres, farmed per tractor, 52; per farm, 355
Adams Farm, 186n, 194
Adams Farm of North Dakota and Iowa, 194-202
Adams, John Quincy, 194, 196
Adams, Margot, 200n
Adams Ranch, size, 194; sold, 201
Adams, Robert B., 200
Adams, R. T., 359, 373n
Adams, W. P., 194-196, 198n, 200
Adams, W. P. II, 200
Adding machine, 176n
Administrative personnel at Adams Ranch, 197
Adversity, 287, 396, 425
Aerial spraying, 112, 129
After harvest, 339-351
Aggressiveness, lack of, 468
Agricultural Chemical Company, 323
Agricultural colleges, 406, 409
Agricultural Credit Corporation, 193
Agricultural extension, 490
Agricultural loan officers, 476n
Agricultural Outlook Conference, 461
Agricultural publications, 490
Agricultural specialist, 491
Agriculture, depression in, 407
Agrifax, 483n, 484
Agsco Company, Inc., 324
Ahlstrand, Gilbert, 397
Ah-Wa-Go-Da-Ay-Goosh, 159-163
Airplane, at CFC, 128; dusting potatoes, 328; purchased by Hanson, 304
Alcohol, 58

Alfalfa, 263, experimental plots, 297; growing of, 147; in beef ration, 290; interfered with corn, 288; largest producer of, 290; loss of, 294; obsolete as hay, 266; production of, 193, 289, 296, 303; profits of, 306; promoted by Hanson, 293; sales of, 289; silage, 440; use of, 447; volume production, 288; yield per acre, 296, 297
Algeria, 101, 425
Allen, Merle, 362, 365, 369n, 370, 375n, 376, 378
Allis Chalmers, 422; tractors, 220, 388, 414
Allotments, 379; for potato marketing, 346; of sugar beets, 358
All State Supply of Willmar, 397
Amana Society, 201
Amb, John, 12
Amenia and Sharon Land Company, 168, 175-179, 186n
Amenia, North Dakota, 177
American Airlines, cattle shipment in C-54, 284
American Beet Sugar Company, 360
American Crystal Sugar Company, 360, 373
American Hereford Association, 200
American Horse Breeders Association, 45, 200
American Meat Institute, 276
American Royal Livestock Show, 200n
American Society of Agricultural Engineers, 422
American Society of Farm Managers and Rural Appraisers, 460
Ames, Iowa, 243 (See also Iowa State University)
Ammonium nitrates, 259
Amortization, 485

Anderson, Dale W., 485, 487, 489
Anderson, Dennis L., 233
Anderson, Gordon, 449, 466, 487, 489
Anderson, Irene (Mrs. Bert Hanson), 281
Anderson, Ray, 247, 271
Anderson, Ruth E., 59ff, 67, 68n, 74, 76, 81, 86, 88
Angus cattle, 192
Anhydrous ammonia, 432
Animal power, farming with, 187
Anticorporation Law of 1932, 228-230; impact of, 229
Appetites, 466
Apple Valley Farm, 244, 248, 275; auction, 243; profits of, 242
Arbuckle, J. N., 15
Arizona, 419; potatoes raised for Jiffy Fry, 353n
Armour and Company, 244, 398
Army worms, 219
Arrowheads, 213
Articulation, in four-wheel drive, 419; need for, 420
Artificial, breeding, 262n; drying of corn, 257
Askegaard, Art, 169n
Askegaard, David, 169n
Askew bonanza, 169n
Askew, Joe, 173
Assets, of Amenia and Sharon Land Company, 175; per farm, 462
Atlantic Hotel, 194
Attitude, change in agricultural, 463, 472; of successful farmers, 217
Auction, 217; by Jarrett, 215; in 1930's, 244; markets, 442; of Dalrymple bonanza, 71; of Lakin farm, 201
Auger, elevator, 125
Aultman-Taylor, 210, 427; capacity of, 41, 115; described, 103; distance per day, 139; grease and dirt, 137; Nebraska test, 29; purchased by Campbell, 103; to haul straw, 106; use of fuel, 119
Austin, Bill, 172
Australia, 425
Automated, cage systems, 393; confined environment, 448; crop production, 296n; equipment, 292, 401; feeding, 293, 299; turkey plant, 397; oilers, 34
Automation, key to production, 300; of farming, 296; use of, 439
Automobiles, 457
Avery combine, 216; thresher, 210; tractor, 23; truck, 23

Awards, 204
Axles, 424

"Backbreakers," 320
Back, Col. H. S., 182n
Back-up management, 491
Bailey, Liberty Hyde, 28, 47, 51ff, 153
Baker, A. C., 192
Baker, Harry Franklin, 191
Balance sheet, 471
Baldwin, George, 186
Baling wire, 122
Ball, A. Gordon, 461
Bank, chains of, 453; competition, 458; country, 475; established, 454; failure of, 43, 216, 458; for cooperatives, 477n; need for, 491; pressure on Romain, 12
Bankers, 395, 463, 490, 491; dealings with Campbell, 102, 143, 151; decline loan, 18, 222, 327, 436; demand inventory, 144; disservice of, 475; imaginative, 470; judgment of, 458; lack of agricultural, 392; oppose beef cow business, 261; options of, 477; refuse to finance tractors, 31ff; relations with, 403
Banking, historical setting, 450-458
Bank of North Dakota, 229, 237
Bankruptcy, 226, 407, 444, 452, 455, 457, 486, avoided; 392; of Dalrymple, 170, 173; of Sargent County bank, 455; of Swift Falls Creamery, 398
Banks, Clarence, 284, 285, 287, 293, 307, 309
Barbecued buffalo, 160
Barbed wire, 103
Bargaining group, 352
Barley, increased yield of, 322; in ration, 440
Barn, burned, 196n; cost of, 195
Barnes County, North Dakota, 208
Barnesville, Minnesota, 352
Barrel O' Fun, 351n
Bartlett, Boyd C., 427
Bauder, Frank, 471
Bauer, Bill, 219
Bearcat, 414n
Beardon soil, 313
Beef, a basic industry, 474; consumption of, 266, 276; feeding, 290; produced per acre, 295, 300, 447
Beef calves, 441
Beef cattle, 89, 202, 224; purchased, 106, 384
Beet, acreage, 368, 379; allotments,

368n; drill, 362, 366, 370, 372, 374; growers, 361, 373, 377; harvest crew, 372; harvester, 374, 375n, 376; industry, 359; lifter, 363, 367, 375; loaders, 372n; plants, 360, 361; planter (beet drill), 374; production of, 359, 362, 378; rate of harvest, 372; row, 366n; seed, 359, 362; thinners, 377; tops, 369

Beets, commercially grown, 358, cost of production, 371; government payment of, 368; income from, 371; in Red River Valley, 379; irrigated, 369; piled in field, 365; price of, 361, 368; profitability of, 358, 361; struggle to produce, 359, 361, 368; yield of, 359, 369

Bell housing, 413
Bell, Viv, 242, 244, 267
Beeson, W. M., 266
Bemidji, Minnesota, 391
Bemidji State College, 391
Benson & Quinn, 232
Benson County, North Dakota, 208
Benson, Minnesota, 398
Bergquist farm, 368
Bergstrom, Carl, 426
Berlin Township, 170
Berthold bank, 216
Best track-type tractor, 31
B.I.A. (*See* Bureau of Indian Affairs)
Bianchi, Mr., 284
Bids, 443
Big Bertha, 411
Bigday, Henry, 139
"The Big Dry," 9ff, 431
Big 4 tractor, 23, 35
Big Horn County, 441
Big Horn Mountains, 107
Bigness, criticized, 459
Big Valley Brand, 393
Billings Junior Chamber of Commerce, 432, 433
Billings, Montana, 433, 443
Bilsborrow, George W., 312
Binder, bypassed by Russians, 161; cost of, 42, 68; hitched in tandem, 37, 119; modified to windrow harvest, 117; on Adams Farm, 195; purchase of, 15
Bin pilers, 338n
Bird Hat, Freddie, 159
Bird in Ground, Adam, 159
Birth rates, 277
Bison, South Dakota, 209
Black leg, 315, 316
Black rust, 212

Blacks, 185n (*See also* Negroes)
Blacksmith, forge, 413; shops, 105, 214, 388, 412
Blackstone Hotel, 176
Blackwelder Company, 373
Blanchard bonanza, 49
Blight, 328
Blizzard, 468
Block farming, 113
Blocking, of beets, 362, 370
Blue Earth County, Minnesota, 287, 304
Boarding halls, 79
Bobcat loader, 388; mechanical scoop shovel, 343; skid loader, 445
Bogart, Earnest, 28
Bogestad, John, 82, 317n, 349
Bohlen, Joe, 469n
Boles, Bob, 129n
Bolley, Dr. Henry L., 315
Bonanza farms, 330, 462n; demise of, 70ff, 97; described, 167, 168; financial returns, 70; horse rental, 22; in Red River Valley, 56; working conditions, 135
Bonjer, A. L., 196
Bonuses, 139
Bookkeepers, 191
Boom psychology, 452
Boone, Iowa, 242
Boorman, Robert, 432, 433
Boss Harrow, 318, 319
Bourgois, Ervin, 380
Brainerd, Minnesota, 399
Brakes, lack of, 414
Brand names, of potatoes, 342, 348
Breaking plow, 190
Breckenridge, Minnesota, 368n
Breeders, 388
Breeders Gazette, 282
Breeding, for performance, 442
Brekke brothers, 376
Bresnahan, Ellery, 380
Brezhnev, L., 274
Bridge, Frances G., 201
Bridges restrict machinery, 329
Bridge, William O., 201
Brimmer, Governor Andrew, 460
Britton, South Dakota, 214, 228, 234
Brockton, Montana, 104, 106
Broiler, production of, 383, 384, 387
Brokers, bypassed, 400; economy of, 349; of potatoes, 346
Brome grass, 259
Brood cows, 202, 265n
Brooding battery, 390

Brown Farm, 193 (*See also* Terminal Farm)
Brown, Henry, 250
Brown, Larry, 323
Brush, clearing of, 190
Buckeye trencher, 291
Budget, 478; need for, 438, 439
Buffalo bones, 213
Buffalo-Pitts, 12
Buffalo River, 319n
Buildings, cause of losses, 194; cost of, 83; erection of, 137, 392; maintenance of, 199
Bulk, loading, of potatoes, 332, 340; shipping, 341
Bull, Bess (Mrs. Tom Campbell), 99
Bull, Mr., 99
Bull, sold to Guatemala, 284
Bulls, production of, 448
Bull tractor, 33
Bunkhouses, 79, 199
Burbidge, Arden, 317n, 325, 328, 343, 352
Bureau of Indian Affairs, 100, 113, 145, 212
Burke County, North Dakota, 236
Burroughs Company, 176n
Burroughs, Wise, 263
Business administration, 468
Business is fun, 394
Bus, used by CFC, 137
Buttermaker's helper, 394
Butter, price of, 282; produced on Adams Ranch, 199
Buxton, North Dakota, 224; bank, 454
Buying, policy of, 443

Cab, tractor, 424
California, 146, 419, 457
Calves, production of, 448
Campbell, Bess (Mrs. Tom), 99, 114
Campbell Farming Corporation, 95, 143; income of, 109; size of operation, 115ff
Campbell, Hardy W., 7, 100
Campbell, Minnesota, 353
Campbell, Phil, 460
Campbell, Thomas D., Jr., aerial spraying, 129; and Crow Indians, 100, 160ff; consultant to governments, 162; described 95, 96, 99, 101, 114, 129, 131, 133; early life, 97-99; employed by Torrance, 99; finances, 102, 108, 109, 143; Fort Peck closed, 111; government payments, 145, 149, 155; grain wagon train, 127; machin-
ery and equipment 96, 103, 114, 122, 124, 133, 152; marketing, 157, 158, 159n; operating cost, 140, 173n; political activity, 97, 100, 112, 149, 154, 162; predictions of, 427; resentment toward, 159ff; Russian dealings, 160ff, 22ln, 272; seeks land, 141, 146; size of farm, 100, 129ff, 150; views of labor, 132, 133, 135, 138, 170
Campbell, Thomas D., Sr., 97, 98, 143
Camp system, 104, 106, 107, 126, 137
Canada, 146, 267
Canadian, beet growers, 371; farmers, 426; wheat surplus, 425
Capital, availability of, 449; creation of, 481; expansion of, 270; how raised, 451; invested per worker, 463; lack of an excuse, 466; limits on borrowing, 203; need for, 440, 462, 473; on bonanzas, 167; profits of, 488; reliability of, 460; shortage of, 481; substitute for labor, 203, 299, 471; use of, 5-6, 459, 471
Capital investment, 302, 473; determines farm size, 270; in agriculture, 273, 276
Capitalized mechanized farm, 464
Capital Trust and Savings Bank of St. Paul, 72
Cargill, 232
Carlisle, William, 178
Carload shipments, 193; of Ryan potatoes, 342; of turkeys, 400
Carlson, Andrew, 81
Carlson, Clarence, 317n
Carlson, Ron, 317n
Carpenters, 191
Carroll County, Iowa, 244
Carroll, Iowa, 252
Carson County, South Dakota, 209
Carter, Montana, 11
Case, combines, 122; thresher, 208; tractors, 38, 119, 123, 221, 230, 418
Cash, demanded on turkey sales, 398
Cash flow, 391, 470, 471, 478, 479, 481; generated by Kiene, 66, 68, 72, 82, 83, 87; generated by Schwartz, 223; generated by Young, 224
Cash rent, 145
Cass-Cheney-Dalrymple bonanza, 170 (*See also* Dalrymple bonanza)
Cass County, North Dakota, 170
Casselton, North Dakota, 95, 173, 175
Cass, George, 170
Caterpillar Tractor Company, 105, 144n; sales to Russia, 221; tractors,

190, 221, 326 (*See also* Tractor, track-type)

Caterpillars, capacity of, 50; Flaat purchase of, 46; impact of, 90; Kiene purchase of, 87; replacement of, 92; Steiger purchase of, 412; to break land, 44; to load manure, 370

Cattle, acquisition cost of, 442; air shipment of, 284; boom of 1973, 437; feeding, 178, 265, 293, 346, 369, 432, 434, 435, 454n, 477; marketing, 442; price of, 226, 254, 260, 261; production of, 193; ration, 257, 259; sale of, 235, 443

Cattlefax, 444

Cavert, W. L., 407

Cederholm, Jonas, 60, 185

Cellulose, in livestock feeding, 259, 262, 265, 446

Census, agricultural, 88

Central Livestock Association, 223

Central National Bank of Chicago, 471, 477

Certified seed, 188, 315, 342; direct marketing of, 348; grain, 343; potatoes, 320, 322n, 343

Chaff, 261, 440, 446

Chaffee, Carrie T., 177, 178

Chaffee, E. W., 175, 177

Chaffee, H. F., 177

Chaffee, H. L., 176, 178

Chaffee, Lester, 178

Chain banks, 471

Chain drive, 408, 420

Chain stores, 348, 400, 402

Challenge, as seen by Jarrett, 228; in size, 486; lack of, 345; of farming, 174, 230, 449, 468, 469; need for, 307, 347, 403; search for, 282

Challenger, four-wheel drive, 409

Chamber of Commerce, 360

Change, in agriculture, 50; caused by mechanical power, 407

Charge accounts, 452, 455

Charles City, Iowa, 23

Charter Gas Engine Company, 23

Chase Manhattan Bank, 436, 438, 474

Chaska, Minnesota, 359, 360

Chavez, Senator Dennis, 147, 148

Cheap food, 159, 268, 448

Chemicals, companies experiment with, 297; cost of control, 265, 371; for insect control, 377; for weed control, 363, 377; way to progress, 377

Cheney, Benjamin, 170

Chevrolet trucks, 127, 286

Chicago bankers, 143

Chicago Board of Trade, 194

Chicago, Illinois, 176, 194

Chicago Northwestern Railway, 244

Chicago World's Fair, 185n

Chickens, 217, 382ff; chores, 382; number needed, 393; pies, 393; production of, 383, 387; profits of, 387; size of flock, 391

Chisel plow, 114

Chokio, Minnesota, 189

Chopped straw, 440

Christianson, Earl, 416, 445

Christian Science Monitor, 267

Chrystal, John, 249, 465, 475

Churning butter, 395

Cincinnati Post, 158

Civilian Conservation Corps, 223

Civil War, 4

Clarion Webster soil, 284, 304, 426

Clarke Equipment Company, 445

Clausen, Ed, 474

Clay County, Minnesota, 312

Cletrac, 31, 123

Climax, Minnesota, 158

Clover, 199, 263 (*See also* Hay)

Clutch, adoption of, 388

Coal, consumption of in plowing, 34

Coffee grinder, to make flour, 12, 179

Coil spring harrow, 445

Cold confinement, 299

Cold storage locker plants, 396

Coleharbor, North Dakota, 208

Collective farms, 161, 221

Columbus, North Dakota, 220

Combine, 3; advantages of, 335n; basis for expansion, 325; capacity of, 115; cost of in 1931, 83; efficiency of, 184, 227; experimental, 211; for windrow harvest, 117; labor saver, 86, 87; owned by CFC, 123, 162; purchased by Engen, 229; purchased by Schwartz, 223; record set at CFC, 124

Comfrey, Minnesota, 411

Commercial agriculture, 458; banks, 473, 474; farmers, 48; fertilizer, 253, 282

Commercial fertilizer, 252-258

Commission, amount of, 397; earned, 395; firms, 400; growth of, 395

Commuting, 472

Component parts, 413, 417, 421, 423, 424

Comstock, Minnesota, 169n

Concentrate ration, 307

Concrete, North Dakota, 99

Confined feeding, of cow herd, 295, 299; of turkeys, 401
Connecticut Mutual Life Insurance Company, 436
Conservation, 111
Conservatism, 268
Consignment, 398
Consolidated Mining and Smelting Company, 254
Consolidation, at Tilneys, 203; in England, 296
Construction, by Kiene, 83
Consultant, 163
Consumers, demand cheap food, 448; eating habits, 159, 400; profit from agricultural progress, 49, 324, 407
Consumption, of beef, 266; of processed potatoes, 351
Containerized beef, 444
Continental Bank of Chicago, 471, 474
Continental Commercial Bank, 116
Continental Oil Company, 119
Continuous corn, 297
Contract for deed, 196
Contractual arrangements, 349, 350, 351, 363ff, 367, 371, 384, 385, 392, 401, 417, 434, 441, 477
Convenience foods, 351, 354
Coombs, Eugene, 466, 482, 484, 489, 490
Cook, Charles Willard, 197
Cooke, Jay, 179
Cooks, at Camp 4, 137; on bonanza farms, 169, 180
Cooling system, 408
Coon Rapids Enterprise, 247n, 271, 272
Coon Rapids, Iowa, 241-244, 251, 254, 258, 270, 466
Cooperative competition, 396
Cooper, Thomas, 406
Corn, acres, 252; amount purchased, 403; borer, 243, 258; cobs, 199, 258, 259, 261, 263, 446; combine, 253n, 266; continuous, 255, 264, 266; cribs, 198, 259; cultivating, 211; deficit area, 403; drying, 252, 257; efficient crop, 264; horsepower cost, 406; husking, 282, 290; hybrid (*See* Hybrid corn); labor requirements, 250; monoculture, 255 (*See also* Continuous corn); need for, 311; picking, 211, 243, 250, 252, 290; planter, 211, 281, 291; price of, 192, 227, 244, 250, 251, 258, 282, 403; production of, 193, 199, 210, 227, 253, 265, 269, 384, 389, 402, 439; repurchased from government,

287; sale of, 300; sealed for government loan, 200; seed, 297; shelling, 199, 290; silage, 266, 447; stalks, 250n, 261, 264, 265; stover, 261 (*See also* Roughage); sweet, 297, 300; waste of, 446; yields, 203, 245n, 246, 249, 251n, 253, 255, 268, 287, 297
Corn Belt, 261, 418, 426
Corn cobs and cows, 258-267
Cornell University, 99, 409, 422
Corporation, formed by Kiene, 66; Manor Farms, 178
Corps of Engineering, 101
Correspondent banks, 474, 475, 478
Cost, of beet harvesters, 375n; of contract labor, 367; of dressing chickens, 391; of first Steiger, 414; of gain, 307; of hand vs. machine harvesting, 374; of horses, 22; of machinery, 318, 319; of pickup, 395; reductions, 117ff, 402, 439; of tractor, 386
Cost accounting, 204, 382
Cost of production, 170, 485; by Campbell, 103, 154; concern of progressive farmer, 416; increase of, 315; on Dalrymple bonanza, 173; of beets, 371; of eggs, 394; in grain, 134; of potatoes, 334, 340, 341, 344; reduced, 118, 251
Cothren, Len, 110, 141
Cotton chopper, 377n
Coulter, Iowa, 253
Country banks, growth of, 459, 473, 475, 476, 479
Country elevators, 441
Country Life Commission, 430
Country store, 453
Courthouse auctions, 230
Cover crop, 337n
Coverdale, John, 254
Cow, cost of, 15, 66, 217, 219, 224; Engens, 218; forage eater, 296; herds, 241, 261, 262, 284, 288, 431, 447; numbers in Iowa, 265; productivity of, 299; saved the farm, 214
Cow barns, 319, 414
Cowboy Hall of Fame, 305
Cow chips, 213
Crawler tractors, 419, 420
Cream and egg stations, 244, 454
Creamery board, 395; profits, 397
Cream of Wheat, 99
Credit needs, in agriculture, 413, 436, 452-454, 460, 463, 471, 473, 478, 479, 483
Critics, abundance of, 414, 416, of

four-wheel drive, 414n, of Jarrett, 228, of mercy wheat program, 158
Croak, Jerry, 244n
Crop dryer, 129
Crookston, Minnesota, 353, 359, 372, 416
Crookston Winter Shows, 416
Croonquist, Jack, 353
Crop, failure of, 11, 17, 43, 108, 194, 219, 220, 221, 224, 227, 229, 312, 412, 431; Hansons', 300; horse consumption, 48; in Russia, 274
Cropland, displacement of, 407
Crop production on Great Plains, 100
Crop rotation, 272
Crop-share renting, 169
Crop yields, 203
Crossbreeding, 275
Cross cultivation, 366, 371, 377
Crow Indian Reservation, 104
Crow Indians, 110, 132, 138ff, 140, 159
Crystal, North Dakota, 314
Cudigan gear harvester, 314, 320
Culling, 469
Cultivating, cost of, 371; field rate of, 414n; of beets, 362
Cultivator, 3, 318, 366, 372, 373n
Cummins Diesel, 424
Curnow, Les, 122, 428
Custom combining, 123, 223; crews, 123n; work, 33, 39, 42, 182, 289, 291, 415

Daellenbach, Oswald, 409
Dahl, Jack, 178
Dahlman, potato equipment manufacturer, 334
Dairy, barn, 413; cows, 384; farmer, 477; herds, 391; Jarretts, 215; Schwartz, 223
Dakota Chief potatoes, 352
Dalrymple, auction, 71; bonanza, 168, 173
Dalrymple bonanza, 170-175
Dalrymple, John S. III, 174, 380
Dalrymple, John, Sr., 171, 173
Dalrymple, Oliver, 44, 49, 95ff, 170, 183
Dalrymple, William, 170, 171, 173
Damart Turkeys, Inc., 389
Darnsteadt & Reed, 123n
Davenport, North Dakota, 219
Davison, Earl, 380
Day labor, 454
Dealers, 417, 425
Dean, John S., 317n
Death loss, in feedlot, 300n

Death, taxes, change, 301
de Baca, Dr. Robert, 267, 275
Debt, 286, 461, 469, 470, 481
Deere and Company, 418, 419, 421, 427
Deere and Webber, 282
Deere Plow Works, 419
Deer River, Minnesota, 390
Deer River Theater, 390
Defective components, 424; in first Steiger, 415
Deferred payment, of grain, 232
Deitchler, D. E., 438
Delco light plant, 190
Delinquent, interest, 73; loans, 457n
Demonstration plots, 249, 255
Depreciation, 471
Depression, 153, 221, 226, 250, 407
Deserted farms, 11
Desexing chicks, 390
Des Moines, Iowa 243
"Despotism of custom," 247
Detasseling, 251, 254
Detroit diesel, 414
Detroit four-wheel drive, 409
Detroit Tractor Company, 409
Di-calcium phosphate, 294
Dickerson, L. A., 479
Diesel, tractor, 370
Dining hall, 89ff
Direct buying, 441
Direct marketing, 144, 174, 236, 286, 348, 400, 402
Direct selling, 231-236
Dirt, on potatoes, 320, 331, 333
Disability insurance, 397
Disc, 83, 319
Discing, 410
Disease, of potatoes, 315, 327, 343; of turkeys, 307
Diversified farming, 180, 184, 186, 188, 217, 382, 385
Dockage, 232
Dodge touring car, 137
Donahue, George, 480, 483
Donaldson, Captain H. W., 71
Dos Passos, John, 257
Double, cropping, 265; driving sheave, 408; work shifts, 120; tree, 318
Double-action hydraulic cylinders, 414
Douglas, Roy, 348, 352
Dowigac grain drill, 321
Drags, 198 (*See also* Harrow)
Drainage, 71, 254, 410
Drills, 68
Driscoll, James, 360
Driscoll, Keith, 317n, 327, 345, 380

Driscoll, Ray, 317n, 345, 380
Drought, 43, 108, 144, 175, 179, 251, 286, 456, 457, 468; in Kansas, 262
Drought-resisting corn, 248
Drum, David G., 431, 434, 437, 439, 442, 443, 444, 449
Drum Farm and Garden Center, 432
Drum, Frances F. (Mrs. Jay), 431
Drum, Jay Gould, 430, 431
"Drum Starvation Center," 432
Drum, William, 435
Drying plant, 251
Dryland farming, 7
Dry weather, 431
Ducks, 382
Dues, 81
Dugout, 12, 213n
Duluth, potato market, 350
Dumping, of potatoes, 347
Dump rake, 3, 289
Dunn, Eddie, 489n
Dusting, of potatoes, 328
Duty free, 361
Dwight Farm, 187-189
Dwight, Jeremiah W., 187
Dwight, John W., 187n

Earl B. Olson Farms, 398, 401, 402
Early Ohio potatoes, 352
Early years, 242-245
East Grand Forks, Minnesota, 343, 352, 358, 359, 360, 361, 372
Eastman, E. R., 27
Eaton, Cyrus, 448
Eckmire, John L., 428
Economics, of cow confinement, 299; of farming, 492
Economy, decline of, 453; of four-wheel drives, 427; of horses, 49; of scale, 19, 52, 92, 168, 192, 317, 331, 338, 344, 349, 360, 377, 383, 385, 486, 492; of steam plowing, 35; strengthening of, 459
Edinburg, North Dakota, 99, 342
Education, 430, 451, 463
Efficiency, in agriculture, 296, 303, 402, 406; of confined barn, 299; of four-wheel drive, 415, 418; optimum on farms, 461; per man, 269; test of, 385
Egg and cream route, 395
Eggs, collection of, 382; cost of, 393; money, 382; processors, 393; purchase of, 382
Eight-horse hitch, 394
Eisenhower, Dwight, 157, 162, 271
Elbow Lake, Minnesota, 416

Electricity, 406; value of, 285, 340, 406
Electronic thinner, 363, 377
Elevator, Adams farm, 195, 198; bonanza farm, 169, 174; Dwight farm, 188; fire, 171n; John Deere, 125; Larimore's, 183; potato, 332; profits of, 403; purchase of, 453; Schwartz and Engen farms, 233; Velo's, 389
Elk Valley Farm, 183, 360 (*See also* Larimore bonanza)
Ellendale, North Dakota, 185
Elm River, 182n
Elson, Howard, 476
Emerson, Iowa, 201
Employees, capacity of, 438; loyalty of, 60, 89; number of, 86, 355, 385, 393, 402; of Hansons, 286; standards of, 439
Employment agencies, 192
Engen, Otto, 207, 213, 214, 216, 218, 219, 222, 228, 231, 233, 237
Engen, Mrs. Otto, 214, 228, 237
Enger, Fingal, 179, 429
Engine, 200 HP, 414
Engineering, needed in farming, 115
Engineers, promote tandem hitch, 410
England, farm consolidation, 296
Enzyme action, 393
Equipment, lack of, 413; size of, 114; transportation of, 73; used with Caterpillars, 121
Equity, loss of, 457
Equity position, 475
Ercoupe, 304
Erickson, Erick, 51
Erosion, 217
Espe loader, 372n
Espe Machine Company, 373n
Esprit de corps, 439
Euclid, Minnesota, 425
Eureka, California, 140
Europe, honeymoons in, 176
Evelo, Inc., 389
Ewing, S. A., 263
Expansion, 457, 483; by Hall brothers, 337; methods of, 270; of crop land on CFC, 144; of facilities, 402; of farm, 390; opposition to, 464; reason for, 394; to lower cost, 460
Expense account, of Kiene, 80
Expenses, farm, 462; of beet production, 371; trucking, 287n
Experimentation, by Dalrymples, 173; by Garst, 245, 268; by Hanson, 297ff; in feedlot, 295; Tilney Farms, 203; with tractor, 419

Experimenting with power, 122-129
Export, beef, 444
Extension service, 314

Faar, Nels, 454
Factory farms, 167 (*See also* Bonanza)
Failure, at Two Legging Creek, 107; of homesteaders, 10; rate of, 452
Fairdale, North Dakota, 453
Fair return, 486
Fairview Farm, 194 (*See also* Adams Farm)
Fairview Junction, 195
Fallow, on CFC, 111, 144
Family, budget, 484; car, 395; closeness, 487; farms, 384, 422, 485; influence, 491n; interest of, 178; needs, 484; problems, 479; support, 485
Family-sized turkey unit, 386
Fargo Clay soil, 219, 319; problems with, 71, 73
Fargo Foundry, 369
Fargo-Moorhead, 319
Fargo, North Dakota, 12, 312
Farmall tractor, 31, 39, 46, 211, 252, 325, 326, 366, 370
Farm, assets, 462n; children, 382, 429; crash of 1921, 172; credit systems, 470, 473, 477, 481, 492; crisis of 1920's, 74ff; debts, 462n; failures, 8, 107, 143, 187, 431; labor, 130; management, 244; mortgages, 244; numbers, 5, 74ff, 402, 430, 457, 458; organizations, 207, 303, 480, 491n; problems, 46; programs, 153, 478; service elevator, 399; size, 6, 9, 18, 26, 70, 149, 270, 281, 346, 384, 407, 429, 460, 461, 464, 465, 469; subsidies, 148; suppliers, 477; work, 99; workers, 473
Farm Bureau, 245, 282
Farm Credit Act, 476
Farm Equipment Institute, 282n, 408
Farmers, abandon farms, 218n; attitude, 154; avoid roughage feeding, 446; beginning, 470; capital needs of, 460; classes of, 5; faith of, 456; large, 492; leave the land, 221, 226, 228, 229, 455, 459; on loan boards, 473; optimism of, 452, 455; resist change, 408; skeptical, 45, 430; surplus of, 157; waiting to buy, 425; want fourwheel drive, 424; want power and speed, 419; wife, 451, 479
Farmers and Merchants Bank, 475
Farmers and Merchants National Bank

of Hatton, 13
Farmers Feed Store, 287
Farmer's Holiday Association, 303
Farmers Home Administration, 481
Farmers State Bank, Schleswig, Iowa, 474
Farmfest, U.S.A., 1972, 305
Farming, a God given right, 466; a good business, 430; a way of life, 151; challenge of, 430, 449; described, 405; freedom of, 430; profitable for Garnaas, 14; stability of, 466; success in, 39
Farm Journal, 262
"Farrow to Finish," 448
Fat cattle packer buyers, 443
Fattening ration, 290
Fawkes, J. W., 21
F.B.I., 135
Fear, 466; causes farmers to join NFO, 303; of change, 480; of failure, 489; of growth, 491
Fecher, Iver, 12
Federal Crop Insurance, 478
Federal Farm Board, 154
Federal farm program, 247
Federal Housing Administration, 478, 479
Federal Intermediate Credit Banks, 481
Federal Land Bank, 213, 214, 229, 237, 284, 342, 449, 454, 457, 466, 476, 487, 489
Federally insured loans, 478
Federal Reserve Board, 460
Federal Security Administration, 147
Feed, byproducts, 257; costs reduced, 293, 295, 392; deficit area, 236; for horses, 22, 407; formulation, 402; line established, 396; loans, 454; mill, 399; payment, 391; processing, 392, 399, 402; purchase of, 384; quality of, 389; requirements reduced, 384; value, 277
Feeder cattle, 218, 290, 293, 384, 474; pigs, 384
Feeders Supply, 438
Feedlot, at Kienes, 89; heifers, 261; Shinrone Farm, 202; T-Bone Feeders, 437; profitable, 434; size of, 440
"Feedlot approach," 261
Feeds and Feeding, 288
Felt, Tom, 432
Fence, construction of, 190; on Adams Farm, 195, 198; on wheat lands, 101
Fermentation pits, 393
Fertility program, at Hansons, 297; of

Adams Ranch, 200

Fertilizer, attachments, 281; balance of, 253; blending plant, 324; dealer in Iowa, 253n; effect of, 322; equipment, 282n; experiments with, 203; increased use in Iowa, 255; on pastures, 256; opposition to, 323; pioneer user of, 297; potential of, 265; reluctance to adopt, 184; spreading, 321; used by Kiene, 86, 259, 321, 337, 362

Field, appearance of, 490; choppers, 264; fires, 142; man, 361

Financial, lever, 284, 464, 469; losses, 371; planning, 413; record of Adams Ranch, 200; statement, 478

Financing, by Hanson, 285, 302; lack of, 353, 432; of binder, 15; of CFC, 143; of Kiene Farms, 66ff, 68, 91; of T-Bone Feeders, 435ff; of tenant, 10; of tractors, 220

Finch, Lindley, 471, 474, 479

Fire, 142, 196, 387

Firesteel, South Dakota, 209

First agricultural revolution, 1850-1910, 4-5

First National Bank, 116

First National Bank of Barnesville, 482

First National Bank of Fergus Falls, 482

Fisher, Minnesota, 334, 359, 360, 364, 367

Fite, Gilbert, 450

Flaat Farms Supply Company, 324

Flaat harvester, 333

Flaat, Ole A., 317n, 337, 354; built large storage, 341; expands, 326, 327; labor force, 330, 338; pioneer fertilizer user, 323; potato harvest, 330ff; size of operation, 331; use of airplane, 328

Flaat, O. Lowell, 317n

Flack, Dr. Duane, 440n

Flail, 161, 179

Flaking plants, 352

Flax, contracting of, 232; price of, 42, 192, 208, 220, 229, 232; production of, 211, 249

Fleishauer, Lloyd, 197n

Florance, Edward, 72

Florance family, 88

Flores, Jim, 392

Flour, made by coffee grinder, 12

Flour mill, 222, 391

Flying Dutchman, gang plow, 14

Flying Farmers, 304

Flywheel, 33

Folson, Nels, 313, 315, 316

Food, allotments, 221; at Camp 4, 137ff;

decreasing cost, 407; destroyed, 16; needs, 446; production, 189; reserve, 267; supply, 277

Forage, eater, 296; choppers, 447; ration, 299

Ford, Henry, 23, 30ff, 422

Ford Motor Company, 423

Fordson, 30, 44, 45, 64, 126, 211, 325, 422, 447

Ford tractors, 412

Foreclosure, 214

Foreign, acceptance, 425; dignitaries, 185n

Forest land, 190

Forks, to load beets, 364

Formaldehyde solution, 315

Forman, North Dakota, 455

Fort Berthold Indian Reservation, 208

Fort Peck Indian Reservation, 104, 106, 107, 110

Fort Smith, Montana, 147n

Fort Totten, North Dakota, 454

Fortune, 268

Fort Yates, North Dakota, 221

"Forty-niners," 141

Foto-pak bags, 342

Foundation, seed, 66, 245, 251, 316; stock, 66

Four-H, buyers, 192, 316

Four point program, 478

Four-wheel-drive tractors, 31, 41, 73, 182, 339, 388, 408ff, 445, 490; advantages of, 416; background, 405-407; bending action, 414; capacity of, 415, 426; cost of, 414; discontinued, 422; experimental, 421; inevitability of, 426; market for, 423, 425; need for, 310; number on farms, 427; outperforms, 420; public display, 415; sales of, 92, 122, 418; surplus of, 425; walk-out, 414

Fragmentation, of agriculture, 474; of the bonanzas, 182

France, four-wheel-drive market, 425

Frank Farms, Inc., 380

Frazee, Frances (Mrs. Jay Drum), 431

Frazee store sold, 72

Freighting teams, 16

Free enterprise, 4, 148, 301

Free, family labor, 451; homestead, 465

Freeman, Orville, 459n

Free Silver, 452

Freezing, of beets, 368; of hybrid seed, 257

Freight rates, 235, 435, 444

French fries, 351, 353, 354

Friesen, John P., 364, 378, 380; changes in beet production, 377; pioneer Manitoban beet grower, 363; production cost, 371; rate of beet harvest, 372, 374n, 375n
Froelich, John, 23
Frontier, at Coleharbor, North Dakota, 40; a safety valve, 429
Forst, 175
Frozen, corn in Russia, 274n; turkey, 400
Fryers, 388
Fuel, consumption reduced, 119, 426; costs, 87; savings, 410, 423
Funk Brothers Company, 257n
The Furrow, 282
Future Farmers of America, 467
Future of agriculture, 258
Future projections, 471

Gager, Stewart D., 474, 485, 488, 491
Garden City, Kansas, 241
Garman, James A., 409, 410
Garnaas, Alf L., 454n
Garnaas, Levor B., 14, 454
Garrison, North Dakota, 313
Garst & Thomas Company, 241, 246, 248, 250, 251ff, 257, 259, 272n
Garst Company, 252n
Garst, David, 247, 253n, 264, 270, 276, 277
Garst, Dr. Michael, 242
Garst Edward, 242, 244n
Garst Elizabeth (Mrs. Roswell), 241, 243, 244, 270
Garst Farms, 241, 264; staff, 275ff
Garst, Jonathan, 242, 243, 253, 256, 266, 268, 277
Garst, Roswell (Bob Garst), 241-246, 248-250, 252ff, 256-258, 261, 263-268; and Russians, 270, 272, 273, 275; corn crusader, 276, 277, 278, 293n, 296, 297, 301; experiments with corn, 446
Garst, Stephen, 253n, 275
Garst, Warren, 242n
Gascoyne, North Dakota, 16
Gasoline engine, need for, 24
Gas Traction Company, 23, 35
Geese, 382
Geiser Peerless steam tractor, 34
General Motors Diesel, 419
General Tractor Corporation, 409
Generator, on tractor, 386
Gerlaugh, Paul, 259, 264
German, prisoners pick potatoes, 330; purchases Dalrymple home farm, 171

Ghost town, 234
G. I. Bill, 391
Gilby, North Dakota, 360
Gildford, Montana, 11
Gitalow, Mr., 271
Gjervold, Duane R., 481, 483, 486, 489, 492
Glenfield, North Dakota, 194n
Glidden, Earl, 380
Glinz, David, 428
Glyndon, Minnesota, 417
Goals, 466
Gold mining, by CFC, 140
Golf course, on Grandin farm, 181
Gondola cars, 364
Goodsell, Wylie D., 464
Goodwin, Bertha, 242
Goose River, 182n
Gordon, Arlo J., 380
Gordon, Fred, 104
Governesses, 177
Governmental interference, 354
Government, loan, 225; payment, 145, 247, 368; program, 107, 111, 155, 200, 226, 233, 246, 252, 259, 286, 300, 307, 435
Grab roll, 375
Grading, of potatoes, 320; plant, 251
Grafton, North Dakota, 314, 342, 352
Graham Hohme digger, 122
Grain, acres of, 193; auger, 225n; competition for, 267; drill, 3, 415; elevator, 3; fed per day, 438; hauling, 51, 127; holding, 80, 188; marketing of, 441; price of, 15, 17, 66, 72, 77, 81, 83, 228, 435, 440; production in N.D., 440, 446; wagon trains, 124, 125, 210
Grain Terminal Association, 232
Grand Forks auditorium, 330
Grand Forks, North Dakota, 97, 143, 315, 330, 348, 352, 391, 418
Grandin, Bert, 180
Grandin brothers, 179
Grandin, Charles, 180
Grandin Farms, 179-182
Grandin, J. L., 180, 182, 182n
Grant, Don, 319n
Grant, Eugene, 313
Granules, 351
Grapevine Ranch, 148
Grasshoppers, 43, 108, 112, 211n, 224, 322, 325
Grazing, 113, 193, 250n, 259, 264ff
Great Northern Railroad, 331
Great Northern roundhouse, 353
Great Plains, 414n; dryland farming, 7;

leads in four-wheel drive, 426
Greeley, Horace, 21
Greenback, 452
Greenberg, Art, 352, 417
Griffiths, Jerry, 347n
Grinder, 36
Gross, assets, 469; income, 485; sales, 469
Gross, Neal, 248
Grosvenor, Wallace, 177n
Ground ear corn, 284, 290, 294
Grover, Edward, 313
Growth, danger in, 491; ease of, 491; industry, 480; rate, 467
Grundy Center, Iowa, 283
Guide wheel, 34
Gulstrand, Fred, 79n
Gummer, F. A., 11, 44
Gunderson, Peter, 14
Guyan, George, 391
Guy, William, 178
Gwinner, North Dakota, 388, 445

Hagboe, B. L., 13
Hagen, Cliff, 317n, 336, 350
Hagen, Thorbald, 329
Hagen, Vern, 317n, 327, 329, 336-338, 341, 345, 350, 352, 353, 372
Hail, 218, 468
Haley, North Dakota, 17
Half tracks, 127
Hall, Bernice, 314n
Hall brothers, 314n, 327, 329, 336-338, 341, 342, 349
Hall, J. G. and Sons, 345, 380
Hall, Johann G., 324, 328
Hallock, Minnesota, 69
Hall's Super Pack, 342
Hall, William, 314n, 317n, 345, 350, 354
Halversons, 314, 315
Hamburg, North Dakota, 390
Hamilton, Robert E., 477
Hampson, Robert J., 52
Hampton, Iowa, 253
Hand, cutting of potato seed, 339; file, 413; harvest of beets, 365ff; labor of beets, 358; labor of potatoes, 354; picking, 332; planting of potatoes, 317; power, 340; tiling, 291
Hanes, Tom, 104, 125
Hansman binder hitch, 37
Hanson, Bert A., 250n, 281-286, 290-309
Hanson, Honnace A., 281
Hanson, Irene (Mrs. Bert), 282, 285, 287, 292
Hanson, Russell, 383, 396, 397, 403

Harderson, Bill, 10
Hardin, Montana, elevator, 142, home of CFC, 95, 104, 107, 125, 137, 145, 146, 158, 159, 441
Hardship, 144
Harkins, Mrs. Myrtle, 288, 302
Harris Power Horse, four-wheel-drive tractor, 409
Harrow, 198, 210n, 318, 371, 414n (*See also* Drags)
Hart, C. W., 23
Hartness, Donald, 448
Hart-Parr, 23, 33, 80, 123, 217, 220
Hart, Tom, 104, 110
Harvest, of beets, 363-365; crew, 374; equipment, 298; labor, 368, 376; limits of, 367; of potatoes, 318, 329; season, 337
Harvestore, 293, 294, 307-309
Hashbrown potatoes, 351
Hatching, eggs, 387; turkeys, 399
Hatton, North Dakota, 429
Haug, Ole, 453
Hauling, beets, 364ff; coal, 214; grain, 214; manure, 369
Havana, North Dakota, 455
Havana State Bank, 455
Hay, 79, 195; baled, 293; baler, 289; baling, 289; hauling, 227; machinery, 117, 243, 289; obsolete, 260, 263
Haybuster, stacker, 234
Haylage, 294, 295, 299, 307
Headers, 3, 210
Headrich, Emil, 123
Health, problems, 448
Heath, Hosea, 252
Hebrides Islands, 253
Heider, tractor, 281
Heifers, 236
Heline, John, 384, 385ff
Henak, Elizabeth, 243 (*See* Garst, Elizabeth)
Hendrum, Minnesota, 452
Henry, Alexander, 312
Henry, Jean, 161
Henthorne, Max, 438, 440, 442
Hereford cattle, 191, 200
Hermanson, Pete, 384ff
Hertz, Carol (Mrs. Milt), 468
Hertz, Gotthold, 468
Hertz, Milt, 467ff
Hi Bred Corn Company, 250 (*See also* Garst & Thomas Company)
Higbee, Edward, 471
Highmore, South Dakota, 275
Hilde, Carl, 60, 78

Hill dropping, 370
Hill Farm, 69, 71-73, 81
Hillhead, South Dakota, 454, 455
Hill, James J., 56, 63, 71, 321
Hill, Walter, 71
Hilstad, George, 49
Hinckley, Minnesota, 287
Hinge, 414
Hired labor, 66, 204, 218, 224
Hitches, for discs and drills, 209, 366
 (*See also* Hansman binder hitch)
Hobo, 285
Hoeing, hand, 318, 362
Hog, breeding, 282; farmer, 477
Hogs, competition to man, 267; dis-
 eases in, 287; gestation period, 282n;
 price of, 66, 77, 192, 244, 254, 282;
 production of, 178, 193, 202, 214, 218,
 224, 281, 387, 448
Holding company, 436
Holds, Ernest, 159
Holly Sugar Plant, 441
Holmstrom, Mr., 180
Holt, Caterpillars, 31, 119, 125, 126,
 127, 211n; combines, 122, 220
Homestead, 12, 16ff, 216, 386
Homestead Act, 5, 8, 100, 450
Homesteader, 429; capital needs of,
 462; speculation of, 6-8; shack, 17
Hoople, North Dakota, 313, 324, 329n,
 342
Hoover, Herbert, 100, 162
Hopkinton, Massachusetts, 195
Hopper cars, 343n
Horn, Paul, Jr., 317n, 380
Horn, Paul, Sr., 317n, 327, 350
Horn Sales, Inc., 350
Horseless farming, 38, 52, 209, 213
Horsepower, cost compared, 25, 68;
 farming, 169, 451; in row crops, 410;
 mechanical, 406; needed in four-
 wheel drive, 421; of diesel engines,
 419; of tractors, 379; on farms, 5, 27,
 422, 427; per tractor, 405, 413; vs.
 tractor power, 49
Horsepower sweep, 183
Horses, 24, 25, 46, 90, 182n, 195-198,
 202, 210, 242, 386, 405, 406; decline
 of, 47ff; destroy crop, 43; difficult to
 handle, 46; cost of, 25, 48, 49, 66, 68,
 77, 219, 220, 245, 325n, 406, 407; destroyed by fire, 196; in war, 243; per
 farm, 26; productivity of, 49, 107, 211,
 406
Horse trading, 22, 27, 38, 69, 169, 218,
 394

Hough loaders, 429
Hours, in agriculture, 405; worked by
 horses, 406
House, built over boundary line, 216;
 conditions of, 490; cost of, 13, 181,
 213n, 294
Household budget, 472
Housewives, 320, 329n
Housing, for labor, 136ff, 330, 377
Howell, Herb, 488
Hubbard, Wilson H., 71
Huber, Hans, 211n, 419
Hudson's Bay Company, 312
Huisinga, Al, 399
Humboldt County, Iowa, 246
Humboldt Farming Company, 56, 87
Humphrey, Senator Hubert H., 162n,
 274
Humus, 255
Huntley, Montana, 436, 437, 441
Hunt, Mr., 86
Huron, South Dakota, 475
Husking, champions, 250n; corn, 283
Hussey, Obed, 21
Hvidsten, Alfred, 314, 320, 324, 327,
 334
Hvidsten, Earl, 317n, 346, 351, 380
Hvidsten, Ralph, 317n, 349
Hybrid corn, developed by Wallace,
 243; experiments, 245, 248, 251, 252;
 farmers refuse to buy, 247, 248; sales,
 250, 252, 270, 273
Hyde, Art, 309n
Hydraulic components, 414; hoists,
 375ff
Hydro-mechanical transmissions, 425

Ice house, 181, 199
Illinois, 426
Illinois Trust Company of Chicago, 194
Immigrants, 242, 403
Implements, large, 420
Imports, of sugar, 368
Income, alternative, 397; fluctuations,
 457; from alfalfa, 290; from silage,
 439; from teaching, 468; improve-
 ment of, 459; of Kiene, 74, 92
Income tax, 92, 232, 301
Incorporate, reasons to, 389
Incubator, 3, 390
Indebtedness, permanent, 460, 463
Independence, of farming, 469
Indian Bureau (*See* Bureau of Indian
 Affairs)
Indian, labor, 225; production rates,
 250; reservation, 101; scrip, 179;

trails, 190
Indians, 189, 320, 454; employed by Youngs, 209; families hired to clear land, 190; rent to CFC, 145; sold land, 192
Industrial, age, 377; agriculture, 150; loaders, 420; mechanical power, 451
Industrialization, 451
Inefficiency, 291
Inflation, 464, 490
Innovation, 4, 486; advantages of, 223, 254; reduces cost, 217
Innovative farmers, 248, 411; spirit, 269
Insecticides, 337
Insects, cause losses, 371; controlled by chemicals, 377
Insolvent, farmers, 458
Institutions, 402
Insulate, 483
Insurance agencies, 454n, 475
Integration, deemed necessary, 399; in farming, 175, 178; of beef business, 262, 437; secret of success, 437
Interest, charged by Kiene, 66; income, 91; payment of, 75ff, 474; rate of, 172, 213, 468
Intermediate credit needs, 477
Intermediate-term loans, 477
Internal Revenue Service, 201, 232
International Harvester Company, 185n, 373, 374; carload shipments, 196; dealership, 431; equipment sale, 211, 252; experimental beet harvester, 373; field tests, 374n; power takeoff, 31; stock, 194; tractor, 45, 120ff, 200, 217-219, 223, 386, 387; trucks, 228, 365
International Livestock Show, 178n, 200n, 283
International Trade Council, 275
International Workers of the World, 131
Inventions, 412, 417, 445
Inventory, 199
Investment, cost of Harvestore, 307; per farm, 473; returns, 487
Investors, 434
Iowa, cow herd size, 265; corn yields, 253; leading poultry state, 383; opened to settlement, 197
Iowa Farm Bureau, 254
Iowa State College, 245 (*See also* Iowa State University)
Iowa State Fair, 200n
Iowa State University, 247, 248, 263, 266, 267, 409, 484, 488
Iron Age planters, 326

Iron mines, 169, 413
Irrigation, 437, 476; at Elk Valley Farm, 184; at Two Legging Creek, 107; by Campbell, 147; for future, 276; need for, 439; of beets, 368, 369; of potatoes, 344
Isolation, impact on price, 440n

Jacobson, Reuben W., 481, 489-491
Jails, contract for beets, 361
Jamaicans, 330
Jamestown, North Dakota, 231
Jardine, W. M., 3
Jarrett Farms, 234, 235
Jarrett, Lena (Mrs. Ray), 227
Jarrett, Ray S., 207, 214-216, 218, 222, 231; comments on weather, 215; comments on failure, 226; marketing experience, 233; produces sheep, 226; reflections on farm growth, 237-238
Jarretts, move to South Dakota, 214
Jarrett stockyards, 235
Jefferson, Ben, 159
Jefferson, Calvin, 159
Jefferson, Iowa, 426
Jefferson, Thomas, 243
Jennie-O Foods, 400
Jennings, Ted, 275
"Jenny" age, 480
Jesness, O. B., 458
Jet age, 480
Jiffy Fry, 353, 354
Jobs, 336, 403
John Deere, experiments for, 298; plow, 64; products engineers, 418; test officials, 419; tractors, 31, 211n, 216, 227, 281, 325, 326, 386, 418, 420, 422, 423
Johnson, Adolph, 87
Johnson, Dr. Jim, 401
Johnson, Jack, 447
Johnson, Lyndon B., 202, 459, 473
Johnson, Walter E., 231
Jordan, Montana, 9, 430
Jorgenson, O. M., 466, 486, 487
Julius Sisch, beet loader, 372
Just, Kenneth, 482

Kampgrounds of America, Inc., 433
Kansas, 146, 426
Keeps, Joe, 109
Keil, Donald, 470
Keller brothers, 343n, 445
Keller, Cyril, 388
Keller, Louis, 388

Kenmare, North Dakota, 216
Kennedy Baseball Association, 91
Kennedy, Minnesota, 314
Kennedy Trading Company, 58, 64, 65, 67, 74, 76, 91
Key inventions, 3-4
Key to success, 423
Kharkov, wheat, 146, 162n
Khrushchev, Nikita, 268, 270-273
Kiene, Ell, 61, 88-91, 347
Kiene Farms, 63, 65, 76, 88, 91
Kiene, Frank, 64ff, 82, 88, 89, 91; early life, 57, 58; potato acreage, 314; potato marketer, 347; traits of, 92ff
Kidder, South Dakota, 234
Kiki Camel ducks, 244
Kindred Farmers Elevator, 228n
Kindred, North Dakota, 228n
Kingman Farm, 170n
King of Spuds, Inc., 352
Kingsley, Truman, 448
K.I.S.S., 295
Kittson County Enterprise, 88
Klefstad, Gilman, 454, 455, 486
Klingenberg, Dan, 436, 438
Knapp, Wallace, 252
Knows Ground, Joe, 159
Koebbe, Art, 107, 109
Kohler engine, 388
Kopac, Ed, 145
Kosygin, Aleksei, 274
Krabbenhoft, Norman, 365, 378
KRM Ranch, 437
Kroeker, Abram, 51
Kroeker brothers, 339, 341, 350, 354, 356n
Kroeker, Don, 317n, 355
Kroeker, Walter, 317n, 327, 335, 346, 355
Kruer, Robert, 418
Krylor, Mikhail, 162
Kube, William, 197

Labor, capacity of, 37, 50ff, 65, 78, 89, 90, 134, 183, 191, 195, 199, 215, 224, 241, 250, 269, 286, 289, 296, 300, 301, 318n, 322, 329, 335n, 337, 338, 340, 342, 343, 345n, 350, 363-366, 369, 375, 376, 379, 388, 393, 401, 405, 440n, 448, 464n, 488; cost of, 5, 9, 22, 73, 78ff, 134, 169, 180, 186ff, 192, 199, 203, 250, 276, 293, 299, 319, 329, 336, 363, 374, 377, 385, 388, 401, 410, 427; force, 5, 34ff, 41, 44, 46, 50, 169n, 188, 211, 298, 325, 371, 384, 386, 407, 471; in agriculture, 5, 70, 133ff, 170n,

227, 317, 422, 450, 451, 462, 464, 471, 487; problems of, 159, 204, 209, 286ff, 292, 393, 439, 448; supply of, 4, 169, 183, 291, 292, 328, 332, 334, 335, 338, 370, 371, 397, 412, 429, 461
Ladd, Gerry, 123n
La Jolla grant, 146
Lake, N. R., 482, 484, 488
Lakin, Charles E., 201
Lakin Ranch, 201
Lambie, J. W., 331, 333
Land, 450, 466; appreciation of, 302; competition for, 84, 218, 461; clearing by Indians, 190; contracts, 15, 171, 173; cost of, 11, 15, 18, 40-42, 64, 82, 83, 88, 147, 148, 168, 171, 172, 179n, 185, 186, 195-197, 201, 212, 213, 220, 223, 225, 226, 229, 230, 242, 244, 281, 285, 287, 300, 387, 412, 452, 455, 457; released by tractor use, 48
Landa, North Dakota, 194n
Land of Plenty, 408
Lane, Franklin K., 100, 102
Lang Machinery Works, 419
Lang Museum, 419
Lantern, carried ahead of tractor, 325; used in potato storage, 318n
Large-scale equipment, 203, 234
Large-scale farming, 13, 56, 201, 204, 316, 317, 358, 361, 488
Larimore bonanza, 168, 183 (*See also* Elk Valley Farm)
Larimore, Jameson II, 183, 184
Larimore, Jameson III, 184, 317n
Larimore, North Dakota, 183, 184, 360
Larimore's Elk Valley Farm, 183-185
Larson, Frank, 466, 471, 474, 475, 486, 488, 489
Larson, Hans, 314, 315
Larson, Lars, 314, 315
Larson, Lyall, 384, 385
Larson, Ole D., 452
Larson, Paul A., 457
Larson, Winslow, 61
Laser beam, 291n
Lawsuit, 465
Laying hens, 387, 393
L.B.J. Ranch, 202
Leach and Gamble, of Wahpeton, 454
Leaf roll, 316
Leasing, rate of, 398; of land, 173
Leazanby, William, 180, 181
Lee, Herman H., 380
Legumes, 263, 266
Lemmon, South Dakota, 209
Leonard-Crosset and Riley, Inc., 348

Letellier, Manitoba, 364
Lethbridge, Alberta, 26
Lever brakes, 408
Lewisville, Minnesota, 202
Liberal credit, 452
Liberty bonds, 65
Liberty State Bank, 455
Life, 155
Life insurance companies, 457n
Life style, 180ff
Light plant, 212
Lights, on tractor, 31, 33, 386
Lime, use of, 298
Lincoln, Abraham, 21
Line of credit, 485
Liquidation, of farms, 459
Litchfield, Minnesota, 398
Litchfield Processing Plant, 398
Little Big Horn River, 96
Litters, of pigs, 282
Live power takeoffs, 425
Livery stables, 319, 369
Livestock, feed, 222, 444, 456; production, 88, 179, 191, 204, 223, 235, 270, 311; train, 235
Livestock expositions, 283
Living expenses, 176, 451, 484-486
Loaders, capacity of, 325, 365, 367
Loading, beets, 364; ramps, 364; stations, 361; time, 341, 343, 371, 372
Loan, companies, 218; limits, 473, 476; officer, 471; operating, 102; refused, 398, 438
Loans, 390; bank, 456; increase of, 474; losses on, 455; on stored wheat, 155; personal, 453; size of, 309, 342n, 436; to employees, 59
Lock, Luther, 123n
Lockwood, potato equipment manufacturer, 334
Loftsgard, L. D., 360n
Log cabin, 12
Lohr, J. M., 72
Long, H. D., 352
Loss, from farm operation, 200; in cattle feeding, 306; in value by baling, 289n; on sale of Schermerhorn Farm, 193
Losses, caused by drought, 144; causes of, 387; from farming, 173, 282; on Baldwin Farm, 187; on farm sales, 456; on farms in 1930's, 194; on turkeys, 400, 401; suffered by Olson, 397, 399; sustained at CFC, 146
Lubkeman, Henry, and son, 253
Lumber, camps, 169; cost of, 15

Lumberjacks, 320
Lund, Melvin, 44
Lund, Oscar, 44
Lysfjord, Charles, 84, 88, 317n
Lysfjord, Conrad, 60, 65, 79, 85ff

Machinery, cost of, 19, 52, 103, 110, 118, 171, 199, 213, 318, 325, 447, 462; discount on, 115, 196; leasing of, 52, 492; manufacturers, 105, 421, 426, 477; on collectives, 161; use of, 6, 7, 10, 146, 203, 377
Machine shop, 129
MacLeish, Archibald, 96
Mack truck, 127
Maddox, Mrs. Ola B., 104, 110
M & D Feedlots, 438
Mahnomen County, Minnesota, 192
Mahnomen, Minnesota, 189, 190, 191
Mahnomen Telephone Company, 190
Maids, 177
Mail route, 11
Malin, James C., 6
Mallony, DeWitt, 253
Management, 73, 103ff, 167, 171, 180, 202, 204, 301, 309, 314, 358, 385, 403, 430, 437, 438, 443, 466, 474, 477, 479, 482, 486; backup, 491-492; contract, 170, 180; of time, 491n; secondary, 331; value of, 84, 181, 183, 331, 479, 488
Managers, 443, 487, 490
Man hours, 463
Manila, Iowa, 170
Manitoba, 358, 383; restrictions on farm size, 346, 356n
Manitoba Potato Board, 355
Manitoba Sugar Company, 361, 363, 371
Mankato, Minnesota, 282
Manor Farms, 178
Manpower, 473
Man with the hoe, 377
Manufacturing of farm implements, 27ff, 417, 445, 478
Manure, 255, 277, 298, 300, 319, 369ff, 388, 393, 439, 448
Mapleton, Minnesota, 410
Mapleton, North Dakota, 169n
Marbeet harvester, 373
Marble, Fred, 436
Marginal land, 192
Market, dependency of, 407; free, 355; management, 350, 355; potential, 410, 416, 423
Marketing, 84ff, 185, 231, 234, 276, 286,

342, 346, 347, 349, 350, 385, 387, 389, 392, 398, 402, 425, 441, 443, 444; agreements, 350, 385; information, 423, 442
Marietta Creamery, 397
Marietta, Minnesota, 397
Marmalejos, Jesus, 365
Marshall County agent, 324
Marshall County, Minnesota, 334
Marshall, Merle W., 475
Mason City, Iowa, 360
Massey-Ferguson, 201
Massey Harris Company, 23, 408
Mass production, of four-wheel drives, 422-425; of food by Russians, 161
Masters, Walt, 198n
Mathews, John L., 196
Matskevich, V. V., 270
Mattson, Grant, 52, 492
Mayville Stock Farm, 180
McBride, Rev. James, 403
McCabe Brothers Grain Company, 69, 84ff
McCarter, Bill, 155
McCleary, David, 71
McCormick-Deering, 31, 36, 252, 454
McCormick Harvester Company, 465
McGibinery, 136
McIntosh, South Dakota, 222, 225
McLaughlin, South Dakota, 209, 210, 212, 218n, 231; bank of, 212
McLean County, North Dakota, 208
McNary, J. W., 185, 186
Meals, cost of, 79
Meat, 138, 181, 199
Mechanical harvester, 321
Mechanics, on farms, 105, 336, 337n
Mechanization, 47, 105, 130, 133, 168, 325, 326, 332, 339, 342, 343, 354, 358, 359, 363, 367, 371ff, 373, 377, 388, 430, 451, 462
Meek, Henry, 79
Mekinock, North Dakota, 454
Melgaard, L., 58
Mellies, Virgil, 380
Melroe, Edward G., 444, 445
Melroe, Lester W., 444-448
Melroe Manufacturing Company, 388, 445
Melroe, Olaf, 444
Melrosa Ranch, 448
Melrose Produce Company, 399
Men, Machines, and Land, 408
Merchants, 477
Merger, 492
Merl, North Dakota, 453

Metropolitan Life Insurance Company, 185
Metzer, George, 221
Meyer, Bill, 172
Meyers, Bennet, 158n
Mexican-Americans, 330
Michigan, 359
Michigan State University, 409
Midgarden, A. N., 314-316, 321, 322
Midland National Bank, 398
Migrant workers, 377
Migration from the farm, 483
Mikoyan, Mr., 271
Miles City, Montana, 431
Miles, Governor, 147
Military surplus equipment, 121ff
Milk cows, 387, 413; price of, 224
Milking, 394, 243; machines, 197
Miller, John, 188
Miller, Leon, 434, 435, 439, 443
Miller, Mrs. John, 188
Millet, 227
Millies, William, 410
Milling, 222
Mill, John Stuart, 247
Milton, North Dakota, 453
Milwaukee Railroad, 235, 243
Minneapolis Chamber of Commerce, 298, 305
Minneapolis, Minnesota, 176, 192, 398
Minneapolis tractor, 29, 63, 169n
Minneapolis Tribune, 235, 239
Minnesota, 167, 169, 216, 305, 351, 358, 419
Minnesota Agricultural High School, 394
Minnesota Certified Double A, 393
Minnesota Flying Farmers Association, 304
Minnesota Mutual Life Insurance Company, 65
Minnesota Potato Growers Exchange, 92
Minnesota State Fair, 283
Minnesota State Prison, 61
Minnesota Sugar Company, 359, 360
Minot Credit Company, 219, 220, 237
Minot, North Dakota, 219, 230
Missionary, 189
Missouri River, 106
Moberg, E. F., 85
Model A Ford, 219
Model T Ford, 42, 68, 107, 111, 115, 210, 365, 373n
Moe, Ed, 392
Mogul tractor, 40, 44, 208

Moisture cycle, 112, 114

Molasses, 263

Moline, 31

Monarch tractor, 123

Monastery, 361

Money, 390, 430, 470

Monfort of Colorado, 436, 440n

Monogerm seed, 376

Montana, 8, 383, 411, 419, 426, 435, 440n, 457

Montana bankers, 466; ranchers, 437; sheepherders, 193

Montana Beef Industries, 436-438, 443, 444

Montana Farming Corporation, 106-109, 143

Moore brothers, 33

Moore, Nettie, 195

Mooreton, North Dakota, 194

Moorhead, Minnesota, 312, 313, 317, 392

Morgan, J. P., 102, 107, 108, 143

Morocco, 425

Morris, John, 283

Morris, Minnesota, 189, 394

Mortgage, holders, 178, 452; use of, 8, 73, 177n, 196, 219, 457, 468

Moscow, U.S.S.R., 221

Motivation, of farmers, 207, 231, 301, 437

Mott, North Dakota, 467

Mountain Pocket, Joe, 159

Mounted cultivator, 373n; plow, 281

Mount Moriah, Missouri, 180

Mountrail County, North Dakota, 208

Moyer, Art, 236

MRS, tractor, 423

Mueller, E. W., 459

Mules, 24, 196, 405 (*See also* Horses)

Multi-carload shipments, 441

Multi-germ seed, 362

Multiple hitch, 26, 39, 42, 45; management, 492

Mundstock, C. W., 313

Murdock Cooperative Creamery, 395

Murdock, Minnesota, 394

Murray Brothers and Ward Land Company, 172

Muskeg, 419

Mustard, 110, 362

Myhro, Dean, 390-394

Myhro, Lillian (Mrs. Dean), 390, 394

Myhro Pullet Farms, 392

Nash, North Dakota, 315, 348

National Advisory Commission on Food and Fiber, 473

National Farm Equipment Manufacturers Association, 296

National Farmers' Organization, 303

National Field Research, 204

National Flying Farmers, 304

National Implement Company, 409

National Life Insurance Company, 84n, 456

Native sod, 225

Natural competitors, 489

Nebraska Society of Farm Managers and Rural Appraisers, 302

Nebraska Tests, for tractors, 29

Negative, attitude, 207; balance, 466, 469; cash flow, 475

Negroes, 185n

Nelson four-wheel drive, 408

Net income, 484; worth, 91, 467, 469, 470, 477, 478

Nevada, 457

Newark, South Dakota, 234

New Deal, 148

New Era Tractor Farms, 211, 212, 220, 224

New Mexico, 146

New Way Harvester, 211n

New York bankers, 143

New York Times, 271

New York World Telegram, 235

Nichols-Shepard, 211n

Nicollet County, Minnesota, 281

Night shift, for beet harvest, 376n

Nitrogen, 253; cost of, 263; priority of, 254; test plots, 255; value of, 256, 264, 266

Nixon's Salute to Agriculture, 305

Noble Foundation Farms, 26

Nordling, Oscar, 60, 65, 78, 79n, 85

North Africa, 100

Northcote, Minnesota, 69

North Dakota, 358, 383, 419, 435, 457; leading big tractor state, 426; settlement of, 8; site of bonanzas, 167

North Dakota Agricultural College, 30

North Dakota Agricultural Experiment Station, 315, 321

North Dakota Mill and Elevator, 391

North Dakota State Seed Department, 352

North Dakota State University, 489n

North Dakota Wool Pool, 178n

Northern Dakota Railroad Company, 99

Northern Pacific Land Grant, 195

Northern Pacific Railroad, 167, 208, 331

Northwest Bank Corporation, 327n,

436, 456, 485
Northwestern Implement Company, 69
Northwestern University, 243
Northwest Farm Manager's Association, 191
Norton, Charles D., 102
Norwegian immigrants, 444
Notes, 179, 220, 221
Novosibirsk, Russia, 221
Noy, Bill, 306, 307ff
Noy, Shirley (Mrs. Bill), 307-309
Nutrition, 401
Nyhus, Ellef, 12

Oats, price of, 188, 192, 244
Odebolt, Iowa, 194, 197, 198
Off-the-farm income, 462, 483, 488, 492
Offutt, Ronald D., Jr., 317n, 341n, 351; largest potato grower, 337; potato harvesters, 338; sale of farm, 351n
Offutt, Ronald D., Sr., 317n, 337, 351
Ohio Agricultural Experiment Station, 259
Oil, deposits, 189; drips on tractor, 34
Okinawa, 433
Oklahoma, 189
Old Coyote, Barney, 138, 139
Old Coyote, Henry, 137, 138
Old Coyote, John, 159
Old Elk, George, 159
Oliver, plow, 44; tractors, 211, 220, 221, 412
Olmstead, four-wheel drive, 408
Olsen, Pete, 66
Olson, Dorothy (Mrs. Earl), 403, 404
Olson, Earl B., early years, 394; enters feed processing, 399; expands, 395, 397, 398, 401; faces adversity, 396; income of, 398, 399, 400; nation's largest turkey grower, 394; outlook on life, 403
Olson, Olaf, 394
Olson, Ole, 84n, 194, 456
Omelyanenko, Demetri, 162
Onida, South Dakota, 233
Onions, 313, 318, 322, 330, 332
Onstad, E. G., 155
Opal beet harvester, 374
Open-pollinated corn, 251
Operating cost, 146, 173, 192, 193; loss, 468
Opportunity, 451; to expand, 308; in farming, 465, 468
Opposition, to expansion, 464
Optimism, 452, 458
Organizations, 469, 470

Outlook, Montana, 10
Outstanding Young Farmer, 467, 469, 476
Overby, Mrs. Ole, 430
Overexpansion, 450, 452, 458, 473
Overhead hopper, 328
Overline loans, 478
Overproduction, of turkeys, 399
Overstocking, of horses, 406
Owahe Grain Company, 233
Owatonna Manufacturing Company, 298
Owen, F. V., 377n
Owens, Johnny, 105, 116, 122, 125, 131, 136, 137, 139, 141, 142
Oxen, cost of, 213

Packer buyers, 444
Packers, acquisition costs, 443
Packing plants, 443
Page, William R., 45
Pakosh, Peter, 422, 423
Parity income, 459n; ratio, 145
Park River, North Dakota, 314
Parr, C. H., 23
Parsons, Thomas D., 195, 196
Partnerships, 435, 437, 478
Part-time farmer, 485
Pasture, 256, 264, 266
Paul Horn Farms, Inc., 350
Paynesville, Minnesota, 399
Payroll, at CFC, 140; delayed payments, 61; size of, 401
Pazandak, Ferd, 34, 36, 37, 38
Peas, production of, 300
Peavy Elevator, 194n
Peck Hatchery and Feed Store, 390
Peddling, of potatoes, 350
Pelican Rapids, Minnesota, 389
Pembina, North Dakota, 312
Pension, 472
Peoria, Illinois, 221
Per capita income, 457
Percheron horses, 281
Performance Registry International, 299
Perham, Minnesota, 351
Perham Potato Farm, 338, 351n
Perkins County, South Dakota, 209
Permanent employees, 140
Pesticides, 337
Peters, Robert, 231
Peterson, Charles, expands, 326, 327; fertilizer user, 319, 321, 322; first potato crop, 317, 318; sugar beet producer, 368, 369; truck gardener, 313
Peterson, Harold T., 380

Peterson, Henry, 317, 326, 364, 380; labor supply, 338; marketing, 348; onion grower, 330n, 332n; potato production, 328, 331-335; production cost, 371; sugar beet grower, 368; use of fertilizer, 349n, 370

Peterson, Leslie W., 459ff, 473-475, 478, 479

Phosphate, 253, 321, 373

Picker-sheller, 258

Pickup truck, 390, 395

Pigeon grass, 362

Pigs, 220

Piling stations, 376

Pinches, Harold, 47

Pioneer farmers, 462

Pioneer tractor farmers, 32-34

Pitchfork, 300

Planting, of beets, 373, 376; of potatoes, 328

Plow, cost of, 319; size of, 29, 34, 413, 425; steel, 3; steam lift, 34

Plow down, 321

Plowing 14, 410; at night, 33; cost of, 11, 219, 426; methods of, 47; rate of, 29, 199ff, 425, 426; with four-wheel drive, 120n; with horses and oxen, 217

Politics, in agriculture, 153; of Garst, 246ff

Polk County, Minnesota, 376

P and O breaking plow, 41

"Pool," 92 (*See also* Minnesota Potato Growers Exchange)

Pool buying, 184

Poor, farm bankers, 481; losers, 489; management, 451, 475, 491

Popcorn, 197, 198

Poplar, Montana, 104, 110

Population, on farms, 150

Populist, 452

Pork, production of, 384

Portable grain bin, 210; generator, 337n

Poss, Dr. Peter, 401

Postal money orders, 217

Potash, 253

Potato, acreage, 89, 337, 338, 342, 344, 345, 356, 359; brokers, 342; commercial production, 312; consumption of, 342, 353; cost of production, 76, 319, 328; demand for, 328; digger, 318, 319, 326, 331; diseases of, 315, 327; diversification of, 184; duster, 328; early growers, 313; fields, 336, 337; grader, 340; harvest, 334, 335n, 341; harvest crew, 318-320, 329, 331, 332;

harvesters, 90, 314, 320, 331-337, 341, 344; income of, 326; irrigation of, 344; largest farm, 330; loading of, 343; machinery, 344; marketing problems, 90; planters, 319, 326, 331, 344; price of, 10, 66, 76, 83, 318, 322, 327, 347; problems of production, 76; processed, 351-354; processing plants, 330, 351n, 353; profit potential of, 314; sacks, 341; sale of, 333, 354; seed, 339; sorting crew, 320, 340, 341; shipment of, 318, 320, 333, 340, 353; storage, 340; surplus, 346; table stock, 354; truck boxes, 336; warehouse, 191, 314; wash plant, 332, 342; weeder, 319; yield of, 318, 324

Poults, 397

Power, cost of, 49, 50ff; divider, 415; lack of, 412; machinery, 265; need for, 419, 446; plants, 137, 221, 224; search for, 412

Powerbuilt four-wheel drive, 409

Power, J. B., 56, 464

Power takeoff, 31, 219, 325

Prairie Farms, Inc., 91

Prairie hay, 447

Prairie Provinces, 414n

Precipitation, 181, 246

Precision tools, 413; production, 422

Press drills, 223

Presto light, 39

Price, collapse of, 73; high, 445; of alfalfa, 288; of beets, 361; of cattle, 369; of corn, 287, 288; of equipment, 19, 68, 83, 424; of farm products, 66, 76ff, 81, 244, 282, 456, 473; of grain, 15, 83, 84ff; of hay, 79; of land, 83, 171ff, 185, 195, 322, 455; of potatoes, 83, 314, 318, 322; of wheat, 108, 145, 322; problems of, 77; regulation of, 354

Primitive methods of growing beets, 361-366

Probstfield, R. M., 312, 430, 462

Processed turkey, 400, 401

Production, increase of, 425, 471; of beets, 359; of horses, 406; of Jennie-O plants, 399, 400; of Offutt enterprise, 351; per acre, 295; per combine, 124; per employee, 300, 330, 335n, 438; regulation of, 154, 352, 354; requirement for corn, 269

Production cost, 485, 488; corn vs. cows, 265; for potatoes, 344, 345n; of corn, 269

Production Credit Association, 436, 466n, 470, 474, 477, 481

Production designer, 422
Production line, 342
Productivity, increase of, 5, 18; of labor in beets, 378; of U.S., U.S.S.R., and India compared, 273; per man, 115
Products engineer, 418
Proebstel, Steve, 470
Profiles of some modern poultry farmers, 384-386
Profit and loss, 175, 478
Profits, during 1918-1919, 322; from beets, 358, 367, 369n; from consolidation, 204; from corn vs. alfalfa, 288; from corn vs. pasture, 264ff; from direct marketing, 232, 235, 350; from farming, 130, 242, 285, 403, 445; from livestock, 204, 223, 224, 294; from 1928 to 1954, 387; from potato production, 313, 314, 326, 336; from turkey production, 386, 391, 397; increased by use of machinery, 203; in 1930's, 83; in 1926, 197, 199; of Baldwin Farm, 186ff; of CFC, 144; of elevator, 389; of Elk Valley bonanza, 183; of Farmfest, 305; of government corn program, 287; of Grandin No. 1, 181; of Kiene Farms, 66ff; of Larimore bonanza, 183; of purchasing grain, 384; of Schermerhorn Farm, 193; of seed contracting, 232; of threshing rig, 227; of Tilney Farms, 202; potential, 355, 446, 480; reason for Dwight Farms, 188
Profit sharing plan, of CFC, 110
Progressive farmers, 306, 415, 416, 418, 427, 468
"Prophet of the Future," 115
Prosperity, 444
Protein, 259
Protest, 361
Proving up, 472
Psychology of buying, 442
Public domain, 472

Quack grass, 45, 386, 456
Quality, control, 421; components, 423; of potatoes, 352; of Shadybrook, 308; production, 393
Quebec, 420
Quotas, 355

Radio, on farms, 92
Radishes, 368
Railroad, shipments, 441; siding, 364
Railroads, enter Iowa, 197; need for, 359

Rain, 227, 455
Ralston Purina Company, 383
Ranchers, 434
Ranch land, 88
Raskob, John Jacob, 146
Rate, of change, 301; of gain, 295
Ratio, family living to gross income, 484
Ration, 294, 438, 440
R.E.A. (*See* Rural Electrification Administration)
Real estate, 75, 97, 463
Reaper, 3
Rear mounted loader, 388
Reconstruction Finance Corporation, 454
Record keeping, 152, 179, 307, 387, 388, 403, 468, 471, 472, 482, 483, 485
Recycling, 277, 448
Red Lake Falls, Minnesota, 412
Red Pontiac potatoes, 352
Red River, 319n; used for irrigation, 368
Red River Hatchery, 392
Red River Valley, 99ff, 167, 179ff, 311, 336, 337n, 376, 453; early tractors in, 33ff; farm credit needs, 462; large farms, 56; potato region, 312, 314, 317, 343, 347ff, 352; sugar producer, 358, 379
Reed, Walter R. family, 176-178, 186n
Reese, Herbert, 50
Reeves, Charley C., 61
Refinancing, 73
Refrigeration, lack of, 137
Registered livestock, 66, 281
Reitan, Oscar, 33, 36
Reitan, T. S., 33
Reitan, Walter, 33, 184
Remodeling, 181
Removing the ache from haying, 292-301
Renault, 419
Rennsch, Oscar, 424
Rent, of horses, 22, 77ff; rates on land, 15, 81, 84, 181, 208, 209, 468
Rental units, 174, 233, 470
Rented land, of Indians, 101, 145
Repossessed, tractors, 220; land, 172, 173, 180, 196
Republic trucks, 44
Reque, John H., 380
Research staff, 421
Reserve, for bad debts, 467, 470
Resistance to change, 25, 248
Rest home, 191
Retail selling, 402

Return to capital, 207, 402, 488
Retz, Richard, 463
Revolt, 162n
Revolutions in agriculture, 48, 426
Rice, Jimmy, 312n
Richland County, North Dakota, 188, 195
Riis, Herman, 250n
Rinehart, Mary Roberts, 124, 127
Rippey, Charles, 248
Risks, in agriculture, 6, 395, 399, 469, 470, 479, 486, 491
Risk takers, 385, 463, 489
Roach, Pat, 105, 117, 125, 136, 137, 138, 141
Road contractors, 416
Roads, 36, 43, 44, 126, 137, 190, 303
Robinson, Arthur, J., 44, 189, 192, 193
Robinson, Elwyn B., 472
Robinson, Roy, 422, 423
Rockefeller, John D., 177
Rock phosphate, 298
Rocks, cleared from fields, 213
Roguing, 316n
Roll bars, 424n
Romain, Clarence, 46
Romain, Fred, 11ff
Room and board, cost of, 138, 169
Roosevelt, Franklin D., 149n, 243, 247
Roosevelt, Theodore, 430
Root cellars, 340 (*See also* Potato, storage)
Ross, Armin, 373
Ross, Ferdinand, 359, 360
Ross, Walter J., sugar beet pioneer, 362, 364-368, 370, 372-374, 377, 378, 380
Rotation, of crops, 111ff, 199, 345, 388; of harvest shifts, 376n
Rothsay, Minnesota, 385, 388
Roto-beater, 375
Roughage, efficiency of, 261, 440, 446, 447; equipment, 446
Row crop farming, 269, 311, 410, 418, 425
Row finders, 375n
Rubber corn roll, 253
Rubber tires, on tractors, 39, 46, 146, 370, 409
Rumely tractor, 23, 29, 42, 210, 211n, 412
Ruminant animal, 267, 440, 446
Rupp, Robert, 460, 469
Rural Electrification Administration, 190
Rush River Township, 170
Russell, Minnesota, 224

Russia, trade with, 148, 273, 425; Campbell mission to, 160, 162
Russian, farms, 221, 270, 418, 425; immigrants, 376; thistle, 185, 225, 431
Rust, in grain, 183, 220, 223
Ruthruff, rail siding, 365
Ryan brothers, 328, 341, 342, 348, 353, 354 (*See also* Ryan, Gerald, Jr., and Ryan, Thomas)
Ryan, Bryce, 248
Ryan, Charles L., 360
Ryan, Dennis, 348
Ryan, Gerald, Jr., 317n, 327, 342n, 346, 348, 380
Ryan, Gerald, Sr., 342n, 348
Ryan, Thomas, 317n, 342, 348, 380
Rye, price of, 214

Sabe, J. Odell, 18
Sabe, John, 16
Sabe, Oscar, 18
Sabin, Charles H., 102
Sac County, Iowa, 197
Sacks, in potato harvest, 320, 333
Sage brush, 106
St. Joseph, Manitoba, 363, 364, 372
St. Paul, Minnesota, 304
St. Pierre farm, 190
St. Xavier, Montana, 160
Salary, for farm manager, 185; of Kiene, 81; of Olson, 395, 397
Sale, of Shadybrook Farm, 306; of surplus machinery at CFC, 121
Sales, gross, 461; restricted by law, 356n; success of, 422; techniques, 249
Salesmen, 255
Salt, in ration, 294
Samson Iron Horse, four-wheel drive, 408
Sanborn, North Dakota, 208
"Sand dunes," 412
Sang, Sam, 79n
San Margerita Ranch, 141, 146
Sargent County, North Dakota, 218
Sarle, C. F., 407
Savings, 488
Savisky, Dr. Helen, 376
Savisky, Dr. V. F., 376
Scab, in potatoes, 315
Schermerhorn cottage, 191n
Schermerhorn Farms, 44, 87, 189-194
Schermerhorn, J. B., 189, 191, 192, 194
Scherweits, 172
Schmidt, Bill, 411
Schmidt, Gene, 411

Schnad, Erwin, 119n, 145
School, children as potato pickers, 320, 329n; consolidation, 451; rental of land, 15
School board, 468
Schroeder, Ernest, 312n
Schroeder, Henry, 312, 313, 319n, 321, 340, 347
Schroeder, Henry E., 312n
Schroeder, Robert, 312n
Schroeder, Stephen, 312n
Schutt, Charles, 170, 173
Schutt, George, 172
Schwartz, Earl, 216, 220, 222, 223, 231, 233, 236
Schwartz, Jacob, 220
Scientific farming, 276; poultry production, 383
Scoop shovels, 388
Scott, John W., Jr., 317n, 380
Scott, John W., Sr., 89, 317n, 327, 353, 360, 380, 417
Scott, W. M., 226
Screenings, 223, 233
Scrip, 202
Sears, Roebuck, 322
Seattle World's Fair, 433
Sebens, Edward, 196, 197
Sebens, William P., 194n, 196, 197
Secretary of Agriculture, 3
Secretary of Interior, 100 (*See also* Lane, Franklin K.)
Seed, corn, 241, 396; loans, 220, 222; plants, 233; potential of, 265, 456; segmented, 372
Seeding, of beets, 370
Selkirk, Lord, 312
Selley, Roy, 201
Semitrailer trucks, 127, 289, 338, 400
Serum, Mrs. Orlando, 180
Settlement of Red River Valley; 312
Severson, Al, 392
Shadybrook Farm, 284-287; visitors to, 285, 291, 298, 306, 308
Shadybrook Monarch, 284
Share renting, 181, 306
Sharp, Andrew, 194n
Sheep, 178, 193, 195, 198, 217, 223, 235, 236, 327, 369
Sheepherders, 193
Shepherd, Montana, 433, 435, 437
Sheyenne, North Dakota, 14, 454
Shinrone Farms, 201, 202
Shipping cost of wheat, 145n
Shocking, 118
Shoestring potatoes, 351

Shorthorn herd, 283
Short-term loans, 457
Shovels, for grain, 125
Shrubbery, 191
Sickle, 161
Sickness, 283
Silage, production of, 202, 293, 322, 438, 439, 440
Simms Machine Shop, 334
Simplot, J. R., 354
Sinner, Al, 172, 173
Sinner, George, 380
Sinner, John B., 170
Sinner, William, 380
Sioux City stockyards, 202
Sisson, Francis, 102
Six-horse hitch, 406
Size of farm unit, 111, 355, 386, 390, 393, 442
Skinner System, for irrigation, 368
Skolness, Arthur, 380, 417
Skyberg, Adolph, 334
Skyberg, Herman, 334
Slattery, Floyd, 148, 155
Sledge hammer, 413
Sleeping quarters, 196
Sliding shed, for potato sorting, 320
Slip clutch, 410
Small farm, 32, 78, 79, 97, 391, 488
Smith, A. O., 296, 298, 477n
Smith, Thomas R., 462, 465n, 469, 474, 478, 479, 484-489, 491
Snack foods, 351
Snare, A. A., 87
Snow, Don, 123n
Social legislation, 377
Social resentment, 236-239
Socorro, New Mexico, 146
Sod breaking, 34, 40, 100ff, 208, 215, 217, 229, 236
Sod buster, 430
Soil analysis, 323; bank, 78, 156; compaction, 421; lime, 253; of Red River Valley, 312, 358
Solberg, Peder L., 453, 454
Sonstegard, Donald, 399
Sorghum, seed, 241
Sorvik, Ben, 86
South Dakota, 221, 383, 419
Southern Minnesota Conservation Association, 305
South Saint Paul, Minnesota, 223
South Saint Paul Union Stockyards, 235, 289
Soviet leaders visit CFC, 124
Soviet Minister of Agriculture, 270

Soviet Premier, 268
Soybeans, 267, 403, 418
Specialists, 351n
Speculators, 186
Speed, of tractors, 419
Spicer 500; transmission, 423
Spike pitcher, 364
Spindle tuber, 316
Spoilage, of crops, 293, 294
Spraying, aerial, 129
Spray trucks, 143
Stablemen, 177
Stacker, use of, 447
Stalin, Joseph, 161, 273
Standard of living, 484
Standard Oil, 35
Standing Rock Indian Reservation, 209
Stanley, Mountrail County, North Dakota, 230
Star auto, 137n
Stasiuk, Mike, 422,-424
State Bank of Vernon Center, 302n
State Potato Show, 76
State Seed Department of North Dakota, 316
Steam, power, 3, 21, 22, 63, 181, 197, 216; thresher, 98
Steele County, North Dakota, 12, 179, 188
Steering, of tractor, 34, 414, 420
Steering stick, 415
Steers, sale of, 219, 236
Steiger brothers, 412-414, 417, 423, 425, 445
Steiger, Douglas, 412, 413, 415, 418
Steiger, John, 411, 412, 414, 414n
Steiger Manufacturing Company, 417, 418, 423
Steiger, Maurice, 412, 413, 415
Steiger, Mrs. John, 412
Steiger tractors, 120n, 414, 418, 423, 425, 447, 448n
Stephen, Minnesota, 314
Stillman, James, 102
Stillwater, Oklahoma, 304
Stinson, 304
Stoa, A. E., 239
Stock market crash, 216
Stockton, California, 220
Stockyards, 191, 235
Stockyards National Bank of Chicago, 32
Stoneberg, Everett, 488
Storage, of grain, 440, 441; of potatoes, 320, 339, 341, 344, 348, 351
Storage batteries, 387

Stoves, for potato storage, 340
Straw, uses of, 195, 214, 261, 340, 369, 446, 447
Strike, cost of, 386
Strip farming, 113
Stroer, Joe H., 336
Strohm, John, 246, 268, 270, 271
Stutz Bearcat, 142
Subsidiary enterprises, 90, 175
Subsidy, government, 155
Subsistence farming, 311, 382, 480
Success, factors of, 207, 231, 402, 432
Successful Farming, 262
Successor to a crusader, 306-310
Sugar (*See* Beets)
Sugar Act, 367, 368, 371
Sulerud, Allen C., 181
Sullivan, James T., 361
Sullivan, Minnesota, 361
Sully County, South Dakota, 233
Sundberg, B., 57, 64
Sundberg, J. E., 64
Sundberg, Mrs. B. E., 64
Sundberg, Roy, 86, 87, 89
Sunflowers, 198, 243
Support price (*See* Subsidy)
Surplus, investment, 473; labor, 473; production, 18, 46, 48, 156, 407
Survey crews, 9
Sutton, Edmund, 32
Sutton, John, Sr., 32
Swamp land, 190
Swather, self propelled, 184
Swath pickup, 445
Sweet clover, 197
Swenson, J. W., 317n, 338, 341n, 351
Swift and Company, 189
Swift County Bank, 398
Swift Falls Cooperative Creamery, 395-398
Swift Falls, Minnesota, 395, 399
Swine (*See* Hogs)
Syndicates, 478

Table stock potatoes, 342, 351
Tagus Bank, 214, 216
Tagus, North Dakota, 213, 214, 219
Talley, H. R., 406
Tandem tractor hitch, 409, 410
Tank wagon, 35, 119
Tardieu, Andre, 100
Tare, 375ff
Tariff, 148, 154
Taxes, delinquent, 85, 147, 173, 188; discriminating, 177; effect on farm profits, 70; exceed income, 194; in-

come, 74; paid by Kiene, 80; on La Jolla lands, 147; real estate, 15, 66, 80, 84, 183, 186n; relief for farmers, 157

Taylor, J. R. ("Punk"), 101n, 104ff, 107, 110, 116, 122, 125, 131, 132, 141, 144n

T-Bone Feeders, 433-436; 438

Technological Revolution, 247

Technology, 269, 277

Telephone, 190

Temperance, 92

Tenant farming, 10, 176, 183, 187

Tenants, 184, 204

Terminal elevators, 232

Terminal Farm (Brown Farm), 193

Test, of tractors, 29; plots, 246; weight of wheat, 194n

Testing, of cream, 395

Texans, at CFC, 140n

Texas, 426; cattle shipped to, 106

Thermal efficiency, 406

Thief River Falls, Minnesota, 192, 413

Thinning, of beets, 362, 366, 371, 374, 377

Thistles, 386

Thomas, Bertha, 246

Thomas, Charles, 246, 250, 253

Thomas, Charolois herd, 246n

Thomas, Ken, 465

Thompson, A. B., 316

Thompson, Joe, 314, 317n, 321, 322, 325, 327, 347, 354

Thompson, Louis M., 267

Thompson, Tom, 13-14, 314, 318

Thorpe, Walton, 226

Three-point hitch, 409, 412, 420, 425

Three-wheeled vehicle, 388 (*See also* Bobcat)

Threshing, 227, 335n; by flail, 179; by horsepower, 183; by tractor, 23; for a strawpile, 220; machine, 3, 12, 41, 64, 116, 117, 184, 217, 227

Tibert, Budd, 317n, 334, 335n, 337, 340, 417

Tibert, L. E., 314, 316, 327, 334

Tidoute, Pennsylvania, 179

Tiger, tractor, 414n

Tile, 291ff

Tiling, 319n, 255; machine, 291n

Tillage systems, at CFC, 112ff

Tilney family, 203

Tilney farms, 202-204, 380

Timber Lake, South Dakota, 209

Time, 159

Time required to, bale hay, 289; feed cattle, 299; produce corn, 269n

Tires, size of, 414, 420, 424

Titanic, 176, 177

Titan tractors, 63, 216

Tolley, Clarence, 179n

Tompson brothers, 116

Toppers, beet, 367, 369, 372

Topping, cost of, 363, 371, 372

Torrance, J. S., 99, 100, 102

Total ration, 295

Tower silos, 293

Towns, abandoned, in Iowa, 244

Township constable, 303

Trace minerals, 294

Traction, 413

Tractor, bankrupts owner, 45; cabs, 281; cost of, 30ff, 43, 63, 68, 83, 245, 325, 386; driver, 139; early users, 36ff, 196, 226, 281, 324, 426; field laboratory, 419; fuel, 35; impact of, 38, 45, 48ff, 70, 87, 120ff, 200, 281, 321, 325, 329, 366, 386, 407, 422, 447, 457, increased size, 49, 52, 359, 411, 413, 419; invention of, 23; loans, 31ff; manufacturers of, 27ff; mechanical problems, 33, 35, 40-42; modifications, 162n; mounted cultivator, 373; named, 23n; number of, 27ff, 300, 326, 355, 405ff; on collectives, 221; opposition to, 31ff; plowing, 243n; predictions of, 51, 52-53; pulling contest, 411, 417; rental of, 87; replaced, 119ff; revolution, 407; sales, 118; small size, 30ff; steam, 34; test, 29, 30; track-type, 31, 113, 121, 326, 421, 425

Tradition, 442, 443

Transient workers, 135

Transit in bond, 361n

Transmission drop box, 424

Transportation, cost on large farms, 128, 149

Tree planting, 180

Trench silos, 202, 440

Trench storage, for potatoes, 320, 344

Trites, Fillmore, 392

Trowbridge, Hugh, 380

Truck, cost of, 83, 287n, 344, 376; drivers, 376n; early use, 127, 366, 378; need for, 329, 335, 339, 367, 375, 441; number of, 89, 92, 128, 338, 376, 393, 402, 406; uses of, 332, 333

Truck gardener, 312

Trucking business, 51, 90, 227, 286, 291, 350, 395, 400

Truman, Harry, 158, 159

Tuberculosis, 99

Tumbleweed, 110

Turkey, 382, 384, 387, 391, 397; buildings, 387, 388; center, 399; farms 383; marketing of, 387, 398; products, 400; processing plants 397, 398; production, 383, 387, 394, 399, 401, 402
Turner, Frederick Jackson, 429
T.V. dinners, 393
Tvede, Mrs. Harold, 82
Tweedy, Robert, 420, 422
Twin City tractors, 43, 80, 212
Two Legging Creek, 107
Tyler, Mrs. Anne Dwight, 188

Underlee, C. H., 45, 380
Underlee, Leslie, 380
Underlee, Nolan, 380
Under Secretary of Agriculture, 105
Union Central Life Insurance Company, 219
Union State Bank of Thief River Falls, 413
Union Stockyards, St. Paul, 290
Union Stockyards, West Fargo, 236
United Nations delegates, 275
United States, agricultural advantages, 267
United States Department of Agriculture, 147, 407, 461; studies, 100; subsidy payments, 156; sugar deficit, 317, 368
United States Farming Corporation, 152
United States Soil Conservation Service, 146
University of Illinois, 406
University of Minnesota, 406
University of Moscow, 161
University of North Dakota, 98
Unloader, bottom 293; dump, 367
Urea, used in cattle ration, 260-263, 265, 266
Urevig, Edgar M., 203
U.S.S.R. (*See* Russia)

Valentine, Mr., 71
"Valley Bee" tractor, 417
Van Hook, North Dakota, 208, 231
Van Nice, Mr., 218n, 221, 225
Vaughan, Montana, 437
Vegetables, 9
Velo, Eddie, 343n, 386-389, 445
Velo, Leola, 387-389
Velo, Martin, 386, 387
Velomix, Inc., 389
Velos, 386-390
Vern Hagen Farms, 343
Vernon Center, Minnesota, 285, 301, 306

Versatile Manufacturing, Ltd., 414n, 422, 423
Versatile, tractors, 122, 423-425, 447
Vertical, expansion, 270; integration, 383
Veterinary work, 402
Vial, Mr., 245
Victory garden, 330n
Viker, Bruce, 182
Viker, D. E., 181, 182
Vine turner, 319
Visitors, to Shadybrook, 298, 306
Vitamin A, 260
Volume, 402; buying, 169, 351; importance of in farming, 285, 287, 309, 335, 445, 482, 487; selling, 169, 351, 356, 400, 490
Voss, North Dakota, 314

Wages, 135ff, 135n; earned by Myhro, 390; in Russia, 221; of A. J. Robinson, 190; of farm labor, 5, 9, 15, 65, 66, 78, 85ff, 128, 133ff, 139, 180, 214, 224, 319, 371, 394
Wagner four-wheel drive, 412, 413, 416, 418, 420, 422, 423
Wagner, L. D., 359
Wagon hitch, 37ff
Wagons, in agriculture, 16, 198; train, 38
Wahpeton, North Dakota, 188n
Walfer, Henry, 63
Walker, Ludwig, 36
Walking plow, 317, 318
Wallace, Henry A., develops hybrid corn, 46, 199, 243, 245, 247, 248, 257n, 283; Secretary of Agriculture, 199, 457
Wallace, John, 433
Wallander, Greg, 438
Wallis Bear tractor, 23
Wall, William, 159
Walsh County Agricultural School, 352
Walsh County, North Dakota, 314, 321, 418
Walsh, John, 250n
Walter A. Woods Harvester Company, 195
Warehouse, crew, 330; largest, 331 (*See also* Storage)
Wardner, Dr. A. C., 323
War Food Administration, 330, 332
Warren, Floyd Darroll, 130, 145
Warren, G. F., 23, 429, 473
Washing, clothes, 219
Water flume system, 341

Water, problems of securing, 17, 106, 218; pump, 3, 181; table, 456; used in steam engines, 34
Watson, Ralph, 123n
W.C.T.U., 177
Wealth, accumulation of, 452
Weaning weights, 469
Weather, 144, 215, 221, 371, 486
Webster County, Iowa, 246
Weeds, 46, 111, 112, 129, 195, 200, 223, 311, 362, 366, 371, 377
Weiland, Gregory, 120n, 425
Weiland, Henry, 428
Welle, Nick, 391
Wemple, Rex, 122, 125, 131, 133n, 136, 137, 139, 141, 142
West Central Turkey Cooperative, 389
Western Livestock Reporter, 236
West Fargo, North Dakota, 219, stockyards, 197, 236, 369
West River Country, 228
Wet cycle, 100, 412
Whalen, Fredric E., 81
Wheat, direct sales of, 144; in ration, 440; price of, 10, 108, 145, 158, 173, 188, 192, 214, 220, 222, 225, 322; production of, 145, 170, 183, 194n, 195, 199, 220, 311; purchase, 162n, 274; seed, 222; stockpiles of, 425; to feed the needy, 156
Wheat-flax-weed crop, 117
Wheat monoculture, 179, 180, 313
Wheelbarrow, 324
Wheeler, H. C., 197, 198
Wheeler, Ralph, 253
Wheeler Ranch, 198n
Wheeler, Senator Burton K., 149
Wheel lifter, 375
Whetstone, Leo, 123n
White Earth Indian Reservation, 190
White Earth Lake, 191n
White House, 305
White, Owen P., 153
Whitman, Ward, 122n, 234n
Wiedemann, Ewald, 313
Wiese brothers, 411
Wiese, Donald, 411
Wife, influence of, 376n, 490
Wigand, Carl, 359
Wije, Reuel, 188
Wilber family, 285
Wilder, Jay, 374
Wild oats, 362
Wilkin County, Minnesota, 312, 313
Williams, Don, 276
Williams, W. H., 23n

Willmar Grain Terminal, 399
Willmar, Minnesota, 394, 397, 399
Willmar Poultry and Egg Company, 399
Willson, George O., 15
Willson, Percy, 231
Wilson, M. L., 9, 26, 105
Wilson, R. S., 180
Wilson, Woodrow, 100
Wind chargers, 287
Wind erosion, 337n
Windrow, 372; harvest method, 117
Windstorms, 387
Winkler, Manitoba, 335
Winnebago, Minnesota, 283
Winnipeg, Manitoba, 361
Winter wheat, 225
Wisconsin lumber camps, 169
Wolf, Bob, 159
Woodward, Dr. Ray, 437
Wool, price of, 226
Work, load, 349; sheet, 139
Working, conditions, 134ff; habits of Indians, 138n, 139
World ploughing contest, 305
"World's biggest farm," 102
World War I, 70ff, 243, 314, 451
World War II, 224, 254, 328, 330, 359, 451
Worms, 371
Worthington, Dr. Wayne H., 418-421
Woven wire, 198
Wright, Charles R., 183

Yaggie, Leo, 360n, 368n, 377n
Yaggie, Robert, 360n, 377n, 378
Yearbook of Agriculture, 1926, 406
Yellowstone County, beef production, 440, 441
Yields, in corn production, 245n, 249, 252, 255, 256, 268; increase with fertilizer, 184, 246, 251, 337; increase in potatoes, 318, 319, 337, 343; in sugar beets, 359, 365, 367, 369, 374n, 377, 378; of alfalfa, 296; of wheat, 145, 313
Young brothers, 209, 220
Young Brothers Implement Company, 211n, 221
Young, Dwight, 209, 221
Young, Edmond, 208
Young, Eugene, 208-209; innovative farmer, 41, 138n, 207, 208, 212, 216, 221-225, 232, 237
Young, George, 209, 220, 221, 225
Young, Guilford, 208, 221
Young, Senator, Milton, 157

Zachary, Mr., 106
Zeigler, M. A., 60

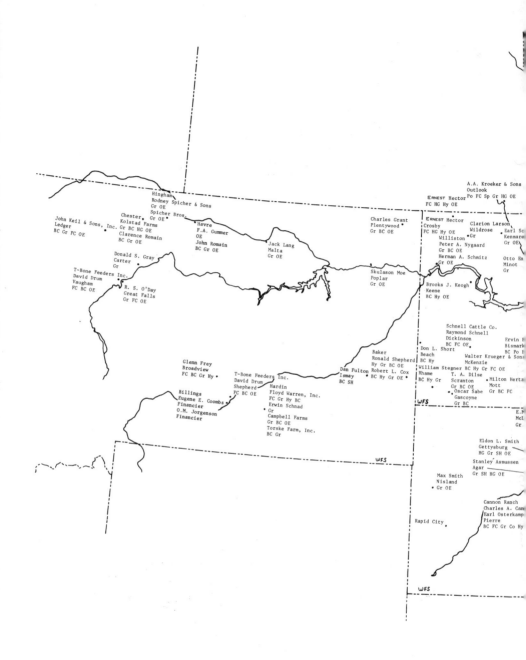

Hingham
Rodney Spicher & Sons
Gr OE
Spicher Bros.
Gr OE

Chester
Kolstad Farms
Gr BC HG OE
Clarence Romain
BC Gr OE

John Keil & Sons, Inc.
Ledger
BC Gr FC OE

Havre
F.A. Gummer
OE

John Romain
BC Gr OE

Jack Lang
Malta
Gr OE

Donald S. Gray
Carter
Gr

T-Bone Feeders Inc.
David Drum
Vaughan
FC BC OE

R. S. O'Day
Great Falls
Gr FC OE

Skulason Moe
Poplar
Gr OE

Charles Grant
Plentywood
Gr BC OE

ERNEST Hector
FC HG Hy OE

ERNEST Hector
Crosby
FC HG Hy OE
Williston
Gr BC OE

Peter A. Nygaard
Gr BC OE

Herman A. Schmitz
Gr OE

A.A. Kroeker & Sons
Outlook
Po FC Sp Gr HG OE

Clarion Larson
Wildrose
Gr

Earl Sc
Kenmare
Gr OE

Otto En
Minot
Gr

Brooks J. Keogh
Keene
BC Hy OE

Schnell Cattle Co.
Raymond Schnell
Dickinson
BC FC OE

Ervin I
Bismark
BC Po I

Walter Krueger & Sons
McKenzie

T. A. Dilse
Scranton
Gr BC OE

Milton Hertz
Mott
Gr BC FC

Oscar Sabe
Gascoyne
Gr BC

Don L. Short
Beach
BC Hy

William Stegner BC Hy Gr FC OE
Rhame
BC Hy Gr

Baker
Ronald Shepherd
Hy Gr BC OE
Robert L. Cox
BC Hy Gr OE

Dan Fulton
Ismay
BC SH

Glenn Frey
Broadview
FC BC Gr Hy

T-Bone Feeders Inc.
David Drum
Shepherd
FC BC OE

Hardin
Floyd Warren, Inc.
FC Gr Hy BC
Erwin Schnad
Gr
Campbell Farms
Gr BC OE
Torske Farm, Inc.
BC Gr

Billings
Eugene E. Coombs
Financier
O.M. Jorgenson
Financier

WFS

E.
McL
Gr

Eldon L. Smith
Gettysburg
HG Gr SH OE

Stanley Asmussen
Agar
Gr SH HG OE

Max Smith
Nisland
Gr OE

Cannon Ranch
Charles A. Can
Earl Osterkamp
Pierre
BC FC Gr Co Hy

Rapid City

WFS

WFS

IOWA, MINNESOTA, MONTANA, NORTH DAKOTA,
SOUTH DAKOTA, AND SOUTHERN CANADA